Microbial
Endophytes

BOOKS IN SOILS, PLANTS, AND THE ENVIRONMENT

Soil Biochemistry, Volume 1, edited by A. D. McLaren and G. H. Peterson

Soil Biochemistry, Volume 2, edited by A. D. McLaren and J. Skuji š

Soil Biochemistry, Volume 3, edited by E. A. Paul and A. D. McLaren

Soil Biochemistry, Volume 4, edited by E. A. Paul and A. D. McLaren

Soil Biochemistry, Volume 5, edited by E. A. Paul and J. N. Ladd

Soil Biochemistry, Volume 6, edited by Jean-Marc Bollag and G. Stotzky

Soil Biochemistry, Volume 7, edited by G. Stotzky and Jean-Marc Bollag

Soil Biochemistry, Volume 8, edited by Jean-Marc Bollag and G. Stotzky

Soil Biochemistry, Volume 9, edited by G. Stotzky and Jean-Marc Bollag

Soil Biochemistry, Volume 10, edited by Jean-Marc Bollag and G. Stotzky

Organic Chemicals in the Soil Environment, Volumes 1 and 2, edited by C. A. I. Goring and J. W. Hamaker

Humic Substances in the Environment, M. Schnitzer and S. U. Khan

Microbial Life in the Soil: An Introduction, T. Hattori

Principles of Soil Chemistry, Kim H. Tan

Soil Analysis: Instrumental Techniques and Related Procedures, edited by Keith A. Smith

Soil Reclamation Processes: Microbiological Analyses and Applications, edited by Robert L. Tate III and Donald A. Klein

Symbiotic Nitrogen Fixation Technology, edited by Gerald H. Elkan

Soil–Water Interactions: Mechanisms and Applications, Shingo Iwata and Toshio Tabuchi with Benno P. Warkentin

Soil Analysis: Modern Instrumental Techniques, Second Edition, edited by Keith A. Smith

Soil Analysis: Physical Methods, edited by Keith A. Smith and Chris E. Mullins

Growth and Mineral Nutrition of Field Crops, N. K. Fageria, V. C. Baligar, and Charles Allan Jones

Semiarid Lands and Deserts: Soil Resource and Reclamation, edited by J. Skuji š

Plant Roots: The Hidden Half, edited by Yoav Waisel, Amram Eshel, and Uzi Kafkafi

Plant Biochemical Regulators, edited by Harold W. Gausman

Maximizing Crop Yields, N. K. Fageria

Microbial Endophytes

edited by

Charles W. Bacon
Richard B. Russell Center, Agricultural Research Service
U.S. Department of Agriculture
Athens, Georgia

James F. White, Jr.
Cook College, Rutgers University
New Brunswick, New Jersey

CRC Press
Taylor & Francis Group
Boca Raton London New York

CRC Press is an imprint of the
Taylor & Francis Group, an **informa** business

First published 2000 by Marcel Dekker, Inc

Published 2019 by CRC Press
Taylor & Francis Group
6000 Broken Sound Parkway NW, Suite 300
Boca Raton, FL 33487-2742

© 2000 by Taylor & Francis Group, LLC
CRC Press is an imprint of Taylor & Francis Group, an Informa business

First issued in paperback 2019

No claim to original U.S. Government works

ISBN 13: 978-0-367-44741-0 (pbk)
ISBN 13: 978-0-8247-8831-5 (hbk)

**Visit the Taylor & Francis Web site at
http://www.taylorandfrancis.com**

**and the CRC Press Web site at
http://www.crcpress.com**

Preface

It is clear that many plants harbor endosymbiotic microbes within their tissues as symptomless intercellular infections. The symptomless nature of these infections is not absolute, but rather it is the apparent nonpathogenic to weakly pathogenic nature that characterizes this assemblage of organisms. These endosymbiotic microbes usually remain undetected until some investigator dissects plant tissue and subjects it to the scrutiny of laboratory analysis, usually microscopy. The presence of endophytic microbes impacts the ecology of host plants, often increasing fitness traits, and animals and humans may also be affected if they consume infected plants. Many types of endosymbionts have been found in plants. In its leaves a plant may harbor bacteria active in fixing nitrogen, as well as fungi that act as feeding deterrents to herbivores; in its roots a plant may harbor mycorrhizae and bacteria. The plant and its microbial partner form a symbiotic unit that figures in the evolution of the plant species. The impetus for the development of symbiotic associations is the increased ability of the symbiotic unit to survive, compete, and reproduce. Indeed, most of these symbiota have become so interdependent during evolution that the independent existence in nature of each component organism devoid of the other is prevented.

There is great potential to utilize symbiotic associations such as those described in this book. Through the use of these symbioses we may be able to produce crops with minimal use of water, fertilizers, and pesticides. Therefore, it is critical that we develop an understanding of the biology of the symbioses.

Recent interest has focused on endophytic microbes for their pharmaceutical potential. Fungal endophytes *Taxomyces andreanae* and *Pestalotiopsis microspora*, and several other fungi isolated from the bark of yew trees are potential new sources of the anticancer drug taxol. The clavicipitaceous grass endophytes are known to produce indole derivatives and other products that are active as plant hormones, antifungal agents, hallucinogens, vasoconstrictors, etc. Many other endosymbiotic microbes have not been investigated for their pharmaceutical value. Since endosymbiotic microbes must interact biochemically with host tissues to obtain nutrients, overcome host defenses, and defend host tissues, it is likely that many endophytes produce secondary metabolites which perform key roles related to the survival of the microbe and symbiotic unit. The endophytic habitat is a niche that merits continued exploration.

This book represents a compilation of writings that deal with the morphology, physiology, ecology, genetic-molecular biology, and evolution of endophytic microbes. The initial impetus for this project developed from a symposium at the conference of the Second International Symbiosis Society at Woods Hole, Massachusetts. The major objective of the symposium was to develop and present evolutionary strategies of symbiotic endophytes, primarily grass endophytes. The excitement generated by the participants stimulated us to invite others to contribute a series of papers, in addition to those presented at Woods Hole, which focused on the major aspects of endophytic microorganisms, emphasizing the non-grass endophytes. The resulting topics are by no means exhaustive but they are varied enough for students and investigators to gain a perspective on these endosymbionts and to develop a base from which deeper investigations into these curious associations may be launched.

Charles W. Bacon
James F. White, Jr.

Contents

Contributors

Charles W. Bacon, Ph.D.   Toxicology and Mycotoxin Research Unit, Richard B. Russell Research Center, Agricultural Research Service, U.S. Department of Agriculture, Athens, Georgia

David P. Belesky, Ph.D.   Appalachian Farming Systems Research Center, Agricultural Research Service, U.S. Department of Agriculture, Beaver, West Virginia

Thomas L. Bultman, Ph.D.   Division of Science, Truman State University, Kirksville, Missouri

Michael J. Christensen   Grasslands Research Centre, New Zealand Pastoral Agriculture Research Institute Limited (AgResearch), Palmerston North, New Zealand

Ann B. Gould, Ph.D.   Department of Plant Pathology, Cook College, Rutgers University, New Brunswick, New Jersey

Peter Herman, Ph.D.   Department of Biology, New Mexico State University, Las Cruces, New Mexico

Shung-Chang Jong, Ph.D.   Department of Microbiology, American Type Culture Collection (ATCC), Manassas, Virginia

D. Y. Kobayashi, Ph.D.   Department of Plant Pathology, Cook College, Rutgers University, New Brunswick, New Jersey

Gretchen A. Kuldau, Ph.D.   Toxicology and Mycotoxin Research Unit, Richard B. Russell Research Center, Agricultural Research Service, U.S. Department of Agriculture, Athens, Georgia

Marion M. Kyde, Ph.D.   The Tulgey Wood, Ottsville, Pennsylvania

Geoffrey A. Lane, Ph.D.   Grasslands Research Centre, New Zealand Pastoral Agriculture Research Institute Limited (AgResearch), Palmerston North, New Zealand

Dariusz P. Malinowski, Ph.D.   Texas A&M University, Vernon, Texas

Christopher O. Miles, D.Phil.   Ruakura Research Centre, New Zealand Pastoral Agriculture Research Institute Limited (AgResearch), Hamilton, New Zealand

Joseph B. Morton, Ph.D.   Division of Plant and Soil Sciences, West Virginia University, Morgantown, West Virginia

John Charles Murphy, M.S.   Division of Science, Truman State University, Kirksville, Missouri

J. D. Palumbo, Ph.D.   Department of Plant Pathology, Cook College, Rutgers University, New Brunswick, New Jersey

Ponaka V. Reddy, Ph.D.   Amersham Pharmacia, Biotech, Inc., Piscataway, New Jersey

Michael D. Richardson, Ph.D.   Department of Horticulture, University of Arkansas, Fayetteville, Arkansas

Christopher L. Schardl, B.S., M.S., Ph.D.   Department of Plant Pathology, University of Kentucky, Lexington, Kentucky

Jan Schmid, Ph.D.   Institute of Molecular BioSciences (IMBS), Massey University, Palmerston North, New Zealand

Martin J. Spiering, Ph.D.   Institute of Molecular BioSciences (IMBS), Massey University, Palmerston North, New Zealand

Jeffrey K. Stone, Ph.D.   Department of Botany and Plant Pathology, Oregon State University, Corvallis, Oregon

Edith L. Taylor, Ph.D.   Department of Ecology and Evolutionary Biology and Museum of Natural History and Biodiversity Research Center, University of Kansas, Lawrence, Kansas

Thomas N. Taylor, Ph.D.   Department of Ecology and Evolutionary Biology and Museum of Natural History and Biodiversity Research Center, University of Kansas, Lawrence, Kansas

Jan S. Tkacz, Ph.D.   Natural Products Drug Discovery, Merck Research Laboratories, Rahway, New Jersey

James F. White, Jr., Ph.D.   Department of Plant Pathology, Cook College, Rutgers University, New Brunswick, New Jersey

Heather H. Wilkinson, B.A., M.A.T., Ph.D.   Department of Plant Pathology and Microbiology, Texas A&M University, College Station, Texas

Dennis Wilson, Ph.D.   Department of Zoology, Arizona State University, Tempe, Arizona

Ida E. Yates, Ph.D.   Toxicology and Mycotoxin Research Unit, Richard B. Russell Research Center, Agricultural Research Service, U.S. Department of Agriculture, Athens, Georgia

I

ASSOCIATION IN STEMS AND LEAVES

1
An Overview of Endophytic Microbes: Endophytism Defined

Jeffrey K. Stone
Oregon State University, Corvallis, Oregon

Charles W. Bacon
Richard B. Russell Research Center, Agricultural Research Service, U.S. Department of Agriculture, Athens, Georgia

James F. White, Jr.
Cook College, Rutgers University, New Brunswick, New Jersey

1. INTRODUCTION

To microbes, land plants present a complex, spatially and temporally diverse ecological habitat. Symbiotic associations between microorganisms and plants are ancient and fundamental, and many examples of complex and highly specific symbioses between plants and microbes have been described in detail. Endophytic microbes are an intriguing group of organisms associated with various tissues and organs of terrestrial and some aquatic plants, and are the subject of increasing interest to mycologists, ecologists, and plant pathologists. In general, "endophytic" infections are inconspicuous, the infected host tissues are at least transiently symptomless, and the microbial colonization can be demonstrated to be internal, either through histological means, by isolation from strongly surface disinfected tissue, or, most recently, through direct amplification of fungal nuclear DNA from colonized plant tissue. Infections of land plants by endophytes are ubiquitous, having been found throughout a broad range of host orders, families, and genera worldwide, and representing a diverse array of terrestrial and aquatic habitats.

Biologically and ecologically, endophytes represent a diversity of nutritional modes from biotrophic parasites to interim or facultative saprotrophs, and associations with their hosts span the continuum from biotrophic mutualists and benign commensals to nectotrophic, antagonistic pathogens. Ecologically, interactions between host plants, pathogens, and herbivores are mediated by endophytes. The species composition and distribution of endophytic associations within and among hosts, and the reciprocal ecological effects of endophyte colonization on host fitness and the composition of plant communities, are problems of common interest to endophytologists and plant ecologists. Research interests in endophytes are correspondingly diverse and include community analysis, anatomical-histological relationships, organismal biology, biodiversity studies, population ecology and evolutionary biology, and ecological interactions among endophytes, hosts, and herbivores. Endophytism may be an important factor in microbial speciation and biodiversity. Accumulating evidence suggests that endophytes represent a large reservoir of genetic diversity and a rich source of heretofore undescribed species. Practical applications of endophytes include potential biological control agents, sources of novel metabolites for medicine, plant protection, and industrial uses, and as research model systems for investigation of host–parasite interactions and evolution in natural systems.

2. DEFINITIONS

The biological and ecological diversity of endophytes is reflected in the varying emphasis and heterogeneity of concepts among researchers concerned with studying them. Often the terms ''endophyte'' and ''endophytic'' are used with particular meaning by different workers and for particular groups of hosts and microbes. Workers investigating asymptomatic fungal infections of grasses caused by species of Clavicipitaceae and those investigating other microbes, such as endophytic bacteria, have adopted the terms ''endophyte'' and ''endophytic,'' and application of the same terms to different systems has contributed to some confusion and controversy. Contemporary application of the terms is not always consistent or accepted among all workers (Petrini, 1991; Wennström, 1994; Wilson, 1995b), although as commonly used the terms generally apply to microbes capable of symptomless occupation of apparently healthy plant tissue (Chanway, 1996).

    Important papers published during the past 20 years or so by various authors have stressed particular concepts of endophytism. These have varied in degree of inclusiveness depending on whether primarily organismal interactions or descriptive ecology of endophyte assemblages was emphasized, and on whether the research methodology was primarily histological or based on cultural isolations. The use of ''endophyte'' and ''endophytic'' in these varied contexts has contributed to a sense of ambiguity in the application of the terms.

    Researchers with a primarily organismal focus have attempted to characterize endophytes in terms of particular ecological or physiological attributes, often emphasizing aspects of host specificity and adaptation. Examples of studies that have focused attention on particular mechanisms of host recognition, attachment, and initial infection include studies on selective adhesion and germination behavior of the fungus *Discula umbrinella* on leaves (Toti et al., 1992) and of the fungus *Hypoxylon fragiforme* on *Fagus sylvatica* (Chapela et al., 1990; Chapela and Haggman,1991; Chapela and Bielser, 1993).

Alternatively, investigators whose primary interest is in descriptive ecology, biodiversity studies, or community analysis broadly include all microbes found inhabiting host tissue, often including ruderal saprobes and species familiar to plant pathologists as latent, quiescent, or opportunistic pathogens among lists of endophytes. Much of the early interest in endophyte infections emphasized their symptomless nature and invited comparison with mutualisms such as mycorrhizal associations. Over the past several years, recognition of the heterogeneity of endophyte–host interactions has gradually broadened so that mutualism is not commonly considered to be an assumed feature of endophyte interactions. Interactions other than mutualistic interactions in endophytes are recognized and anticipated.

Among the definitions proposed for the term endophyte are "fungi colonizing living plant tissue without causing any immediate, overt negative effects" (Hirsch and Braun, 1992). This definition includes virtually the entire spectrum of symbiotic interactions in which fungi and plants participate: parasitism, commensalism, and mutualism (Bills, 1996). This definition, however, fails to include prokaryotic microbes, such as bacteria and blue-green algae, or endophytic vascular plants (Fisher et al., 1992; Mauseth et al., 1985). A more inclusive definition of "endophyte" should stress the symptomless nature of the infection on the host without limiting the term to any particular group of organisms. Latent and quiescent pathogens are endophytes as are mutualistic microbes and benign commensals. Petrini (1991) considers the term endophyte to be purely topographical: "Endophytes colonize symptomlessly the living, internal tissues of their host, even though the endophyte may, after an incubation or latency period, cause disease." This latter definition is broad enough to include virtually any microbe or vascular plant that colonizes the internal tissues of plants. A synopsis of several major groups of endophytic microbes is given below.

3. ENDOPHYTIC CLAVICIPITACEAE

Researchers primarily concerned with endophytic fungi of grasses use "endophyte" to differentiate asymptomatic, systemic, seedborne endophytes, *Neotyphodium* spp., as well as pathogenic species of Balansiae (Clavicipitaceae) from the strictly epibiotic parasites of pooid and bambusoid grasses in the tribe Balansiae of the Clavicipitaceae. There are only two genera, *Epichloe*, with anamorphs in *Neotyphodium*, and *Balansia*, with *Ephelis* anamorphs, that contain endophytic species (i.e., systemic internal colonization of host plants). Pathogenic endophyte species may cause partial or complete sterilization of hosts (choke diseases) due to the production of a fungal stroma on the flowering culms of the host. The degree of effect on host reproduction varies with species. Clavicipitaceous endophytes are widespread in grasses (White, 1987) and frequently enhance resistance of their hosts to stresses (Clay, 1988).

4. OTHER SYSTEMIC FUNGAL ENDOPHYTES

Species of *Acremonium* have been isolated from asymptomatic rice plants (Ramanan et al., 1996). Although they are relatively commonly isolated, little is known of the impact of these endophytes on host plants. Similarly, an endophyte identified as *Pseudocercosporella trichachnicola* was found to be widespread in the warm season grass *Trichachne insularis* (White et al., 1990). The histological features of the endophytic mycelium of *P. trichachnicola* are similar to those of the clavicipitaceous endophytes: mycelium is intercellular, longitudinally oriented, unbranching, and present in leaf sheaths, culms, and seeds. It is unknown as to whether an external stage is ever produced by *P. trichachnicola*. The impact that this endophyte has on its host is also completely unknown although *T. insularis* has been reported to be toxic under some circumstances (White and Halisky, 1992).

Few studies have examined graminaceous hosts for nonsystemic, nonseedborne fungi. Dominant nonsystemic endophytes of these hosts are generally either familiar epiphytes such as *Alternaria alternata*, *Cladosporium* spp., and *Epicoccum purpurascens*, together with pathogens typical of grass hosts. Riesen and Close (1987) found barley (*Hordeum vulgare*) leaves in New Zealand infected primarily by the pathogen *Didymella phleina*, *Alternaria* spp., and *Stemphyllium botryosum*. Riesen and Sieber (1985) found *Phaeosphaeria* (*Stagonospora*) *nodorum*, a common leaf and culm blotch, the most common of 196 endophytic colonists of winter wheat in Switzerland. *Alternaria alternata*, *Cladosporium tenuissimum*, and *Epicoccum purpurascens* were the dominant endophytes of rice (*Oryza sativa*) (Fisher 1992) and maize (Fisher et al., 1992), occurring together with latent pathogens such as *Phoma sorghina*, *Fusarium equiseti*, *F. oxysporum*, *F. graminearum*, and *Ustilago* sp. Leslie et al. (1990) found frequent infection of maize and sorghum (*Sorghum bicolor*) with at least one species of the *Fusarium* section Liseola, primarily *F. moniliforme*. Both hosts were frequently colonized concurrently by several *Fusarium* species. In symptomless maize plants infected by *F. moniliforme* intercellular hyphae occur throughout the host plant (Bacon and Hinton, 1996). Cabral et al. (1993) investigated endophytes of

Juncus spp. in Oregon and used both culture methods and direct microscopy to document unique patterns of internal colonization of leaves and culms by *Alternaria alternata*, *Cladosporium cladosporioides*, *Drechslera* sp., and *Stagonospora innumerosa*.

5. FUNGAL ENDOPHYTES OF DICOTS

Recent research has demonstrated that the genetic conversion of the common pathogen *Colletotrichum magna* to an asymptomatic endophyte may involve mutation in a single gene (Freeman and Rodriguez, 1993). If endophytism is so easily achieved, it might have developed independently in many groups of fungi. Endophytic fungi in dicotyledonous plants are surprisingly common (Table 1). A large proportion of the fungal genera frequently encountered as endophytes of woody perennials are inoperculate discomycetes. An endophytic habit (inconspicuous occupation of healthy tissue) may be widespread in the Rhytismatales (Livsey and Minter, 1994), and in the Phacidiaceae and Hemiphacidiaceae (Leotiales). *Fabrella tsugae* is commonly found fruiting in late winter on the oldest needles of several species of *Tsuga*, where its appearance coincides with normal senescence. *Lophodermium* spp., conspicuous on senescent and abscised conifer needles and recognizable in culture by their anamorphs and culture morphology, are among the most common endophytic isolates from *Abies*, *Picea*, and *Pinus*. Species of *Rhytisma*, such as *R. punctata* on *Acer grandifolia*, and *Coccomyces* may have similar endophytic niches in broad-leaved trees and shrubs. Appearance of fungal fruit bodies on leaves usually coincides with senescence, but maturation and release of ascospores coincides with bud opening and leaf emergence, a pattern typical of "latent pathogens." Anamorphs of *C. stellatus* are isolated at high frequency from healthy leaves of *Mahonia nervosa* (Petrini et al., 1982). Genera of Leotiales recurrently isolated as endophytes include *Tiarosporella*, *Ceuthospora* (Phacideaceae), *Pezicula*, *Dermea*, *Mollisia* (Dermateaceae), and *Chloroschypha* (Leotiaceae). Dothidiales, such as *Phyllosticta* anamorphs of *Guignardia* spp. and *Hormonema* anamorphs of *Dothiora* and *Pringsheimia*, Diaporthales, such as *Phomopsis* spp., and various Hypocreales are also ubiquitous as endophytes.

Xylariaceae represent an exceptionally speciose group of endophytes (Petrini and Petrini, 1985; Petrini et al., 1995; Petrini, 1986). The greatest biodiversity in Xylariaceae is found in the tropics, where they are especially common and diverse in living plants, dead wood, and plant litter (Petrini et al., 1995), but diversity is also quite high among endophytic Xylariaceae infecting temperate zone hosts (Brunner and Petrini, 1992; Petrini and Petrini, 1985). While distribution of field collections might lead to the conclusion that many xylariaceous species are host-specific, the relatively common recovery of anamorphic states in

Table 1 Endophytic Mycobiota Reported for Various Hosts Worldwide

Hosts	Tissue/organ	No. species	Comments	Location	Ref.
Abies alba	Branch bases	44	17 Common	Germany and Poland	Kowalski and Kehr 1992
		50	2 Endemic	Switzerland	Sieber 1989
	Twigs	4	2 Common	Switzerland	Carroll et al., 1974
	Needles	4	—	Switzerland	Ahlick and Sieber 1996
Abies amabilis	Roots Needles	4	—	Oregon and Washington	Carroll and Carroll 1978
Abies concolor	Needles	5	2 Endemic	Oregon and Washington	Carroll and Carroll 1978
Abies grandis	Needles	6	1 Endemic	Oregon and Washington	Carroll and Carroll 1978
Abies lasiocarpa	Needles	4	3 Endemic	Oregon and Washington	Carroll and Carroll 1978
Abies magnifica	Needles	5	2 Endemic	British Columbia	Sieber and Dorworth 1994
Acer macrophyllum	Leaves, twigs	9			
Acer pseudoplatanus	Branch bases,	28	16 Common	Germany, Poland	Kowalski and Kehr 1992
		5	5 Endemic		Pehl and Butin 1994
		22		Germany	
Araceae	leaves	29	2 New species, *Anthostomella aracearum*, *Chaetosphaeria endophytica*	Guyana	Dreyfuss and Petrini 1994

Host	Tissue	No.	Notes	Location	Reference
Arctostaphylos uva-ursi	Leaves,	176	23 Common	Switzerland	Widler and Muller 1984
	leaves, twigs,	13	—	Oregon	Petrini et al., 1982
		35	29 Common	Switzerland	Widler and Muller 1984
	roots	14	8 Common	Switzerland	Wilder and Muller 1984
Betula nana	Stems	1	1 New species, *Myrothecium gronlandicum*	Greenland	Bohn 1993
Betula papyrifera	Roots	5	Aquatic hyphomycetes	Nova Scotia	Sridhar and Barlocher 1992a,b
Betula pendula	Branch bases	23	14 Common, 3 endemic	Germany, Poland	Kowalski and Kehr 1992
Bromeliaceae	Foliage	16	1 New species, *Chaetosphaeria endophytica*	Guyana	Dreyfus and Petrini 1984
Carpinus betulus	Branch base	29	17 Common 3 endemic	Germany, Poland	Kowalski and Kehr 1992
Carpinus caroliniana	Bark	155	11–12 species per tree, 5 Basidiomycetes	New Jersey	Bills and Polishook 1991a
Castanea sativa	Stems	14	10 Common	Switzerland	Bissegger and Sieber 1994
Calocedrus decurrens	Foliage	15	—	Oregon	Petrini and Carroll 1981
Chamaecyparis lawsoniana	Foliage	18	1 Basidiomycete (*Luellia*)	Oregon	Petrini and Carroll 1981
Chamaecyparis thyoides	Leaves and twigs	88	8–12 Species per tree, 1 new species: *Mycoleptodicus* sp.	New Jersey, West Virginia	Bills and Polishook 1991b, and 1992
Dryas octopetala	Leaves	4	—	Switzerland	Fisher et al., 1995
	leaves	23		Switzerland	Fisher et al., 1995
Eucalyptus globulus	Stems	41	9 Basidiomycetes	Uruguay	Bettuci and Saravay 1993

Table 1 Continued

Hosts	Tissue/organ	No. species	Comments	Location	Ref.
Eucalyptus viminalis	Leaves	23	36 Epiphytic species	Argentina	Cabral 1985
Euterpe oleracea	Leaves	57	21 Common, 1 new genus with 3 species: *Idriella species.*	Brazil	Rodrigues 1994
Ericaceae (7 species)	Leaves	18	1 New species: *Cryptocline arctostaphyii*	Switzerland	Petrini 1984, and 1985
Ericaceae (6 species)	Leaves, twigs	22	—	UK	Petrini 1984
Fagus sylvatica	Branch bases	37	23 Common, 3 endemic	Germany, Poland	Kowalski and Kehr 1992
	Stems	18	—	UK	Chapela and Boddy 1988a,b
	Twigs				Petrini and Fisher 1988
	Leaves	18	—	UK	Sieber and Hugentobler 1987
	Leaves	21	—	Switzerland	Sieber and Hugentobler 1987
	Leaves	64	1 Basidiomycete	Switzerland	Pehl and Butin 1994
	Root	19	—	Germany	Ahlick and Sieber 1996
		12	*Cryptosporiopsis radicicola, Phialocephala fortinii*	Germany, Switzerland	
Fraxinus excelsor	Branch bases	36	18 Common, 1 endemic	Germany, Poland	Kowalski and Kehr 1992
Gaultheria shallon	Leaves	13	—	Oregon	Petrini et al., 1982
Hordeum vulgare	Leaves	14	*Didymella phleina*	New Zealand	Riesen and Close 1987
Juncus spp. (3 species)	Leaves	6	—	Oregon	Cabral et al., 1993

Juncus bufonius	Leaves	14	—	Oregon	Cabral et al., 1993
Juniperus communis	Leaves	114	—	Switzerland	Petrini and Müller 1979
Juniperus occidentalis	Leaves	6	—	Oregon	Petrini and Carroll 1981
Larix decidua	Branch bases	27	17 Common, 1 endemic	Germany, Poland	Kowalski and Kehr 1992
Licuala ramasayi	Roots	51		Germany	Kehr 1995
	Leaves	11	1 New species: *Idriella licualae*	Australia	Rodrigues and Samuels 1990
Mahonia aquifolia	Leaves	9	—	Oregon	Petrini et al., 1982
Mahonia nervosa	Leaves	6	—	Oregon	Petrini et al., 1982
Manikara bidentata	Leaves	23	—	Puerto Rico	Lodge et al., 1996
Orchidaceae	Leaves	21	—	Guyana	Dreyfuss and Petrini 1984
Oryza sativa	Leaves and roots	30	—	Italy	Fisher 1992
Picea abies	Branch bases	30	21 Common, 1 endemic, 2 new species: *Phialocephala compacta, P. scopiformis*	Germany	Kowalski and Kehr 1992
	Twigs	58	—	Poland	Sieber 1989
		85	*Tryblidiopsis pinasti*	Switzerland	Barklund and Kowalski 1996
	Roots	120	25 Common	Sweden	Holdenrieder and Sieber 1992
		20	*Phialocephala fortinii*	Germany	Ahlick and Sieber 1996
Picea breweriana	Needles	2	2 Endemic	Germany, Switzerland / Oregon and Washington	Carroll and Carroll 1978

Table 1 Continued

Hosts	Tissue/organ	No. species	Comments	Location	Ref.
Picea engelmannii	Needles	6	6 Endemic	Oregon and Washington	Carroll and Carroll 1978
Picea excelsa	Needles	29	8 Common	Switzerland	Carroll et al., 1974
Picea glauca	Roots	9	Aquatic hyphomycetes	Nova Scotia	Sridhar and Barlocher 1992a,b
Picea mariana	Roots	97	—	Ontario, Canada	Summerbell 1989
Pinus sitchensis	Needles	10	1 Endemic	Oregon and Washington	Carroll and Carroll 1978
Pinus attenuata	Needles	3	1 Endemic	Oregon and Washington	Carroll and Carroll 1978
Pinus banksiana	Needles	6	25 Epiphytic species	Quebec	Legault et al., 1989
Pinus contorta	Needles	4	1 Endemic	Oregon and Washington	Carroll and Carroll 1978
Pinus densiflora	Needles	9	—	Japan	Hata and Futai 1995
Pinus lambertiana	Needles	3	1 Endemic	Oregon and Washington	Carroll and Carroll 1978
Pinus monticola	Needles	1	—	Oregon and Washington	Carroll and Carroll 1978
Pinus nigra	Needles	15	4 Common	France	Carroll et al., 1974
Pinus ponderosa	Needles	7	3 Endemic	Oregon and Washington	Carroll and Carroll 1978
Pinus resinosa	Needles	8	37 Epiphytic species	Quebec	Legault et al., 1989
Pinus sylvestris	Branch bases	28	18 Common, 2 endemic species	Germany	Kowalski and Kehr 1992
	Stems	18	*Pezizella pulvinata* var. *lignicola*	Poland	Petrini and Fisher 1988
	Roots	15	*Phialocephala fortinii*	Finland, Germany, and Switzerland	Ahlick and Sieber 1996

Pseudotsuga menziesii	Needles	9	2 Endemic, 1 new species	Oregon and Washington	Carroll and Carroll 1978
Pteridium aquilinum	Needles	20	—	Switzerland	Carroll et al., 1974
	Various organs	61	6 Common, 1 new species: *Stagonospora pteridicola*	UK	
Quercus ilex	Leaves	50	27 Common	UK, Switzerland, Spain	Fisher et al., 1994
	Twigs	7	—		Fisher et al., 1994
Quercus petraea	Leaves	49	13 Common	Austria	Halmschlager et al., 1993
	Twigs	20	8 Common	Austria	Halmschlager et al., 1993
Quercus robur	Branch bases	31	23 Common, 2 endemic	Germany	Kowalski and Kehr 1992
	Leaves	25	—	Poland	Pehl and Butin 1994
	Twigs	36	12 Common	UK	Petrini and Fisher 1990
Salicornia perennis	Stems	31	—	UK	Petrini and Fisher 1986
Salix fragilis	Twigs	33	9 Common	UK	Petrini and Fisher 1986
Sequoia sempervirens	Needles	3	—	California	Carroll and Carroll 1978
	Needles	26	—	California	Espinosa-Garcia and Langenheim 1990
Stylosanthes guianensis	Needles	12	—	France	Carroll et al., 1974
	Leaves	12	—	Brazil	Periera et al., 1993
Suaeda fruticosa	Leaves	7	—	UK	Fisher and Petrini 1987
Taxus baccata	Stems	9	—	UK	Fisher and Petrini 1987

Table 1 Continued

Hosts	Tissue/organ	No. species	Comments	Location	Ref.
Taxus brevifolia	Needles	6	1 Common	Switzerland	Carroll et al., 1974
Thuja brevifolia	Foliage	5	4 Endemic	Oregon and Washington	Carroll and Carroll 1978
Tilia cordata	Bark Leaves	17	1 Basidiomycete (*Luellia*), 1 new genus: *Taxomyces*	Oregon and Montana	Strobel et al., 1993
Triticum aestivum	Leaves, culms, glumes, roots, seeds	163	*Phaeosphaeria nodorum*	Switzerland	Riesen and Sieber 1985
Tsuga heterophylla	Needles	10	9 Endemic	Oregon and Washington	Carroll and Carroll 1978
Tsuga mertensiana	Needles	8	3 Endemic	Oregon and Washington	Carroll and Carroll 1978
Ulex europeaus	Stems	20	—	UK	Fisher et al., 1986
Ulex galii	Stems	21	—	UK	Fisher et al., 1986
Umbellularia californica	Leaves	5	—	—	Petrini et al., 1982
Various hosts (36 supp)	Leaves and roots	56	—	—	Dreyfuss and Petrini 1984
Various orchid species (51 species)	Roots	67	—	Costa Rica	Richardson and Currah 1995
Various hosts	Shoots, roots	31	6 Desert halophytes	Manitoba	Mushin and Booth 1987
Zea mays	Leaves, stems, and roots	1	*Fusarium moniliforme*	Global	Bacon and Hinton, 1996
Zea mays	Leaves, stems	23	—	UK	Fisher et al., 1992

culture from diverse substrates suggests otherwise (Petrini et al., 1995). Certain hosts or substrata evidently provide specific requirements for formation of teleomorphic states of Xylariaceae. The broad distribution of Xylariaceae, both as endophytes and as saprobes, together with their well-documented ability to produce a variety of bioactive metabolites points to a significant but as yet unelucidated ecological role (Petrini et al., 1995).

6. FUNGAL ENDOPHYTES OF LICHENS

Lichen thalli may be colonized by endophytic fungi. Endophytes have been isolated at high frequencies from lichen thalli (Petrini et al., 1990). A sampling of 17 fruticose lichen samples yielded 506 different fungal taxa, the majority of which (306) were isolated only once (Petrini et al., 1990). A more exhaustive study of only two lichen species revealed differences between fungal assemblages but a similar level of biodiversity on two species from a common site. Most isolates were not representative of lichenicolous fungi, but included many genera and species known from various substrata. The high level of fungal diversity may have been due to the highly porous and heterogeneous nature of the lichen thalli. Roots and other tissues of various tropical epiphytes have been examined by Dreyfuss and Petrini (1984) and Richardson and Currah (1995).

7. ENDOPHYTIC FUNGI OF BRYOPHYTES AND FERNS

Bryophyta present another habitat for endophytes. Comprehensive studies of the endophyte assemblages occurring in these hosts are not common. Weber (1995) rediscovered *Selenospora guernisacii*, an inconspicuous discomycete, associated with mosses in northwestern North America. Döbbler (1979) has reported pyrenocarpous and pezizalean parasites of mosses in Europe. Intracellular associations between achlorophyllous gametophytes of hepatics and pteridophytes and various fungi are apparently widespread (Ligrone et al., 1993; Ligrone and Lopes, 1989; Pocock and Duckett, 1984).

Associations between endophytic fungi and nonvascular plants, such as the hornwort *Phanoceros laevis* (Ligrone, 1988), are known from histological studies. Ascomycetes, basidiomycetes, and zygomycetes have been reported to form associations with a variety of nonvascular hosts, in a range of cytological specializations from simple to complex (Pocock and Duckett, 1984, 1985a, b). Schmid and Oberwinkler (1993) coined the term "lycopodioid mycothallus interaction" to recognize the distinct nature of the association between fungal endophytes and the achlorophyllous gametophytes of *Lycopodium clavatum*. "Endomycothalli" is a general term for the fungal colonization of hepatics (Ligrone et al., 1993).

Endophytes of roots of fern sporophytes are well known (Boullard, 1957,

1979). Most endophytes of terrestrial ferns are considered to be arbuscular mycorrhizae, although there are also numerous reports of septate hyphae in pteridophyte roots (Boullard, 1957; Schmid et al., 1995). Few comprehensive surveys of fungi colonizing fern roots exist. Roots of the fern *Pteridium aquilinum* are colonized by a variety of fungi, including zygomycetes (*Absidia cylidrospora, Mortierella* spp.), several anamorphic ascomycetes, and a sterile basidiomycete (Petrini et al., 1992). Several species of tropical, arboreal epiphytic ferns were found to have roots similarly colonized by an undetermined ascomycete. Epidermal and cortical cells were invaded in a manner resembling ericoid mycorrhizae, with hyphal coils occupying the epidermal and outer cortical cells (Schmid et al., 1995).

8. ENDOPHYTIC FUNGI OF TREE BARK

Bark of trees represents a fruitful niche for fungi that have the capacity to colonize it. Many species of fungi asymptomatically inhabit living bark on twigs and small branches of coniferous and broad-leaved trees, but almost nothing is known of their biology. Resinous young bark of conifers, such as Douglas fir, as well as several smooth-barked deciduous trees such as *Alnus* are frequently colonized by *Arthopyrenia plumbaria* and other members of the Arthopyreniaceae including *Mycoglaena subcoerulescens* (= *Winteria coerulea*) and *Mycoglaena* sp. (''*Pseudoplea*''). In eastern North America *Arthonia impolita* is ubiquitous on young bark of *Pinus strobus*. *Vestigium felicis*, an unusual coelomycete with ''cat's paw''–shaped conidia, is known only from young living twigs of *Thuja plicata* in the Pacific Northwest (Pirozynski and Shoemaker, 1972). Other bark endophytes, primarily ascomycetes, that fruit on recently dead twigs still attached to otherwise healthy trees include members of the Rhytismataceae, *Lachnellula* spp. (Hyaloscyphaceae), *Pezicula* and *Mollisia* (Dermateaceae). *Tryblidiopsis pinastri*, a common circumboreal species that occurs on *Picea spp.*, and *Discocainia treleasei*, on *P. sitchensis*, are representative. Both fruit in abundance in the spring on twigs that have been dead for less than a year and thus must be suspected of routinely colonizing bark of living twigs. Bark-colonizing endophytes may have colonization strategies similar to that of some foliar endophytes that inconspicuously colonize healthy young tissue and fruit only on necrotic tissue. *Colpoma* spp. and *Tryblidiopsis pinastri* are frequently found fruiting on recently killed twigs of oaks and spruces, respectively, but are also frequently isolated as endophytes from healthy inner bark. Other species whose biology appears to be similar are *Therrya pini* and *T. fuckelii* on *Pinus* spp., *Coccomyces strobi* on *P. strobus* L., *Coccomyces heterophyllae* on *Tsuga heterophylla*, and *Lachnellula ciliata* and *L. agasizii* on *P. menziesii* and *Abies* spp., respectively. Insect parasites, such as *Beauveria bassiana* and *Verticillium lecanii*, and *Paecilomyces farinosus* have been isolated from living bark (Bills and Polishook, 1991b) and are not uncommon as endophytes of foliage. The endophytic occurrence of insect parasites has prompted the suggestion that bark may provide an interim substrate for saprobic growth (Carroll, 1991).

9. FUNGAL ENDOPHYTES OF XYLEM

Xylem-colonizing endophytes are a distinct guild of endophytes comprising mainly xylariaceous species such as *Hypoxylon* and related genera, Diaporthales (e.g., *Phragmoporthe, Amphiporthe, Phomopsis*), Hypocreales (*Nectria* spp.), a few basidiomycetes (e.g., *Coniophora*), and infrequently other species more typical of the periderm mycobiota (Bassett and Fenn, 1984; Boddy et al., 1987; Chapela and Boddy, 1988a). In general, species diversity and abundance are low compared to bark, shoot, and foliar endophytes. Chapela and associates (1990, 1991, 1993) have demonstrated that the specialized host attachment and recognition in *Hypoxylon* spp. and the novel germination response triggered in *H. fragiforme* in response to host-specific monolignol glucosides are apparently host-specific recognition mechanisms in these fungi.

Living branches of oak and beech (Chapela and Boddy, 1988b; Boddy, 1992), *Alnus* spp. (Fisher and Petrini, 1990), conifers (Sieber, 1989; Kowalski and Kehr, 1992), and aspen (Chapela, 1989) are frequently endophytically colonized by a mycobiota characteristic for each host. Colonization of host tissue by these fungi occurs initially as disjunct infections that remain quiescent in healthy wood. The higher water content of functional sapwood serves to prevent active invasion and/or colonization, but active colonization resumes in response to decreased water content caused by host death, stress, or injury (Chapela and Boddy, 1988b). Active growth and eventual sporulation is a response to drying of the substrate (Chapela, 1989; Boddy, 1992). Xylem and foliar endophytes have comparable life history strategies that involve establishing early infections followed by a prolonged period of interrupted growth, which enables immediate invasion and saprobic exploitation of the substrate at the onset of physiological stress or senescence. Facultative pathogens, such as *Hypoxylon mammatum* on aspen, as well as many wood decay fungi apparently have adopted this strategy of early endophytic occupation.

10. FUNGAL ENDOPHYTES OF ROOTS

Endomycorrhizae include the well-known arbuscular mycorrhizae (Glomales). These fungi are known to be beneficial to plants by enhancing absorption of soil nutrients such as phosphorus. Arbuscular mycorrhizae are obligate symbionts of plants but form resistant spores that may lie dormant in soil until roots of a susceptible host are near. In plants bearing arbuscular mycorrhizae, an intercellular mycelium permeates the cortex of infected roots.

Many orchids contain endophytic fungi of genus *Rhizoctonia*. Orchid seeds are small and lack sufficient nutrients to fuel orchid embryo development. The endophytic fungus grows out of the seeds after dispersal and enzymatically degrades the bark or other substrate to supply nutrients for the developing orchid embryo (Harley and Smith, 1983; Petrini and Dreyfuss, 1981).

Roots of trees are colonized by a variety of endophytes. Considerable overlap occurs between soil fungi, saprobic rhizosphere fungi, fungal root pathogens, and endophytes, although certain taxa appear to be isolated recurrently and preferentially as symbionts from living roots. Nonmycorrhizal microfungi isolated from serially washed mycorrhizal roots of *Picea mariana* (Summerbell, 1989) were primarily *Mycelium radicis atrovirens*, mycelia sterilia, and penicillia. Holdenrieder and Sieber (1992) compared populations of endophytic fungi colonizing roots of *Picea abies* in relation to site and soil characteristics. Of the 120 taxa recovered, *Mycelium radicis atrovirens*, *Penicillium* spp., *Cylindrocarpon destructans*, and *Cryptosporiopsis* sp. were the most commonly isolated.

Phialocephala fortinii, *P. dimorphospora*, *P. finladia*, *Oidiodendron* spp., *Geomyces* sp., and *Scytalidium vaccinii* (Dalpe et al., 1989) are common components of a guild of endophytes forming associations with roots of alpine ericoid and other perennial hosts. *Mycelium radicis atrovirens*, generally regarded as a heterogeneous taxon, is the name commonly applied to sterile dematiaceous isolates. Because of their unique morphology particularly in association with ericoid hosts, roots colonized by these fungi are sometimes termed ericoid mycorrhizae, although the fungi have apparently a much broader host range (Stoyke and Currah, 1991; Stoyke et al., 1992). Endophytes with dark, septate hyphae dominated the mycobiota isolated from the fine roots or several forest trees and shrubs in Europe and western Canada. A large proportion of these isolates proved to be *Phialocephala fortinii*, a root-inhabiting fungus with an apparently very broad host and geographic distribution (Ahlick and Sieber, 1996). In addition to the ericoid hosts *Phialocephala fortinii* and *Mycelium radicis atrovirens* are commonly isolated from roots of *Fagus sylvatica* (Ahlick and Sieber, 1996), *Abies alba*, *Picea abies*, *Pinus sylvestris*, *P. resinosa*, *P. contorta*, and several additional alpine perennial hosts (Wang and Wilcox, 1985; Holdenrieder and Sieber, 1992; Stoyke et al., 1992; O'dell et al., 1993). Root morphology is described as ectomycorrhizal, ectendomycorrhizal, pseudomycorrhizal, nonmycorrhizal, to possibly pathogenic interactions (Wilcox and Wang, 1987). Differentiation of species is based on morphotypes, but because these are not very informative a current focus is to use biochemical or genetic methods for distinguishing host- or site-specific strains. This is exemplified by the restriction fragment length polymorphism (RFLP) analyses of sterile *P. fortinii* isolates from various alpine hosts by Stoyke et al. (1992) and on the E-strain mycorrhizal fungi, a relatively uniform morpho group that produces chlamydospores on and within infected roots (Egger and Fortin, 1990; Egger et al., 1991). Hyphae in roots appear *Rhizoctonia*-like with "monilioid hyphae" and frequently produce a loose weft of mycelium on the outer root surface. Root colonization is relatively extensive, but intracellular colonization of outer cortical cells is limited. Coiled or branched hyphae and intracellular microsclerotia may be present. In association with conifer hosts ectomycorrhiza-like structures are formed, i.e., intercellular colonization resembles Hartig net (Wilcox and Wang, 1987; O'dell et al., 1993). In culture, the fungi characteristically have septate, thick-walled, dark-pigmented hyphae and are frequently sterile or very slow to sporulate.

Penicillium nodositatum, an endophyte of *Alnus glutinosa* and *A. incana*, forms specialized root associations, "myconodules," which resemble actinorrhizae. Myconodules are similar to the ericoid mycorrhizae, being confined to the outer cortical layer. Cortical cells invaded by the fungus are eventually killed as the hyphae expand to occupy the entire cell with a highly branched and convoluted hyphal mass (Cappellano et al., 1987). *Penicillium janczewskii* Zaleski was reported from similar myconodules on *A. glutinosa* (Valla et al., 1989). Although *P. nodositatum* was first described from root nodules of *Alnus incana* in France (Valla et al., 1989), it has since been recovered together with *Aspergillus tardus* as a foliar endophyte from *Linnea borealis* in Oregon, suggesting the existence of a broader ecological and geographic range for this fungus. Other *Penicillium* and *Aspergillus* spp. have been recovered as endophytes from various hosts in Oregon and from *Sorbus* spp. in Germany, and included more common rhizosphere species such as *P. expansum*, *P. westlingii*, *P. pinophilum*, *P. citreonigrum*, *A. sydowii*, and *A. terreus*. Summerbell (1989) reported 20 species of *Penicillium* from roots of *Picea mariana*, of which *P. spinulosum* and *P. montanense* were frequently encountered. Endophytic penicillia and aspergilli are apparently widespread and heterogeneous.

11. FUNGAL ENDOPHYTES OF GALLS AND CYSTS

Complex interactions between host plants, insects and other invertebrates, and internal fungi often occur in abnormally developed tissues of plants. Galled plant tissues may harbor fungal populations distinct from those of normal tissue. Gall midges (Lasiopterini and Asphondyliidi), for example, introduce a variety of coelomycetous fungi to the galls which become a food source for the developing larvae (Bissett and Borkent, 1988). Cecidiomyid midge galls on Douglas fir needles often support heavy fruiting of the *Meria* anamorph of *Rhabdocline parkeri*, giving the appearance of a fungal disease (Stone, 1988). Other fungi invading galls may be saprobes, insect parasites, or inquilines (organisms inhabiting insect galls not parasitizing the gall maker but otherwise utilizing the gall tissue for food) (Wilson, 1995a).

Phialocephala sp. and *Leptostoma* sp. were the most common endophytic fungi in both healthy and galled needles of *Pinus densiflora*, but needles galled by *Thecodiplosis japonensis* were preferentially colonized by *Phomopsis, Pestalotiopsis, Alternaria alternata*, and an unidentified coelomycete (Hata and Futai, 1995). Secondary invasion of leaf galls by foliar endophytes has been repeatedly reported in connection with incidence of larval mortality (Carroll, 1988; Pehl and Butin, 1994; Butin, 1992; Halmschlager et al., 1993). Wilson (1995a) compared fungal populations of leaves and galls of three host–insect pairs and found that the fungus species colonizing cynipid wasp galls on *Quercus garryana* and *Q. agrifolia* were typical of the endophyte species on those hosts, i.e., galls were secondarily invaded by foliar endophytes. Fungi isolated from galls of the aphid *Pemphigus betae* on cottonwood leaves (*Populus angustifolia*), however, were not found as endophytes of cottonwood, and included probable saprobic penicillia, *Cladosporium cladosporioides*, and *Verticillium lecanii*, which may act either as insect or mycoparasite. An unusual *Lophodermium* sp. confined to galls of the midge *Hormomyia juniperina* on *Juniperus* foliage but distinct from the endophytic *L. juniperinum* is mentioned by Cannon and Hawksworth (1995).

12. PROKARYOTIC ENDOPHYTES OF PLANTS

Prokaryotic endophytes of plants are sometimes thought of as less common or less important than fungal endophytes. This, however, is far from correct as it is clear that healthy plant tissues are frequently colonized by endophytic, intracellular prokaryotes (Hollis, 1951; Kloepper et al., 1992). Bacterial endophytes may be limited to xylem, as in infections by *Clavibacter xyli*, an endophyte commonly encountered in grasses such as *Cynodon dactylon* (Metzler et al., 1992). They may also be distributed in ground tissues of roots, stems, and leaves, as seen in endophytic infection of sugarcane by *Acetobacter diazotrophicus* (Cavalcante and Dobereiner, 1988). Some prokaryotic endophytes are known to significantly impact on the ecology of their host plants. For example, it has been reported that sugarcane has been cultivated continuously for 100 years in parts of Brazil without addition of nitrogen; instead, nitrogen is supplied by the endophytic nitrogen-fixing bacterium *Acetobacter diazotrophicus* (Bodey et al., 1991). Nitrogen-fixing blue-green algae, such as *Nostoc*, are also known to become endophytic in hornworts, cycads, rice, and many other plants. In these symbioses, nitrogen-fixation is the basis for a mutualistic association between plant and microbe (Chanway, 1995).

REFERENCES

Ahlick, K. and Sieber, T. N. (1996). The profusion of dark septate endophytic fungi in non-mycorrhizal fine roots of forest trees and shrubs. *New Phytologist* **132**, 259–270.

Bacon, C. W. and Hinton, D. M. (1996). Symptomless endophytic colonization of maize by *Fusarium moniliforme*. *Canadian Journal of Botany* **74**, 1195–1202.

Barklund, P. and Kowalski, T. (1996). Endophytic fungi in branches of Norway spruce with particular reference to *Tryblidiopsis pinastri*. *Canadian Journal of Botany* **74**, 673–678.

Bassett, E. N. and Fenn, P. (1984). Latent colonization and pathogenicity of *Hypoxylon atropunctatum* on oaks (*Quercus alba, Quercus mailandica, Quercus velutina*). *Plant Disease* **68**, 317–319.

Bettucci, L. and Saravay, M. (1993). Endophytic fungi of *Eucalyptus globulus*: a preliminary study. *Mycological Research* **97**, 679–682.

Bills, G. F. (1996). Isolation and analysis of endophytic fungal communities from woody plants. *In* S. C. Redlin and L. M. Carris (eds.), *Endophytic Fungi in Grasses and Woody Plants. Systematics, Ecology, and Evolution*. APS Press, St. Paul, MN, pp. 31–65.

Bills, G. F. and Polishook, J. D. (1991a). Microfungi from *Carpinus caroliniana*. *Canadian Journal of Botany* **69**, 1477–1482.

Bills, G. F. and Polishook, J. D. (1991b). A new species of *Mycoleptodiscus* from living foliage of *Chamaecyparis thyoides*. *Mycotaxon* **43**, 453–460.

Bills, G. F. and Polishook, J. D. (1992). Recovery of endophytic fungi from *Chamaecyparis thyoides*. *Sydowia* **44**, 1–12.

Bissett, J. and Borkent, A. (1988). Ambrosia galls: the significance of fungal nutrition in the evolution of the Cecidomyiidae (Diptera). *In* K. A. Pirozynski and D. L. Hawksworth (eds.), *Coevolution of Fungi with Plants and Animals*. Academic Press, London, pp. 203–225.

Bissegger, M. and Sieber, T. N. (1994). Assemblages of endophytic fungi in coppice shoots of *Castanea sativa*. *Mycologia* **86**, 648–655.

Boddey, R. M., Urquiaga, S., Reis, V. and Dobereiner J. (1991). Biological nitrogen fixation associated with sugarcane. *Plant & Soil* **137**, 111–117.

Boddy, L. (1992). Development and function of fungal communities in decomposing wood. *In* G. C. Carroll and D. T. Wicklow (eds.), *The Fungal Community: Its Organization and Role in the Ecosystem*, 2nd ed. Marcel Dekker, New York, pp. 749–782.

Boddy, L. and Griffith, G. S. (1989). Role of endophytes and latent invasion in the development of decay communities in sapwood of angiospermous trees. *Sydowia* **41**, 41–73.

Boddy, L., Bardsley, D. W. and Gibson, O. M. (1987). Fungal communities on attached ash branches. *New Phytologist* **107**, 143–154.

Bohn, M. (1993). *Myrothecium groenlandicum* sp. nov., a presumed endophytic fungus of *Betula nana* (Greenland). *Mycotaxon* **46**, 335–341.

Bose, S. R. (1947). Hereditary (seed-borne) symbiosis in *Casuarina equistifolia* Forst. *Nature* **159**, 512–514.

Boullard, B. (1957). La mycotrophie chez les ptéridophytes. Sa fréquence, ses charactères, sa signification. *Botaniste* **41**, 5–185.

Boullard, B. (1979). Consideration sur la sybiose chez les ptéridophytes. *Syllogeus* **19**, 1–59.

Boursnell, J. G. (1950). The symbiotic seed-borne fungus in the Cistaceae. *Annals of Botany* **24**, 217–243.

Brunner, F. and Petrini, O. (1992). Taxonomy of some *Xylaria* species and xylariaceous endophytes by isozyme electrophoresis. *Mycological Research* **96**, 723–733.

Butin, H. (1992). Effect of endophytic fungi from oak on the mortality of leaf inhabiting gall insects. *European Journal of Forest Pathology* **22**, 237–246.

Cabral, D. (1985). Phyllosphere of *Eucalyptus viminalis*: dynamics of fungal populations. *Transactions of the British Mycological Society* **85**, 501–511.

Cabral, D., Stone, J. K. and Carroll, G. C. (1993). The internal mycobiota of *Juncus* spp.: microscopic and cultural observations of infection patterns. *Mycological Research* **97**, 367–376.

Cavalcante, V. A. and Dobereiner, J. (1988). A new acid tolerant nitrogen-fixing bacterium associated with sugarcane. *Plant and Soil* **108**: 23–31.

Cannon, P. F. and Hawksworth, D. L. (1995). The diversity of fungi associated with vascular plants: the known, the unknown and the need to bridge the knowledge gap. *Advances in Plant Pathology* **11**, 277–302.

Cappellano, A., deQuartre, B., Valla, G. and Moiroud, A. 1987. Root nodule formation by *Penicillium* sp. on *Alnus glutinosa* and *Alnus incana*. *Plant and Soil* **104**, 45–51.

Carroll, F. E., Müller, E. and Sutton, B. C. (1974). Preliminary studies on the incidence of needle endophytes in some European conifers. *Sydowia* **29**, 87–103.

Carroll, G. C. (1986). The biology of endophytism in plants with particular reference to woody perennials. *In* N. J. Fokkema and J. Van den Heuvel (eds.), *Microbiology of the Phylloplane*. Cambridge University Press, Cambridge, pp. 205–222.

Carroll, G. C. (1988). Fungal endophytes in stems and leaves: from latent pathogen to mutualistic symbiont. *Ecology* **69**, 2–9.

Carroll, G. C. (1991). Fungal associates of woody plants as insect antagonists in leaves and stems. *In* P. Barbosa, V. A. Krishnik, and C. G. Jones (eds.), *Microbial Mediation of Plant–Herbivore Interactions*. John Wiley and Sons, New York, pp. 253–271.

Carroll, G. C. (1995). Forest endophytes: pattern and process. *Canadian Journal of Botany* **73**, 1316–1324.

Carroll, G. C. and Carroll, F. E. (1978). Studies on the incidence of coniferous needle endophytes in the Pacific Northwest. *Canadian Journal of Botany* **56**, 3034–3043.

Chanway, C. P. (1996). Endophytes: they're not just fungi! *Canadian Journal of Botany* **74**, 321–322.

Chapela, I. H. (1989). Fungi in healthy stems and branches of American beech and aspen: a comparative study. *New Phytologist* **113**, 65–75.

Chapela, I. H. and Boddy, L. (1988a). Fungal colonization of attached beech branches. I. Early stages of development of fungal communities. *New Phytologist* **110**, 39–45.

Chapela, I. H. and Boddy, L. (1988b). Fungal colonization of attached beech branches. II. Spatial and temporal organization of communities arising from latent invaders in bark and functional sapwood, under different moisture regimes. *New Phytologist* **110**, 47–57.

Chapela, I. H., Petrini, O. and Petrini, L. E. (1990). Unusual ascospore germination in *Hypoxylon fragiforme*: first steps in the establishment of an endophytic symbiosis. *Canadian Journal of Botany* **68**, 2571–2575.

Chapela, I. H. and Hagmann, L. (1991). Monolignol glucosides as specific recognition messengers in fungus-plant symbioses. *Physiological and Molecular Plant Pathology* **39**, 289–298.

Chapela, I. H. and Bielser, G. (1993). The physiology of ascospore eclosion in *Hypoxylon fragiforme*: mechanisms in the early recognition and establishment of an endophytic symbiosis. *Mycological Research* **97**, 157–162.

Christensen, M. A. (1981). Species diversity and dominance in fungal communities. *In* D. T. Wicklow and G. C. Carroll (eds.), *The Fungal Community, Its Organization and Role in the Ecosystem*. Marcel Dekker, New York, pp. 201–232.

Clark, E. M., White, J. F. Jr. and Patterson, R. M. (1983). Improved histochemical techniques for the detection of *Acremonium coenophialum* in tall fescue and methods of in vitro culture of the fungus. *Journal of Microbiological Methods* **1**, 149–155.

Clay, K. (1988). Fungal endophytes of grasses: a defensive mutualism between plants and fungi. *Ecology* **69**, 10–16.

Dalpe, Y., Litton, W. and Sigler, L. (1989). *Scytalidium vaccinii* sp. nov., an ericoid endophyte of *Vaccinium angustifolium* roots. *Mycotaxon* **35**, 371–377.

Döbbler, P. (1979). Untersuchungen an moosparasitischen Pezizales aus der Verwandtschaft von *Octospora*. *Nova Hedwigia* **31**, 817–864.

Dreyfuss, M. and Petrini, O. (1984). Further investigations on the occurrence and distribution of endophytic fungi in tropical plants. *Botanica Helvetica* **94**, 34–40.

Egger, K. and Fortin, J. A. (1990). Identification of E-strain mycorrhizal fungi by restriction fragment analysis. *Canadian Journal of Botany* **68**, 1482–1488.

Egger, K., Danielson, R. M. and Fortin, J. A. (1991). Taxonomy and population structure of E-strain mycorrhizal fungi inferred from ribosomal and mitochondrial DNA polymorphisms. *Mycological Research* **95**, 866–872.

Espinosa-Garcia, F. J. and Langenheim, J. H. (1990). The endophytic fungal community in leaves of a coastal redwood population—diversity and spatial patterns. *New Phytologist* **116**, 89–97.

Fisher, P. J., Anson, A. E. and Petrini, O. (1986). Fungal endophytes in *Ulex europaeus* and *Ulex gallii*. *Transactions of the British Mycological Society* **86**, 153–193.

Fisher, P. J., Anson, A. E. and Petrini, O. (1984a). Novel antibiotic activity of an endophytic *Cryptosporiopsis* sp. isolated from *Vaccinium myrtillus*. *Transactions of the British Mycological Society* **83**, 145–187.

Fisher, P. J., Graf, F., Petrini, L. E., Sutton, B. C. and Wookey, P. A. (1995). Fungal endophytes of *Dryas octopetala* from a high polar semidesert and from the Swiss Alps. *Mycologia* **87**, 319–323.

Fisher, P. J. and Petrini, O. (1987). Location of fungal endophytes in tissues of *Suaeda fruticosa*: a preliminary study. *Transactions of the British Mycological Society* **89**, 246–249.

Fisher, P. J. and Petrini, O. (1990). A comparative study of fungal endophytes in xylem and bark of *Alnus* species in England and Switzerland. *Mycological Research* **94**, 313–319.

Fisher, P. J. and Petrini, O. (1992). Fungal saprobes and pathogens as endophytes of rice (*Oryza sativa* L.). *New Phytologist* **120**, 137–143.

Fisher, P. J., Petrini, O. and Lappin Scott, H. M. (1992). The distribution of some fungal and bacterial endophytes in maize (*Zea mays* L.). *New Phytologist* **122**, 299–305.

Fisher, P. J., Petrini, O., Petrini, L. E. and Sutton, B. C. (1994). Fungal endophytes from the leaves and twigs of *Quercus ilex* L. from England, Majorca, and Switzerland. *New Phytologist* **127**, 133–137.

Fisher, P. J., Webster, J. and Petrini, O. (1991). Aquatic hyphomycetes and other fungi in living aquatic and terrestrial roots of *Alnus glutinosa*. *Mycological Research* **95**, 543–547.

Fisher, P. J. and Punithalingham, E. (1993). *Stagonospora pteridiicola* sp. nov., a new endophytic coelomycete in *Pteridium aquilinum*. *Mycological Research* **97**, 661–664.

Fisher, P. J., Spooner, B. M. and Petrini, O. (1987). *Pezizella pulvinata* var. *lignicola* var. nov., and endophyte of the xylem of *Pinus sylvestris*. *Mycological Research* **89**, 593–596.

Fisher, P. J. and Webster, J. (1991). Aquatic hyphomycetes and other fungi in living aquatic and terrestrial roots of *Alnus glutinosa*. *Mycological Research* **95**, 543–547.

Freeman, S. and Rodriguez, R. J. (1993). Genetic conversion of a fungal plant pathogen to a nonpathogenic, endophytic mutualist. *Science* **260**, 75–78.

Halmschlager, E., Butin, H. and Donaubauer, E. (1993). Endophytische Pilze in Blättern un Zwiegen von *Quercus petraea*. *European Journal of Forest Pathology* **23**, 51–63.

Harley, J. L. and Smith, S. E. (1983). *Mycorrhizal Symbiosis*. Academic Press, New York.

Hata, K. and Futai, K. (1995). Endophytic fungi associated with healthy pine needles and needles infested by the pine needle gall midge, *Thecodiopsis japonensis*. *Canadian Journal of Botany* **73**, 384–390.

Hirsch, G. and Braun, U. (1992). Communities of parasitic microfungi. *In* W. Winterhoff (ed.), *Handbook of Vegetation Science, Vol. 19, Fungi in Vegetation Science*. Kluwer Academic Dordrecht, pp. 225–250.

Holdenrieder, O. and Sieber, T. N. (1992). Fungal associations of serially washed healthy non-mycorrhizal roots. *Mycological Research* **96**, 151–156.

Hollis, J. P. (1951). Bacteria in healthy potato tissue. *Phytopathology* **41**, 320–366.

Kehr, V. (1995). Endophytische Pilze der Lärche (*Larix decidua* Mill.) und ihr Einfluß auf phenolische Inhaltstoffe der Lärchenwurzel. Diplomarbeit der Fakultät für Biologie der Universtität Tübingen, March 1995, Braunschweig.

Kloepper, J. W., Schippers, B. and Bakker, P. A. H. M. (1992). Proposed elimination of term endorhizosphere. *Phytopathology* **82**, 726–727.

Kowalski, T. and Kehr, R. D. (1992). Endophytic fungal colonization of branch bases in several forest tree species. *Sydowia* **44**, 137–168.

Kowalski, T. and Kehr, R. D. (1995). Two new species of *Phialocephala* occurring on *Picea* and *Alnus*. *Canadian Journal of Botany* **73**, 26–32.

Legault, D., Dessureault, M. and Laflamme, G. (1989). Mycoflore des aguilles de *Pinus banksiana* et *Pinus resinosa*. I. Champignons endophytes. *Canadian Journal of Botany* **67**, 2052–2060.

Leslie, J. F., Pearson, C. A. S., Nelson, P. E. and Tousson, T. A. (1990). *Fusarium* spp. from corn, sorghum, and soybean fields in the central and eastern United States. *Phytopathology* **80**, 343–350.

Ligrone, R. (1988). Ultrastructure of a fungal endophyte in *Phanoceros laevis* (L.) Prosk. (Anthocerotophyta). *Botanical Gazette* **149**, 92–100.

Ligrone, R. and Lopes, C. Cytology and development of a mycorrhiza-like infection in the gametophyte of *Conocephalum conicum* (L.) Dum. (Marchantiales, Hepatophyta). *New Phytologist* **111**, 423–433.

Ligrone, R., Pocock, K. and Duckett, J. G. (1993). A comparative ultrastructural study of endophytic basidiomycetes in the parasitic achlorophyllous hepatic *Cryptothallus mirabilis* and the closely allied photosynthetic species *Aneura pinguis* (Metzgeriales). *Canadian Journal of Botany* **71**, 666–679.

Livsey, S. and Minter, D. W. (1994). The taxonomy and biology of *Tryblidiopsis pinastri*. *Canadian Journal of Botany* **72**, 549–557.

Lodge, D. J., Fisher, P. J. and Sutton, B. C. (1996). Endophytic fungi of *Manilkara bidentata* leaves in Puerto Rico. *Mycologia* **88**, 733–738.

Maranová, L. and Fisher, P. J. (1991). A new endophytic hyphomycete from alder roots. *Nova Hedwigia* **52**, 33–37.

Mauseth, J. D., Montenegro, G. and Walckowiak, A. M. (1985). Host infection and flower formation by the parasite *Tristerix aphyllus* (Loranthaceae). *Canadian Journal Botany*. **63**, 567–581.

McCutcheon, T. L., Carroll, G. C. and Schwab, S. (1993). Genotypic diversity in populations of a fungal endophyte from Douglas-fir. *Mycologia* **85**, 180–186.

Metzler, M. C., Zhang, Y. P. and Chen, T. A. (1992). Transformation of the gram-positive bacterium *Clavibacter xyli* subsp. *cynodontis* by electroporation with plasmids from the IncP incompatibility group. *Journal of Bacteriology* **174**, 4500–4503.

Moore-Landecker, E. (1982). *Fundamentals of the Fungi*. Prentice-Hall, Englewood Cliffs, NJ.

Muhsin, T. M. and Booth, T. (1987). Fungi associated with halophytes of an inland salt marsh, Manitoba, Canada. *Canadian Journal of Botany* **65**, 1137–1151.

Muhsin, T. M., Booth, T. and Zwain, K. H. (1989). A fungal endophyte associated with desert parasitic plant. *Kavaka* **17**, 1–5.

Noble, H. M., Langley, D., Sidebottom, P. J., Lane, S. J. and Fisher, P. J. (1991). An echinocandin from an endophytic *Cryptosporiopsis* sp. and *Pezicula* sp. in *Pinus sylvestris* and *Fagus sylvatica*. *Mycological Research* **95**, 1439–1440.

O'dell, T. E., Massicote, H. B. and Trappe, J. M. (1993). Root colonization of *Lupinus latifolius* Agardh. and *Pinus contorta* Dougl. by *Phialocephala fortinii* Wang & Wilcox. *New Phytologist* **124**, 93–100.

Pehl, L. and Butin, H. (1994). Endophytische Pilze in Blättern von Laubbäumen und ihre Beziehungen zu Blattgallen (Zoocecidien). *Mitteilungen aus der Biologischen Bundesanstalt für Land- und Forstwirtschaft* **297**, Berlin.

Pereira, J., Azevedo, J. and Petrini, O. (1993). Endophytic fungi of *Stylosanthes*: a first report. *Mycologia* **85**, 362–364.

Peterson, R. L. and Howarth, M. J. (1981). Interactions between a fungal endophyte and gametophyte cells in *Psilotum nudum*. *Canadian Journal of Botany* **59**, 711–720.

Petrini, L. E. and Petrini, O. (1985). Xylariaceous fungi as endophytes. *Sydowia* **38**, 216–234.

Petrini, O. (1984). *Cryptocline arctostaphyli* sp. nov., an endophyte of *Arctostaphylos uva-ursi* and other Ericaceae. *Sydowia* **37**, 238–241.

Petrini, O. (1985). Wirtsspezifität endophytischer Pilze bei einheimischen Ericaceae. *Botanica Helvetica* **95**, 213–238.

Petrini, O. (1987). Endophytic fungi of alpine Ericaceae. The endophytes of *Loiseleuria procumbens*. *In* G. A. Laursen, J. F. Ammirati, and S. A. Redhead (eds.), *Arctic and Alpine Mycology*, Vol. 2. Plenum Press, New York, pp. 71–77.

Petrini, O. (1986). Taxonomy of endophytic fungi in aerial plant tissues. *In* N. J. Fokkema and J. van den Heuvel (eds.), *Microbiology of the Phyllosphere*. Cambridge University Press, Cambridge, pp. 175–187.

Petrini, O. (1991). Fungal endophytes of tree leaves. *In* J. H. Andrews and S. S. Hirano (eds.), *Microbial Ecology of Leaves*. Springer-Verlag, New York, pp. 179–187.

Petrini, O. and Carroll, G. C. (1981). Endophytic fungi of some Cupressaceae in Oregon. *Canadian Journal of Botany* **59**, 629–636.

Petrini, O. and Dreyfuss, M. (1981). Endophytische Pilze in Epiphytischen Araceae, Bromeliaceae, und Orchidaceae. *Sydowia* **34**, 135–148.

Petrini, O. and Müller, E. (1979). Pilzliche Endophyten, am Beispeil von *Juniperus communis* L. *Sydowia* **32**, 224–251.

Petrini, O. and Fisher, P. J. (1986). Fungal endophytes of *Salicornia perennis*. *Transactions of the British Mycological Society* **87**, 647–651.

Petrini, O. and Fisher, P. J. (1988). A comparative study of fungal endophytes in xylem and whole stem of *Pinus sylvestris* and *Fagus sylvatica*. *Transactions of the British Mycological Society* **91**, 233–238.

Petrini, O. and Fisher, P. J. (1990). Occurrence of fungal endophytes in twigs of *Salix fragilis* and *Quercus robur*. *Mycological Research* **94**, 1077–1080.

Petrini, O., Fisher, P. J. and Petrini, L. E. (1992). Fungal endophytes of bracken (*Pteridium aquilinium*) with some reflections on biological control. *Sydowia* **44**, 282–293.

Petrini, O., Hake, U. and Dreyfus, M. M. (1990). An analysis of fungal communities isolated from frutcose lichens. *Mycologia* **82**, 444–451.

Petrini, O., Petrini, L. E. and Rodrigues, K. F. (1995). Xylariaceous endophytes: an exercise in biodiversity. *Fitopatologia brasiliera* **20**, 531–539.

Petrini, O., Stone, J. and Carroll, F. E. (1982). Endophytic fungi in evergreen shrubs in western Oregon: a preliminary study. *Canadian Journal of Botany* **60**, 789–796.

Petrini, O., Sieber, T. N., Toti, L. and Viret, O. (1992). Ecology, metabolite production, and substrate utilization in endophytic fungi. *Natural Toxins* **1**, 185–196.

Pirozynski, K. A. and Shoemaker, R. A. (1972). *Vestigium*, a new genus of Coelomycetes. *Canadian Journal of Botany* **50**, 1163–1165.

Pocock, K. and Duckett, J. G. (1984). A comparative ultrastructural analysis of the fungal endophytes in *Cryptothallus mirabilis* Malm. and other British thalloid hepatics. *Journal of Bryology* **13**, 227–233.

Pocock, K. and Duckett, J. G. (1985a). Fungi in hepatics. *Bryological Times* **31**, 2–3.

Pocock, K. and Duckett, J. G. (1985b). On the occurrence of branched and swollen rhizoids in British hepatics: their relationships with the substratum and associations with fungi. *New Phytologist* **99**, 281–304.

Polishook, J. D., Dombrowski, A. W., Tsou, N. N., Salituro, G. M. and Curotto, J. E. (1993). Preussomerin D from the endophyte *Hormonema dematioides*. *Mycologia* **85**, 62–64.

Ramanan, B. V., Balakrishna, P. and Suryanarayanan, T. S. (1996). Search for seed borne endophytes in rice (*Oryza sativa*) and wild rice (*Porteresia coarctata*). *Rice Biotechnology Quarterly* **27**, 7–8.

Rayner, M. C. (1915). Obligate symbiosis in *Calluna vulgaris*. *Annals of Botany* **29**, 96–131.

Rayner, M. C. (1929). The biology of fungus infection in the genus *Vaccinium*. *Annals of Botany* **43**, 55–70.

Richardson, K. A. and Currah, R. S. (1995). The fungal community associated with the roots of some rainforest epiphytes of Costa Rica. *Selbyana* **16**, 49–73.

Riesen, T. K. and Close, R. C. (1987). Endophytic fungi in propiconizole-treated and untreated barley leaves. *Mycologia* **79**, 546–552.

Riesen, T. K. and Sieber, T. (1985). *Endophytic Fungi In Winter Wheat* (*Triticum aestivum* L.). Mikrobiologisches Institut, Swiss Federal Institute of Technology, Zurich.

Rodrigues, K. F. (1994). The foliar fungal endophytes of the Amazonian palm *Euterpe oleracea*. *Mycologia* **86**, 376–385.

Rodrigues, K. F. and Samuels, G. J. (1990). Preliminary study of endophytic fungi in a tropical palm. *Mycological Research* **94**, 827–830.

Rodrigues, K. F. and Samuels, G. J. (1992). *Idriella* species endophytic in palms. *Mycotaxon* **48**, 271–276.

Roll-Hansen, F. and Roll-Hansen, H. (1980a). Microorganisms which invade *Picea abies* in seasonal stem wounds. II. Ascomycetes, fungi imperfecti, and bacteria. *European Journal of Forest Pathology* **10**, 396–410.

Rollinger, J. L. and Langenheim, J. H. (1993). Geographic survey of fungal endophyte community composition in leaves of coastal redwood. *Mycologia* **85**, 149–156.

Ruscoe, Q. W. (1971). Mycoflora of living and dead leaves of *Nothofagus truncata*. *Transactions of the British Mycological Society* **56**, 463–474.

Schmid, E. and Oberwinkler, F. (1993). Mycorrhiza-like interaction between the achlorophyllous gametophyte of *Lycopodium clavatum* L. and its fungal endophyte studied by light and electron microscopy. *New Phytologist* **124**, 69–81.

Schmid, E., Oberwinkler, F. and Gomez, L. D. (1995). Light and electron microscopy of a host–fungus interaction in the roots of some epiphytic ferns from Costa Rica. *Canadian Journal of Botany* **73**, 991–996.

Schulz, B., Wanke, U., Draeger, S. and Aust, H.-J. (1993). Endophytes from herbaceous plants and shrubs: effectiveness of surface-sterilization methods. *Mycological Research* **97**, 1447–1450.

Schulz, B., Sucker, J., Aust, H. J., Krohn, K., Ludewig, K., Jones, P. G. and Döring, D. (1995). Biologically active secondary metabolites of endophytic *Pezicula* species. *Mycological Research* **99**, 1007–1015.

Sherwood-Pike, M., Stone, J. K. and Carroll, G. C. (1986). *Rhabdocline parkeri*, a ubiquitous foliar endophyte of Douglas-fir. *Canadian Journal of Botany* **64**, 1849–1855.

Sieber, T. N. (1989). Endophytic fungi in twigs of healthy and diseased Norway spruce and white fir. *Mycological Research* **92**, 322–326.

Sieber, T. N. and Dorworth, C. E. (1994). An ecological study of endophytic fungi in *Acer macrophyllum* in British Columbia; in search of candidate mycoherbicides. *Canadian Journal of Botany* **72**, 1397–1402.

Sieber, T. N. and Hugentobler, C. (1987). Endophytishche Pilze in Blättern und Äten gesunder und geschädigter Buchen (*Fagus sylvatica* L.). *European Journal of Forest Pathology* **17**, 411–425.

Sieber, T. N., Sieber-Canavesi, F. and Dorworth, C. E. (1991). Endophytic fungi of red alder (*Alnus rubra*) leaves and twigs in British Columbia. *Canadian Journal of Botany* **69**, 407–411.

Sridhar, K. R. and Bärlocher, F. (1992a). Endophytic aquatic hyphomycetes of roots of spruce, birch, and maple. *Mycological Research* **96**, 305–308.

Sridhar, K. R. and Bärlocher, F. (1992b). Aquatic hyphomycetes in spruce roots. *Mycologia* **84**, 580–584.

Stone, J. K. (1988). Fine structure of latent infections by *Rhabdocline parkeri* on Douglas fir. *Canadian Journal of Botany* **66**, 45–54.

Stoyke, G. and Currah, R. S. (1991). Endophytic fungi from the mycorrhizae of alpine ericoid plants. *Canadian Journal of Botany* **69**, 347–352.

Stoyke, G., Egger, K. and Currah, R. S. (1992). Characterization of sterile endophytic fungi from the mycorrhizae of subalpine plants. *Canadian Journal of Botany* **70**, 2009–2016.

Strobel, G., Stierle, A. D. and Hess, W. M. (1993). *Taxomyces andreane*, a proposed new taxon for a bulbilliferous hyphomycete associated with pacific yew (*Taxus brevifolia*). *Mycotaxon* **57**, 71–80.

Summerbell, R. C. (1989). Microfungi associated with the mycorrhizal mantle and adjacent microhabitats within the rhizosphere of black spruce. *Canadian Journal of Botany* **67**, 1085–1095.

Toti, L., Viret, O., Horat, G. and Petrini, O. (1993). Detection of the endophyte *Discula umbrinella* in buds and twigs of *Fagus sylvatica*. European Journal of Forest Pathology **23**, 147–152.

Valla, G., Cappellano, A., Hugueney, R. and Moiroud, A. (1989). *Penicillium nodositatum* Valla, a new species inducing myconodules on *Alnus* roots. Plant and Soil **114**, 142–146.

Viret, O. and Petrini, O. (1994). Colonization of beech leaves (*Fagus sylvatica*) by the endophyte *Discula umbrinella* (teleomorph: *Apiognomonia errabunda*). *Mycological Research* **98**, 423–432.

Wang, C. J. K. and Wilcox, H. E. (1985). New species of ectendomycorrhizal and pseudo-mycorrhizal fungi: *Phialocephala finlandia*, *Chloridium paucisporum*, and *Phialocephala fortinii*. *Mycologia* **77**, 951–958.

Weber, N. S. (1995). Western American Pezizales. *Selenaspora guernisacii*, new to North America. *Mycologia* **87**, 90–95.

Wennström, A. (1994). Endophyte—misuse of an old term. *Oikos* **73**, 535–536.

White, J. F., Jr. (1987). Widespread distribution of endophytes in the Poaceae. *Plant Disease* **71**, 340–342.

White, J. F., Jr. and Halisky, P. M. (1992). Association of systemic fungal endophytes with stock-poisoning grasses. *In* L. F. James, R. F. Keeler, E. M. Bailey, P. R. Cheeke, and M. P. Hegarty (eds.), *Poisonous Plants.* Iowa State University Press, Ames, pp. 574–578.

White, J. F., Jr., Morrow, A. C. and Morgan-Jones, G. (1990). Endophyte–host associations in forage grasses. XII. A fungal endophyte of *Trichachne insularis* belonging to *Pseudocercosporella. Mycologia* **82**, 218–226.

White, J. F., Jr. and Owens, J. R. (1992). Stromal development and mating system of *Balansia epichloë*, a leaf-colonizing endophyte of warm-season grasses. *Applied and Environmental Microbiology* **58**, 513–519.

Widler, B. and Müller, E. (1984). Untersuchungen über endophytische Pilze von *Arctostaphylos uva-ursi* (L.) Sprengel (Ericaceae). *Botanica Helvetica* **94**, 307–337.

Wilcox, H. E. and Wang, C. J. K. (1987). Mycorrhizal and pathological associations of dematiaceous fungi in roots of 7-month-old tree seedlings. *Canadian Journal of Forest Research* **17**, 884–889.

Wilson, D. (1995a). Fungal endophytes which invade insect galls: insect pathogens, benign saprophytes, or fungal inquilines? *Oecologia* **103**, 255–260.

Wilson, D. (1995b). Endophyte—the evolution of a term, and clarification of its use and definition. *Oikos* **73**, 274–276.

Wilson, D. and Carroll, G. C. (1994). Infection studies of *Discula quercina*, an endophyte of *Quercus garryana. Mycologia* **86**, 635–647.

2

The Rhynie Chert Ecosystem: A Model for Understanding Fungal Interactions

Thomas N. Taylor and Edith L. Taylor
Museum of Natural History and Biodiversity Research Center,
University of Kansas, Lawrence, Kansas

1. INTRODUCTION

Fungi were an important component of ancient ecosystems where they functioned as decomposers of lignin and cellulose much as they do today. While few would challenge this assumption, our understanding of the complexity of fungal interactions in ancient ecosystems has been greatly hampered by an inability to sufficiently document fungi in time and space (Taylor and Taylor, 1997). Both fossil specimens (Taylor, 1993) and divergence times based on molecular sequence data (e.g., Bruns et al., 1991) suggest that many fungal groups have a long geological history. As a result of current molecular techniques it is hypothesized that the origin of eukaryotic kingdoms, including the fungi, took place about 1 billion years ago, and that terrestrial fungi diverged from their chytrid ancestors during the Cambrian (approximately 550 million years ago) (Berbee and Taylor, 1993). Despite the fact that well-preserved cyanobacteria are known from many Proterozoic deposits, no fungal remains have been recovered from rocks of this age.

The earliest fossils that might represent fungi occur as traces in Cambrian reefs (Kobluk and James, 1979) and in the shells of chitinozoa (Grahn, 1981). These fossils have been suggested as evidence of endolithic fungi, but just as likely represent the activities of algae. To date the first evidence of terrestrial fungi comes from the Early Silurian in the form of narrow, branched septate filaments, some of which may contain chains of conidia (Pratt et al., 1978). Similar hyphae, some showing perforate septae and flask-shaped appendages like those produced by imperfect stages of some Ascomycetes, have also been reported from the Upper Silurian of Sweden (Sherwood-Pike and Gray, 1985). Both of these early records were based on acid maceration of sediment and, while important in documenting the existence of terrestrial fungi as early as the Silurian, provide no information about the interactions of these fungi in the ecosystems in which they lived.

Today the Lower Devonian (Edwards et al., 1992) or perhaps the Upper Silurian (Tims and Chambers, 1984) marks the first appearance of plants with vascular tissue. However, it is increasingly clear that the cells in the conducting strand of some of these organisms are quite unlike the tracheids of vascular land plants today. Although the occurrence of dispersed spores in rocks as early as the mid-Ordovician has been used to document the presence of vascular plants far earlier than the Upper Silurian, some of these spores have been interpreted as the products of plants at a bryophytic grade of evolution (Taylor, 1995). Irrespective of the affinities of these early land plants, none are sufficiently preserved to provide any information about potential interactions with fungi. To date the best example of an ecosystem that is preserved almost in its entirety is the 400-million-year-old Lower Devonian Rhynie chert flora.

2. RHYNIE CHERT FUNGI

The biological significance of the Rhynie chert Lagerstätten was first demonstrated by the British researchers Kidston and Lang in 1917, who initiated a series of papers that described and illustrated the flora of the Rhynie chert. In addition to exquisitely preserved land plants, these authors also recorded the existence of a number of fungal remains (Fig. 1a) (Kidston and Lang, 1921). While the macroplants have served as anatomical and morphological prototypes of early terrestrial plants, only recently has the fungal component in this ecosystem been examined in detail. It is now apparent that the Rhynie chert ecosystem not only contained a diverse mycoflora but that within this ancient community are numerous examples of fungal interactions with the other organisms preserved within the chert.

The Rhynie chert site contains several plant-bearing beds that are preserved as siliceous sinters and that are regarded as approximately 400 million years old (Pragian) based on palynomorph assemblages (Richardson, 1967). Sedimentological and paleobiological evidence suggests that this ancient ecosystem consisted of a series of freshwater lakes that were periodically desiccated (Trewin and Rice, 1992). The organisms growing in these lakes and along the shores were preserved in silica that had its origin from airborne ash formed as a result of the extensive volcanic activity in the area. The small plants grew on a sandy substrate and because of their relatively small size were preserved in growth position. Also

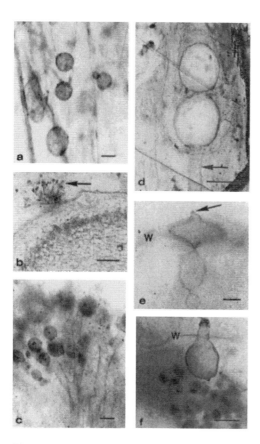

Figure 1 (a) Section of Rhynie chert stem showing several chlamydospores and hyphae. Bar = 20 μm. (b) *Palaeoblastocladia* thallus (arrow) extending from the surface of a stem. Bar = 200 μm. (c) *Palaeoblastocladia* thallus showing terminal zoosporangia. Bar = 20 μm. (d) Portion of *Palaeonitella* showing two enlarged (hypertrophied) cells of axis. Compare with normal cell (arrow). Bar = 200 μm. (e) *Krispiromyces* showing rhizomycelium extending into cell lumen. Note zoosporangium with discharge cap (arrow) extending above spore wall (W). Bar = 10 μm. (f) *Milleromyces* zoosporangium with discharge tube extending through cell wall (W). Note dark granular bodies. Bar = 10 μm.

preserved in these ephemeral pools were cyanobacteria, eubacteria, and algae, which were closely associated with the fungi in a variety of biological interactions. This Early Devonian locality thus affords a view of a 400-million-year-old terrestrial ecosystem in which a variety of biological interactions can be accurately documented.

The age of the Rhynie chert organisms is important because it provides an opportunity to examine fungal diversity early in the terrestrialization of the earth. Equally significant is the preservation of these organisms as siliceous permineralizations, making it possible to resolve delicate structures such as flagella on zoospores (Taylor et al., 1992a). It is the intent of this chapter to survey a variety of these fungal associations with terrestrial plants, algae, cyanobacteria, eubacteria, and other fungi, and to examine these symbiotic associations in relation to the divergence of major fungal lineages.

2.1. Saprobes

The degradation and decomposition of plant and animal tissues is certainly the most widespread but least documented interaction involving fungi in the fossil record. The general absence of information on this critical component of the carbon cycle is no doubt related to our inherent bias of selecting specimens that are well preserved and complete. Few paleobiologists collect specimens that are only partially preserved, and as a result there are few examples of fungal saprophytism recorded from the paleontological record. In addition to the preservation of whole organisms, the Rhynie chert also contains abundant degraded plant material, some of which represents the activities of fungi. With few exceptions, however, it is impossible to determine which fungi were responsible for these saprophytic activities.

Today all major groups of fungi contain representatives that are saprobes, but in the Rhynie chert the most ubiquitous saprobes are various chytridiomycetes (Taylor et al., 1992b). These include both holocarpic and eucarpic forms that are associated with most of the Rhynie chert plants. For example, chytrids are found attached to the rhizoids of *Nothia*, within the cells of multicellular algae, and on the stems of land plants. Some of the fossil chytrids possess well-defined discharge tubes; in others this diagnostic feature is absent. In addition to occurring as small pustules on the stem surfaces of several plants, chytridiaceous fungi have been found within cortical tissues, inside and outside of several land plant spores, attached to cellular gametophytes, and on the surface and inside of several other fungi.

While it is impossible to identify the specific chytrids that functioned as saprotrophs, other, more complex aquatic fungi in the Rhynie chert possess features that can be directly compared with extant forms known to utilize nonliving organic matter. One of these is *Palaeoblastocladia*, a form that occurs as tufts of branching hyphae (Fig. 1b) arising from the stomata and beneath the cuticle of *Aglaophyton* (Remy et al., 1994a). This fossil fungus is morphologically identical to modern forms of *Euallomyces* in which there are two types of identical mature thalli: one bearing globose terminal zoosporangia (Fig. 1c), and the other, pairs of gametangia. *Palaeoblastocladia* is of special interest because it is identical to one of the few modern species of fungi that possesses a true alternation of generations. Since modern members of the Blastocladiales are saprobes, and since the fossils arise from beneath the epidermis and from stomata, it appears highly probable that *Palaeoblastocladia* was a saprobe as well.

There are both direct and indirect examples of saprotrophic activity in fossils that are geologically younger than those in the Rhynie chert. These include wood cell walls with areas of lignin and cellulose that have been selectively degraded and appear as solution troughs on tracheid walls, or in the form of hyphae within conducting elements (Stubblefield et al., 1985). The activities of other saprobes sometimes result in easily identified symptoms such as the characteristic spindle-shaped areas in wood formed by pocket rot fungi (Stubblefield and Taylor, 1986). None of these features has been identified to date in the land plants of the Rhynie chert, certainly in part because these organisms had not yet evolved the capacity to produce secondary tissues in the form of wood. Also, recent research on the conducting elements of the Rhynie chert plants suggests that many of these elements were not tracheids, and thus the cell walls may not have been constructed of lignin and hemicellulose. Nevertheless, even if the conducting elements of these fossil plants were constructed of cellulose and lignin it does not necessarily follow that the enzyme systems were identical to modern ones, regardless of the morphological similarity of the fossils with their modern counterparts. It is quite probable that as land plants evolved increasingly complex growth forms that required new cells and tissue systems the biochemical pathways involved in enzyme synthesis were also undergoing change in the various fungal groups. Thus, in the absence of certain biopolymers such as lignin it is likely that the Rhynie chert saprobes were only capable of cellulose degradation, and therefore some of the symptoms that can be documented later in geological time are simply not present. Finally, it is highly probable that some fungal groups have changed nutritional modes during their evolution. For example, White and Taylor (1989a) have suggested that the zygomycetes may have been the principal cellulose and lignin decomposers during the Triassic, and that their role may have shifted to one of mycorrhizal symbionts as the basidiomycetes and ascomycetes radiated. Perhaps in the Lower Devonian it was the chytridiomycetes that in fact were the chief decomposers.

2.2. Parasites

Among the Rhynie chert fungi are several types that were parasitic on land plants. Some of the holocarpic chytrids that are ubiquitous in the chert samples were no doubt parasites of a number of land plants and algae; however, demonstrating this interaction in the fossil record is difficult. Lacking a distinct and measurable host response it is impossible to differentiate parasitism from a number of other fungal interactions such as the activities of saprobes and other fungi functioning in mycorrhizal associations. However, a highly visible host response is present in the charophyte *Palaeonitella* (Taylor et al., 1992b). Normal internodal cells in the axis of this charophyte measure up to approximately 75 μm in diameter (Fig. 1d). In some *Palaeonitella* axes a few of the internodal cells are enlarged and exceed 300 μm in diameter. These cells represent a host response to the fossil fungus *Krispiromyces*, a small disc-shaped thallus characterized by a cap-like discharge papilla (Taylor et al., 1992b). Figure 1e shows the fungus zoosporangium extending outside the cell wall and the rhizomycelium within the lumen of the cell. There are a number of modern chytrids that are parasitic on freshwater algae, including charophytes, and some of these are morphologically similar to *Krispiromyces* (Sparrow, 1960).

The developmental response of the fossil charophyte to the presence of the fungus is particularly interesting. Cell enlargement of this type is referred to as hypertrophy and a similar pattern has been reported in extant algae including some species of *Chara* (Karling, 1928). What is perhaps most interesting about *Krispiromyces* and the interaction of this Devonian fungus with *Palaeonitella* is the fact that as early as 400 million years ago some parasitic fungi had already evolved the molecular and physiological signals to affect growth and development of the host cells. Finally, this host response is identical to that produced by some parasitic fungi today. It remains impossible to determine whether the fossil parasite responsible for the hypertrophied cells ultimately functioned as a pathogen in the Rhynie chert ecosystem.

Another example of a host response in the Rhynie chert involves the endobiotic and holocarpic fungus *Milleromyces*, which also attacked *Palaeonitella* (Taylor et al., 1992b). Because of its small size, entire stages in the life history of this fossil fungus are preserved, including the production of a zoosporangium with an elongate discharge papilla extending out from the cell wall (Fig. 1f). An interesting report by Karling (1928) suggests that in modern *Chara* the parasite may be responsible for the formation of large starch grains in the plastids of the host. In some of the nonhypertrophied cells of *Palaeonitella* that also contain zoosporangia of *Milleromyces* are dark granular bodies (Fig. 1f). Whether the dark bodies in the fossil are starch grains is problematic, as is whether or not they were formed in response to a parasitic fungus. Nevertheless, they appear to be consistently associated with many *Palaeonitella* cells infected by chytrids and therefore potentially represent another example of a host response.

2.3. Mycoparasitism

Although mycoparasitism is defined as parasitism of one fungus by another (Hawksworth et al., 1995), the nutritional mode and degree of biological interaction may extend over a wide range of conditions. Within the Rhynie chert ecosystem are several examples of mycoparasitism that include both necrotrophic and biotrophic associations. All of the fossil examples that have been documented to date include mycoparasites associated with chlamydospores (Fig. 2a) and thin-walled vesicles of arbuscular mycorrhizae (Hass et al., 1994). Epibiotic mycoparasites are represented by a number of chytrid thalli, each with extensive rhizoidal systems that extend into the lumen of the chlamydospore. Some fossil mycoparasites are morphologically similar to species of *Spizellomyces* (Barr, 1980), a modern chytrid that is a mycoparasite of some zygomycetous chlamydospores.

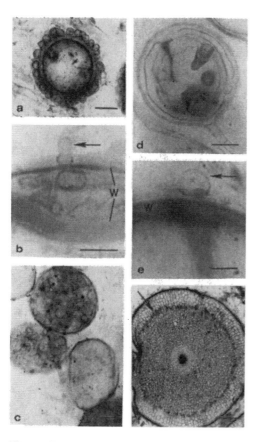

Figure 2   (a) Chlamydospore covered with chytrid thalli. Bar = 50 μm. (b) Section through multilayered chlamydospore wall (W) showing position of mycoparasite between wall layers. Arrow indicates discharge tube of parasite. Bar = 5 μm. (c) Two chlamydospores with endobiotic mycoparasites. Bar = 10 μm. (d) Chlamydospore containing several papillae. Bar = 10 μm. (e) Chytrid zoospore (arrow) on chlamydospore wall (W) with papilla extending into spore lumen. Bar = 5 μm. (f) Transverse section of *Aglaophyton* axis showing histology and position of narrow zone of arbuscules (arrow). Bar = 0.5 μm.

Other fossil mycoparasites occur between the wall layers of chlamydospores (Fig. 2b), often with the discharge tube of the zoosporangium extending through the outer wall and the rhizomycelium extending inside the spore. Still other mycoparasites occur entirely within the lumen of spores and vesicles (Fig. 2c). These endobiotic forms are often characterized by irregular aseptate hyphae that fill the spore lumen. The absence of a host response in some of the fossil chlamydospores strongly suggests that these mycoparasites were necrotrophs.

Other mycoparasites in the Rhynie chert involved biotrophic hosts. The most obvious example of this interaction is represented by chlamydospores with papillae (= callosities, lignitubers, or conical projections) extending into the lumen from the inner surface of the spore wall (Fig. 2d). In section view these papillae are often elongate with a tapered end and constructed of a series of concentric, convex layers (Fig. 2d). Extending the length of the projection is a narrow infection canal. Such structures have been found in a variety of modern endogonaceous chlamydospores that are identical to those in the fossil spores (Boyetchko and Tewari, 1991). They represent a definite host response by a living chlamydospore in which spore wall material is periodically synthesized and deposited around the invading papilla, presumably as a defense against the parasite. On a few fossil spores a chytrid zoospore or thallus is positioned directly over the papilla on the outer surface of the spore wall (Fig. 2e), suggesting that the chytrid was the organism responsible for the host infection. In some spores there may be more than 20 papillae along the inner surface of the spore wall, indicating the high concentration of chytrids in the environment and perhaps the susceptibility of the chlamydospores.

Since each chlamydospore can germinate to form hyphae that infect potential host roots in the form of endomycorrhizae, a reduction in the number of chlamydospores as a result of mycoparasitism may significantly impact the development of succeeding populations of host plants in the ecosystem. While it would be an impossible task to determine the impact of such mycoparasites on endomycorrhizal associations in the fossil record, such interactions would have played an important role in an ecosystem that was probably nutrient deficient.

2.4. Mycorrhizae

Today fungi are associated with the roots and underground organs of more than 240,000 plant species where they form a variety of mycorrhizal associations. Although this degree of interaction is often termed mutualistic, i.e., both plant and fungus derive some benefit from the symbiosis, under some conditions the fungus may become parasitic. The most widespread of these interactions involves approximately 50 species of zygomycetous fungi belonging to the Glomales, a group that has recently been suggested to be polyphyletic (Stürmer and Morton,

1997). The fact that these arbuscular mycorrhizae occur today in approximately 80% of all vascular plant species (Bonfante and Perotto, 1995) provides compelling evidence that this symbiosis is not only ancient but may have initially contributed to the colonization of the land by plants (Pirozynski and Malloch, 1975).

While it is perhaps difficult to attempt to characterize physiological interactions in fossil plants, the presence of the arbuscule defines the function of this association because it is through this highly branched hyphal network that carbon and phosphorus exchange takes place. *Glomites* is the name used for the endomycorrhizal fungus that occurs in both the aerial and underground stems of the Rhynie chert plant *Aglaophyton* (Remy et al., 1994; Taylor et al., 1995). In addition to aseptate intercellular hyphae and multilayered spores in the cortex (Fig. 3a), *Glomites* produced numerous arbuscules. The arbuscules occur in files of small, specialized cortical cells that extend throughout the axial system of the plant except the apical meristems (Fig. 2f). Each arbuscule is highly branched with the ultimate branch tips less than 0.5 μm in diameter (Fig. 3b). Like the condition in living plants containing arbuscular mycorrhizae, a thickened region or collar forms in the cell wall at the point where the arbuscule trunk penetrates the host cell. This structure is common in infected extant plant cells and is interpreted as a host response to the invading fungus. Although the host cell wall is breached in extant plants, the plasma membrane is not disrupted but rather invaginates around the arbuscule, compartmentalizing the fungus in the apoplast.

Endomycorrhizal fungi of the *Glomites* type are now known in two of the Rhynie chert plants (*Aglaophyton* and *Rhynia gwynne-vaughani*), and the morphological similarity in all stages of endomycorrhizal development in the fossils is remarkably similar to that in certain extant plants. There are several illustrations of hyphal coils, loops, and swellings in the other Rhynie chert plants (Kidston and Lang, 1921) that may represent the arbuscules of another fossil endomycorrhiza. Some of these are more similar to a *Gigaspora* type than those produced by *Glomites*. They are morphologically identical to what have been termed coralloid arbuscules in some extant ferns, such as *Pteridium* (Boullard, 1979) and *Ophioglossum* (Schmid and Oberwinkler, 1996). They appear in the youngest infected cells and are quite similar to some of the coiled hyphae and vesicles observed in the cells of other fossils containing endomycorrhizae (Phipps and Taylor, 1996). If these fossil structures are in fact arbuscules of different endomycorrhizae, they demonstrate that not only were there different forms present by the Early Devonian but that the fungi possessed the signaling mechanisms that enabled them to coexist in the same host. Finally, there is increasing evidence that the free-living gametophytes of the Rhynie chert plants also may have been mycorrhizal (H. Hass, personal communication).

It should be noted that *Aglaophyton* is not a vascular plant but rather combines features of both bryophytes and vascular plants (Edwards, 1986). It is

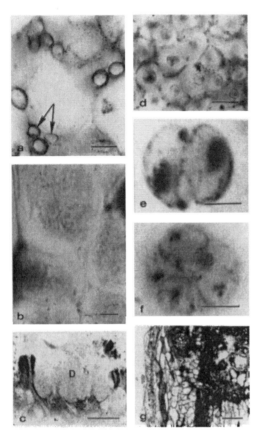

Figure 3 (a) Transverse section of the stem cortex of *Aglaophyton* showing *Glomites* hyphae (arrows) extending between the cortical cells. Bar = 20 μm. (b) Several cells containing highly branched arbuscules. Bar = 20 μm. (c) Section through a portion *Winfrenatia* thallus showing a depression (D). Bar = 0.5 μm. (d) A portion of the hyphal net in the depression in Fig. 3c. Note unicell in center of each lacuna. Bar = 20 μm. (e) Two photobiont cells of *Winfrenatia*. Bar = 5 μm. (f) Multicelled stage of the *Winfrenatia* photobiont. Bar = 5 μm. (g) Portion of *Rhynia gwynne-vaughanii* stem showing wound area and possible hyperplasia region in cortex (arrow). Bar = 200 μm.

known that *Aglaophyton* was leafless and that photosynthesis was accomplished in green aerial stems containing stomata. *Glomites* was not restricted to the absorbing organs of the plant as is typical in most extant plants, but rather arbuscules occur in a distinct zone that extends continuously from just beneath the growing tips of both the underground and aerial axes. When compared with endomycorrhizal infection in modern plants, *Aglaophyton* appears to have had a larger percentage of fungal tissue in the form of arbuscules relative to the photosynthetic tissue of the host. One might hypothesize that this mycorrhizal association functioned in increasing phosphorus uptake in nutrient-low substrates, which would have been common in the wet substrate of the Rhynie chert ecosystem. Another hypothesis is that the high ratio of fungus to photosynthetic tissue is due to the fact that the fungus was initially parasitic on the host and that during the evolution of endomycorrhizae a physiological equilibrium became established between the host and fungus resulting in the mycorrhizal association. The possibility also exists that in this ecosystem the arbuscular fungi were parasites that eventually killed their hosts.

One of the many inferred benefits of endomycorrhizae to higher plants is protection from soil pathogens and other organisms that attack the roots. Perhaps in the Early Devonian we are viewing an early stage in the fungus–host relationship in which a zygomycetous endophyte that was initially parasitic on the host has now evolved to protect the host from other parasites. What is perhaps most interesting is that during the early colonization of the land there were apparently a number of zygomycetous fungi that possessed the necessary genetic plasticity to coevolve with several grades of terrestrial plants in a variety of symbiotic associations.

2.5. Lichens

Lichens represent another interaction in which there is a stable symbiosis between a fungus and one to several photosynthetic autotrophs. In this association a characteristic thallus results that is morphologically distinct from that of either symbiont. Although lichens today represent a conspicuous component of the ''flora,'' there is almost nothing known about this interaction from the fossil record (Taylor and Taylor, 1993). To date the oldest and most completely known lichen is *Winfrenatia*, a cyanolichen, from the Rhynie chert (Taylor et al., 1997). The thallus is constructed of superimposed layers of aseptate hyphae that form numerous open depressions on the upper surface of the thallus (Fig. 3c), each approximately 1.0 mm deep. Hyphae extend into each depression and form a net-like structure. The lichen photobiont consists of solitary unicells, each approximately 10–16 μm in diameter and located within the spaces formed by the hyphal net (Fig. 3d). Each cell is surrounded by an extracellular sheath like that in extant cyanobacteria. Both within and above each depression of the thallus are cyanobacterial cells in various stages of division (Fig. 3e, f). While the photobiont is most comparable to the extant cyanobacteria *Gloeocapsa*, *Chroococcus*, and *Chroococcidiopsis*, the affinities of the fungus would appear to be most closely related to the zygomycetes.

Each of the thallus depressions represents a different stage in the development of the lichen. In some depressions the hyphal net is weakly organized and many of the unicells are not in contact with the fungus. In other regions the entire depression may be filled with hyphae with a few moribund unicells at the base. It is hypothesized that *Winfrenatia* represents a symbiosis in which the fungus parasitizes some, but not all, of the photobiont cells. Some of the cyanobacterial cells are stimulated to divide and produce new daughter cells that settle on the hyphal mat and initiate the formation of new depressions at the thallus margin. It is suggested that there is a physiological balance between the fungus and photobiont since new cells are produced that maintain the carbon source for the fungal partner. The benefit to the photobiont may be the availability of new niches and perhaps protection from invertebrate herbivory through the production of lichen acids or secondary metabolites.

3. CONCLUSION

There is a continuing discussion regarding the evolution of nutritional modes of fungi in their relationships with terrestrial plants. Some have argued that fungal parasites of land plants have evolved from saprophytic ancestors as a result of increasing nutritional specialization with increased dependence on the host (e.g., Lewis, 1973). An alternative view suggests that parasitism in terrestrial fungi is the more primitive state, and that increasing nutritional versatility as a necrotroph or saprobe is the derived condition (e.g., Cooke and Whipps, 1980). A perhaps more flexible approach to this question is offered by Heath (1987) who suggests that there is no single universal pattern for the evolution for nutritional modes in fungi, that each fungus–host relationship represents a spectrum of highly complex interactions each of which in turn can be modified by a variety of biological and ecological factors (Johnson et al., 1997). Within the Rhynie chert are examples of saprobes and parasites that illustrate fungus–host interactions that are identical to those in modern ecosystems and that involve a wide range of types of interactions. Present are a variety of chytrids that functioned as saprobes and parasites, as well as zygomycetes that were endomycorrhizal. The host responses that have been documented to date in the Rhynie chert are also identical to those in many modern fungus–plant interactions, including hypertrophy and the formation of papillae in chlamydospores. One host response that may have been overlooked in the Rhynie chert plants is hyperplasia—overproduction of host cells in response to a parasite. Kidston and Lang (1921) describe and illustrate necrotic areas in some stems of *Aglaophyton*, some of which consist of enlarged patches of parenchyma cells in the cortex of the axis, often associated with necrotic zones. In other regions these authors found hemispherical projections extending from the epidermis, or from beneath stomata. Prior to the report of free-living gametophytes in the Rhynie chert (Remy, 1982), some of these stem surface features were interpreted as sex organs (Lemoigne, 1968). In *Psilophyton*, another Early Devonian plant, extensive tissue development beneath the epidermis was interpreted as wound tissue formed in response to arthropod activity (Banks, 1981). It may be possible that some of these examples of abnormal tissue proliferation are examples of hyperplasia caused by fungal interactions (Fig. 3g).

Despite the antiquity of the Rhynie chert fungi, there is little new information about how various nutritional modes evolved in the major groups of fungi. Although Ascomycetes appear to predate the Rhynie chert fungi, to date there is only a single example from the Lower Devonian cherts (Taylor et al., 1999). Basidiomycetes, which are known as early as the Upper Devonian (Stubblefield et al., 1985), also appear to be absent from the Rhynie chert. Today both groups represent the principal decomposers of cellulose and lignin, but it appears that during the Early Devonian the principal saprobes were chytrids and possibly Zygomycetes. The Zygomycetes are especially interesting because today the group contains not only saprobes and parasites but the symbionts that form endomycorrhizal associations. While the zygomycetous affinities of the Rhynie chert endomycorrhizae are unequivocal, the potential relationship of these fungi with cyanobacteria in the Rhynie chert lichen symbiosis is less certain. Modern lichens generally have Ascomycetes as the mycobiont, in some instances Basidiomycetes; *Geosiphon* is the only modern symbiosis that includes a zygomycete with the cyanobacterium *Nostoc*. It is clear that during the evolution of terrestrial ecosystems there was also a concomitant evolution in the nutritional modes of the various fungal groups. For example, fossil evidence indicates that by Early Devonian time the Zygomycetes had achieved the complex enzyme systems necessary to enter into an endomycorrhizal mutualism, and also perhaps weakly parasitic interactions with cyanobacteria. These nutritional modes demonstrated by the Rhynie chert fossils are also consistent with molecular evidence based on nucleotide substitutions which suggest that the Glomaceae potentially diverged from the other terrestrial fungi 410–580 million years ago (Berbee and Taylor, 1993; Simon et al., 1993).

There are several arthropods known from the Rhynie chert (e.g., Kevan et al., 1975; Rolfe, 1985), but only indirect evidence that any were herbivores. In fact, most evidence suggests that the early land arthropods were predators. One of the many benefits ascribed to mycorrhizal plants is that the fungal infection enhances the host defenses. One idea suggests that plant families with a high incidence of mycorrhizal species also have a high proportion of specialist herbivores rather than generalist feeders, and that the loss of mycorrhizal infection in some species may have been due to the fact that the infection did not offer a selective advantage in herbivore resistance. Thus, if mycorrhizae evolved before the herbivores, this might explain the fact that most insects today are specialist feeders. The discovery of modern-appearing endomycorrhizae in the Lower

Devonian provides strong evidence that this fungal–plant symbiosis may in fact have predated terrestrial herbivorous arthropods and therefore was responsible in directing the evolution of herbivory.

Based on the Rhynie chert fungi it is highly probable that the Zygomycetes played an important role in the evolution of early terrestrial ecosystems. One highly specialized group of these fungi that have not been discovered in the chert thus far are the Trichomycetes, obligate zygomycetous fungi that today live in the hindgut of many aquatic arthropods. To date the only report of a fossil trichomycete comes from the Triassic of Antarctica (White and Taylor, 1989a), and it will be interesting to see if the complex interactions established within the other Zygomycetes during the Lower Devonian also include symbioses with other organisms such as arthropods.

The significance of the Rhynie chert fungi is far more important than the first appearance of fungal structures and identification of major groups, but rather is the unique picture these ancient fungi provide about a range of interactions with other organisms. While paleontological evidence can address questions relative to morphology and structure–function relationships in fossil ecosystems, there are a variety of equally significant questions necessary for understanding biological interactions focused at the genetic and molecular levels that can only be addressed using extant organisms. For example, there is compelling evidence that signal molecules such as flavonoids, isoflavonoids, cytokinins, phytohormones, and phenolic compounds that are produced by the host stimulate hyphal development in extant mycorrhizal systems (Harrison, 1997). The potential for differential expression of the genes encoding fungal enzymes by the host throughout the development of the symbiosis also has been suggested for ectomycorrhizal symbioses (Cairney and Burke, 1994). Similarly, the mechanism of phosphate and carbon transfer between the fungus and host remain fruitful areas in mycorrhizal research. While the answers to these questions cannot be directly addressed in ancient symbioses such as those presented in the Rhynie chert, the complexity of these molecular interactions today further underscores the fact that such genetic mechanisms are no doubt very ancient. The opportunity to characterize in fossil specimens host responses and symbiont interactions that are identical to those in modern organisms will perhaps someday make it possible to calibrate the evolution of the molecular systems that were responsible for these ancient interactions.

ACKNOWLEDGMENTS

This study was supported in part by the National Science Foundation (OPP-9614847) awarded to Edith L. Taylor and Thomas N. Taylor. The authors are indebted to the late Winfried Remy for his enthusiasm, generosity, and longstanding commitment to research on the Rhynie chert organisms, and to Hans Kerp and Hagen Hass for their continued cooperation and support of work on the Rhynie chert ecosystem.

REFERENCES

Banks, H. P. (1981). Peridermal activity (wound repair) in an early Devonian (Emsian) trimerophyte from the Gaspé Peninsula, Canada. *Palaeobotanist* 28–29, 20–25.

Barr, D. J. S. (1980). An outline for the reclassification of the Chytridiales, and for a new order, the Spizellomycetales. *Canadian Journal of Botany* 58, 2380–2394.

Berbee, M. L. and Taylor, J. W. (1993). Dating the evolutionary radiations of the true fungi. *Canadian Journal of Botany* 71, 1114–1127.

Bonfante, P. and Perotto, S. (1995). Strategies of arbuscular mycorrhizal fungi when infecting host plants. *New Phytologist* 130, 3–21.

Boullard, B. (1979). Considération sur la symbiose fongique chez les Ptéridophytes. *Syllogéus* 19, 1–59.

Boyetchko, S. M. and Tewari, J. P. (1991). Parasitism of spores of the vesicular-arbuscular mycorrhizal fungus, *Glomus dimorphicum. Phytoprotection* 72, 27–32.

Bruns, T. D., White, T. J. and Taylor, J. W. (1991). Fungal molecular systematics. *Annual Review of Ecology and Systematics* 22, 525–564.

Cairney, J. W. G. and Burke, R. M. (1994). Fungal enzymes degrading plant cell walls: their possible significance in the ectomycorrhizal symbiosis. *Mycological Research* 98, 1345–1356.

Cooke, R. C. and Whipps, J. M. (1980). The evolution of modes of nutrition in fungi parasitic on terrestrial plants. *Biological Review* 55, 341–362.

Edwards, D., Davis, K. L. and Axe, L. (1992). A vascular conducting strand in the early land plant *Cooksonia. Nature* 357, 683–685.

Edwards, D. S. (1986). *Aglaophyton major,* a non-vascular land plant from the Devonian Rhynie chert. *Botanical Journal of the Linnean Society* 93, 173–204.

Grahn, Y. (1981). Parasitism on Ordovician chitinozoa. *Lethaia* 14, 135–142.

Harrison, M. J. (1997). The arbuscular mycorrhizal symbiosis: an underground association. *Trends in Plant Science* 2, 54–60.

Hass, H., Taylor, T. N. and Remy, W. (1994). Fungi from the Lower Devonian Rhynie chert: mycoparasitism. *American Journal of Botany* 81, 29–37.

Hawksworth, D. L., Kirk, P. M., Sutton, B. C. and Pegler, D. N. (1995). *Ainsworth and Bisby's Dictionary of the Fungi,* 8th ed. International Mycological Institute, Cambridge University Press, Cambridge.

Heath, M. C. (1987). Evolution of parasitism in the fungi. In A.D.M. Rayner, C.M. Brasier, and D. Moore (eds.), *Evolutionary Biology of Fungi.* Cambridge University Press, Cambridge, pp. 147–160.

Johnson, N. C., Graham, J. H. and Smith, F. A. (1997). Functioning of mycorrhizal associations along the mutualism-parasitism continuum. *New Phytologist* 135, 575–585.

Karling, J. S. (1928). Studies in the Chytridiales III. A parasitic chytrid causing cell hypertrophy in *Chara. American Journal of Botany* 15, 485–495.

Kevan, P. G, Chaloner, W. G. and Saville, D.B.O. (1975). Interrelationships of early terrestrial arthropods and plants. *Palaeontology* 18, 391–417.

Kidston, R. and Lang, W. H. (1921). On Old Red Sandstone plants showing structure, from the Rhynie chert bed, Aberdeenshire. Part 4. Restorations of the vascular cryptogams, and discussion of their bearing on the general morphology of the Pteridophyta and the organisation of land-plants. *Transactions of the Royal Society of Edinburgh* 52, 831–854.

Kobluk, D. R. and James, N. P. (1979). Cavity-dwelling organisms in Lower Cambrian patch reefs from southern Labrador. *Lethaia* 12, 193–218.

Lemoigne, Y. (1968). Observation d'archégones portés par des axes de type *Rhynia gwynne-vaughanii* Kidston et Lang. Existence de gamétophytes vascularisés au Dévonien. *Comptes Rendus des Séances des l'Académie des Sciences, Paris* 266, 1655–1657.

Lewis, D. H. (1973). Concepts in fungal nutrition and the origin of biotrophy. *Biological Review* 48, 261–278.

Phipps, C. J. and Taylor, T. N. (1996). Mixed arbuscular mycorrhizae from the Triassic of Antarctica. *Mycologia* 88, 707–714.

Pirozynski, K. A. and Malloch, D. W. (1975). The origin of land plants: a matter of mycotrophism. *BioSystems* 6, 153–164.

Pratt, L. M., Phillips, T. L. and Dennison, J. M. (1978). Evidence of non-vascular land plants from the early Silurian (Llandoverian) of Virginia, U.S.A. *Review of Palaeobotany and Palynology* 25, 121–149.

Remy, W. (1982). Lower Devonian gametophytes: relation to the phylogeny of land plants. *Science* 215, 1625–1627.

Remy, W., Taylor, T. N. and Hass, H. (1994). Early Devonian fungi: a blastocladalean fungus with sexual reproduction. *American Journal of Botany* 81, 690–702.

Remy, W., Taylor, T. N., Hass, H. and Kerp, H. (1994). Four hundred-million-year-old vesicular arbuscular mycorrhizae. *Proceedings of the National Academy of Sciences USA* 91, 11841–11843.

Richardson, J. B. (1967). Some British Lower Devonian spore assemblages and their stratigraphic significance. *Review of Palaeobotany and Palynology* 1, 111–129.

Rolfe, W. D. I. (1985). Early arthropods: a fragmentary record. *Philosophical Transactions of the Royal Society of London* B 309, 207–218.

Schmid, E. and Oberwinkler, F. (1996). Light and electron microscopy of a distinctive VA mycorrhiza in mature sporophytes of *Ophioglossum reticulatum*. *Mycological Research* 100, 843–849.

Sherwood-Pike, M. A. and Gray, J. (1985). Silurian fungal remains: probable records of the class Ascomycetes. *Lethaia* 18, 1–20.

Simon, L., Bousquet, J, Levesque, R. C. and Lalonde, M. (1993). Origin and diversification of endomycorrhizal fungi and coincidence with vascular land plants. *Nature* 363, 67–69.

Sparrow, F. K. (1960). *Aquatic Phycomycetes.* University of Michigan Press, Ann Arbor.

Stubblefield, S. P. and Taylor, T. N. (1986). Wood decay in silicified gymnosperms from Antarctica. *Botanical Gazette* 147, 116–125.

Stubblefield, S. P., Taylor, T. N. and Beck, C. B. (1985). Studies of Paleozoic fungi. V. Wood-decaying fungi in *Callixylon newberryi* from the Upper Devonian. *American Journal of Botany* 72, 1765–1774.

Stürmer, S. L. and Morton, J. B. (1997). Developmental patterns defining morphological characters in spores of four species of *Glomus*. *Mycologia* 89, 72–81.

Taylor, T. N. (1993). Fungi. *In* M.J. Benton (ed.), *The Fossil Record 2*. Chapman & Hall, London, pp. 9–13.

Taylor, T. N. and Taylor, E. L. (1993). *The Biology and Evolution of Fossil Plants*. Prentice-Hall, Englewood Cliffs, NJ.

Taylor, T. N. and Taylor, E. L. (1997). The distribution and interactions of some Paleozoic fungi. *Review of Palaeobotany and Palynology* 95, 83–94.

Taylor, T.N. , Hass, H. and Kerp, H. (1997). A cyanolichen from the Lower Devonian Rhynie chert. *American Journal of Botany* 84, 992-1004.

Taylor, T. N., Hass, H. and Remy, W. (1992a). Devonian fungi: interactions with the green alga *Palaeonitella*. *Mycologia* 84, 901–910.

Taylor, T. N., Remy, W. and Hass, H. (1992b). Fungi from the Lower Devonian Rhynie chert: Chytridiomycetes. *American Journal of Botany* 79, 1233–1241.

Taylor, T. N., Remy, W., Hass, H. and Kerp, H. (1995). Fossil arbuscular mycorrhizae from the Early Devonian. *Mycologia* 87, 560–573.

Tims, J. D. and Chambers, T. C. (1984). Rhyniophytina and Trimerophytina from the early land flora of Victoria, Australia. *Palaeontology* 27, 265–279.

Trewin, N. H. and Rice, C. M. (1992). Stratigraphy and sedimentology of the Devonian Rhynie chert locality. *Scottish Journal of Geology* 28, 37–47.

White, J. F. and Taylor, T. N. (1989a). Triassic fungi with suggested affinities to the Endogonales (Zygomycotina). *Review of Palaeobotany and Palynology* 61, 53–61.

White, J. F. and Taylor, T. N. (1989b). A trichomycete-like fossil from the Triassic of Antarctica. *Mycologia* 81, 643–646.

3

Biotrophic Endophytes of Grasses: A Systematic Appraisal

James F. White, Jr.
Cook College, Rutgers University, New Brunswick, New Jersey

Ponaka V. Reddy
Amersham Pharmacia, Biotech, Inc., Piscataway, New Jersey

Charles W. Bacon
Richard B. Russell Agricultural Research Center, Agricultural Research Service, U.S. Department of Agriculture, Athens, Georgia

1. INTRODUCTION

Endosymbiosis is an important process of biological evolution where simple cells combine to produce more complex cells. Through this process major evolutionary change has occurred giving rise to the eukaryotic kingdoms of organisms from prokaryotes. The process of endosymbiosis is ongoing and occurs within the fungi. The major groups of terrestrial fungi (Ascomycetes, Basidiomycetes, and Zygomycetes) are believed to have evolved from an aquatic chytridiomycete that associated with early terrestrial plants, either as a saprophyte or as a parasite of living plants (Pirozynski and Malloch, 1975). The earliest land plant fossils contain fungal remains that may be interpreted as chytridiomycetous and zygomycetous in nature (Kidston and Lang, 1921; see Chap. 2 in this volume). These Early Devonian fungal remains have been interpreted as endosymbionts (Pirozynski and Malloch, 1975), but are probably saprophytes due to the degraded condition of host tissues in which mycelium is encountered (Kidston and Lang, 1921). It is evident that some time after plants colonized land, symbiotic associations began to develop between plants and terrestrial fungal groups. Fossils containing roots from the Triassic were found to contain fungi (Zygomycetes) of the ecologically important symbiotic association, known as arbuscular mycorrhizae (Stubblefield, et al., 1987). Even earlier Devonian fossil plants have been found to contain arbuscle-like structures, suggesting that the arbuscular mycorrhizae may have a much earlier origin (see Chap. 2 in this volume). It is unknown when or from what progenitor the Ascomycetes and Basidiomycetes evolved. It seems probable that the ancestor to these groups was zygomycetous and that their evolution was influenced by their association with land plants since they evolved as mycorrhizae, pathogens, and saprophytes of plant materials. It is within the Clavicipitaceae (Ascomycetes) that another type of fungal endosymbiont evolved. The family Clavicipitaceae contains members that are biotrophic parasites of insects, fungi, and plants. Within the groups that are parasitic on grasses, many are episymbiotic (superficial on host tissues), while some are endosymbiotic (colonize internal tissues of the host). In this chapter we will examine biological aspects of episymbiotic and endosymbiotic members of the Clavicipitaceae.

2. TAXONOMY AND EVOLUTION

The Clavicipitaceae are believed to have evolved from within the order Hypocreales (Rehner and Samuels, 1995; Spatafora and Blackwell, 1993), an order that includes the families Clavicipitaceae, Hypocreaceae, Nectriaceae, etc. Fungi of the family Clavicipitaceae (Ascomycotina) may be recognized by several features, including perithecia that undergo *Epichloe*-type centrum development rather than *Nectria*-type development; ascospores that are filamentous and many-celled rather than ellipsoidal and few-celled; and asci that generally possess a pronounced apical thickening (White, 1993, 1997). Some species of Clavicipitaceae are obligate parasites of insects (e.g., species of *Cordyceps* [Fr.] Link); others are obligate parasites of grasses and sedges (e.g., species of *Balansia* Speg., *Claviceps* Tul, and *Epichloe* [Fr.] Tul.). Spatafora and Blackwell (1993) examined molecular phylogenetic relationships of several members of the Clavicipitaceae and concluded that host affiliation was a reliable predictor of relatedness within the family. This seems to confirm earlier classification of the majority of the graminicolous Clavicipitaceae into the subfamily Clavicipitoideae (Gäumann, 1926; Diehl, 1950). Diehl (1950) further subdivided the subfamily Clavicipitoideae into tribes Clavicipiteae, Balansieae, and Ustilaginoideae. The tribe Clavicipiteae contained *Claviceps*; Balansieae contained *Balansia*, *Epichloe*, and other genera; and Ustilaginoideae contained conidial states classified in genera *Shropshiria* Stevens, *Munkia* Speg., *Neomunkia* Petrak, and *Ustilaginoidea* Bref. The separation of the Clavicipitoideae into tribes as defined by Diehl (1950) seems of questionable value, although this question is not the focus of critical evaluation in this study.

3. THE GENERA

The taxonomic knowledge of the plant-infecting Clavicipitaceae is not well developed. One species, *Epichloe bertonii* is known to infect dicotyledonous plant species in the American tropics (Diehl, 1950); however, the majority of the species worldwide are graminicolous, infecting predominantly grasses. Species of genus *Epichloe* (Fr.) Tul. and its conidial state in *Neotyphodium* Glenn, Bacon & Hanlin (= *Acremonium* sect. *Albolanosa* Morgan-Jones and W. Gams) are commonly encountered in cool-season grasses of North America and Europe. Many *Neotyphodium* spp. have lost the capacity to develop the external *Epichloe* state and produce no conspicuous fruiting structures on plants. Among these endosymbionts are the economically important endophytes *N. coenophialum* and *N. lolii*, widespread in tall fescue (*Festuca arundinaceae*) and perennial ryegrass (*Lolium perenne*), respectively, two important species of turf and forage grasses. Genera *Balansia*, *Parepichloe*, and *Myriogenospora* tend to infect warm-season grasses. Species of *Balansia* are most numerous in the Americas but may also be found in Asia; species of *Parepichloe* are limited to African and Asian tropics; species of *Myriogenospora* are found exclusively in the Americas.

4. DEVELOPMENT OF ENDOPHYTISM

Two genera of the Clavicipitaceae contain endosymbionts, including *Balansia* and *Epichloe*. In the genus *Balansia*, the step to endosymbiosis was made in a group of species (*B. claviceps* subclade) that are limited to the Americas (Diehl, 1950). Endosymbiotic *B. claviceps* and *B. obtecta* were thus derived from an episymbiotic ancestor. All species of the graminicolous *Epichloe* are endosymbionts. Phylogenetic evidence suggests that episymbiotic genus *Parepichloe* and the graminicolous *Epichloe* diverged from a common ancestor that was likely also episymbiotic (White and Reddy, 1998), the genus *Epichloe* acquiring the endophytic habit, and the genus *Parepichloe* retaining the ancestral episymbiotic habit. The occurrence of an episymbiotic stage in the life cycles of some *Epichloe* and *Neotyphodium* species may be a holdover from the episymbiotic habit (White et al., 1996).

4.1. Balansia

Endophytes of *Balansia* may be derived from epibiotic species of that genus. This is suggested since endophytic species *B. obtecta*, *B. claviceps*, *B. discoidea*,

B. granulosa, and *B. strangulans* form a subclade (Fig. 1) that arises from the more deeply rooted branch containing the epibiotic species *B. ambiens*. Previous work has shown that endophytic species of *Balansia* are limited to the Americas (White, 1994). This may be explained in that endophytism apparently evolved only in the *B. claviceps* subclade that is limited to the Americas. The *B. asperata* subclade contains species native both to Asia and the Americas; thus it is not exclusively Asian.

4.2. Epichloe and Parepichloe

A suite of morphological, ecological, and molecular data (Fig. 2) suggest that two separate lines of evolution were followed by *Epichloe* and *Parepichloe*. That species of *Epichloe* and *Parepichloe* occur on two phylogenetically separate groups of hosts (i.e., *Epichloe* on cool-season grasses and *Parepichloe* on warm-season grasses) suggests that the two genera have at least partly coevolved with host groups. It is notable that the tropical and subtropical *Parepichloe* species produce ascomata that are black, while the temperate *Epichloe* have ascomata that are light yellow to orange, a feature seen in other representatives of the Hypocreales (Rogerson, 1970; Spatafora and Blackwell, 1993). One may specu-late that dark pigmentation in *Parepichloe* is an adaptation for coping with solar radiation. Where solar radiation was intense, as in parts of Africa and Asia, dark pigmentation was selected for, whereas in temperate regions solar radiation was not a selective factor and pigmentation remained light. A phialidic anamorph classified in *Neotyphodium* forms on young stromata in *Epichloe*, but no ana-morphs have been reported for *Parepichloe*. Anamorphs could have been lost early in the evolution of *Parepichloe*, or alternatively, these may be ephemeral and simply not observed. The structure of stromata in *Epichloe* is surprisingly consistent. Stromata always include the inflorescence primordium and the leaf sheath of a leaf (the stromal leaf) whose blade emerges from the apex of the stroma (Fig. 3a). In *Parepichloe* several variations are seen in stromal structure. In *Parepichloe cinerea*, the stroma forms only on the inflorescence and no stromal leaves are included in the stroma (Fig. 3b). In *Parepichloe sclerotica* and *P. oplismeni*, the stroma surrounds a primordial tiller and attached to the adaxial side of a stromal leaf that is longer than the stroma (Fig. 3b). In *Parepichloe cynodontis*, the stroma is similar to those of *P. sclerotica* and *P. oplismeni*, except that all stromal leaves are shorter than the stroma.

5. ADVANTAGES OF ENDOPHYTISM

It is probable that among the benefits to the fungal symbiont resulting from a switch from episymbiosis to endosymbiosis are the following: (1) greater access

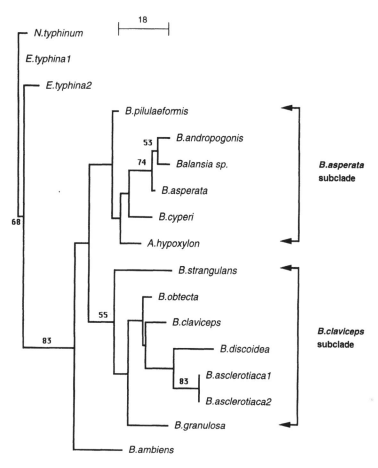

Figure 1 One of three most parsimonious trees (length = 197; consistency index = 0.761; homoplasy index = 0.239; retention index = 0.745) resulting from maximum parsimony analysis of an ITS1 dataset using the heuristic search option of PAUP 3.1.1. Numbers above the branches indicate the percentage bootstrap support for that branch from heuristic bootstrap analysis (500 bootstrap replicates). Scale bar indicates the number of nucleotide changes per unit of measure. *Epichloe* and *Neotyphodium* species were employed in the out group. GenBank accession numbers and specimen data for each sequence included in the tree are as follows: *Atkinsonella hypoxylon* (U78051), *Balansia andropogonis* (U89370), *B. ambiens* (AFO65612), *B. asclerotiaca* (U89367, U89368), *B. asperata* (U89375), *B. claviceps* (U89366), *B. cyperi* (U89369), *B. discoidea* (AFO65614), *B. granulosa* (AFO65613), *B. obtecta* (U77965), *B. pilulaeformis* (AFO65611), *Balansia* sp. (U89373), *Epichloe typhina* (L07133, L07131), *Neotyphodium typhinum* (L07134).

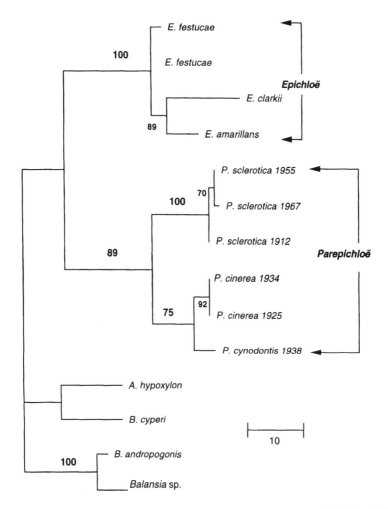

Figure 2   One of six equally parsimonious trees (length = 148; CI = 0.8) resulting
from maximum parsimony analysis of an ITS1 dataset using the heuristic search option
of PAUP 3.1.1. Numbers above branches indicate the percentage bootstrap support for
that branch. GenBank accession numbers for each sequence included in the tree are as
follows: *Atkinsonella hypoxylon* (U78051), *Balansia andropogonis* (U89370), *B. cyperi*
(U89369), *B. discoidea* (U89373), *Epichloe festucae* (L07134, L07133), *E. clarkii*
(U57666), *E. amarillans* (U57664), *Parepichloe cinerea* (AF003989, AF003987), *P. cynodontis* (AF003988), *P. sclerotica* (AF003984, AF003985, AF003986).

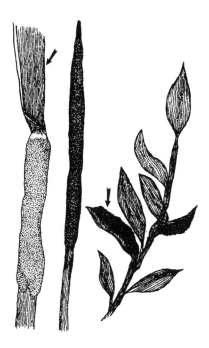

Figure 3 (left) Stroma of *Epichloe baconii* showing apical stromal leaf (arrow), ×9. (center) Stroma of *Parepichloe cinerea* showing absence of apical stromal leaf, ×9. (right) Branch of *Oplismenus* sp. showing stroma of *Parepichloe oplismeni* (arrow) on upper surface of leaves, ×9.

to nutrients; (2) protection from desiccation; (3) protection from surface-feeding insects; and (4) protection from parasitic fungi and the competition of other microbes.

5.1. Greater Access to Nutrients

In episymbiotic fungi, as exemplified by *Myriogenospora atramentosa*, mycelium is limited to the upper surfaces of leaves. In order to produce the stroma (the fungal reproductive structure), a substantial amount of nutrients must be obtained from the host tissues. Nutrient acquisition is accomplished by the episymbiont by modifying tissues of the plant so that nutrients flow to mycelium on the surfaces of those tissues. Epidermal cells in close association to stromata are swollen (hypertrophied) and lack the waxy cuticle that normally prevents escape of water and nutrients from the leaf (White and Glenn, 1994). Because the cuticle is absent on the leaf, the episymbiont may extract nutrients from the leaf across the modified epidermal layer. In endosymbionts, the extraction of nutrients from the host is enhanced since mycelium is distributed among the internal cells of the plant tissues (White and Owens, 1992). Thus both internal and external tissues of the host may be modified to enhance flow of nutrients to the stromal mycelium (White et al., 1997).

5.2. Protection from Desiccation

Episymbionts perennate as mycelium on the surfaces of plants. Meristem tillers seem to be a region of high concentration of perennating mycelium (Leuchtmann and Clay, 1988). It is reasonable to expect that the episymbiont is vulnerable to desiccation if conditions become dry. In desert environments only endosymbionts, such as the asymptomatic *Neotyphodium* spp., are known to occur. Endosymbiotic species of *Balansia*, *Epichloe*, and *Neotyphodium* may be protected from the effects of a desiccating environment since they are contained within the moist internal tissues of the plant host.

5.3. Protection from Fungus-Feeding Insects, Parasitism, and Competition of Microbes

The surfaces of plants are scoured by insects, such as fungus mites and snails, that actively search out and consume fungal mycelium and spores. Epibiotic fungi likely fall prey to many mycophagous organisms. In addition, the air contains spores of numerous species of fungi, such as *Trichoderma* spp., that have the capacity to parasitize mycelium of other species of fungi. These propagules likely find their way to the plant surfaces where epibionts may be parasitized. Many fungi and bacteria are known to colonize the surfaces plants. These microorganisms may actively compete with episymbiotic fungi for nutrients and space. Endosymbionts escape all of the potentially deleterious interactions that might occur on the surfaces of plants.

6. FORMATION OF STROMATA ON PLANTS

In order to reproduce sexually, endosymbionts and episymbionts must produce stromata on host plants (Fig 3). The stroma generally consists of a mixture of both fungal and host tissues, interfacing in such a way as to facilitate flow of nutrients from living host tissues into the fungal component of the stroma

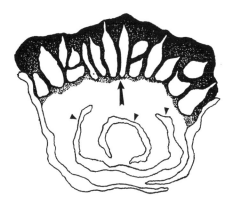

Figure 4 Stroma of *Parepichloe sclerotica* showing relationship of plant and fungal tissues. Illustration shows leaves enclosed in stroma (arrowheads) and a densely pigmented ascomatal layer (arrow) containing perithecia, ×9.

(Fig. 4). The fungal component of the stroma is composed of differentiated mycelia in which three tissues may be distinguished. The fungal component includes interstitial tissues found in spaces between the inflorescence primordium and leaves embedded within the stroma; a fungal cortex layer composed of a tight palisade of parallel hyphae that appears to function to impede loss of water and nutrients to the environment; and an outer hymenium on which spermatia are produced and following fertilization is converted to the ascomal stroma in which perithecia (Fig. 4) and asci containing ascospores (Fig. 5) differentiate.

7. MODIFICATION OF PLANT TISSUES

Plant tissues within the stroma are generally modified, apparently to maximize the flow of nutrients into the mycelium. Mesophyll cells within stromata of *Epichloe* may be hypertrophied (White et al., 1993). Epidermal cells trapped within the stroma show modification to the extent that it softens and fails to function as a barrier layer. Additionally, mycelium is often observed within vascular tissues within the stroma. Evaporation of water has been demonstrated to occur rapidly from the mycelial surface of the stroma. Rapid evaporation of water from the stroma may be an important mechanism whereby nutrients are delivered to the mycelium of the stroma. Enhanced flow of water into the stroma to replace that lost by evaporation may facilitate the transfer sugars and other compounds in solution to the mycelium of the stroma (White and Camp, 1995).

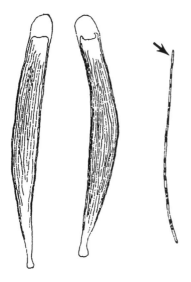

Figure 5 Asci and ascospore (arrow) of *Epichloe bertonii* showing typical apically thickened ascus tip of Clavicipitaceae and filamentous multiseptate ascospore, ×9.

8. SEXUAL REPRODUCTION

Stromata provide the foundation on which the fungal reproductive cells are differentiated. In the endosymbiotic *Balansia obtecta*, the stroma has been modified into a sclerotium. Stromata of *B. obtecta* germinate in the spring to form gametangia consisting of clusters of receptive hyphae and cup-shaped conidiomata bearing conidia, often classified in form genus *Ephelis*. Ephelidial conidia function as spermatia in a heterothallic mating process. Perithecia in stalked ascomatal stromata develop when spermatia of one mating type are transferred to receptive hyphae of the opposite mating type. It has been suggested that ascospores infect hosts by infection of florets (Diehl, 1950). In *Epichloe*, sexual reproduction similarly involves a heterothallic mating system. Stromata are initially covered with a layer of spermatia or *Neotyphodium* conidia and receptive hyphae (White and Bultman, 1987). Spermatia are vectored between opposite mating types of the stroma by symbiotic flies, classified in the genus *Phorbia*. Female flies visit several stromata, feeding on conidia and mycelium. Conidia are not digested and instead accumulate in the digestive tract of the fly. Females deposit eggs on stromata and in the process deposit spermatia on stromata by defecating on them. After perithecia develop, ascospores are produced that are ejected from asci and are believed to also infect grasses at the florets. Plants developing from infected seeds contain the endophyte within the embryo.

9. BENEFITS TO HOSTS

Clavicipitaceous endosymbionts are so common in many grasses that it is reasonable to expect that they impart some benefit to their hosts. Numerous studies have suggested that beneficial effects on grasses may include enhanced insect and nematode resistance, fungus disease resistance, enhanced vegetative growth, and increased drought tolerance (Clay et al., 1985; West et al., 1990; Bacon and De Battista, 1990). Enhanced insect resistance has been proposed to be the result of endosymbiont-produced alkaloids, such as peramine, which serve as antifeeding agents. Enhanced vegetative growth may be the result of auxins that are produced by the endosymbionts (Porter et al., 1979). The mechanism of increased drought tolerance is believed to involve the increased ability of meristems to recover after drought conditions; however, the mechanism is not clearly understood (West et al., 1990). One recent study (Clarke et al., 1997) has also demonstrated endosymbiont-enhanced resistance in fine fescue grasses to the dollar spot disease, caused by the fungal pathogen *Sclerotinia homeocarpa*. It has been suggested that enhanced resistance to fungal pathogens could be the result of antifungal compounds produced by endophytes (White and Cole, 1985). Koshino et al. (1989) described several fungitoxic sesquiterpenes and sterols from an endophyte (*Epichloe typhina*) of timothy grass (*Phleum pratense*) that were demonstrated to inhibit development of fungal pathogens of the grass. Whatever mechanisms account for the benefits to hosts, they translate to increased persistence and competitive capacity of endosymbiont-containing grasses in natural populations (Clay, 1988). It has been reported that pastures containing a low percentage of endosymbiont-infected plants when first established will, over a few years, drastically increase in the percentage of infected plants (Lewis and Clements, 1986; Latch et al., 1987). The selective advantage of endosymbiont-infected grasses is most pronounced when grasses are subjected to environmental stresses (Clarke et al., 1997).

10. TOXIC SYNDROMES

Due to the production of ergot alkaloids and other secondary compounds by Clavicipitaceae, host plants are often rendered toxic to mammalian herbivores. Several well-studied toxic syndromes may develop in mammals feeding on forage grasses containing endosymbiotic species of *Neotyphodium*. Cattle that consume tall fescue containing the endosymbiont *N. coenophialum* may develop a condition known as "fescue foot." Due to the vasoconstrictive properties of the ergot alkaloids produced by species of *Neotyphodium*, over a long period of time the flow of blood is reduced to the body extremities of animals and consequently hooves and tails may rot off. In perennial ryegrass infected by *N. lolii*, the production of lolitrems by the endosymbiont results in a condition whereby animals may develop uncontrollable spasms, a condition called "ryegrass staggers." Animals may show toxic effects until they are removed from toxic forage. Another grass, called sleepy grass (*Achnatherum robustum*), contains a *Neotyphodium* endophyte that produces lysergic acid amide (Petroski et al., 1992), an alkaloid that has the effect of causing horses that consume a relatively small quantity of the grass (1% of body weight) to sleep for 2–3 days. In South America several grass species are infected by a *Neotyphodium* endosymbiont that causes a condition commonly known as "tembladera." This condition is similar to ryegrass staggers but is often fatal to poisoned animals.

ACKNOWLEDGMENTS

This research was supported by NSF DEB 96-96041. We are grateful to Rachna Patel (Rutgers University) for illustrations in this chapter.

REFERENCES

Bacon, C. W. and De Battista, J. (1990). Endophytic fungi of grasses. *In* D. K. Avora, B. Rai, K. G. Mukerji, and G. R. Knudsen (eds.), *Soil and Plants*. Marcel Dekker, New York, pp. 231–256.

Clarke, B. B., White, J. F., Jr., Funk, C. R., Jr., Sun, S. and Huff, D. R. Enhanced resistance to dollar spot in endophyte-infected fine fescues. *Plant Disease*. In press.

Clay, K. (1988). Fungal endophytes of grasses: a defensive mutualism between plants and fungi. *Ecology* 69, 10–16.

Clay, K., Hardy, T. N. and Hammond, A. M. Jr. (1985). Fungal endophytes of grasses and their effects on an insect herbivore. *Oecologia* 66, 1–5.

Diehl, W. W. (1950). Balansia *and the Balansiae in America*. USDA Monograph No. 4, U.S. Government Printing Office, Washington, D.C.

Gäumann, E. A. (1926). *Vergleichende Morphologie der Pilze*. Gustav Fischer. Jena.

Kidston, R. and Lang H. W. (1921). On the old red sandstone plants showing structure, from the Rhynie chert bed, Aberdeenshire. *Transactions of the Royal Society of Edinburgh* 52, 855–902.

Koshino, H., Terada, S., Yoshihara, T., Sakamura, S., Shimanuki, T., Sato, T. and Tajimi, A. (1989). A ring B aromatic sterol from stromata of *Epichloe typhina*. *Phytochemistry* 28, 771–772.

Latch, G. C. M., Potter, L. R. and Tyler, B. F. (1987). Incidence of endophytes in seeds from collections of *Lolium* and *Festuca* species. *Annals of Applied Biology* 111, 59–64.

Leuchtmann, A. and Clay, K. (1988). *Atkinsonella hypoxylon* and *Balansia cyperi*, epibiotic members of the Balansieae. *Mycologia* 80, 192–199.

Lewis, G. C. and Clements, R. O. (1986). A survey of the ryegrass endophyte (*Acremonium lolii*) in the U.K. and its apparent ineffectuality on a seedling pest. *Journal of Agricultural Science Cambridge* 107, 633–638.

Petroski, R. J., Powell, R. G. and Clay, K. (1992). Alkaloids of *Stipa robusta* (sleepy grass) infected with an *Acremonium endophyte*. *Natural Toxins* 1, 84–88.

Pirozynski, K. A. and Malloch, D. W. (1975). The origin of land plants: a matter of mycotrophism. *BioSystems* 6, 153–164.

Porter, J. K., Bacon, C. W. and Robbins, J. D. (1979). Lysergic acid amide derivatives from *Balansia epichloe* and *Balansia claviceps* (Clavicipitaceae). *Journal of Natural Products* 42, 309–314.

Rehner, S. A. and Samuels, G. J. (1995). Molecular systematics of the Hypocreales: a telomorph gene phylogeny and the status of their anamorphs. *Canadian Journal of Botany* 73, S816–S823.

Rogerson, C. T. (1970). The hypocrealean fungi (Ascomycetes, Hypocreales). *Mycologia* 62, 865–910.

Spatafora, J. W. and Blackwell, M. (1993). Molecular systematics of unitunicate perithecial ascomycetes: the Clavicipitales–Hypocreales connection. *Mycologia* 85, 912–922.

Stubblefield, S. P., Taylor, T. N. and Trappe, J. M. (1987). Vesicular-arbuscular mycorrhizae from the Triassic of Antarctica. *American Journal of Botany* 74, 1904–1911.

West, C. P., Oosterhuis, D. M. and Wullschleger, S. D. (1990). Osmotic adjustment in tissues of tall fescue in response to water deficit. *Environmental and Experimental Botany* 30, 149–156.

White, J. F., Jr. (1993). Endophyte-host associations in grasses. XIX. A systematic study of some sympatric species of *Epichloe* in England. *Mycologia* 85, 444–455.

White, J. F., Jr. (1994). Endophyte-host associations in grasses. XX. structural and reproductive studies of *Epichloe amarillans* sp. nov. and comparisons to *E. typhina*. *Mycologia* 86, 571–580.

White, J. F., Jr. (1997). Perithecial structure in the fungal genus *Epichloe*: an examination of the clavicipitalean centrum. *American Journal of Botany* 84, 170–178.

White, J. F., Jr., Bacon, C. W. and Hinton, D. M. (1997). Modifications of host cells and tissues by the biotrophic endophyte *Epichloe amarillans* (Clavicipitaceae; Ascomycotina). *Canadian Journal of Botany* 75, 1061–1069.

White, J. F., Jr. and Bultman, T. L. (1987). Endophyte-host associations in forage grasses. VIII. Heterothallism in *Epichloe typhina*. *American Journal of Botany* 74, 1716–1721.

White, J. F., Jr. and Camp, C. 1995. A study of water relations of *Epichloe amarillans* White, an endophyte of the grass *Agrostis hiemalis* (Walt.)B.S.P. *Symbiosis* 18, 15–25.

White, J. F., Jr. and Cole, G. T. (1985). Endophyte–host associations in forage grasses. III. *In vitro* inhibition of fungi by *Acremonium coenophialum*. *Mycologia* 77, 487–489.

White, J. F., Jr. and Glenn, A. E. (1994). A study of two fungal epibionts of grasses: structural features, host relationships, and classification in the genus Myriogenospora Atk. (Clavicipitales). *American Journal of Botany* 81, 216–223.

White, J. F., Jr., Glenn, A. E. and Chandler, K. F. (1993). Endophyte-host associations in grasses. XVIII. Moisture relations and insect herbivory of the stromal leaf of *Epichloe typhina*. *Mycologia* 85, 195–202.

White, J. F., Jr., Martin, T. I. and Cabral, D. (1996). Endophyte-host associations in grasses. XXIII. Conidia formation by *Acremonium* endophytes in the phylloplanes of *Agrostis hiemalis* and *Poa rigidifolia*. *Mycologia* 88, 174–178.

White, J. F., Jr. and Owens, J. R. (1992). Stromal development and mating system of *Balansia epichloe*, a leaf-colonizing endophyte of warm-season grasses. *Applied Environmental Microbiology* 58, 513–519.

4

Hybridization and Cospeciation Hypotheses for the Evolution of Grass Endophytes

Christopher L. Schardl
University of Kentucky, Lexington, Kentucky

Heather H. Wilkinson
Texas A&M University, College Station, Texas

1. INTRODUCTION: ENDOPHYTE DIVERSITY

Grass symbionts of the genus *Epichloe*, including related asexual fungi, can be important for biological protection of their hosts under stresses imposed biotically (e.g., herbivores and parasites) or abiotically (e.g., drought). The symbionts can be categorized as three types. The most benign (nonpathogenic) type encompasses symbionts that only transmit vertically by systemic infections of seeds and tend to be highly beneficial to their hosts. In these cases, the symbiont is asexual (thus taxonomically classified in genus *Neotyphodium*). In contrast, the most antagonistic *Epichloe* species only transmit horizontally (i.e., contagiously), completely suppress host seed production, and are obligately sexual. In the third type, balanced (pleiotropic) symbiosis, the symbiont aborts some host inflorescences, is transmitted both horizontally and vertically, and, like the benign type, can greatly enhance host fitness. In balanced symbiosis, both host and fungus have a sexual stage in addition to their asexual reproductive capabilities (Fig. 1).

Diversity of endophytes is manifested not only in their types of symbiosis, host specificities, and morphologies, but in the types of benefits they confer to their hosts. For example, there is considerable variation in the profiles and levels of antiherbivore alkaloids produced in grass–*Epichloe*/*Neotyphodium* symbioses. No less than four classes of alkaloids are produced by the endophytes or only

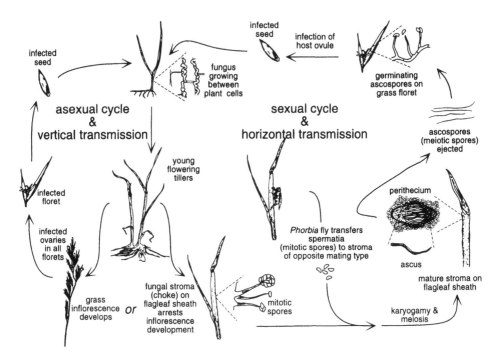

Figure 1 An example of balanced symbiosis involving coordinated life cycles of *Epichloe festucae* (MP-II) and one of its host grasses, *Festuca rubra*. The alternative life cycles (clockwise and counterclockwise loops) occur on different tillers of a symbiont-infected plant. In the symbiont's asexual life cycle (clockwise loop) it systemically infects the seeds produced on the mother plant and is thus transmitted vertically through successive host generations. The symbiont's sexual cycle (counterclockwise loop) is initiated by a fungal growth (stroma) that envelops a young flowering tiller and prevents the inflorescence from maturing. Following fertilization and maturation, meiotic spores are ejected and mediate horizontal (contagious) transmission. Antagonistic *Epichloe* species sterilize most or all flowering tillers of their host plants and are never (or perhaps rarely) transmitted vertically. (From Schardl et al. 1997.)

in symbiota (Siegel and Bush, 1996). Two of the classes—saturated aminopyrrolizidines (''lolines'') and pyrrolopyrazines (peramine)—have no close chemical relatives known elsewhere. These alkaloids protect against herbivory by insects. The other two classes—ergotoxins and indolediterpenoid tremorgens—are potent antimammalian neurotoxins known from other fungi. In some symbiota none of these alkaloids have been detected. However, loline alkaloids often exceed 1% dry weight in symbiota of meadow fescue (*Festuca pratensis*) and *Neotyphodium uncinatum* (= *Acremonium uncinatum*) (Siegel and Bush, 1996); and ergotoxins approach 0.3% dry weight in *Achnatherum inebrians* (drunken horse grass) with a *Neotyphodium* sp. endophyte (Miles et al., 1996). Other host benefits attributable to the endophytes but for which the mechanistic bases are unknown include enhanced drought tolerance (Arechavaleta et al., 1989) and resistance to nematodes (Kimmons et al., 1990). The evolutionary diversity of endophytes reflects the several types and diverse benefits of these symbionts.

In this chapter we discuss molecular phylogenetic evidence and other information attesting to the diversity of the very evolutionary processes underlying *Epichloe/Neotyphodium* symbioses with grasses.

2. SYMBIOTIC LIFE HISTORIES

Symbioses of grasses with fungi of the genus *Epichloe* are among the most intimate associations known. Many species of cool-season grasses (subfamily Pooideae) are hosts to sexual *Epichloe* or asexual *Neotyphodium* species (Clay and Leuchtmann, 1989; White, 1987). A grass plant infected with an *Epichloe* or *Neotyphodium* species typically maintains the symbiosis throughout its life, and often transmits its symbiont vertically (Fig. 1): seeds produced by the infected mother plant bear infections as a direct result of clonal propagation of the symbiont in the ovule (Philipson and Christey, 1986). In contrast, the sexual cycle of an *Epichloe* species requires horizontal transmission because it only occurs on host flowering tillers whose development is arrested by the fungal stroma, the ectophytic structure on which mating can occur (Chung et al., 1997). This stage is called grass choke disease. Again, in balanced grass-*Epichloe* symbioses most flowering tillers do not elaborate stromata but instead produce healthy, symbiont-infected seeds. In other symbioses the *Epichloe* strain completely suppresses host seed set and must be transmitted horizontally (Chung et al., 1997).

3. HOST SPECIFICITY

Most species of *Epichloe* and *Neotyphodium* are host-specific. Extensive surveys of grass–fungus symbioses have been conducted over the past decade, and mating tests have identified biological species (Schardl et al., 1997; Schardl and Tsai, 1992; White, 1993; White, 1994). These are grouped as mating populations MP-I through MP-IX, which largely correspond to recently described or revised morphospecies. Thus, MP-I is composed primarily of *Epichloe typhina* sensu stricto, MP-II of *Epichloe festucae*, MP-IV of *Epichloe amarillans*, and MP-V of *Epichloe baconii*. Biological species MP-III and MP-VI through MP-IX have also been taxonomically described and named (Table 1).

The biological species of *Epichloe* closely follow both host taxonomic divisions and geographic divisions (Schardl et al., 1997). For example, two species

Table 1 *Epichloe/Neotyphodium* Species and Host Interactions

Species	Mating population	Host tribe	Transmission strategy	Symbiotic character	Occurrence[a]	Continent of origin	References
E. typhina	MP-I	Aveneae	Horizontal	Antagonistic	Rare	Eurasia	1–3
E. typhina	MP-I	Poeae	Horizontal	Antagonistic	Rare-Moderate	Eurasia	1–4
E. typhina	MP-I	Brachypodieae	Horizontal	Antagonistic	Rare	Eurasia	1–3
E. clarkii	MP-I	Poeae	Horizontal	Antagonistic	Rare	Eurasia	1–4
E. festucae	MP-II	Poeae	Mixed	Balanced	Common	Eurasia	2–3, 5–6
E. elymi	MP-III	Triticeae	Mixed	Balanced	Rare-Moderate	N. America	2–3, 7
E. amarillans	MP-IV	Aveneae	Mixed	Balanced	Rare-Moderate	N. America	2–3, 8–9
E. baconii	MP-V	Aveneae	Horizontal	Antagonistic	Rare	Eurasia	2–4
E. bromicola	MP-VI	Bromeae	Horizontal	Antagonistic-Benign	Rare-Moderate	Eurasia	1–3
E. sylvatica	MP-VII	Brachypodieae	Mixed	Balanced	Common	Eurasia	1–3
E. glyceriae	MP-VIII	Meliceae	Horizontal	Antagonistic	Rare	N. America	2–3, 7
E. brachyelytri	MP-IX	Brachyelytreae	Mixed	Balanced	Rare	N. America	2–3, 7
N. coenophialum	N/A[b]	Poeae	Vertical	Benign	Common	Eurasia	10
N. lolii	N/A	Poeae	Vertical	Benign	Moderate	Eurasia	11
N. lolii × *E. typhina*	N/A	Poeae	Vertical	Benign	Rare	Eurasia	12–13
N. uncinatum	N/A	Poeae	Vertical	Benign	Common	Eurasia	3
Neotyphodium spp.	N/A	Stipeae	Vertical	Benign	Common	N. America	14–15

[a] Estimated occurrences are: rare, less than 1% host plants possess the endophyte; moderate, 1–50% possess the endophyte; and common, more than 50% possess the endophyte.
[b] N/A = not applicable.

1. Leuchtmann, A., and Schardl C. L. (1998). Mating compatibility and phylogenetic relationships among two new species of *Epichloe* and other congeneric European species. *Mycological Research* 102, 1169–1182.

2. Schardl, C. L., Leuchtmann, A., Chung, K.-R., Penny, D. and Siegel, M. R. (1997). Coevolution by common descent of fungal symbionts (*Epichloe* spp.) and grass hosts. *Molecular Biology and Evolution* 14, 133–143.

3. Leuchtmann, A. (ETH Zürich, Switzerland) personal communications, and personal observations of the authors.

4. White, J. F., Jr. (1993). Endophyte-host associations in grasses. XIX. A systematic study of some sympatric species of *Epichloe* in England. *Mycologia* 85, 444–445.

5. Leuchtmann, A., Schardl, C. L., and Siegel, M. R. (1994). Sexual compatibility and taxonomy of a new species of *Epichloe* symbiotic with fine fescue grasses. *Mycologia* 86, 802–812.

6. Bazely, D. R., Vicari, M., Emmerich, S., Filip, L., Lin, D., and Inman, A. (1997). Interactions between herbivores and endophyte-infected *Festuca rubra* from the Scottish islands of St. Kilda, Benbecula, and Rum. *Journal of Applied Ecology* 34: 847–860.

7. Schardl, C. L., and Leuchtmann, A. (1999). Three new species of *Epichloë* with North American grasses. *Mycologia* 91, 95–107.

8. White, J. F., Jr., and Chambless, D. A. (1991). Endophyte-host associations in forage grasses. XV. Clustering of stromata-bearing individuals of *Agrostis hiemalis* infected by *Epichloë typhina*. *American Journal of Botany* 78, 527–533.

9. White, J. F., Jr. (1994). Endophyte-host associations in grasses. XX. Structural and Reproductive studies of *Epichloe amarillans* sp. nov. and comparisons to *E. typhina*. *Mycologia* 86, 571–580.

10. Zabalgogeazcoa, I., Garcia-Ciudad, A., and Garcia-Criado, B. (1997). Endophytic fungi in grasses from semiarid grasslands in Spain. In: N. S. Hill, and C. W. Bacon, (eds). *Neotyphodium*/grass interactions. Plenum Press, New York, pp. 89–91.

11. Lewis, G. C., Ravel, C., Naffaa, W., Astier, C., Charmet, G. (1997). Occurrence of *Acremonium* endophytes in wild populations of *Lolium* spp. in European countries and a relationship between level of infection and climate in France. *Annals of Applied Biology* 130, 227–238.

12. Christensen, M. J. Leuchtmann, A., Rowan, D. D. and Tapper, B. A. (1993). Taxonomy of *Acremonium* endophytes of tall fescue (*Testuca arundinacea*), meadow fescue (*F. partensis*), and perennial rye-grass (*Lolium perenne*). *Mycological Research* 97, 1083–1092.

13. Kuldau, G. A., Tsai, H.-F., Schardl, C. L. (1999). Genome sizes of *Epichloë* species and anamorphal hybrids. *Mycologia* (in press).

14. Kaiser, W. J., Bruehl, G. W., Davitt, C. M., and Klein, R. E. (1996). Acremonium isolates from *Stipa robusta*. *Mycologia* 88, 539–547.

15. White, J. F., Jr., and Morgan-Jones, G. (1987). Endophyte-host associations in forage grasses. VII. *Acremonium chisosum*, a new species isolated from *Stipa eminens* in Texas. *Mycotaxon* 28, 179–189.

of *Epichloe* are associated with *Agrostis* and related genera (tribe Aveneae), but *E. baconii* is Eurasian and *E. amarillans* is North American. Specificities of the nine known biological species of *Epichloe* are shown in Table 1. Only MP-I is not confined to hosts within a single tribe. However, MP-I exhibits considerable genetic variation, encompasses at least two morphospecies (A. Leuchtmann personal communication), and may include several incipient host-specific species for which genetic barriers to mating have not yet fully evolved.

Phylogenetic evidence suggests that selection on the grass–*Epichloe* systems maintains balanced (''pleiotropic'') symbioses in which each symbiotic partner exhibits both sexual and asexual reproduction. As will be discussed later, balanced symbiosis is associated with a pattern of common descent of grass taxa (tribes, genera, or species) and *Epichloe* species.

Even in the case of *E. typhina*, a species associated with a wide taxonomic range of hosts, individual isolates can be highly host-specific. For example, isolates from perennial ryegrass (*Lolium perenne*) and orchard grass (*Dactylis glomerata*) are typically more compatible with their own than with each others hosts. Genetic analyses employing such *E. typhina* strains suggest that host specificity is under complex control involving several genes and likely gene interactions (Chung et al., 1997).

Specificity of grass–*Neotyphodium* symbioses can be manifested at several levels. First, natural symbioses exhibit no apparent host cellular response, whereas artificial movement of endophytes to new host species is often associated with an accumulation of osmiophilic (perhaps phenolic) material in the intercellular matrix surrounding endophyte hyphae (Koga et al., 1993). In some such novel associations, host tillers can develop necrotic crowns (Christensen, 1995). In others the endophytic hyphae can invade vascular bundles and become closely associated with phloem companion cells, a phenomenon associated with stunting of the plants (Christensen et al. 1997). Thus, host specificity is typical of the asexual as well as sexual endophytes.

4. EVOLUTIONARY CONSIDERATIONS

4.1 Evolutionary Implications of Host Specialization

Host specialization in the *Epichloe/Neotyphodium* group clearly fails to adhere to predictions based on a coevolutionary arms race. It has been suggested that hosts should evolve broad compatibility with mutualists, whereas the coevolutionary arms race would lead to a high degree of specialization in host–pathogen interactions (Law, 1985; Law and Lewis, 1983). Contrary to this expectation, *Epichloe* species with more mutualistic character—such as *E. festucae*—are associated with only a single genus of host, whereas the relatively antagonistic MP-I is associated with seven host genera spanning three tribes (Schardl et al., 1997).

It should be noted, however, that although the species has a broad host range, individual isolates can be host-specific (Chung et al., 1997). Host specificity is also a feature of the benign *Neotyphodium* species. Despite the fact that tall fescue (*Festucae arundinacea* var. *genuina*), meadow fescue, and perennial ryegrass are so closely related that they can be routinely mated to form female-fertile hybrids (Chandrasekharan et al., 1972), their endophytes exhibit various levels of incompatibility with any but their native hosts (Christensen, 1995; Christensen et al., 1997; Koga et al., 1993).

Much of the underlying speculation on evolutionary trajectories is based on teleological or adaptive arguments, while little is known of the physiological or genetic requirements for symbiotic specialization or effective mutualism. Consideration should be given to the possibility that at least some symbioses may have a mechanistic connection between mutualism and specialization at the species or intraspecies level. If so, then selection for mutualism would concomitantly select for specialization. Indeed, the grass–*Epichloe* system is not the only one in which such a link is empirically suggested. Associations between distinct lineages of the legume *Amphicarpaea bracteata* and *Bradyrhizobium* sp. vary in their relative mutualistic effectiveness manifested in formation of nitrogen-fixing nodules and enhancement of plant growth. Such variation, whereby more specific associations are more effective, is suggestive of coadaptation, though common descent (i.e., parallel divergence) was not indicated (Wilkinson et al., 1996). Coadaptation may also be important for grasses and endophytes to contribute maximum mutual benefit at several levels: e.g., stability of the associations in vegetative plants and seeds, proper plant growth regulator levels for optimal development, and appropriate levels and types of stress tolerance to counterbalance costs of alkaloid production and other fitness-enhancing mechanisms. We speculate that those particular cases where host and fungus both display their sexual states require such a high degree of coadaptation that, once this situation evolved, it was maintained primarily by common descent of host and symbiont lineages (Schardl et al., 1997).

4.2 Benefits of Sex to Symbionts

Schardl (1996) argues that maximal mutual benefit between a grass and its *Epichloe* symbiont may require that the symbiont be capable of both vertical and horizontal transmission in that host. The underlying rationale is that sexual capability may be beneficial for both host and symbiont. Part of the benefit may be related to dissemination of diaspores. Grass seeds can disperse to new locales whereas tillers and stolons have a more spacially restricted dissemination capability. In the case of *Epichloe* species, meiotic spores (ascospores) can colonize host individuals other than the maternal descendants of an infected host. However, dissemination may not be the only or even primary benefit of the sexual cycle (after all, fungal dissemination is often more efficient via asexual rather than sexual diaspores). Accumulating empirical evidence supports a direct benefit of sexual recombination upon an organism's genome, rendering a recombining genome more fit, on average, than a clonally propagated genome (Muller, 1964). Therefore, maximal mutual benefit is expected in balanced symbioses in which many host tillers produce symbiont-infected seeds, but some flowering tillers are sacrificed to the sexual state of the fungus.

5. CLONES AND MUTATIONAL MELTDOWN

Muller (1964) stated the hypothesis that clonal lineages are subject to relentless accumulation of marginally deleterious mutations that ratchet their fitness downward. This hypothesis, known as Muller's ratchet, leads to the prediction of "mutational meltdown" (Lynch, 1996) and extinction of clonal lineages. Rice (1994) dramatically demonstrated this effect by an approach based on the fact that *Drosophila melanogaster* chromosomes recombine in females but not in males. Nonrecombining chromosomes were maintained through 35 generations, while chromosomes of the same origin were allowed to recombine in a separate series of 35 generations. The recombined and nonrecombined chromosomes were then introduced into a *D. melanogaster* line and their effects on fitness measured. A significant reduction in fitness was attributable to the nonrecombined chromosomes.

Reduced fitness of clonal lineages poses special problems for organelles that arose endosymbiotically and maintain functional genetic material. It is important to note, however, that the equivalent of sexual recombination between organellar genomes can occur in some organisms (such as mitochondrial genome recombination in fungi; Chung et al., 1996; Hintz et al., 1988). Even in these cases, however, there is considerably more clonality of organellar genomes than nuclear genomes of sexual eukaryotes. Studies by Moran (1996) and Lynch (1996) strongly suggest that Muller's ratchet has been in operation in endosymbiotic bacteria and organellar genomes.

The particular dilemma of organelles with mutable but nonrecombining genomes has a parallel in clonal grass endophytes, but a fundamental difference is that grasses do not depend on endophytes for basic cellular processes as many eukaryotes depend on cytoplasmic organelles. If a grass plant were to lose its endophyte it might survive well in a fostering environment. Even so, stresses and competition with symbiotic conspecifics may cause the demise of nonsymbiotic plant lineages. In those instances when the hosts are highly dependent on their endophytes, an endophyte's ability to escape Muller's ratchet would be important for the long-term survival of its host lineage. In fact, even asexual endophytes possess a means of escaping the ratchet: Phylogenetic evidence presented later in this chapter indicates that many asexual endophytes have a history of interspecific hybridization with *Epichloe* species.

6. INTERDEPENDENCE

As stated above, the analogy of asexual endophytes with endosymbiotic organelles—as clonally propagated, maternally inherited entities—does not extend to the level of interdependence. Dependence of a host species on an endophyte species appears to be the exception, not the rule. High infection frequencies suggest a degree of dependence of the host on its endophyte(s) (Clay, 1990). *Festuca rubra* is commonly symbiotic with *E. festucae* in many populations (Funk et al., 1994). *Brachypodium sylvaticum* is essentially always symbiotic with *Epichloe* sylvatica or closely related asexual lineages (Bucheli and Leuchtmann, 1996). A Europe-wide seed collection of meadow fescue (from Ian D. Thomas, Welsh Plant Breeding Station, Aberystwyth) was approximately 70% endophyte-infected (M. R. Siegel, personal communication). Yet despite these dramatic examples, many pooid grass species are either not known to possess *Epichloe/Neotyphodium* endophytes or are rarely infected.

Although ecological dependence of hosts on endophytes appears to be rare, it is possible that these symbioses are significant in the evolution of the Pooideae. This would be the case if the symbionts provide their hosts with unique functionality and adaptability. That is, for the endophytes to influence speciation of their hosts they need not be ubiquitous. Instead, they need only facilitate host speciation by improving chances to occupy new environments or ecological niches. Indications that they can do so come from two systems that have been extensively studied due to their agricultural importance. The perennial ryegrass–*N. lolii* association naturalized to New Zealand is under the selection of an introduced insect, the Argentine stem weevil (*Listronotus bonariensis*). This efficient herbivore forces dependence of perennial ryegrass on the anti-insect alkaloids produced by its endophyte (Rowan and Latch, 1994). The other example is tall fescue in North America, where its spectacular success in the temperate transition zone (otherwise typically occupied by warm-season C4 grasses) may largely be due to endophyte-enhanced tolerance to heat, drought, and other stresses (Arechavaleta et al., 1989; Schardl and Phillips, 1997; West, 1994).

Another indication of the importance of endophytes in the evolution of the Pooideae is that these symbioses span the phylogenetic diversity of the grass subfamily. Of the 10 tribes in the subfamily, members of 8 tribes are known hosts of *Epichloe* species, *Neotyphodium* species or both (Table 1). Only tribes Nardeae and Diarrheneae lack reported endophytes. Inspection of phylogenetic relationships among the tribes (Davis and Soreng, 1993) indicates that host clades have diverged from one another early in pooid radiation. For example, relative to the more recent radiations of tribes Aveneae, Poeae, Triticeae, and Bromeae, tribes Brachyelytreae, Brachypodieae, Meliceae, and Stipeae appear deeply rooted.

Although a proposed influence of endophytes on pooid evolution is highly speculative, it is a possible explanation for phylogenetic patterns suggestive of *Epichloe*–grass cospeciation. Likewise, the possibility that even the rare cases of host dependence on endophytes are evolutionarily relevant may explain the preponderance of interspecific hybrids among clonal endophyte lineages. In the next two sections we will discuss the phylogenetic evidence first for cospeciation and then for interspecific hybridization.

7. COSPECIATION

Diversity within and between species can be studied in detail by analysis of isozyme and DNA sequence polymorphisms, allowing inferences about evolutionary patterns and possible cospeciation with the grasses. A particularly informative region for comparisons of species within the *Epichloe/Neotyphodium* genus has been the intron-rich β-tubulin gene (*tub2*). Because *E. typhina* and its sexual relatives were each found to possess only a single *tub2* copy (Byrd et al., 1990; Schardl et al., 1994), the evolution of this copy may approximate the evolution of the species. For the most part, this expectation has been borne out. For sexual species the *tub2* gene tree is largely concordant with that generated from a comparably informative region of nuclear rDNA (Schardl et al., 1997). The one exception is the placement of sequences from MP-VIII (*Epichloe glyceriae* from Glyceria striata). There is such a dramatic discordance between MP-VIII *tub2* and rDNA phylogenies as to suggest an origin of this species by interspecific hybridization. This conclusion must be considered tentative, however, pending phylogenetic analysis of additional genes.

Another underlying pattern, again with an exception, was that *tub2* sequences from members of the same *Epichloe* biological species tended to group together. The exception was the relationship of MP-I and MP-VII *tub2* sequences, which suggested that MP-I was paraphyletic to MP-VII (Fig. 2). Again, additional gene trees are required to elucidate these relationships.

The possibility of cospeciation can be addressed for eight of the nine *Epichloe* biological species. MP-I is excluded from this consideration because its host range nearly spans the phylogenetic diversity of the Pooideae. Other species are specific for single species, genera, or groups of genera within a tribe. The relationships of six tribes—Brachyelytreae, Brachypodieae, Aveneae, Poeae, Bromeae, and Triticeae—mirror those of the seven *Epichloe* biological species MP-II through MP-VII, and MP-IX (Fig. 3). In the *tub2* gene tree, only MP-VIII and its host, *Glyceria striata* (tribe Meliceae), do not fit the mirror phylogeny. However, the rDNA sequence (recall that this is discordant with *tub2* for MP-

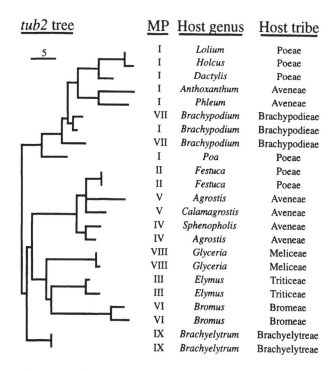

tub2 tree	MP	Host genus	Host tribe
	I	*Lolium*	Poeae
	I	*Holcus*	Poeae
	I	*Dactylis*	Poeae
	I	*Anthoxanthum*	Aveneae
	I	*Phleum*	Aveneae
	VII	*Brachypodium*	Brachypodieae
	I	*Brachypodium*	Brachypodieae
	VII	*Brachypodium*	Brachypodieae
	I	*Poa*	Poeae
	II	*Festuca*	Poeae
	II	*Festuca*	Poeae
	V	*Agrostis*	Aveneae
	V	*Calamagrostis*	Aveneae
	IV	*Sphenopholis*	Aveneae
	IV	*Agrostis*	Aveneae
	VIII	*Glyceria*	Meliceae
	VIII	*Glyceria*	Meliceae
	III	*Elymus*	Triticeae
	III	*Elymus*	Triticeae
	VI	*Bromus*	Bromeae
	VI	*Bromus*	Bromeae
	IX	*Brachyelytrum*	Brachyelytreae
	IX	*Brachyelytrum*	Brachyelytreae

Figure 2 Phylogeny of genus *Epichloe* spp. *tub2* sequences. Biological species of *Epichloe* are listed as mating populations (MP) I-IX. The gene tree was generated by maximum parsimony on the aligned sequences of the 5′-portion of *tub2*, in which nearly all variation is within about 400 positions in introns 1, 2, and 3. Shown is the single most parsimonious tree by exact search (branch-and-bound algorithm). All nucleotide substitutions were weighted equally, and each alignment gap regardless of length was weighted as one nucleotide substitution. The length of each horizontal branch is proportional to the inferred number of character changes on that branch (numbered bar = 5 changes); vertical lines have no numerical value and serve only to separate clades.

VIII) place MP-VIII isolate 277 in the expected position to mirror host phylogeny. The mirror host and *Epichloe* phylogenies hold for all balanced symbioses, suggesting a greater propensity for coevolution by common descent among balanced associations (Schardl et al., 1997).

8. INTERSPECIFIC HYBRIDIZATION

Also of considerable interest are the origins of highly beneficial asexual species such as *Neotyphodium coenophialum*, *N. lolii*, and *N. uncinatum*. In their isozyme

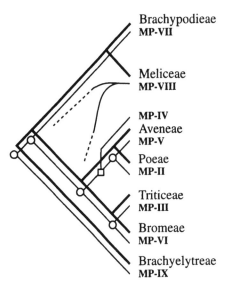

Figure 3   Proposed common descent of grasses and *Epichloe* species. Thick lines represent evolutionary history of tribes within the Pooideae, estimated from data of Davis et al. (1993); thin lines represent evolutionary history of *Epichloe* species (excluding MP-I) estimated from data of Schardl et al. (1997). Circles represent postulated cospeciation events; the square represents a symbiont speciation without concomitant host speciation. The origin of MP-VIII associated with host tribe Meliceae is unclear but may involve hybridization.

studies of sexual and asexual species, Leuchtmann and Clay (1990) noted that most sexual isolates had single-locus isozymes, but that multiband patterns in many asexual species suggested two or more copies of some isozyme loci. This was interpreted as a likely indication of aneuploidy in the asexual species, but euploidy (the haploid state in this fungal group) in sexual *Epichloe* species. Later, Christensen et al. (1993) noted considerable diversity of tall fescue endophytes, all of which gave multiband allozyme patterns. In contrast, *N. lolii* genotypes, though variable, did not exhibit the multiband patterns. *N. uncinatum* gave primarily, but not exclusively, single-band patterns. DNA sequences of noncoding portions of β-tubulin (*tub2*) and nuclear rRNA (rDNA) genes were determined by Tsai et al. (1994) to help elucidate the origins of the diverse tall fescue symbionts, and by Schardl et al. (1994) to elucidate the origin of a rare and distinct genotype isolated from perennial ryegrass in southern France (Fig. 4). These were the first documentations, at the genetic level, of interspecific hybrid fungi.

The hypothesis of interspecific hybridization in the evolution of many asexual endophytes continues to be supported with the expanded survey and phyloge-

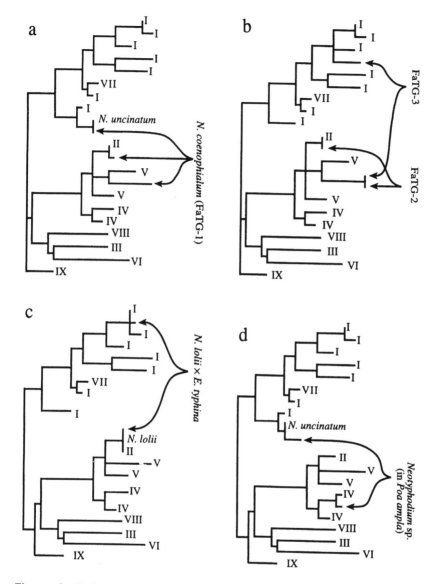

Figure 4 Phylogenetic relationships of asexual (*Neotyphodium*) endophytes with *Epichloe* species, as inferred by *tub2* gene phylogenies. Gene trees were determined as in Fig. 2, but for each *Epichloe* species only one isolate from each host genus was included. Arrows indicate the multiple *tub2* genes in the endophyte taxa indicated. The left edge of each phylogram indicates the midpoint root, which shifts slightly depending on which taxa are included in the analysis. Panels a and b indicate relationships of *tub2* gene copies from endophytes of hexaploid tall fescue (*Festuca arundinacea*). Panel c indicates *tub2* gene relationships for two endophytes of perennial ryegrass, *Neotyphodium lolii* and an apparent *N. lolii* × *Epichloe typhina* hybrid. Panel d indicates relationships of the two *tub2* copies in an endophyte of big bluegrass (*Poa ampla*).

netic analysis of *Epichloe* species (Fig. 4). All three taxonomic groups of tall fescue endophytes indicated by Christensen et al. (1993) have *tub2* phylogenies indicative of hybrid origin. One group, FaTG-1 (= *N. coenophialum*) has three *tub2* genes, each related to the single *tub2* copy of a different *Epichloe/Neotyphodium* species: one with identical sequence to the *N. uncinatum* gene, one related to the *E. festucae* (MP-II) gene, and the third related to the *E. baconii* (MP-V) gene (Fig. 4a). Taxonomic group FaTG-2 had *E. festucae-* and *E. baconii*-related *tub2* copies, and FaTG-3 genes were related to those of *E. baconii* and *E. typhina* (MP-I) (Fig. 4b).

At least three hybridization events were required to account for the copy numbers and sequence relationships of *tub2* genes in tall fescue endophytes. Furthermore, the complex hybrid nature of *N. coenophialum* suggested that some or all of the hybridization events were somatic (following vegetative anastomosis) rather than sexual. A possible scenario for the origin of the tall fescue–*N. coenophialum* symbiota, including both host and symbiont hybridizations, is shown in Fig. 5. These symbiota constitute some of the most complex genetic systems known, including three grass nuclear genomes, contributions from at least three fungal species to the symbiont genome, and, of course, the genomes of fungal and plant cytoplasmic organelles.

A rare perennial ryegrass symbiont (taxonomic group LpTG-2) has a genotype consistent with a hybrid origin from *N. lolii* and *E. typhina*. The basis for this conclusion was analysis of 10 loci including *tub2* and *pyr4* gene sequences and electrophoretic profiles of eight polymorphic isozymes (Collett et al., 1995; Schardl et al., 1994). As in the tall fescue endophytes, reanalysis of the *tub2* phylogenies with the larger dataset now available for *Epichloe* spp. *tub2* genes continues to support the inferred hybrid origin of this strain (Fig. 4c). *Neotyphodium lolii* has *tub2* and *pyr4* sequences identical to *E. festucae*, but differences

---

Figure 5 Proposed origin of the tall fescue–*Neotyphodium coenophialum* symbiosis. The likely origin of hexaploid (2n = 6x = 42 chromosomes) tall fescue was the hybridization of meadow fescue (2n = 2x = 14) and *Festuca glaucescens* (2n = 4x = 28) (Humphreys et al., 1995; Xu and Sleper, 1994). Both tall fescue and meadow fescue have well characterized and ubiquitous symbionts, *N. coenophialum* and *Neotyphodium uncinatum*, respectively. Apparently, 6x tall fescue originally inherited the symbiont of its meadow fescue ancestor. Later, a tall fescue–*N. uncinatum* symbiotum was infected with an *Epichloe* species, and the coexisting fungi hybridized and combined nuclei. A repeat of this process led to the double hybrid, *N. coenophialum*. Although *E. baconii* is shown as the first to participate in symbiont hybridization, it is equally likely that *E. festucae* was first.

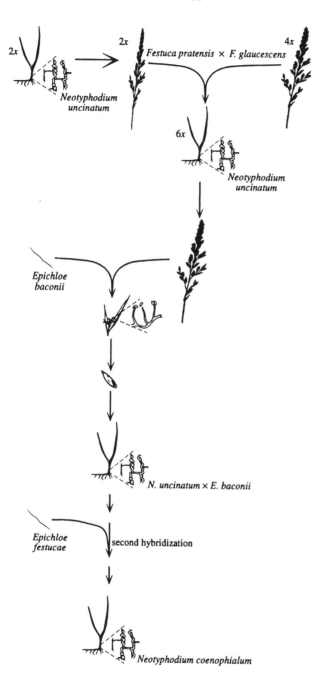

in isozyme polymorphisms and mitochondrial DNA profiles (based on restriction endonuclease digests and electrophoretic analysis). These differences provide a basis for suggesting that *N. lolii*, rather than *E. festucae*, was one of the ancestors of LpTG-2 (Schardl et al., 1994). The other ancestor is a close relative specifically of *E. typhina* associated with perennial ryegrass. Two *E. typhina* isolates from this host have the same *tub2* sequence (within the region spanning the first three introns). This sequence differs from the *tub2*-1 copy of LpTG-2 by only a single nucleotide change. Also, the hybrid has the same rDNA-ITS sequence as *E. typhina* from perennial ryegrass. Thus, the formula that describes LpTG-2 is *N. lolii* × *E. typhina*, where the latter ancestor is a genotype specifically associated with perennial ryegrass.

Another hybrid endophyte is a symbiont of *Poa ampla* plants in Yukon, Canada. This endophyte produces stromata on a small minority of infected plants. However, we have not yet been able to identify a sexually compatible partner in order to demonstrate a biological species relationship of this endophyte. In fact, we have not observed any reaction indicative of mating-type specificity. This is unusual. Most asexual endophytes tested behave either as mat-1 or mat-2 genotypes when used as males on any *Epichloe* sp. tester. Possibly, the *P. ampla* endophyte is asexual, but its stroma expression is a vestige of its ancestral sexual species. It has two *tub2* gene copies, and, as for other endophytes with multiple *tub2* copies, phylogenetic analysis of the copies indicates relationships to different *Epichloe* species (Fig. 4d). This analysis suggested two ancestral species for the *P. ampla* endophyte: one being a relative of either *N. uncinatum* or *E. typhina* from *Poa pratense* (confirmed by rDNA-ITS sequence comparison), and the other being *E. amarillans*.

Interestingly, the two *tub2* sequences in the *P. ampla* endophytes are associated with species on different continents: *N. uncinatum* and *E. typhina* known only in European grasses, and *E. amarillans* known only in North America. The suggestion is that members of the *Epichloe* clades were once in contact with each other, so that their sequence divergence predates their geographic separation. This fits with the observation that there is no biogeographic pattern in the overall *Epichloe* phylogeny which, like the Pooideae, appears to be a circumpolar complex.

The phylogenetic evidence suggests that *N. coenophialum* and the rare perennial ryegrass symbiont LpTG-2 each had an asexual endophyte among its ancestors (*N. uncinatum* and *N. lolii*, respectively). Therefore at least some interspecific hybridization probably occurred somatically, by fusion of fungal hyphae (multinucleate ''cells'') and subsequent fusion of nuclei from the different sources. Many fungi possess vegetative incompatibility systems that can serve as a barrier to such a process (Leslie, 1993), raising the question of whether a vegetative incompatibility system is lacking in *Epichloe/Neotyphodium*. This was addressed by standard methods of complementing nitrate-nonutilizing mutants derived from four different biological species of *Epichloe* (Chung and Schardl, 1997). In fact, complementation was frequently observed both in intraspecific and interspecific pairings. It was confirmed that even instances of interspecific complementation involved formation of heterokaryons (hyphae with nuclei of different origins). Although heterokaryon formation was not common and the heterokaryons were unstable, vegetative incompatibility was not a likely barrier to hybridization. Further experimentation is needed to determine whether nuclear fusion is a common occurrence in *Epichloe/Neotyphodium* heterokaryons.

Even without a vegetative incompatibility barrier, it seems likely that hybridization is a rare phenomenon. Plants possessing endophytes need to become further infected by *Epichloe*, the coinfecting symbionts need to merge and fuse nuclei, and somehow the hybrid nuclei need to be maintained and perhaps outcompete the parental nuclei. The frequent occurrence of hybrid asexual endophytes suggests a selection favoring them. A likely source of selection is the mutational meltdown effect of Muller's ratchet acting on the asexual endophytes. Hybridization with sexually derived endophytes should replaced essential genes that may have suffered reduced functionality by Muller's ratchet.

9. CONCLUSIONS

The diversity of grass–endophyte associations makes this a model system for the study of coevolution of symbiotic partners. As such this system reveals that coevolution of symbiotic partners is likely to be affected by levels of interdependence, modes of transmission, degree of mutualism vs. antagonism, and geographic isolation. Furthermore, these characteristics are frequently linked (e.g., we predict that the most mutualistic associations are those with balanced modes of transmission for the partners).

The only consistent pattern of coevolution by common descent within the *Epichloe* phylogeny involves the endophytes that participate in balanced symbioses. All such species studied apparently share the same pattern of species divergence as their host tribes (see Fig. 3). It is these same types of associations that are likely to provide the greatest mutual benefit to the symbiotic partners. This makes intuitive sense, i.e., those associations with a balance of benefits between partners are more likely to persist and diverge together over time, while associations in which either partner is relatively disadvantaged would be less likely to lead to stable patterns of cospeciation.

Many endophytes that reproduce only through asexually derived propagules disseminated with host seeds have evolved via interspecific hybridization. In these cases, the evolution of the endophyte has been strongly driven by the captive life history it experiences. We predict that over time the accumulation of nonlethal deleterious mutations due to a lack of sexual recombination (Muller,

1964) ultimately causes extinction in those endophyte lines that do not hybridize with other endophytes. Thus, although interspecific hybridization is likely to be a very rare event, many of the asexual endophytes that persist today show evidence of hybrid origins.

The variety of coevolutionary trajectories begs the question of the direction of evolution in grass–endophyte associations. If, as we hypothesize, (1) interspecific hybrids are selected by Muller's ratchet on asexual lineages and (2) balanced symbioses are maintained for long evolutionary time periods by cospeciation, then the purely horizontally transmitted and the purely vertically transmitted associations would likely be derived from balanced symbioses. The simplest argument for balanced symbioses as the ancestral state is that loss of a trait is more likely than a gain. The rise of the purely horizontally transmitted and the purely-vertically transmitted associations would each involve a loss of one form of symbiont transmission from the balanced symbioses. Furthermore, if balanced symbioses have, in fact, cospeciated with their hosts, then this implies that the associations are quite ancient. In contrast, the established phylogenies of many purely vertically transmitted *Neotyphodium* sp. and preliminary data for one purely horizontally transmitted endophyte indicate more recent hybridization events. Thus, the available data support the balanced grass–*Epichloe* symbioses as representing the most ancient condition.

REFERENCES

Arechavaleta, M., Bacon, C. W., Hoveland, C. S. and Radcliffe, D. E. (1989). Effect of the tall fescue endophyte on plant response to environmental stress. *Agronomy Journal* 81, 83–90.

Bucheli, E. and Leuchtmann, A. (1996). Evidence for genetic differentiation between choke-inducing and asymptomatic strains of the Epichloe grass endophyte from *Brachypodium sylvaticum. Evolution* 50, 1879–1887.

Byrd, A. D., Schardl, C. L., Songlin, P. J., Mogen, K. L. and Siegel, M. R. (1990). The b-tubulin gene of *Epichloe typhina* from perennial ryegrass (*Lolium perenne*). *Current Genetics* 18, 347–354.

Chandrasekharan, P., Lewis, E. J. and Borrill, M. (1972). Studies in *Festuca*. II. Fertility relationships between species of sections Bovinae and Scariosae, and their affinities with *Lolium. Genetica* 43, 375–386.

Christensen, M. J. (1995). Variation in the ability of *Acremonium* endophytes of perennial rye-grass (*Lolium perenne*), tall fescue (*Festuca arundinacea*) and meadow fescue (*F. pratensis*) to form compatible associations in three grasses. *Mycological Research* 99, 466–470.

Christensen, M. J., Ball, O. J.-P., Bennett, R. and Schardl, C. L. (1997). Fungal and host genotype effects on compatibility and vascular colonisation by *Epichloe festucae. Mycological Research* 101, 493–501.

Christensen, M. J., Leuchtmann, A., Rowan, D. D. and Tapper, B. A. (1993). Taxonomy of *Acremonium* endophytes of tall fescue (*Festuca arundinacea*), meadow fescue (*F. pratensis*), and perennial rye-grass (*Lolium perenne*). *Mycological Research* 97, 1083–1092.

Chung, K.-R., Hollin, W., Siegel, M. R. and Schardl, C. L. (1997). Genetics of host specificity in *Epichloe typhina*. *Phytopathology* 87, 599–605.

Chung, K.-R., Leuchtmann, A. and Schardl. C. L. (1996). Inheritance of mitochondrial DNA and plasmids in the ascomycetous fungus, *Epichloe typhina*. *Genetics* 142, 259–265.

Chung, K.-R. and Schardl, C. L. (1997). Vegetative compatibility between and within *Epichloe* species. *Mycologia* 89, 558–565.

Clay, K. (1990). Fungal endophytes of grasses. *Annual Review of Ecology and Systematics* 21, 275–295.

Clay, K. and Leuchtmann, A. (1989). Infection of woodland grasses by fungal endophytes. *Mycologia* 81, 805–811.

Collett, M. A., Bradshaw, R. E. and Scott, D. B. (1995). A mutualistic fungal symbiont of perennial ryegrass contains two different *pyr4* genes, both expressing orotidine-5′-monophosphate decarboxylase. *Gene* 158, 31–39.

Davis, J. I. and Soreng, R. J. (1993). Phylogenetic structure in the grass family (Poaceae) as inferred from chloroplast DNA restriction site variation. *American Journal of Botany* 80, 1444–1454.

Funk, C. R., Belanger, F. C. and Murphy, J. A. (1994). Role of endophytes in grasses used for turf and soil conservation. *In*: C. W. Bacon, and J. F. White, Jr. (eds.), *Biotechnology of Endophytic Fungi of Grasses*, CRC Press, Boca Raton, FL, pp. 201–209.

Hintz, W. E .A., Anderson, J. B. and Horgen, P. A. (1988). Nuclear migration and mitochondrial inheritance in the mushroom *Agaricus bitorquis*. *Genetics* 119, 35–41.

Humphreys, M. W., Thomas, H. M., Morgan, W. G., Meredith, M. R., Harper, J. A., Thomas H., Zwierzykowski, Z. and Ghesquire, M. (1995). Discriminating the ancestral progenitors of hexaploid *Festuca arundinacea* using genomic in situ hybridization. *Heredity* 75, 171-174.

Kimmons, C. A., Gwinn, K. D. and Bernard, E. C. (1990). Nematode reproduction on endophyte-infected and endophyte-free tall fescue. *Plant Disease* 74, 757–761.

Koga, H., Christensen, M. J. and Bennett, R. J. (1993). Cellular interactions of some grass/ *Acremonium* endophyte associations. *Mycology Research* 97, 1237–1244.

Law, R. (1985). Evolution in a mutualistic environment. *In*, D.H. Boucher (eds.), *The Biology of Mutualism*. Oxford University Press, New York, pp. 145–170.

Law, R. and Lewis, D. H. (1983). Biotic environments and the maintenance of sex: some evidence from mutualistic symbioses. *Biology Journal of the Linnean Society* 20, 249–276.

Leslie, J. F. (1993). Fungal vegetative compatibility. *Annual Review of Phytopathology* 31, 127–151.

Leuchtmann, A. and Clay, K. (1990). Isozyme variation in the Acremonium/Epichloe fungal endophyte complex. *Phytopathology* 80, 1133–1139.

Lynch, M. (1996). Mutation accumulation in transfer RNAs: molecular evidence for

Muller's ratchet in mitochondrial genomes. *Molecular and Biology Evolution* 13, 209–220.

Miles, C. O., Lane, G. A., Di Menna, M. E., Garthwaite, I., Piper, E. L., Ball, O. J. P., Latch, G. C. M., Allen, J. M., Hunt, M. B., Bush, L. P., Min, F. K., Fletcher, I. and Harris, P. S. (1996). High levels of ergonovine and lysergic acid amide in toxic Achnatherum inebrians accompany infection by an *Acremonium*-like endophytic fungus. *Journal of Agriculture and Food Chemistry* 44, 1285–1290.

Moran, N. A. (1996). Accelerated evolution and Muller's rachet in endosymbiotic bacteria. *Proceedings of the National Academy of Science of the USA* 93, 2873–2878.

Muller, H. J. (1964). The relation of recombination to mutational advance. *Mutation Research* 1, 2–9.

Philipson, M. N. and Christey, M. C. (1986) The relationship of host and endophyte during flowering, seed formation, and germination of *Lolium perenne*. *New Zealand Journal of Botany* 24, 125–134.

Rice, W. R. (1994). Degeneration of a nonrecombining chromosome. *Science* 263, 230–232.

Rowan, D. D. and Latch, G. C. M. (1994). Utilization of endophyte-infected perennial ryegrasses for increased insect resistance. *In*: C. W., Bacon and J. F. White, Jr. (eds.), *Biotechnology of Endophytic Fungi of Grasses*, CRC Press, Boca Raton, FL, pp. 169–183.

Schardl, C. L. (1996). *Epichloe* species: fungal symbionts of grasses. *Annual Review of Phytopathology* 34, 109–130.

Schardl, C. L., Leuchtmann, A., Chung, K.-R., Penny, D. and Siegel, M. R. (1997). Coevolution by common descent of fungal symbionts (*Epichloe* spp.) and grass hosts. *Molecular Biology and Evolution* 14, 133–143.

Schardl, C. L., Leuchtmann, A., Tsai, H.-F., Collett, M..A., Watt, D. M. and Scott, D. B. (1994). Origin of a fungal symbiont of perennial ryegrass by interspecific hybridization of a mutualist with the ryegrass choke pathogen, *Epichloe typhina*. *Genetics* 136, 1307–1317.

Schardl, C. L. and Phillips, T. D. (1997). Protective grass endophytes: where are they from and where are they going? *Plant Disease* 81, 430–437.

Schardl, C. L. and Tsai, H.-F. (1992). Molecular biology and evolution of the grass endophytes. *Natural Toxins* 1, 171–184.

Siegel, M. R. and Bush, L. P. (1996). Defensive chemicals in grass-fungal endophyte associations. *Recent Advances in Phytochemistry* 30, 81–120.

Tsai, H.-F., Liu, J.-S., Staben, C., Christensen, M. J., Latch, G. C. M., Siegel, M. R. and Schardl, C. L. (1994). Evolutionary diversification of fungal endophytes of tall fescue grass by hybridization with *Epichloe* species. *Proceedings of the National Academy of Science*, of the *USA* 91, 2542–2546.

West, C. P. (1994). Physiology and drought tolerance of endophyte-infected grasses. *In*: C. W. Bacon and J. F. White, Jr. (eds.), *Biotechnology of Endophytic Fungi of Grasses*, CRC Press, Boca Raton, FL, pp. 87–99.

White, J. F., Jr. (1987). Widespread distribution of endophytes in the Poaceae. *Plant Disease* 71, 340–342.

White, J. F., Jr. (1993). Endophyte-host associations in grasses. XIX. A systematic study of some sympatric species of *Epichloe* in England. *Mycologia* 85, 444–455.

White, J. F., Jr. (1994). Endophyte-host associations in grasses. XX. Structural and reproductive studies of *Epichloe amarillans* sp. nov. and comparisons to *E. typhina*. *Mycologia* 86, 571–580.

Wilkinson, H. H, Spoerke, J. M. and Parker, M. A. (1996). Divergence in symbiotic compatibility in a legume-*Bradyrhizobium* mutualism. *Evolution* 50, 1470–1477.

Xu, W. W. and Sleper, D. A. (1994). Phylogeny of tall fescue and related species using RFLPs. *Theoretical and Applied Genetics* 88, 685–690.

5

Evidence for *Fusarium* Endophytes in Cultivated and Wild Plants

Gretchen A. Kuldau and Ida E. Yates
Richard B. Russell Research Center, Agricultural Research Service, U.S. Department of Agriculture, Athens, Georgia

1. INTRODUCTION

The anamorphic genus *Fusarium* link is arguably the most important fungal genus agronomically and is known for its plant pathogenic members and for mycotoxin production. *Fusarium* species are defined by the production of fusoid macroconidia (asexual spores) that bear a foot-shaped basal cell (Booth, 1971). *Fusarium* taxonomy and thus the accepted number of species has been in a state of flux since its inception (Wollenweber, 1931; Synder and Hansen, 1945; Booth, 1971; Nelson et al., 1983; Nelson, 1991). However, recent trends are toward an increase in the number of named species via the application of combined efforts in morphology and molecular phylogenetics (e.g., Nirenberg and O'Donnell, 1998; O'Donnell et al., 1998) and through mating studies (e.g., Klittich et al., 1997). Currently, there are well over 35 described species. Teleomorphs (sexual stages) of *Fusarium* are found in *Gibberella*, *Nectria* and *Calonectria* (Booth, 1981).

Members of the genus are found worldwide predominantly in soil and organic matter and, although known for their pathogenicity, many nonpathogenic isolates have been described (Burgess, 1981; Stoner, 1981; Gordon and Martyn, 1997). *Fusarium oxysporum* is the most cosmopolitan species of *Fusarium* in both cultivated and noncultivated soils worldwide (Stoner, 1981). Thus, it is not surprising that it is the most commonly reported *Fusarium* endophyte of both wild and crop plants (e.g., Gordon et al., 1989; Leslie et al., 1990) (Tables 1–3).

Table 1 *Fusarium* Endophytes of Wild Plant Species

Fusarium endophyte	Host plant or family	Method[a]	Pathogen[b]	Ref.(s)
F. acuminatum	*Chenopodium album*	C	?	Gordon, 1959
	Dracocephalum parvifolium	C	?	Ibid.
F. aquaeductuum	*Licuala ramsayi*	C	?	Rodrigues and Samuels, 1990
F. avenaceum	*Capsella bursa-pastoris*	C	No	Jenkinson and Parry, 1994
	Cirsium arvense	C	No	Ibid.
	Galium aparine	C	No	Ibid.
	Matricaria spp.	C	No	Ibid.
	Ranunculus acris	C	No	Ibid.
	Ranunculus repens	C	No	Ibid.
	Rumex obtusifolius	C	No	Ibid.
	Senecio vulgaris	C	No	Ibid.
	Stellaria media	C	No	Ibid.
	Urtica dioica	C	No	Ibid.
	Viola arvensis	C	No	Ibid.
F. culmorum	*Beta vulgaris*	C	No	Jenkinson and Parry, 1994
	Capsella bursa-pastoris	C	No	Ibid.
	Galium aparine	C	No	Ibid.
	Matricaria spp.	C	No	Ibid.
	Ranunculus acris	C	No	Ibid.
	Ranunculus repens	C	No	Ibid.
	Rumex obtusifolius	C	No	Ibid.
	Senecio vulgaris	C	No	Ibid.
	Urtica dioica	C	No	Ibid.
	Viola arvensis	C	No	Ibid.
F. lateritium	*Taxus baccata*	C	?	Strobel et al., 1996
	Ulex europaeus, U. gallii	C	?	Fischer et al., 1986
F. moniliforme	*Cynomorium coccineum*	C	?	Mushin and Zwain, 1989
	Crypripedium calceolus	C	?	Peschke and Volz, 1978
	Cypripedium reginae	C	?	Ibid.
	Yucca sp.	C	?	Bills, 1996

F. oxysporum			
Araceae	C	?	Petrini and Dreyfuss, 1981
Bromeliaceae	C	?	Ibid.
Orchidaceae	C	?	Ibid.
Abutilon theophrasti	C	No	Helbig and Carroll, 1984
Achillea millefolium	C	No	Windels and Kommedahl, 1974
Agropyron trachycaulum	C	No	Ibid.
Amaranthus retroflexus	C	No	Helbig and Carroll, 1984
Ambrosia artemisiifolia	C	No	Ibid.
Ambrosia trifida	C	No	Ibid.
Amorpha canescens	C	No	Windels and Kommedahl, 1974
Andropogon geradi	C	No	Ibid.
Apocynum cannabinum	C	No	Helbig and Carroll, 1984
Artemisia ludoviciana	C	No	Windels and Kommedahl, 1974
Asclepias syriaca	C	No	Helbig and Carroll, 1984
Asclepias tuberosa	C	No	Windels and Kommedahl, 1974
Aster sp.	C	No	Ibid.
Bromus mollis	C	No	Ibid.
Cardiospermum halicacabum	C	No	Haware and Nene, 1982
Chenopodium album	C	No	Helbig and Carroll, 1984
Convovulus arvensis	C	No	Haware and Nene, 1982
Coreopsis tinctoria	C	No	Windels and Kommedahl, 1974
Cyperus rotundus	C	No	Haware and Nene, 1982
Datura stramonium	C	No	Helbig and Carroll, 1984
Daucus carota	C	No	Ibid.
Erigeron canadensis	C	No	Ibid.
Erysimum cheiranthoides	C	No	Gordon, 1959
Euterpe oleracea	C	Yes	Rodrigues, 1996
Gaillardia aristata	C	?	Gordon, 1959
Galium boreale	C	No	Windels and Kommedahl, 1974
Gaura coccinea	C	?	Gordon, 1959
Helianthus annuus	C	No	Ibid.
Ipomoea hederacea	C	No	Helbig and Carroll, 1984
Ipomoea purpurea	C	No	Ibid.

Table 1 Continued

Fusarium endophyte	Host plant or family	Method[a]	Pathogen[b]	Ref.(s)
	Lepidium apetalum	C	?	Gordon, 1959
	Monarda menthaefolia	C	?	Ibid.
	Muhlenbergia racemosa	C	No	Windels and Kommedahl, 1974
	Petalostemum candidum	C	No	Ibid.
	Phleum pratense	C	No	Ibid.
	Phlox pilosa	C	No	Ibid.
	Poa annua	C	No	Ibid.
	Poa pratensis	C	No	Ibid.
	Polygonum pensylvanicum	C	No	Helbig and Carroll, 1984
	Potentilla recta	C	No	Windels and Kommedahl, 1974
	Psoralea agrophylla	C	No	Ibid.
	Solanum carolinese	C	No	Helbig and Carroll, 1984
	Sorghastrum nutans	C	No	Windels and Kommedahl, 1974
	Symphoricarpus occidentalis	C	No	Ibid.
	Thalictrum venulosum	C	?	Gordon, 1959
	Tradescantia bracteata	C	No	Windels and Kommedahl, 1974
	Tribulus terrastris	C	No	Haware and Nene, 1982
	Trifolium pratense	C	No	Windels and Kommedahl, 1974
	Viola sp.	C	No	Helbig and Carroll, 1984
	Xanthium chinense	C	No	Ibid.
	Zizia aurea	C	No	Windels and Kommedahl, 1974
F. oxysporum f. sp. *batatas*	*Cassia tora*	C	No	Armstrong and Armstrong, 1948
	Ipomoea alba	C	No	Clark and Watson, 1983
	Ipomoea hederifolia	C	No	Ibid.
	Ipomoea lacunosa	C	No	Ibid.
	Ipomoea tamnifolia	C	No	Ibid.
F. oxysporum f. sp. *betae*	*Antheum graveolens*	C	No	MacDonald and Leach, 1976
	Brassica nigra	C	No	Ibid.
	Chenopodium alba	C	No	Ibid.

F. oxysporum f. sp. *cepae*	*Oxalis corniculata*	No	C	Abawi and Lorbeer, 1972
F. oxysporum f. sp. *cubense*	*Paspalum fasciculatum*	No	C	Waite and Dunlap, 1953
	Panicum purpurascens	No	C	Ibid.
F. oxysporum f. sp. *cepae*	*Amaranthus graecizans*	No	C	Katan, 1971
	Digitaria sanguinalis	No	C	Ibid.
	Malvus silvestris	No	C	Ibid.
	Oryziopsis miliceae	No	C	Ibid.
F. poae	*Capsella bursa-pastoris*	No	C	Jenkinson and Parry, 1994
	Galium aparine	No	C	Ibid.
	Matricaria spp.	No	C	Ibid.
	Ranunculus acris	No	C	Ibid.
	Salix sp.	?	C	Gordon, 1959
	Viola arvensis	No	C	Jenkinson and Parry, 1994
F. roseum[c]	*Achillea millefolium*	No	C	Windels and Kommedahl, 1974
	Amorpha canescens	No	C	Ibid.
	Artemisia ludoviciana	No	C	Ibid.
	Helianthus annuus	No	C	Ibid.
	Monarda fistulosa	No	C	Ibid.
	Muhlenbergia racemosa	No	C	Ibid.
	Petalostemum candidum	No	C	Ibid.
	Poa pratensis	No	C	Ibid.
F. sambucinum	*Chenopodium album*	?	C	Gordon, 1959
F. semitectum	*Euterpe oleracea*	Yes	C	Rodrigues, 1996
F. solani	*Achillea millefolium*	No	C	Windels and Kommedahl, 1974
	Amorpha canescens	No	C	Ibid.
	Andropogon gerardi	No	C	Ibid.
	Artemesia ludoviciana	No	C	Ibid.
	Asclepias tuberosa	No	C	Ibid.
	Bromus mollis	No	C	Ibid.
	Euphorbia esula	?	C	Gordon, 1959
	Galium boreale	No	C	Windels and Kommedahl, 1974
	Licuala ramsayi	?	C	Rodrigues and Samuels, 1990
	Petalostemmum candidum	No	C	Windels and Kommedahl, 1974

Table 1 Continued

Fusarium endophyte	Host plant or family	Method[a]	Pathogen[b]	Ref.(s)
	Phleum pratense	C	No	Ibid.
	Poa annua	C	No	Ibid.
	Psoralea argophylla	C	No	Ibid.
	Rudbeckia laciniata	C	?	Gordon, 1959
	Solidago rigida	C	No	Windels and Kommedahl, 1974
	Symphoricarpus occidentalis	C	No	Ibid.
	Trifolium pratense	C	No	Ibid.
	Zizia aurea	C	No	Ibid.
Fusarium sp.	_Fagus sylvatica_	C	?	Chapela and Boddy, 1988
	Fraxinus excelsior	C	?	Kowalski and Kehr, 1996
	Tripterygium wildfordii	U	?	Lee et al., 1995
F. subglutinans	_Agropyron trachycaulum_	C	No	Windels and Kommedahl, 1974
F. tricinctum	_Artemesia ludoviciana_	C	No	Ibid.
	Aster sp.	C	No	Ibid.
	Galium boreale	C	No	Ibid.
	Petalostemmum candidum	C	No	Ibid.
	Phleum pratense	C	No	Ibid.
	Psoralea argophylla	C	No	Ibid.
	Solidago rigida	C	No	Ibid.
	Tradescantia bracteata	C	No	Ibid.
	Trifolium pratense	C	No	Ibid.

[a] "C" indicates that the endophyte was identifed by microbiological culturing; U, unknown.

[b] "Yes" indicates the _Fusarium_ endophyte was also a pathogen on the host from which it was isolated; "No" indicates that it was not; "?" indicates that information was not available.

[c] The name _Fusarium roseum_, as well as its cultivars and varieties, has no current taxonomic status. Thus the actual identity of the fungus indicated is uncertain. Prior taxa with this name have been placed under one of the sections Roseum, Arthrosporella, Gibbosum, and Discolor.

Table 2 *Fusarium* Endophytes of Ornamental Plants and Cultivated Trees

Fusarium endophyte	Host plant or family	Method[a]	Pathogen[b]	Ref(s).
F. oxysporum	*Dianthus caryophyllus*	C	No	Postma and Rattink, 1991
	Gladiolus grandiflora	C	Yes/No	Henis and Zilberstein, 1973
F. oxysporum f. sp. *gladioli*	*Gladiolus grandiflora*	C	Yes	Magie, 1971
F. oxysporum f. sp. *lilii*	*Lilium*	M	Yes	Baayen and Forch, 1998
F. oxysporum f. sp. *lycopersici*	*Albizia julibrissin*	C	No	Armstrong and Armstrong, 1948
F. oxysporum f. sp. *pernici-osum*	*Albizia julibrissin*	C	Yes	Gill, 1967
F. oxysporum f. sp. *tulipae*	*Tulipa*	C	Yes	Bergman and Bakker-Van Der Voort, 1979
F. oxysporum f. sp. *vasinfectum*	*Antirrhinum*	C	No	Armstrong and Armstrong, 1948
Fusarium sp.	*Pinus sylvestris*	C	?	Vaartaja and Cram, 1956
	Psuedotsuga menziesii	C	?	Bloomberg, 1966
F. subglutinans f. sp. *pini*	*Pinus* spp.	C	Yes	Storer et al., 1998

[a] "C" indicates that the endophyte was identifed by microbiological culturing; M indicates identification by microscopy.

[b] "Yes" indicates that the *Fusarium* endophyte is also a pathogen on the host from which it was isolated; "No" indicates that it was not; ? indicates that information was not available.

Table 3 *Fusarium* Endophytes of Crop Plants

Fusarium endophyte	Host plant or family	Method[a]	Pathogen[b]	Ref(s).
F. avenaceum	*Hordeum vulgare*	C	?	Riesen and Close, 1987
F. culmorum	*Triticum aestivum*	C/M	Yes	Sieber et al., 1988; Clement and Parry, 1988
F. equiseti	*Gossypium hirsutum*	C	No	Gordon et al., 1989
	Lycopersicon esculentum	C	No	Ibid.
	Oryza sativa	C	Yes	Fisher and Petrini, 1992
F. graminearum	*Triticum aestivum*	C/M	Yes	Sieber et al., 1988; Clement and Parry, 1988
	Zea mays	C	Yes	Fisher et al., 1992
F. moniliforme	*Gossypium hirsutum*	C	No	Armstrong and Armstrong, 1948; Rudolph and Harrison, 1945
	Zea mays	C	Yes	Foley, 1962; Windels and Kommedahl, 1974
F. oxysporum	*Apium graveolens*	C	No	Schneider, 1984
	Citrullus lanatus	C	No	Larkin et al., 1996
	Cucumis sativus	C	No	Paulitz et al., 1987
	Gossypium hirsutum	C	No	Gordon et al., 1989; Rudolph and Harrison, 1945
	Linum usitatissimum	C	No	Nagao et al., 1990
	Lycopersicon esculentum	C	No	Gordon et al., 1989
	Lycopersicon esculentum	C	No	Hallman and Sikora, 1994a,b
	Medicago sativa	C	No	Hancock, 1985
	Phaseolus vulgaris	C	?	Taylor and Parkinson, 1965
	Oryza sativa	C	Yes	Fisher and Petrini, 1992
	Zea mays	C	?	Windels and Kommedahl, 1974; Fisher et al., 1992
F. oxysporum f. sp. *batatas*	*Abelmoschus esculentus*	C	No	Armstrong and Armstrong, 1948
	Citrullus vulgaris	C	No	Hendrix and Nielson, 1958
	Glycine max	C	No	Armstrong and Armstrong, 1948
	Gossypium hirsutum	C	No	Ibid.
	Lycopersicon esculentum	C	No	Hendrix and Nielson, 1958; Armstrong and Armstrong, 1948
	Zea mays	C	No	Ibid.

Fungus	Host	Method[a]	Pathogen[b]	Reference
	Beta vulgaris var *cicla*	C	No	Ibid.
F. oxysporum f. sp. *ciceri*	*Cajanus cajan*	C	No	Haware and Nene, 1982
	Lens esculentus	C	No	Ibid.
	Pisum sativum	C	No	Ibid.
F. oxysporum f. sp. *lycopersici*	*Ipomoea batatas*	C	No	Armstrong and Armstrong, 1948
F. oxysporum f. sp. *melonis*	*Gossypium hirsutum*	C	No	Gordon et al., 1989
	Lycopersicon esculentum	C	No	Ibid.
F. oxysporum f. sp. *phaseoli*	*Ipomoea batatas*	C	No	Hendrix and Nielson, 1958
F. oxysporum f. sp. *vasinfectum*	*Ipomoea batatas*	C	No	Armstrong and Armstrong, 1948
F. roseum[c]	*Medicago sativa*	C	?	Hancock, 1985
	Zea mays	C	?	Windels and Kommedahl, 1974
F. roseum 'Acuminatum'[c]	*Trifolium pratense*	M	Yes	Stutz and Leath, 1983
F. roseum 'Equiseti'[c]	*Zea mays*	M	?	Kommedahl et al., 1979
F. roseum 'Graminearum'[c]	*Zea mays*	M	?	Ibid.
	Zea mays	M	?	Ibid.
F. sambucinum	*Phaseolus vulgaris*	C	?	Taylor and Parkinson, 1965
F. scirpi	*Gossypium hirsutum*	C	?	Rudolph and Harrison, 1945
F. solani	*Gossypium hirsutum*	C	No	Gordon et al., 1989
	Linum usitatissimum	C	No	Nagao et al., 1990
	Lycopersicon esculentum	C	No	Gordon et al., 1989
	Zea mays	C	?	Windels and Kommendahl, 1974
Fusarium sp.	*Arachis hypogaea*	C	No	Haware and Nene, 1982
	Hordeum vulgare	C	?	Maier et al., 1997
	Vigna mungo	C	No	Haware and Nene, 1982
	Vigna radiata	C	No	Ibid.
	Vigna unguiculata	C	No	Ibid.
F. tricinctum	*Zea mays*	M	?	Kommedahl et al., 1979
F. ventricosum[d]	*Zea mays*	C	?	Fisher et al., 1992

[a] "C" indicates that the endophyte was identified by microbiological culturing; "M" indicates identification by microscopy.

[b] "Yes" indicates that the *Fusarium* endophyte was also a pathogen on the host from which it was isolated; "No" indicates that it was not; "?" indicates that information was not available.

[c] *Fusarium roseum*, see explanation in note c of Table 1.

[d] *Fusarium ventricosum* = *Fusarium solani* (Nelson et al., 1981).

Fusarium species are primarily plant-associated fungi. They are known from all of the world's major cereals causing storage rots on these materials and from most, if not all, of the other plants cultivated for food, as well as on feed and fiber crops, ornamentals, and wild species. In addition to causing serious agricultural yield losses, *Fusarium* species produce a variety of toxins known or suspected to be harmful to humans and animals. Examples include the fumonisins, which cause fatal diseases in animals and are correlated with human esophageal cancer, and deoxynivalenol, commonly known as vomitoxin, which can cause feed refusal in animals (Mirocha, 1984; Marasas, 1996). Some species of *Fusarium* are insect pathogens and some occasionally cause human mycoses especially in immunocompromised patients.

While fusaria are not generally known for establishing endophytic colonizations in plants, there is a great deal of evidence that documents such infections in both cultivated and wild plants. As discussed in Chap. 1 of this volume, the generally accepted definition of an endophyte has broadened in recent years (Stone et al., 1999). As defined by Petrini (1991), and emphasized in Chap. 1 and 15 of this volume, we consider any organism that at some point establishes a symptomless infection of the internal tissues of a plant to be an endophyte. In a functional sense, endophytes are microorganisms that grow out of surface-sterilized plant tissue when plated on appropriate media. For the purpose of this chapter, this functional definition has been deemed sufficient evidence to include a *Fusarium* species in our tabulation of endophytes (Tables 1–3). This concept of endophytism includes infections described by plant pathologists as latent infections. Thus, there are examples of *Fusarium* endophytes with short endophytic phases that develop into symptomatic pathogenic infections and also of endophytic infections by nonpathogens that never become symptomatic. One shortcoming of this functional definition, however, is that it does not address the metabolic state of the endophyte. The endophytes identified by isolation from surface-sterilized plant tissue are likely to include metabolically active or quiescent types.

Narrower definitions of endophytes often include or imply specialized and/or mutualistic interactions between the plant and the endophyte. We view these types of associations as one type of endophytic infection. We will describe one example from the biocontrol literature of what appears to be a mutualistic *Fusarium* endophyte of tomatoes. Given the relatively large number of nonpathogenic *Fusarium* endophytes identified to date, it seems probable that there are many undescribed instances of mutualistic associations between plants and *Fusarium* species. We will begin this chapter by discussing methods for collecting evidence of *Fusarium* endophytes, then present the case of *F. moniliforme* and maize as one of the most documented cases of a *Fusarium* endophyte, and finally present the evidence for *Fusarium* endophytes in other crop and wild plant species.

2. METHODS FOR ASSESSING *FUSARIUM* ENDOPHYTE RELATIONSHIPS

Identification of a pathogenic colonization of a host plant by a fungus can often be made by macroscopic field observations based on previous characterization of the host–fungal interaction. However, an endophytic infection requires more analytical tools than visual field inspection. Useful methods for studying endophytic relationships could be placed into one of three broad categories: microbiological, microscopic, and molecular and genetic. Advantages and disadvantages of selected detection methods for microbes in general and fungal endophytes of tree leaves in particular were reviewed by Drahos (1991) and Petrini (1991), respectively. A unique set of information can be acquired from each category of methods and utilizing all categories is essential to understanding the complex nature of host–endophyte relationships.

2.1. Microbiological Methods

The evidence for association of most *Fusarium* species with an endophytic habit has relied primarily on microbiological techniques of fungal isolations from surface-sterilized plant tissues. Standard procedures of fungal endophyte isolation from plant tissues have been amalgamated from microbiology, plant pathology, and plant tissue culture.

A typical isolation procedure would involve some, if not, all of the following steps: (1) plant collection, (2) transport of collected material to the laboratory, (3) water wash of large segments of tissue, (4) surface sterilization, (5) dissection of tissue into 1- to 2-cm segments using aseptic techniques, (5) a second surface sterilization, (6) water rinse, (7) plating the plant tissue on a fungal growth medium, and, finally, (8) reinoculation of the host plant and subsequent reisolation. Any hyphae growing from the tissue segments are presumed to have originated from within the plant segment and the fungus designated as endophytic.

Before tissue is brought into the laboratory, extensive washing in running tap water, followed by three rinses in sterile, distilled water, is useful to remove adhering soil particles or other external debris. The agents most commonly used for surface sterilization have been sodium hypochlorite and ethanol solutions either singly or in combination. Isolation of *Fusarium* species from plant material is enhanced by the use of *Fusarium* selective media (Nash and Snyder, 1962).

Microbiological assessments have the advantages that materials for performing experiments are relatively inexpensive, techniques are simple to perform with appropriate training in aseptic manipulations, and results are available in days. This type of data is a necessary first step for identifying an endophytic fungal colonization and permits taxonomic classification of the fungus involved. Furthermore, microbiological methods may be useful for surveying the extent of fungal colonization within the plant and within plant populations. Disadvantages of microbiological assessments is that information would not be obtained on the cellular and/or intercellular localization of the fungus within the plant or organismal interactions, such as metabolic status of the fungus. Furthermore, identification of an endophytic relationship may be masked by the presence of more rapidly growing microbes and/or may be discredited by speculations that sterilization procedures were inadequate to eliminate all adhering surface contaminants.

2.2. Microscopic Methods

Microscopy is a traditional method used extensively for analysis of host–pathogen cellular interactions. In spite of the availability of light microscopes for centuries and electron microscopes for decades, few reports exist on the analyses of relationships of endophytic *Fusarium* species at the microscopic level (Tables 1–3). Assigning endophytic status to a fungus based on microbiological techniques would be much more conclusive with substantiating microscopic evidence to confirm the in planta localization of the fungus. Since *Fusarium* species rarely conidiate in planta, microscopy of field-collected tissues will only reveal the presence and location of fungal hyphae and does not permit identification of even the fungal genus in most cases. Thus, microscopic studies of particular plant–fungus relationships usually require reconstitution of the endophytic association in the laboratory.

The use of light and electron microscopy reveals the precise internal locations of endophytic fungi and can shed some light on the plant's response to the infection. Generally, assessments are made of whether the fungal infection is intercellular or intracellular, and of the plant cellular anatomy. Differences in plant anatomy from that found in uninfected plants indicate a plant response to infection. The plant response can be probed further by use of stains for particular materials such as lignin or pectin. Comparisons of infected and uninfected plant material by transmission electron microscopy can indicates changes in structure and organization of organelles and development of unusual cell wall, membrane, or inclusion bodies.

Two monocotyledonous families, Liliaceae and Gramineae, account for most of the plants for which a microscopic analysis has been reported to include a *Fusarium* species endophyte (Tables 2 and 3). Morphological and anatomical features of endophytic colonization have been examined for *F. oxysporum* in lily (Baayen and Forch, 1998) and *F. culmorum* and *F. graminearum* in wheat (Clement and Parry, 1998). The applications of microscopy to the analyses of *F. moniliforme* endophyte of maize will be described in a later section.

A major advantage of microscopy is providing details of the cellular features of the endophytic relationship. Microscopy can be used to analyze plant cellular responses during a nonpathogenic in comparison to a pathogenic colonization. However, disadvantages of microscopy include the need for expensive equipment for specimen preparation and visualization, especially at the level of electron microscopy. Likewise, skill, expertise, and aptitude are necessary for specimen preparation and microscope operation. Time required for microscopic studies is a major disadvantage for studies requiring analyses of extensive numbers of samples. A commitment of months or even years would be needed to adequately examine multiple tissues from plants collected in replicated studies.

2.3. Genetic and Molecular Methods

Vegetative compatibility is a genetically controlled mechanism that regulates vegetative heterokaryon formation and is found in most fungi including *Fusarium* species (Glass and Kuldau, 1992; Leslie, 1993). Nitrate-nonutilizing or *nit* mutants are readily generated by selection on chlorate media and have been widely used to assign individual isolates within a species to a vegetative compatibility group (VCG) (Cove, 1976; Puhalla and Spieth, 1985; Glass and Kuldau, 1992; Leslie, 1993). Complementary *nit* mutants are paired on minimal media and those that are able to form stable heterokaryons and are thus from the same VCG grow vigorously, whereas those of different VCGs do not form heterokaryons and grow poorly if at all.

Use of VCG as a marker has been employed primarily in the study of *F. moniliforme*–maize interaction but could be effective with other *Fusarium* endophytes as well. VCG has been used to monitor systemic infection of maize by *F. moniliforme* in field situations (Kedera et al., 1992; Munkvold and Carlton, 1997; Munkvold, McGee, and Carlton, 1997; Desjardins and Plattner, 1998) and to assess the genetic diversity of *Fusarium* species within individual maize stalks (Kedera et al., 1994). Use of VCG or other markers is essential in field and even greenhouse experiments since *F. moniliforme* is ubiquitous and endemic strains can infect previously inoculated plants. One drawback to the use of VCG as a marker in field studies is that sometimes an isolate with the same VCG as the inoculated strain is recovered from uninoculated plants (Munkvold, McGee, and Carlton, 1997). Since VCG identity does not always equate with genetic identity, in such cases the source of the recovered fungus (either experimental or endemic) is ambiguous. Nonetheless, *Fusarium* species studied thus far have such a large number of VCGs that this remains a useful marker.

Another means of marking strains for study is through DNA-mediated transformation with genes that encode products that are directly visible, which can be detected by the addition of appropriate substrates or confer resistance to antibiotics. Fungi carrying such transgenes can often be visualized in planta and, if the genes are active, indicate metabolic activity of the endophyte. Use of multiple antibiotic markers, namely hygromycin and benomyl resistance, in transformants has been used effectively to track the systemic growth of *F. moniliforme* and *F. graminearum* in individual maize stalks (Dickman and Partridge, 1989). This study did not directly address endophytic colonization but clearly showed the potential for use of such markers in studying *Fusarium* endophytes.

Bacteria, fungi, and plants have been transformed with the *Escherichia coli* gene encoding β-glucuronidase (GUS), which is detectable by histochemical and fluorometric enzymatic assays (Jefferson, 1987; Gallagher, 1992). Low endogenous GUS activity in most plants and fungi, and rapid assays for detecting GUS activity, have made this an attractive method for studying plant-fungus interactions. The perennial ryegrass endophyte, *Neotyphodium lolii*, has been transformed with the GUS gene and the transformant was used to determine the distribution of metabolic activity and gene expression of the fungus during colonization of the grass (Herd et al., 1997). The modifications, metabolic activity, and propagation of the GUS reporter gene system are presented in Chap. 12. Similarly, Yates et al. (1999) transformed *F. moniliforme* with the GUS gene and demonstrated in situ GUS expression in maize plants inoculated with the transformed isolate. Green fluorescent protein is another visibly detectable marker recently adapted for expression in filamentous fungi (Spelling et al., 1996; Vanden Wymelenberg et al., 1997; Dumas et al., 1999; Horowitz et al., 1999) that may prove useful in visualizing metabolically active *Fusarium* endophytes in planta.

Molecular approaches using polymerase chain reaction (PCR), random amplified DNA (RAPD), and restriction fragment length polymorphism (RFLP) have been used successfully to differentiate among various *Fusarium* species Species-specific PCR primers for *F. poae* (Parry and Nicholson, 1996) and for *F. culmorum*, *F. graminearum*, and *F. avenaceum* (Schilling et al., Geiger, 1996) based on RAPDs can detect these species in infected plant samples. Murillo et al. (1998) developed primers specific for *F. moniliforme* using the same approach. In similar experiments looking at conifer associated *Fusarium* species, Donaldson et al. (1995) developed primer sets for conserved genes and found that RFLPs associated with these fragments could be used to discriminate among six different *Fusarium* species. The use of random amplified polymorphic DNA (RAPD) has been reported for differentiating mating groups within *Fusarium*, including *F. moniliforme* (Amoah et al., 1996). Although only some of these studies reported in planta detection and none addressed in planta detection and identification of endophytic *Fusarium*, it seems likely that these methods and perhaps even these previously identified primers could be used for this purpose. Recently, microsatellite-based PCR fingerprinting assays have been reported for in planta detection of individual strains of *Epichloe* endophytes (Groppe and Boller, 1997; Moon et al., 1999). Potentially, this methodology could be developed for use with *Fusarium* endophytes to study survival of inoculated fungi, mode of infection, and epidemiology. Advantages of molecular studies often include increased sensitivity, specificity, and speed with which results are produced.

A combination of microbial, microscopic, and molecular techniques can provide direct evidence and conclusive assessment on the tissue specificity and metabolic activity of specific fungal endophyte populations during host plant colonization. Assessments based exclusively on microbial procedures for isolation are indirect evaluations requiring inductive reasoning to support the concept of an endophytic habit. Definite assessment of the endophytic habit requires incorporation of a unique, nonambiguous marker for identifying the fungus in situ. However, limited investigations have utilized markers and molecular tools to study the endophytic relationships of *Fusarium* species with their host plant. The absence of studies of this nature is not for lack of need, but may be due to the recent recognition of an endophytic status for many fungi, including *Fusarium* species.

3. ENDOPHYTIC *FUSARIUM MONILIFORME* AND MAIZE

Other single *Fusarium* species may have been studied more extensively for pathogenic interactions with plants; however, no other single species of the genus has been more thoroughly investigated from an endophytic aspect than *F. moniliforme* in association with maize. The vast majority of maize worldwide is colonized by an asymptomatic infection of *F. moniliforme*; many maize fields have greater than 90% infection and asymptomatic infection is usually more common than symptomatic (Munkvold, McGee, and Carlton, 1997). The symptomatic infections or diseases attributed to *F. moniliforme* are ear and kernel rots, stalk rot, and seedling blight (Futrell and Kilgore, 1969; McGee, 1988). Work on the symptomatic colonization of maize by *F. moniliforme* was reviewed up to the early 1980s by Kommedahl and Windels (1981).

During the 1980s, *F. moniliforme* was determined not only to produce diseases on plants but to cause fatal diseases in animals when contaminated maize and maize products were consumed (Marasas, 1996). These findings led to the identification of a new class of mycotoxins, the fumonisins, that could be isolated from *F. moniliforme* cultures and contaminated grain (Gelderblom et al., 1988). Fumonisin B1 causes equine leukoencephalomalacia and porcine pulmonary edema, and fumonisin levels food are statistically correlated with human esophageal cancer in some parts of the world as reviewed by Norred (1993). The fungus also produces an assortment of other secondary metabolites that have been implicated in both plant and animal toxicoses (Bacon and Williamson, 1992). Concern about the health risks of consuming fumonisins has resulted in increased research on the interaction between *F. moniliforme* and maize.

A worldwide survey has been conducted for *F. moniliforme* and/or fumonisin contamination of maize and maize-based products under the auspices of the Commission of Food Chemistry for the International Union of Pure and Applied

Chemistry (Shepherd et al., 1996). It indicates that *F. moniliforme* occurs ubiquitously in maize-producing areas resulting in worldwide distribution of fumonisins. It is likely though not demonstrated that a substantial percentage of these cases represent endophytic *F. moniliforme*.

3.1. Microbiological Evidence for Endophytic Colonization of Maize *by F. moniliforme*

Microbiological techniques were used by Manns and Adams in 1923 to isolate *F. moniliforme* from apparently sound maize kernels incubated on fungal growth medium following surface sterilization. Koehler (1942) substantiated the results of Manns and Adams (1923) that *F. moniliforme* often grew from mature, sound-appearing kernels. Koehler did a more in-depth analysis by quantifying *F. moniliforme* colonization within the parts of the kernel. The percentage of kernels infected was 76% at the tip cap, 22% in the embryo, and less than 10% in the endosperm.

Many researchers have noted the difficulty in eliminating *F. moniliforme* from kernels by surface sterilization and have thus reported this fungus as a seedborne organism (e.g., Salama and Mishricky, 1973). It appears that complete elimination of *F. moniliforme* from infected kernels requires a heat treatment in addition to surface sterilization, further evidence of the internal nature of the infection (Moore, 1943; Daniels, 1983). In addition to the kernel, other parts of the ear such as the cob pith are endophytically colonized by *F. moniliforme* (Hesseltine and Bothast, 1977).

Not only have the reproductive tissues of kernels been associated with endophytic colonization by *F. moniliforme*; the vegetative tissues of the maize plant have been implicated as well (e. g., Leonian, 1932; Foley, 1962). Foley (1962) surface-sterilized symptomless vegetative as well as reproductive maize tissues and observed that *F. moniliforme* grew from stalks, axillary buds, leaf sheaths, roots, and sound kernels plated on fungal growth medium. Examination of maize tissues at intervals during the growing season determined that the frequency of infection increased with plant age. The microbiological findings of these earlier investigations have been repeatedly verified by many different researchers including Marasas et al. (1979) and Thomas and Buddenhagen (1980).

3.2. Microscopic Evidence for Endophytic Colonization of Maize by *F. moniliforme*

Manns and Adams (1923) provide not only microbiological but also microscopic evidence for the endophytic colonization of maize *by F. moniliforme*. Koehler (1942) validated the early work of Manns and Adams that *F. moniliforme* mycelia were visible with the aid of the microscope in mature, sound-appearing kernels.

The mycelia were reported in both analyses to be concentrated in the tip cap, especially in the cavity between the tip cap and brown-to-black layer covering the scutellum. The development of the fungus from the cap into the embryo and endosperm appeared to be restricted by the black layer. In some cases, the fungal mycelium was also in the cavity around the radicle and plumule. The penetration of the hyphae into the embryo and endosperm was proposed to occur before the dark layer at the base of the scutellum had developed. Earlier reports suggested that development of the black layer during the later stages of kernel maturation formed a barrier to fungal penetration through the tip cap (Johann, 1935).

More recently, ultrastructural characteristics of endophytic colonization of kernels and vegetative tissues were examined by scanning and transmission electron microscopy (Bacon et al., 1992; Bacon and Hinton, 1996). These studies confirmed the work of earlier investigators that endophytic colonization of maize is present in both reproductive and vegetative tissues of the plant (Bacon et al., 1992; Bacon and Hinton, 1996). Scanning electron microscopy demonstrated fungal hyphae located within the pedicel or tip cap of asymptomatic maize (Bacon et al., 1992).

Light and transmission electron microscopy were used to compare cellular distribution in vegetative tissues during symptomless and symptomatic *F. moniliforme* colonization. (Bacon and Hinton, 1996). The results indicated that in symptomless infected plants, hyphae were intercellular only and were present in roots, stem internodes, and leaf sheaths. Symptomless plants remained symptomless throughout the observation period. No evidence was observed at the ultrastructural level for an antagonistic relation of the plant cells to the fungal endophytic colonization. However, plants with disease symptoms had the fungal mycelium both intercellularly and intracellularly, but it was restricted to the cortical tissue and was not present within the vascular bundles. These investigators concluded that *F. moniliforme* is a cortical, not a vascular, rot fungus as did Pennypacker (1981).

In contrast to the ultrastructural observations of Bacon and Hinton (1996), histological studies have demonstrated an apparent defensive reaction to *F. moniliforme* asymptomatic colonization in seedling maize shoots (Yates et al., 1997). Fresh sections of shoots from noninfected maize seedlings did not stain as intensely for lignin as shoots from *F. moniliforme*-infected seedlings. Fixed thin sections confirmed a band of cells with thickened walls located interior to the outermost ring of vascular bundles corresponding to the region of intense staining. Lignin can function as a mechanical barrier for organisms without a lignin-degrading enzyme (Vance et al., 1980). Lignified cells have been described in yellow poplar to contain infection in the bark by *F. solani* (Arnett and Witcher, 1974). Consequently, the lignification detected in stems of young maize seedlings could be indicative of a defense mechanism operative during an asymptomatic, endophytic colonization of maize. However, this defense mechanism should be viewed as being ineffective to *F. moniliforme* since seedlings are infected by it, in spite of this lignification. Alternatively, lignification may be viewed as a positive reaction induced by the association, allowing the seedling to develop structural integrity and barriers to wind and moisture. Certainly, lignification may prevent coinfection of seedlings by other fungi sensitive to lignin barriers.

3.3. Genetic and Molecular Evidence for Endophytic Colonization of Maize by *F. moniliforme*

Kedera et al. (1994) used vegetative compatibility group (VCG) to determine the number of *Fusarium* strains from section Liseola that colonize individual maize plants and to determine if isolates from different tissues of the same plant are identical. Recovered isolates were not subjected to mating tests to determine which species they were; however, previous experience indicates that the majority would be *F. moniliforme*. Disinfected seed were planted in sterilized soil in the greenhouse and in the field. *Fusarium* isolates were recovered from stems and ears, and assignment to a VCG group was based on complementation reactions between *nit* mutants. Use of VCG enabled the observations that asymptomatic maize stalks are ubiquitously infected with *Fusarium* strains from section Liseola usually by more than one individual.

Fusarium moniliforme has been transformed with a plasmid containing the *gusA* gene encoding β-glucoronidase (GUS) and the hygromycin phosphotransferase gene conferring hygromycin resistance (Yates et al., 1999). GUS activity is detectable in planta by histochemical and fluorometric enzymatic assays (Gallagher, 1992; Jefferson, 1987). The observation of in planta GUS activity provides evidence of metabolic activity of *F. moniliforme* (Yates et al., 1999). The presence of in planta mycotoxins is also indicative of fungal metabolic activity. The GUS gene also provides another marker in addition to hygromycin resistance for identifying inoculated strains upon reisolation. This is important since naturally occurring hygromycin-resistant strains have been reported (Yan et al., 1993).

3.4. Mode of Endophytic Infection

Foley (1962) described the path of infection through the plant based on isolation data. *Fusarium moniliforme* was present in the nodes earlier than in internodes and nodes could be infected without any evidence for invasion of the internodal tissue. The investigator proposed that rapid elongation of the stem by activity of the intercalary meristem could account for the temporary freedom from infection of internodal tissue. *Fusarium moniliforme* was also isolated from the epicotyl of the young plant. Another avenue of entrance to nodal tissue is through the leaf sheath attachment; the leaf sheaths are infected early in the life of the plant.

The temporal development of *F. moniliforme* infection in the maize plant has been addressed by at least two studies. Histological observations indicated that the fungus was confined to the basal parts of the stalk until the time flowering at which point a rapid spread of mycelium throughout the plant was seen (Lawrence et al., 1981). A similar time course of infection was reported by Raju and Lal (1976). One interpretation of these results is that the crown and upper parts of the plant are infected by different *F. moniliforme* individuals, an idea that is supported by the report of multiple individuals within one maize stalk (Kedera et al., 1994).

Maize plants may become infected through at least three different routes, including (1) air and/or rain splash (Ooka and Kommedahl, 1977), (2) insect damage (Windels et al., 1976), and (3) seed transmission (e.g., Salama and Mishricky, 1973). The importance of different pathways for kernel infection was recently investigated using vegetative compatibility group as the marker for tracking the movement of the inoculated fungus through the plant (Munkvold, McGee, and Carlton, 1997). Methods analyzed for effectiveness of causing kernel infection included inoculation of the seed before planting, injection of the stalk, injection of the crown, and spraying the silks with spore suspensions. The seed inoculations could be likened to seed transmission; injections, to insect damage; and silk sprays, to air and/or rain splash. The average percentage fungal recovery for three experiments was 7%, 38%, 45%, and 95% of the ears following inoculation of seed, crown, stalk, and silk, respectively (Munkvold, McGee, and Carlton, 1997). These results would indicate silks are the most important pathway of infections. As early as 1942, Koehler was convinced that the silks were an important route for *F. moniliforme* infections of maize kernels. Koehler analyzed different ear tissues for *F. moniliforme* following plating of the kernel parts from open-pollinated maize. With rare exceptions, infections either involved the silks only, or silks plus kernels, or silks plus kernels plus pedicels, or silks plus kernels plus pedicels plus vascular cylinder, or silks plus kernels plus pedicels plus vascular cylinder plus internal tissues at butt of cob.

The impact of insect damage on the incidence of symptomless *Fusarium* in maize kernels was evaluated by Munkvold, Hellmich, and Showers (1997). *Fusarium* colonization was examined on three types of maize plants: (1) a maize hybrid genetically engineered with *Bacillus thuringiensis* genes encoding the endotoxin CrylA(b), an insecticide effective against the European corn borer (ECB), *Ostrinia nubialis*; (2) hybrid plants without the endotoxin producing gene, but sprayed with insecticide, and (3) hybrid plants without the endotoxin producing gene or insecticide spray. The transgenic maize hybrids and insecticide-treated plants exhibited reduced kernel infection by *F. moniliforme*, *F. proliferatum*, and *F. subglutinans*. Expression of CrylA(b) in plant tissues other than kernels did not consistently affect *Fusarium* symptoms of infection. Disease incidence was positively correlated with ECB damage to kernels. These results support earlier studies suggesting a role for insects in transmission of the fungus for ear rot disease (Windels et al., 1976). However, it is still an open question as to whether insects such as European corn borer directly vector *Fusarium* conidia or whether the wounds they incur are sites for infection leading to ear rot. However, these experiment do not present any information on the symptomless infection of the endophytic habit in the maize plant.

Maize fragments overwintered in fields appear to provide another source of viable *F. moniliforme* inoculum. Thickened hyphae in maize stalk debris were observed by Nyvall and Kommedahl (1968) and were hypothesized to be survival structures analogous to chlamydospores. *Fusarium moniliforme* hyphae consistently grew out from debris containing these structures and remained viable for up to 2 years when buried in soil (Nyvall and Kommedahl, 1968). These results were recently confirmed by Cotten and Munkvold (1998) who examined field survival of *F. moniliforme* on maize plant debris in a similar study. Inoculated maize stalk pieces were placed in mesh bags and set on the soil surface or buried at 15 or 30 cm deep and sampled at 1- to 3-month intervals for close to two years. *Fusarium moniliforme* was recovered at equal frequency from surface and buried samples after one year, but more frequently from surface residues thereafter. On the final sampling at 630 days, recovery *of F. moniliforme* was not significantly different between inoculated and noninoculated controls indicating the limit of survival for *F. moniliforme* was near 630 days. Thus, it appears that endophytic stage of *F. moniliforme* persists in maize debris left in fields and may be an important source of inoculum.

3.5. *F. moniliforme*: Friend or Foe of the Maize Plant?

Response of the maize plant to *F. moniliforme* infection depends on a multitude of factors related both to the plant, the fungus, and the environment. Under conditions that promote the *F. moniliforme* diseases ear rot and seedling blight, a range of virulence is observed among different strains (e.g., see Bacon et al., 1994). However, even strains that can be quite virulent are known to be asymptomatic under optimal growth conditions (e.g., see Yates et al., 1997). Suboptimal conditions that promote disease include cool spring planting temperatures, which seems to promote seeding blight, and warm dry post-flowering conditions, which often result in an increase in fumonisin content and ear rot. A reduction in fumonisin in the Iowa maize in 1992 was attributed to a cool, damp growing season so that the plants were not stressed by heat or lack of moisture (Shepherd et al., 1996).

There is some evidence to suggest that the asymptomatic *F. moniliforme* maize association may sometimes be mutualistic. The advantages for the fungus would include a sheltered, moist habitat, vertical (through seed) and horizontal (through debris) transmission mediated by the plant, and a reliable nutrition source. There are several studies that suggest advantages for the maize plant as well. Infection of maize kernels by endophytic *F. moniliforme* protects against infection by *F. graminearum* in experimental studies (Van Wyck et al., 1988). Similarly, Wicklow (1988) observed a negative correlation between *F. moniliforme* and 10 other kernel infecting fungi, including some maize pathogens, in a survey of kernels collected at harvest in North Carolina. Finally, Yates et al. (1997) observed a slight but significant increase in plant weight, shoot height, and shoot diameter in *F. moniliforme*-infected plants 28 days after planting compared with control plants under growth room conditions (Yates et al., 1997). Increased growth and protection against seedborne pathogens are clearly benefits to maize plants, but whether this represents a mutualistic association would require a careful cost accounting for both partners.

4. *FUSARIUM* ENDOPHYTES OF WILD PLANTS

Fusarium moniliforme and many other *Fusarium* species have been identified with an endophytic habit in wild plants, ornamentals, timber trees, and crop plants (Tables 1–3). Peschke and Volz (1978) described *F. moniliforme* from two species of wild orchids, *Crypripedium calceolus* and *C. reginae*, in moist habitats in Michigan (Table 1). *Fusarium moniliforme* was isolated from surface-sterilized leaves, roots, and seeds of *Cypripedium* sp. on repeated occasions. In contrast, endophytic *F. moniliforme* has also been found in the desert plants *Yucca* (Bills, 1996) and *Cynomorium coccineum* (Mushin and Zwain, 1989). In neither case was there mention of a pathogenic relationship between the fungus and plant.

The most commonly reported *Fusarium* endophyte of wild plant species is *F. oxysporum* (Table 1). Although *F. oxysporum* is responsible for serious wilt diseases of crop plants and is found abundantly in soil, it is not reported as a pathogen of wild plant species (Windels and Kommedahl, 1974; Armstrong and Armstrong, 1981; Gordon, 1997). Windels and Kommedahl (1974) found that *F. oxysporum* was the most frequently isolated *Fusarium* from roots of grasses and forbs growing in virgin prairie in Minnesota. This fungus was found in 25 of 28 plant species surveyed. However, *F. solani*, *F. tricintum*, and *F. roseum* were also prevalent in both the grasses and forbs. Members of the Orchidaceae as well as Bromeliaceae and Araceae also host *F. oxysporum* endophytes (Petrini and Dreyfuss, 1981).

Many of the studies reporting asymptomatic *F. oxysporum* infections of wild plants were conducted in agricultural settings with the objective of identifying nonhost carriers of *F. oxysporum* formae (Waite and Dunlap, 1953; Katan, 1971; Haware and Nene, 1982; Clark and Watson, 1983; Helbig and Carroll, 1984). For example, the tomato wilt pathogen *F. oxysporum* f. sp. *lycopersici* has been found in the roots of symptomless weed species collected in fallow tomato fields (Katan, 1971). These isolates were pathogenic to tomato but not to the weeds from which they were isolated, namely, *Amaranthus graecizans*, *Digitaria sanguinalis*, *Malvus silvestris*, and *Oryziopsis miliceae*. This study suggests a balanced commensal relationship between the weed species and the endophytic *F. oxysporum* f. sp. *lycopersici* (Katan, 1971). Dicotyledonous weed species are also known to harbor endophytic *F. avenaceum*, *F. culmorum*, and *F. poae* all of which are pathogenic to winter wheat (Jenkinson and Parry, 1994).

Fusarium endophytes of wild plants are not limited to annual, herbaceous species. Woody, perennial wild plants host endophytic *Fusarium* spp. as well (Table 1). Although the endophytic flora of palms has not been explored in depth and *Fusarium* sp. are not among the dominant components, three species (*F. oxysporum*, *F. aquaeductuum*, and *F. semitectum*) are known (Rodrigues and Samuels, 1990; Rodrigues, 1996). *Fusarium oxysporum* and *F. semitectum* are pathogenic to some palms but these studies did not report on whether the isolates were pathogenic to the hosts they occupied. *Fusarium* species are not the dominant endophytes of trees and shrubs such as *Fraxinus*, *Fagus*, *Ulex*, and *Taxus* but are occasionally isolated (Chapela and Boddy, 1988; Fischer, et al., 1986; Kowalski and Kehr, 1986; Strobel et al., 1996). Endophytic *F. lateritium* isolated from *Taxus baccata* is reported to synthesize the cancer fighting drug taxol (Strobel et al., 1996).

5. *FUSARIUM* ENDOPHYTES OF ORNAMENTALS AND CULTIVATED TREES

Fusarium endophytes have been described from the important timber trees *Pseudotsuga menziesii* (Bloomberg, 1966), *Pinus radiata* (Storer et al., 1998), and *Pinus sylvestris* (Vaartaja and Cram, 1956) (Table 2). *Fusarium subglutinans* f. sp. *pini*, the causal agent of pitch canker disease of pines, has recently been reported as endophytic in 4-month-old seedlings of Monterey pine (*Pinus radiata*) and is also found in the seeds of this species (Storer et al., 1998). The seedborne nature of the fungus is thought to contribute to the epidemiology of the disease (Storer et al., 1998). The pitch canker fungus is also seedborne in cultivated pine in the southeastern United States causing serious disease problems; however, there are no reports regarding asymptomatic infection of tree parts other than seeds for these other pine species (Miller and Bramblett, 1979; Runion and Bruck, 1988). The route of infection of pine seeds by the pitch canker fungus is not known.

Ornamental flower bulbs can harbor latent (symptomless) endophytic infections by forma speciales of *F. oxysporum* that often lead to diseased bulbs and or reduced flowering potential (Magie, 1971; Henis and Zilberstein, 1973; Bergman and Bakker-Van Der Voort, 1979; Baayen and Forch, 1998) (Table 2). A recent histological study of lily bulb infection by *F. oxysporum* f. sp. *lilii* suggested the presence of lily cells involved in transfer of carbohydrates to the fungus. These transfer cells were described as being similar to those formed in mycorrhizal associations (Baayern and Forch, 1998).

The precise environmental cues that result in an end to the symptomless endophytic infection and the start of a symptomatic infection are not known; however, a number of environmental plant stresses appear to trigger this change. Treatment with a high carbon dioxide atmosphere for several days can induce symptomatic infection by *F. oxysporum* f. sp. *gladioli* in gladiolus (Magie, 1971) and treatment with herbicide plant stresses such as paraquat are also known to break latency of endophytic fungi in soybeans (Cerkauskas and Sinclair, 1980, 1982). Plant stress in general may be the cue that induces asymptomatic *Fusarium* infections to become symptomatic. Conditions such as water stress, freezing, transplantation, and defoliation have been cited as factors predisposing plants to disease (Papendick and Cook, 1974; Schoenewiess, 1981; Sinclair, 1991). Some disease development under these circumstances may be due to a shift from endophytic, asymptomatic to pathogenic infection.

6. *FUSARIUM* ENDOPHYTES OF CROP PLANTS

Surveys of crop plants have revealed *Fusarium* endophytes in a number of major grains (Table 3). *Fusarium graminearum* and *F. oxysporum* were found in the stems and leaves of maize (Fisher et al., 1992), but surprisingly, these authors did not report finding endophytic *F. moniliforme* as the predominant *Fusarium* isolated from maize in the United States (Kommedahl and Windels, 1981). In an extensive survey of fungal endophytes in four cultivars of winter wheat, Sieber et al. (1988) reported finding *F. culmorum* and *F. graminearum* in all plant parts. However, they isolated these fusaria more frequently from the roots, culms, and kernels than from the leaves and glumes. Clement and Parry (1998) reported similar results. These three species are pathogens on winter wheat and such infections could result in disease if the appropriate developmental or environmental conditions arose. The pathogens *F. equiseti* and *F. oxysporum* were reported as endophytes of rice in the leaves, leaf sheathes, and roots (Fisher and Petrini, 1992). Studies of fungal infection in the forage crops alfalfa and red clover indicate the presence of latent *Fusarium* pathogens here as well (Stutz and Leath, 1983; Hancock, 1985).

Nonsusceptible crops are frequently carriers of forme speciales of *F. oxysporum* that cause wilt diseases in susceptible hosts (Armstrong and Armstrong, 1948; Hendrix and Nielson, 1958; MacDonald and Leach, 1976; Haware and Nene, 1982; Gordon et al., 1989) (Table 3). For example, in field and greenhouse studies, *F. oxysporum* f. sp. *vasinfectum*, the cotton wilt pathogen, and *F. oxysporum* f. sp. *lycopersici*, the tomato wilt pathogen, were isolated from asymptomatic sweet potato. In the same study, the sweet potato wilt pathogen *F. oxysporum* f. sp. *batatas* was found in symptomless okra, cotton, soybean, tomato, and maize (Armstrong and Armstrong, 1948). Hendrix and Nielson (1958) conducted similar studies with *F. oxysporum* f. sp. *batatas* and added watermelon to the list of nonsusceptible hosts of this fungus. These types of studies often identify other species of *Fusarium* as endophytes as well (Armstrong and Armstrong, 1948; Gordon et al., 1989) (Table 3). In addition to finding weed carriers of *F. oxysporum* f. sp. *betae* (Table 1), MacDonald and Leach (1976) also reported garden beets and Swiss chard as nonsuscept carriers. Cotton and tomato are carriers of *F. oxysporum* f. sp. *melonis* (Gordon et al., 1989) and the nonsuscepts lentil and pea and pigeon pea can harbor *F. oxysporum* f. sp. *ciceri* (Haware and Nene, 1982). The prevalence of nonsuscept crops that harbor asymptomatic, endophytic infection of *Fusarium* wilt pathogens call into question the utility of crop rotation strategies for the control of these pathogens. This phenomenon also underscores the need for further research into the nature of endophytic fungal infections of plants.

Nonpathogenic, endophytic strains of *F. oxysporum* isolated from suppressive soils have been used as biocontrol agents to manage diseases caused by fusaria pathogenic to watermelon, cucumber, celery, and other crops (Schneider, 1984; Paulitz et al., 1987; Postma and Rattink, 1992; Larkin et al., 1996). In each case, these fungi are endophytes of the hosts they protect. Split root experiments with watermelon plants indicate that one of the mechanisms of control is induced systemic resistance (Larkin et al., 1996). However, studies on suppression of *F. oxysporum* f. sp. *apii*–caused celery yellows by nonpathogenic *F. oxysporum* indicated that the primary mechanism in this case was parasitic competition for infection sites (Schneider, 1984). Saprophytic competition for nutrients has also been shown as a mechanism in some systems (see Larkin et al., 1996). Nagao et al. (1990) showed a correlation between the ability of nonpathogenic *F. oxysporum* to colonize roots and the degree of reduction in the percentage of wilted plants.

Endophytic, nonpathogenic *F. oxysporum* can reduce the population of plant parasitic root-knot nematodes in tomato roots (Hallman and Sikora, 1994*a*, *b*). Experiments showed a marked reduction in nematode egg masses and root galls in *F. oxysporum*-treated plants compared to untreated controls (Hallman and Sikora, 1994*b*). Furthermore, culture filtrates of the endophytic *F. oxysporum* are toxic to these nematodes causing inactivation and death (Hallman and Sikora, 1996). The toxic metabolites in the culture filtrate selectively inactivated plant parasitic nematodes and were not harmful to either the mycophagous or bacteriophagous nematodes tested (Hallman and Sikora, 1996). The capacity of some nonpathogenic, endophytic *F. oxysporum* isolates to protect their plant hosts against *Fusarium* wilt pathogens or parasitic nematodes points to the possibility of mutualistic relationships.

7. SUMMARY

The literature review presented herein indicates that *Fusarium* endophytes are both widespread and common within vascular plants, and that these associations range from latent infections to commensal associations to possible mutualistic interactions. Indeed, *Fusarium* species are nearly always found when surface-sterilized plant tissues are plated on appropriate medium. Thus, it is tempting to speculate that endophytism is in fact an important and normal component of the life history of many if not all plant-associated *Fusarium* species. The accumulated evidence also suggests that an endophytic strategy is shared among latent pathogens, nonpathogenic strains, and isolates infecting nonhosts. The genus *Fusarium* may have evolved characteristics that make it especially well suited to occupy internal plant tissues without causing disease.

While many apparent endophytic relationships have been identified, very little is known about the physiology, biochemistry, cell biology, or genetics of these undoubtedly complex interactions. Given the ubiquity and pathogenic and mycotoxigenic potential of the genus *Fusarium*, a deeper understanding of the endophytic lifestyle of these fungi is essential. The relatively well-studied case of *F. moniliforme*, and maize discussed herein is a compelling example of this need for research in this area.

REFERENCES

Abawi, G. S. and Lorbeer, J. W. (1972). Several aspects of the ecology and pathology of *Fusarium oxysporum* f. sp. *cepae*. *Phytopathology* 62, 870–876.

Amoah, B. K., Macdonald, M. V., Rezanoor, N. and Nicholson, P. (1996). The use of random amplified polymorphic DNA technique to identify mating groups in the *Fusarium* section Liseola. *Plant Pathology* 45, 115–125.

Armstrong, G. M. and Armstrong, J. K. (1981). Formae speciales and races of *Fusarium oxysporum* causing wilt diseases. *In* P. E. Nelson, T. A. Tousson, and R. J. Cook (eds.), *Fusarium: Diseases, Biology, and Taxonomy*. Pennsylvania State University Press, University Park, pp. 391–399.

Armstrong, G. M. and Armstrong, J. K. (1948). Nonsusceptible hosts as carriers of wilt Fusaria. *Phytopathology* 38, 808–826.

Arnett, J. D. & Witcher, W. (1974). Histochemical studies of yellow poplar infected with *Fusarium solani*. *Phytopathology* 64, 414–418.

Baayen, R. P. and Forch, M. G. (1998). A biotrophic phase in bulb rot of lilies infected by *Fusarium oxysporum* f.sp. *lilii*. (Abstract) 8th International *Fusarium Workshop*, 17–20 August, CABI Bioscience, Egham, England.

Bacon, C. W. and Hinton, D. M. (1996). Symptomless endophytic colonization of maize by *Fusarium moniliforme*. *Canadian Journal of Botany* 74, 1195–1202.

Bacon, C. W., Hinton, D. M. and Richardson, M. D. (1994). A corn seedling assay for resistance to *F. moniliforme*. *Plant Disease* 78, 302–305.

Bacon, C. W. and Williamson J. W. (1992). Interactions of *Fusarium moniliforme*, its metabolites and bacteria with corn. *Mycopathologia* 117, 65–71.

Bacon, C. W., Bennett R. M., Hinton, D. M. and Voss K. A. (1992). Scanning electron microscopy of *Fusarium moniliforme* within asymptomatic corn kernels and kernels associated with equine leukoencephalomalacia. *Plant Disease* 76, 144–148.

Bergman, B. H. H. and Bakker-Van Der Voort, M. A. M. (1979). Latent infection in tulip bulbs by *Fusarium oxysporum*. *Netherlands Journal of Plant Pathology* 85, 187–195.

Bills, G. (1996). Isolation and analysis of endophytic fungal communities from woody plants. *In* S. C. Redlin and L. M. Carras (eds.), *Endophytic Fungi in Grasses and Woody Plants*. The American Phytopathological Society, St. Paul, MN.

Bloomberg, W. J. (1966). The occurrence of endophytic fungi in Douglas-fir seedlings and seed. *Canadian Journal of Botany* 44, 413–420.

Booth, C. (1981). Perfect states (Teleomorphs) of *Fusarium* species. *In* P. E. Nelson, T. A. Tousson, and R. J. Cook (eds.), *Fusarium: Diseases, Biology, and Taxonomy*. Pennsylvania State University Press, University Park, pp. 446–452.

Booth, C. (1971). *The Genus* Fusarium. Commonwealth Mycological Institute, Kew, Surrey, England.

Burgess, L. W. (1981). General ecology of the Fusaria. *In* P. E. Nelson, T. A. Tousson, and R. J. Cook (eds.), *Fusarium: Diseases, Biology, and Taxonomy*. Pennsylvania State University Press, University Park, pp. 225–235.

Cerkauskas, R. F. and Sinclair J. B. (1980). Use of paraquat to aid detection of fungi in soybean tissues. *Phytopathology* 70, 1036–1038.

Cerkauskas, R. F. and Sinclair J. B. (1982). Effect of paraquat on soybean pathogens and tissues. *Transactions of the British Mycological Society* 78, 495–502.

Chapela, I. and Boddy L. (1988). Fungal colonization of attached beech branches II. Spatial and temporal organization of communities arising from latent invaders in bark and functional sapwood, under different moisture regimes. *New Phytologist* 110, 47–57.

Clark, C. A. and Watson, B. (1983). Susceptibility of weed species of Convolvulaceae to root-infecting pathogens of sweet potato. *Plant Disease* 67, 907–909.

Clement, J. A. and Parry, D. W. (1998). Stem-base disease and fungal colonisation of winter wheat grown in compost inoculated with *Fusarium culmorum*, *F. graminearum*, and *Microdochium nivale*. *European Journal of Plant Pathology* 104, 323–330.

Cotten, T. K. and Munkvold, G. P. (1998). Survival of *Fusarium moniliforme*, *F. proliferatum*, and *F. subglutinans* in maize stalk residue. *Phytopathology* 88, 550–555.

Daniels, B. A. (1983). Elimination of *Fusarium moniliforme* from corn seed. *Plant Disease* 67, 609–611.

Desjardins, A. E. and Plattner, R. D. (1998). Distribution of fumonisins in maize ears infected with strains of *Fusarium moniliforme* that differ in fumonisin production. *Plant Disease* 82, 953–958.

Dickman, M. B. & Partridge, J. E. (1989). Use of molecular markers for monitoring fungi involved in stalk rot of corn. *Theoretical and Applied Genetics* 77, 535–539.

Donaldson, G. C., Ball, L. A., Axelrood, P. E. & Glass, N. L. (1995). Primer sets developed to amplify conserved genes from filamentous ascomycetes are useful in differentiating *Fusarium* species associated with conifers. *Applied and Environmental Microbiology* 61 1331–1340.

Drahos, D. J. (1991). Methods for the detection, identification, and enumeration of microbes. *In* J. H. Andrews and S. S. Hirano (eds.), *Microbial Ecology of Leaves*. Springer-Verlag, New York, pp. 135–157.

Dumas, B., Centis, S., Sarrazin, N. and Esquerre-Tugaye, M.-T. (1999). Use of green fluorescent protein to detect expression of an endopolygalacturonase gene of *Colletotrichum lindemuthianum* during bean infection. *Applied and Environmental Microbiology* 65, 1769–1771.

Fisher, P. J., Anson, A. E. and Petrini, O. (1986). Fungal endophytes in *Ulex europaeus* and *Ulex gallii*. *Transactions of the British Mycological Society* 86, 153–156.

Fisher, P. J. & Petrini, O. (1992). Fungal saprobes and endophytes of rice (*Oryza sativa* L.). *New Phytologist* 120, 137–143.

Fisher, P. J., Petrini, O. and Scott, H. M. L. (1992). The distribution of some fungal and bacterial endophytes in maize (*Zea mays* L.). *New Phytologist* 122, 299–305.

Foley, D. C. (1962). Systemic infection of corn by *Fusarium moniliforme*. *Phytopathology* 52, 870–872.

Futrell, M. C. & Kilgore, M. (1969). Poor stands of corn and reduction of root growth caused by *Fusarium moniliforme*. *Plant Disease Reporter* 53, 213–215.

Gallagher, S. R. (1992). GUS Protocols: *Using the GUS Gene as a Reporter of Gene Expression*. Academic Press, San Diego, CA.

Gelderblom, W. C. A., Jaskiewicz, K., Marasas, W. F. O., Thiel, P. G., Horak, R. M., Vleggaar, R. and Kriek, N. P. J. (1988). Fumonisins-novel mycotoxins with cancer-promoting activity produced by *Fusarium moniliforme*. *Applied and Environmental Microbiology* 54, 1806–1811.

Gill, D. L. (1967). *Fusarium* wilt infection of apparently healthy mimosa trees. *Plant Disease Reporter* 51, 148–150.

Glass, N. L., & Kuldau, G. A. (1992). Mating type and vegetative incompatibility in filamentous ascomycetes. *Annual Review of Phytopathology* 30, 201–224.

Gordon, T. R. and Martyn, R. D. (1997). The evolutionary biology of *Fusarium oxysporum*. *Annual Review of Phytopathology* 35, 111–128.

Gordon, T. R., Okamato, D. and Jacobsen, D. J. (1989). Colonization of muskmelon and nonsusceptible crops by *Fusarium oxysporum* f. sp. *melonis* and other species of *Fusarium*. *Phytopathology* 79, 1095–1100.

Gordon, W. L. (1959). The occurrence of *Fusarium* species in Canada VI. Taxonomy and geographic distribution of *Fusarium* species on plants, insects, and fungi. *Canadian Journal of Botany* 37, 257–290.

Groppe, K. and Boller, T. (1997). PCR assay based on a microsatellite-containing locus for detection and quantification of *Epichloe* endophytes in grass tissue. *Applied and Environmental Microbiology* 63, 1543–1550.

Hallman, J. and Sikora, R. A. (1994a). Influence of *Fusarium oxysporum*, a mutualistic fungal endophyte, on *Meloidogyne incognita* infection of tomato. *Journal of Plant Diseases and Protection* 101, 475–481.

Hallman, J. and Sikora, R. A. (1994b). Occurrence of plant parasitic nematodes and nonpathogenic species of *Fusarium* in tomato plants in Kenya and their role as mutualistic synergists for biological control of root-knot nematodes. *International Journal of Pest Management* 40, 321–325.

Hallman, J. and Sikora, R. A. (1996). Toxicity of fungal endophyte secondary metabolites to plant parasitic nematodes and soil-borne plant pathogenic fungi. *European Journal of Plant Pathology* 102, 155–162.

Hancock, J. G. (1985). Fungal infection of feeder rootlets of alfalfa. *Phytopathology* 75, 1112–1120.

Haware, M. P. and Nene, Y. L. (1982). Symptomless carriers of the chickpea wilt *Fusarium*. *Plant Disease* 66, 250–251.

Helbig, J. B. and Carroll, R. B. (1984). Dicotyledonous weeds as a source of *Fusarium oxysporum* pathogenic on soybean. *Plant Disease* 68, 694–696.

Hendrix, F. F. Jr. and Nielson, L. W. (1958). Invasion and infection of crops other than the forma suscept by *Fusarium oxysporum* f. *batatas* and other formae. *Phytopathology* 48, 224–228.

Henis, Y. and Zilberstein, Y. (1973). Detection of latent *Fusarium* in gladiolus corms. *Journal of Horticultural Science* 48, 189–194.

Herd, S., Christensen, M. J., Saunders, K., Scott, D. B. and Schmid, J. (1997). Quantitative assessment of the in planta distribution of metabolic activity and gene expression of an endophytic fungus. *Microbiology* 143, 267–275.

Hesseltine, C. W. anb Bothast, R. J. (1977). Mold development in ears of corn from tasseling to harvest. *Mycologia* 69, 328–340.

Horwitz, B. A., Sharon, A., Lu, S.-W. Ritter, V., Sandrock, T. M., Yoder, O. C. and Turgeon, B. G. (1999). A G protein alpha subunit from *Cochliobolus heterostrophus* involved in mating and appressorium formation. *Fungal Genetics and Biology* 26, 19–32.

Jefferson, R. A. (1987). Assaying chimeric genes in plants: the GUS gene fusion system. *Plant Molecular Biology Reporter* 5, 387–405.

Jenkinson, P. and Parry, D. W. (1994). Isolation of *Fusarium* species from common broad-leaved weeds and their pathogenicity to winter wheat. *Mycological Research* 98, 776–780.

Johann, H. (1935). Histology of the caryopsis of yellow dent corn, with reference to resistance and susceptibility to kernel rots. *Journal of Agricultural Research* 51, 855–883.

Katan, J. (1971). Symptomless carriers of the tomato *Fusarium* wilt pathogen. *Phytopathology* 61, 1213–1217.

Kedera, C. J., Leslie, J. F. and Claflin, L. E. (1994). Genetic diversity of *Fusarium* section *Liseola* (*Gibberella fuijkuroi*) in individual maize stalks. *Phytopathology* 84, 603–607.

Kedera, C. J., Leslie, J. F. and Claflin, L. E. (1992). Systemic infection of corn by *Fusarium moniliforme*. (Abstract). *Phytopathology* 82, 1138.

Klittich, C. J. R., Leslie, J. F., Nelson, P. E., Marasas, W. F. O. (1997). *Fusarium thapsinum* (*Gibberella thapsina*): A new species in section *Liseola* from sorghum. *Mycologia* 89, 643–652.

Koehler, B. (1942). Natural mode of entrance of fungi into corn ears and some symptoms that indicate infection. *Journal of Agricultural Research* 64, 421–442.

Kommedahl, T. and Windels, C. E. (1981). Root-, stalk-, and ear-infecting *Fusarium* species on corn in the USA. *In* P. E. Nelson, T. A. Tousson, and R. J. Cook (eds.), *Fusarium: Diseases, Biology, and Taxonomy.* Pennsylvania State University Press, University Park, pp. 94–103.

Kommedahl, T., Windels, C. E. and Stucker, R. E. (1979). Occurrence of *Fusarium* species in roots and stalks of symptomless corn plants during the growing season. *Phytopathology* 69, 961–966.

Kowalski, T. and Kehr, R. D. (1996). Fungal endophytes of living branch bases in several European tree species. *In* S. C. Redlin and L. M. Carras (eds.), *Endophytic Fungi in Grasses and Woody Plants.* American Phytopathological Society, St. Paul, MN, pp. 67–86.

Larkin, R. P., Hopkins, D. L. and Martin, F. N. (1996). Supression of *Fusarium* wilt of watermelon by nonpathogenic *Fusarium oxysporum* and other microorganisms recovered from a disease-suppressive soil. *Phytopathology* 86, 812–819.

Lawrence, E. B., Nelson, P. E. and Ayers, J. E. (1981). Histopathology of sweet corn seed and plants infected with *Fusarium moniliforme* and *F. oxysporum. Phytopathology* 71, 379–386.

Lee, J. C., Lobkovsky, E., Pliam, N. B., Strobel, G. and Clardy, J. (1995). Subglutinols A and B: immunosuppressive compounds from the endophytic fungus *Fusarium subglutinans. Journal of Organic Chemistry* 60, 7076–7077.

Leonian, L. H. (1932). The pathogenicity and the variability of *Fusarium moniliforme* from corn. *West Virginia Agricultural Experiment Station Bulletin* 248, 1–16.

Leslie, J. F. (1993). Fungal vegetative compatibility. *Annual Review of Phytopathology* 31, 127–150.

Leslie, J. F., Pearson, C. A. S., Nelson, P. E, & Tousson, T. A. 1990. *Fusarium* spp. from corn, sorghum, and soybean fields in the central and eastern United States. *Phytopathology* 80, 343–350.

MacDonald, J. D. and Leach, L. D. (1976). Evidence for an expanded host range of *Fusarium oxysporum* f. sp. *betae. Phytopathology* 66, 822–827.

Magie, R. O. (1971). Carbon dioxide treatment of *Gladiolus* corms reveals latent *Fusarium* infections. *Plant Disease Reporter* 55, 340–341.

Maier, W., Hammer, K., Dammann, U., Schulz, B. and Strack, D. (1997). Accumulation of sesquiterpenoid cyclohexenone derivatives induced by an arbuscular mycorrhizal fungus in members of the Poaceae. *Planta* 202, 36–42.

Manns, T. F. and Adams, J. F. (1923). Parasitic fungi internal of seed corn. *Journal of Agricultural Research* 23, 495–524.

Marasas, W. F. O. (1996). Fumonisins: History, world-wide occurrence and impact. *In* L. S. Jackson, J. W. De Vries & L. B. Bullerman (eds), *Fumonisins in Food.* Plenum Press, New York, pp. 1–17.

Marasas, W. F. O., Kriek, N. P. J., Wiggens, V. M., Steyn, P. S., Towers, D. K. & Hastie, T. J. (1979). Incidence, geographic distribution, and toxigenicity of *Fusarium* species in South African corn. *Phytopathology* 69, 1181–1185.

McGee, D. C. (1988). *Maize Diseases.* American Phytopathological Society, St. Paul, MN.

Miller, T. and Bramblett, D. L. (1979). Damage to reproductive structures of slash pine by two seed-borne pathogens: *Diplodia gossypina* and *Fusarium moniliforme* var.

subglutinans. In F. Bonner (ed.), *Proceedings of Flowering and Seed Development in Trees: A Symposium.* U.S. Department of Agriculture Forest Service, Southern Forest Experiment Station, pp. 347–355.

Mirocha, C. J. (1984). Mycotoxicoses associated with *Fusarium. In* M. O. Moss and J. E. Smith (eds.), *The Applied Mycology of Fusarium.* Cambridge University Press, Cambridge, pp. 141–155.

Moon, C. D., Tapper, B. A. and Scott, B. (1999). Identification of *Epichloe* endophytes in planta by a microsatellite-based PCR fingerprinting assay with automated analysis. *Applied and Environmental Microbiology* 65, 1268–1279.

Moore, W. C. (1943). New and interesting plant diseases. *Transactions of the British Mycological Society* 2, 20–23.

Munkvold, G. P. and Carlton, W. M. (1997). Influence of inoculation methods on systemic *Fusarium moniliforme* infection of maize plants grown from infected seeds. *Plant Disease* 81, 211–216.

Munkvold, G. P., McGee, D. C. and Carlton, W. M. (1997). Importance of different pathways for maize kernel infection by *Fusarium moniliforme. Phytopathology* 87, 209–217.

Munkvold, G. P., Hellmich, R. L. & Showers, W. B. (1997). Reduced *Fusarium* ear rot and symptomless infection in kernels of maize genetically engineered for European corn borer resistance. *Phytopathology* 87, 1071–1077.

Murillo, I., Cavallarin, L. and San Segundo, B. (1998). The development of a rapid PCR assay for detection of *Fusarium moniliforme. European Journal of Plant Pathology* 104, 301–311.

Mushin, T. M. & Zwain, K. H. (1989). A fungal endophyte associated with a desert parasitic plant. *Kavaka* 17, 1–5.

Nagao, H., Couteaudier, Y. and Alabouvette, C. (1990). Colonization of sterilized soil and flax roots by strains of *Fusarium oxysporum* and *Fusarium solani. Symbiosis* 9, 343–354.

Nash, S. M. & Snyder, W. C. (1962). Quantitative estimations by plate counts of propagules of the bean root rot *Fusarium* in field soils. *Phytopathology* 52, 567–572.

Nelson, P. E. (1991). History of *Fusarium* systematics. *Phytopathology* 81, 1045–1048.

Nelson, P. E., Tousson, T. A. and Marasas, W. F. O. (1983). Fusarium *Species: An Illustrated Manual for Identification.* Pennsylvania State University Press, University Park, PA.

Nirenberg, H. and O'Donnell, K. (1998). New *Fusarium* species and combinations within the *Gibberella fujikuroi* species complex. *Mycologia* 90, 434–458.

Norred, W. P. (1993). Fumonisins-mycotoxins produced by *Fusarium moniliforme. Journal of Toxicological and Environmental Health* 38, 309–328.

Nyvall, R. F. and Kommedahl, T. (1968). Individual thickened hyphae as survival structures of *Fusarium moniliforme* in corn. *Phytopathology* 58, 1704–1707.

O'Donnell, K., Cigelnik, E. and Nirenberg, H. I. (1998). Molecular systematics and phylogeography of the *Gibberella fujikuroi* species complex. *Mycologia* 90, 465–493.

Ooka, J. J. & Kommedahl, T. (1977). Wind and rain dispersal of *Fusarium moniliforme* in corn fields. *Phytopathology* 67, 1023–1026.

Papendick, R. I. and Cook, R. J. (1974). Plant water stress and development of *Fusarium* foot rot in wheat subjected to different cultural practices. *Phytopathology* 64, 358–363.

Parry, D. W. and Nicholson, P. (1996). Development of a PCR assay to detect *Fusarium poae* in wheat. *Plant Pathology* 45, 383–391.

Paulitz, T. C., Park, C. S. and Baker, R. (1987). Biological control of *Fusarium* wilt of cucumber with nonpathogenic isolates of *Fusarium oxysporum*. *Canadian Journal of Microbiology* 33, 349–353.

Pennypacker, B. W. (1981). Anatomical changes involved in the pathogenesis of plants by *Fusarium*. *In* P. E. Nelson, T. A. Tousson, and R. J. Cook (eds.), *Fusarium: Diseases, Biology, and Taxonomy*. Pennsylvania State University Press, University Park, pp. 400–408.

Peschke, H. C and Volz, P. A. (1978). *Fusarium moniliforme* Sheld. association with species of orchids. *Phytoparasitica* 40, 347–356.

Petrini, O. (1991). Fungal endophytes of tree leaves. *In* J. H. Andrews and S. S. Hirano (eds.), *Microbial Ecology of Leaves*. Springer-Verlag, New York, pp. 179–197.

Petrini, O. and Dreyfuss, M. (1981). Endophytische Pilze in Epiphytischen Araceae, Bromeliaceae und Orchidaceae. *Sydowia* 34, 135–148.

Postma, J. and Rattink, H. (1991). Biological control of *Fusarium* wilt of carnation with a nonpathogenic isolate of *Fusarium oxysporum*. *Canadian Journal of Botany* 70, 1199–1205.

Puhalla, J. E. and Spieth, P. T. (1985). A comparison of heterokaryosis and vegetative incompatibility among varieties of *Gibberella fujikuroi* (*Fusarium moniliforme*), *Experimental Mycology* 9, 39–47.

Raju, C. A. and Lal, S. (1976). Relationship of *Cephalosporium acremonium* and *Fusarium moniliforme* with stalk rots of maize. *Indian Phytopathology* 29, 227–231.

Riesen, T. K. and Close, R. C. (1987). Endophytic fungi in Propiconazole-treated and untreated barley leaves. *Mycologia* 79, 546–552.

Rodrigues, K. F. (1996). Fungal endophytes of palms. *In* S. C. Redlin and L. M. Carras (eds.), *Endophytic Fungi in Grasses and Woody Plants*. American Phytopathological Society, St. Paul, MN, pp. 121–132.

Rodrigues, K. A. and Samuels, G. J. (1990). Preliminary report of endophytic fungi in a tropical palm. *Mycological Research* 94, 827–830.

Rudolf, B. A. and Harrison, G. J. (1945). The invasion of the internal structure of cotton seed by certain Fusaria. *Phytopathology* 35, 542–548.

Runion, G. B. and Bruck, R. I. (1988). Effects of thiabendazole-DMSO treatment of longleaf pine seed contaminated with *Fusarium subglutinans* on germinating and seedling survival. *Plant Disease* 72, 872–874.

Salama, A. M. and Mishricky, A. G. (1973). Seed transmission of maize wilt fungi with special reference to *Fusarium moniliforme* Sheld. *Phytopathologische Zeitschrift* 77, 356–362.

Schilling, A. G., Möller, E. M. and Geiger, H. H. (1996). Polymerase chain reaction-based assays for species-specific detection of *Fusarium culmorum, F. graminearum*, and *F. avenaceum*. *Phytopathology* 86, 515–522.

Schoeneweiss, D. F. (1981). The role of environmental stress in diseases of woody plants. *Plant Disease* 65, 308–314.

Schneider, R. W. (1984). Effects of nonpathogenic strains of *Fusarium oxysporum* on celery root infection by *F. oxysporum* f. sp. *apii* and a novel use of the Lineweaver-Burk double reciprocal plot technique. *Phytopathology* 74, 646–653.

Shepherd, G. S., Thiel, P. G., Stockenström, S. and Sydenham, E. W. (1996). Worldwide survey of fumonisin contamination of corn and corn-based products. *Journal AOAC International* 79, 671–687.

Sieber, T., Riesen, T. K., Muller, E. and Fried, P. M. (1988). Endophytic fungi in four winter wheat cultivars (*Triticum aestivum* L.) differing in resistance against *Stagonospora nodorum* (Berk.) Cast. & Germ. = *Septoria nodorum* (Berk.) Berk. *Journal of Phytopathology* 122, 289–306.

Sinclair, J. B. (1991). Latent infection of soybean plants and seeds by fungi. *Plant Disease* 75, 220–224.

Snyder, W. C. and Hansen, H. N. (1945). The species concept in *Fusarium* with reference to Discolor and other sections. *American Journal of Botany* 32, 657–666.

Spelling, T., Bottin, A. and Kahmann, R. (1996). Green fluorescent protein (GFP) as a new vital marker in the phytopathogenic fungus *Ustilago maydis*. *Molecular and General Genetics* 252, 503–509.

Stone, J. K., Bacon, C. W. and White, J. F. Jr. (2000). An overview of endophytic microbes: endophytism defined. *In* C. W. Bacon and J. F. White, Jr. (eds.), *Microbial Endophytes*. Marcel Dekker, New York. In press.

Stoner, M. F. (1981). Ecology of *Fusarium* in non-cultivated soils. *In* P. E. Nelson, T. A. Tousson, and R. J. Cook (eds.), *Fusarium: Diseases, Biology, and Taxonomy*. Pennsylvania State University Press, University Park, pp. 276–286.

Storer, A. J., Gordon, T. R. and Clark, S. L. (1998). Association of the pitch canker fungus, *Fusarium subglutinans* f. sp. *pini* with Montery pine seeds and seedlings in California. *Plant Pathology* 47, 649–656.

Strobel, G. A., Hess, W. M., Ford, E., Sidhu, R. S. and Yang, X. (1996). Taxol from fungal endophytes and the issue of biodiversity. *Journal of Industrial Microbiology* 17, 417–423.

Stutz, J. C. and Leath, K. T. (1983). Virulence differences between *Fusarium roseum* "Acuminatum" and *F. roseum* "Avenaceum" in red clover. *Phytopathology* 73, 1648–1651.

Taylor, G. S. and Parkinson, D. (1965). Studies on fungi in the root region IV. Fungi associated with the roots of *Phaseolus vulgaris* L. *Plant and Soil* 22, 1–21.

Thomas, M. D. and Buddenhagen, I. W. (1980). Incidence and persistence of *Fusarium moniliforme* in symptomless maize kernels and seedlings in Nigeria. *Mycologia* 72, 882–887.

Vaartaja, O. and Cram, W. H. (1956). Damping-off pathogens of conifers and of *Caragana* in Saskatchewan. *Phytopathology* 46, 391–397.

Vance, C. P., Kirk, T. K. and Sherwood, R. T. (1980). Lignification as a mechanism of Disease resistance. *Annual Review of Phytopathology* 18, 259–288.

Vanden Wymelenberg, A. J., Cullen, D., Spear, R. N., Schoenike, B. and Andrews, J. H. (1997). Expression of green fluorescent protein in *Aureobasidium pullulans* and quantification of the fungus on leaf surfaces. *BioTechniques* 23, 686–690.

Van Wyck, P. S., Scholtz, D. J. and Marasas, W. F. O. (1988). Protection of maize seedlings by *Fusarium moniliforme* against infection by *Fusarium graminearum* in the soil. *Plant and Soil* 107, 251–257.

Waite, B. H. and Dunlap, V. C. (1953). Preliminary host range studies with *Fusarium oxysporum* f. sp. *cubense*. *Plant Disese Reporter* 37, 79–80.

Wicklow, D. T. (1988). Patterns of fungal association within maize kernels harvested in North Carolina. *Plant Disease* 72, 113–115.

Windels, C. E., Windels, M. B. & Kommedahl, T. (1976). Association of *Fusarium* species with picnic beetles on corn ears. *Phytopathology* 66, 328–331.

Windels, C. E. and Kommendahl, T. (1974). Population differences in indigenous *Fusarium* species by corn culture of prarie soil. *American Journal of Botany* 61, 141–145.

Wollenweber, H. W. (1931). *Fusarium*- Monographie. Julius Springer, Berlin.

Yan, K., Dickman, M. B., Xu, J.-R. and Leslie, J. F. (1993). Sensitivity of field strains of *Gibberella fujikuroi* (*Fusarium moniliforme* section *Liseola*) to benomyl and hygromycin. *Mycologia* 85, 206–213.

Yates, I. E., Bacon, C. W. and Hinton, D. M. (1997). Effects of endophytic infection by *Fusarium moniliforme* on corn growth and cellular morphology. *Plant Disease* 81, 723–728.

Yates, I. E., Hiett, K. L., Kapczynski, D. R., Smart, W., Glenn, A. E., Hinton, D. M., Bacon, C. W., Meinersmann, R., Liu, S. & Jaworski, A. J. (1999). GUS transformation of the corn fungal endophyte *Fusarium moniliforme*. *Mycological Research* 103, 129–136.

II
ASSOCIATIONS WITH ROOTS
AND BACTERIAL ENDOPHYTES

6

Evolution of Endophytism in Arbuscular Mycorrhizal Fungi of Glomales

Joseph B. Morton
West Virginia University, Morgantown, West Virginia

1. INTRODUCTION

Most terrestrial plants distributed worldwide establish at least one of seven types of mutualistic symbiotic mycorrhizal associations in their roots (Brundrett, 1991; Smith and Read, 1997). The most widespread of these is an endomycorrhizal symbiosis between soilborne fungi (the *mycobiont*) presently classified in the phylum Zygomycota, order Glomales (Morton and Benny, 1990) and approximately 80% of terrestrial plant species (the *phytobiont*) (Trappe, 1987). This symbiosis has been demonstrated to be mutualistic in measurable ways. The most obvious benefit to the mycobiont is a ready supply of carbon within a relatively secure niche in root cortical cells. Jakobsen and Rosendahl (1990) indicate that the mycobiont can capture as much as 20% of the fixed carbon in the phytobiont, mostly by converting glucose to trehalose (Shachar-Hill et al., 1995). The symbiosis is termed "arbuscular" in this chapter because the mycobiont produces specialized arbuscules in root cortical cells that are involved in bidirectional exchange of carbon, phosphorus, and other physiologically important molecules. These arbuscules are considered the key structural evolutionary innovation because they are a conserved feature in all lineages of Glomales (Morton, 1990a). The phytobiont depends to varying degrees on the mycobiont for hyphal uptake and translocation of essential cations in soil which otherwise might not be available from root uptake alone (reviewed in Smith and Read, 1997). The magnitude of dependency often is measured in terms of plant growth responses, which usually are inversely related to plant nutrient (mostly phosphorus) content.

The arbuscular mycorrhizal (AM) symbiosis has received considerable attention by botanists, ecologists, and agriculturalists in recent years because of dramatic responses to plant growth, vigor, and reproductive potential under largely controlled (and therefore somewhat artificial) conditions (Safir, 1987; Smith and Read, 1997). Similar responses in natural settings tend to be more inconsistent or absent, suggesting that growth benefit is only one factor that contributes to the long-term stability of the AM symbiosis. A multitude of other interactions also must be taking place which result in a positive outcome for both symbionts at the scale of the community and above. Clues about the nature of these interactions come from measurement of present day processes, but they rarely are interpreted in an evolutionary context. Genealogical properties of both the mycobiont and phytobiont lineages also play an important role, with causal mechanisms varying with different combinations of host and fungal genotypes operating in specific environments. These historical processes undoubtedly are significant, and so the purpose of this chapter is to elaborate on current hypotheses concerning evolution of the arbuscular mycorrhizal endosymbiosis and link historical patterns and processes to those occurring at the present time.

2. ORIGIN OF THE ARBUSCULAR MYCORRHIZAL SYMBIOSIS

The AM symbiosis became evolutionarily significant when interactions between one or more plant and fungal lineages were sufficiently integrated and heritable to be maintained in offspring that radiated into the divergent lineages recognized today. Three lines of evidence collectively indicate that the AM symbiosis originated at least 400 million years ago and that structure of the mycorrhiza (fungus + root) has not changed appreciably during the intervening years. First, and most directly, was the discovery of finely branched hyphal structures in fossilized roots of *Aglaophyton* from the Rhynie chert that bore striking resemblance to extant arbuscules (Taylor et al., 1995). Cross-sections of fossil spores also approximated morphologies of extant spores of species in the genus *Glomus* (Hass et al., 1994). Second, calculations of age of a small sampling of extant fungal species in five of the six recognized genera were measured using a molecular clock estimate based on nucleotide sequence divergence of 18S rDNA (Simon et al., 1993). Last, distribution of the association appears to be almost universal in early plant taxa (order level and above), with loss of the symbiosis occurring more recently and convergently in families, genera, and species of different plant lineages (Trappe, 1987).

Various hypotheses have been proposed concerning properties of the progenitor mycobiont (common ancestor) in what has been implicitly assumed to be a monophyletic group. Pirozynski and Malloch (1975) suggested that it was a semiaquatic alga and aquatic oomycete "fungus." Morton (1990b), unable to accept an oomycete ancestor, alternatively suggested a saprobic zygomycetous fungus with perhaps a more terrestrial habit. The latter hypothesis is corroborated by affinities of a zygomycete fungus, *Geosiphon pyriforme*, which does not form associations with higher plants but which has morphological and molecular affinities with the *Glomus* lineage in Glomales. The fungus is culturable, albeit with some difficulty, and serves as a host for filaments of *Nostoc* species (usually *N. punctiforme*) in photosynthetic and nitrogen-fixing multinucleate bladders (Kluge et al., 1991). It also forms spores that bear some resemblance to those of a typical *Glomus* species (Schussler et al., 1994). More convincingly, nucleotide sequences in a conserved region of the 18S rRNA gene more closely align with those of glomalean than nonglomalean fungi (Gehrig et al., 1996). Reliance of the mycobiont on photosynthetic carbon assimilation (Kluge et al., 1991), together with indications of relatedness to fungi in Glomales, suggests that a near-obligate to obligate relationship may have existed with an autotrophic prokaryote as an intermediate step in formation of an obligate and integrated endosymbiosis with ancient plants.

The arbuscule is a highly specialized structure that has obvious importance since it is so widely conserved in mycobiont lineages. Ultrastructural, biochemical, and immunological studies indicate its final form (and functional activity) depends on complex interactions with the phytobiont (reviewed by Bonfante and Perotto, 1995), to the extent that the phytobiont contributes important substrates to the interfacial zone (Balestrini et al., 1996). The complexity of steps in this process would suggest that origin of the biotrophic AM symbiosis with the arbuscule as an immediate innovation was not abrupt (saltational). There now is some evidence that intraradical hyphal interfaces may serve as sites of carbon acquisition by the mycobiont and phosphorus efflux to the photobiont (Smith and Smith, 1996). The extent that this mechanism is important evolutionarily to the AM symbiosis is only speculative at this time. The mycobiont in *Pisum* mutants (myc^{-2}) are capable of limited intraradical growth in the absence of functional arbuscules (Gianinazzi-Pearson et al., 1994), but the absence of extraradical hyphal growth and sporulation suggests that hyphal nutrient transfer is not evolutionarily significant in the modern symbiosis. However, its occurrence today as a supplement to arbuscule activity may be a relict process that during origin and integration of the symbiosis was pivotal or even exclusive.

An important question to be resolved is: Did this symbiosis arise only once from a common ancestor or did it arise more than once from separate ancestors? Cladistic analysis of morphological (Morton, 1990a) and 18S rRNA gene sequence divergence (Simon et al., 1993; Simon, 1996) suggests that arbuscular fungi in Glomales arose uniquely from a common ancestor and thus are lineages within a monophyletic group (Fig. 1a, b). Validity of the morphological tree was viewed to be equivocal by Morton (1990a), however, because it depended on homology of only one structure, the arbuscule, and a thorough taxonomic analysis of this character was lacking. The molecular phylogeny was accepted with little reservation because it corroborated morphological patterns at the order level, and significant conflicts at suborder level (Fig. 1a, b) were totally ignored. Additional data when considered together strongly indicate that the arbuscular symbiosis arose twice rather than once (Fig. 1c) and that the current suborders, Glomineae and Gigasporineae, are separate lineages that evolved convergently.

Arbuscule structure and architecture are significantly different between suborders, as first suggested by Brundrett and Kendrick (1990) and currently being corroborated in my laboratory from comparisons of many species in an International Collection of Vesicular-Arbuscular Mycorrhizal Fungi (INVAM) grown on a standard host (*Zea mays*). Arbuscules of fungi in Gigasporineae have a wide trunk hypha (5–12 µm, with most 6–8 µm) from which subsequent branches narrow abruptly to form many fine tips. Hyphal walls always are darkly pigmented with widely used stains such as trypan blue or chlorazol black E. In Glomineae, conversely, arbuscular trunk hyphae generally are thinner (2–8 µm, with most less than 6 µm), with subsequent branches growing incrementally thinner to very fine terminal tips. Hyphal walls are highly variable in intensity of staining reactions, ranging from dark (similar to that of hyphae in gigasporinean fungi) to almost invisible. Data still are being accumulated on another morphological pattern: the extent to which a fungal species produces *Arum* v. *Paris* types of arbuscules (the former being terminal on branches of intercellular hyphae, the latter being terminal on branches of intracellular hyphal coils) (Smith and Read, 1997). Brundrett and Kenrick (1990) conclude that these arbuscular types are

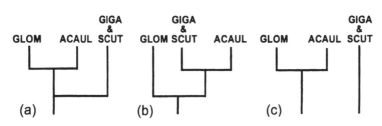

Figure 1   Competing phylogenetic hypotheses of relationships among arbuscular fungi in the genera *Glomus* (GLOM) and *Acaulospora* (ACAUL) of suborder Glomineae and the genera *Gigaspora* (GIGA) and *Scutellospora* (SCUT) of suborder Gigasporineae. (a) Both suborders arising from a common ancestor (Morton, 1990a). (b) Three lineages arising from a common ancestor, but with members of the suborder Gigasporineae more closely related to *Acaulospora* than to *Glomus* (Simon et al., 1993; Simon, 1996). (c) Each suborder arising from a separate ancestor (this chapter).

controlled mainly by host genotype, but considerable variation is elicited by fungal genotypes colonizing the same host. Since these data have yet to be published, the reader may obtain photographic documentation of more than 45 species on INVAM's worldwide web site at http://invam.csf.wvu.edu. With the limited range of possible morphological variation in arbuscule structure and architecture that does not compromise functionality (similarly found among haustoria of biotrophic pathogenic fungi), even the seemingly small differences reported are likely to be phylogenetically significant at the cellular level.

Other fungal structures important in mycorrhizal development also differ between suborders in morphology, stage of life cycles in which they are formed, duration, and function. In Glomineae, all fungal parts appear to be highly infective: intraradical or extraradical spores (widely reported), intraradical vesicles (Biermann and Linderman, 1983), and attached or detached extraradical hyphae (Jasper et al., 1989). Extraradical thin-walled "vesicles" are formed by a few species in *Glomus* that bear some analogy to auxiliary cells in Gigasporineae, and they also are infective (Morton, unpublished). Intraradical vesicles generally increase in abundance after arbuscular colonization is well established, if they form at all. Occurrence and abundance of vesicle development varies considerably with fungal genotype (within and between species) and with environmental conditions (Morton, unpublished). Both intraradical hyphae and vesicles are of long duration and can persist intact in detached roots for 2 years or longer and still be infective. In Gigasporineae, no intraradical vesicles are formed at any time. Instead, clusters of fragile auxiliary cells are formed on hyphae of germinating spores before roots are even contacted and are more abundantly on extraradical hyphae of developing mycorrhizae. They are rarely absent (unlike vesicles), often peaking in abundance during early stages of sporulation and declining thereafter (Franke and Morton, 1994; Morton, unpublished). No evidence has ever been presented to indicate that auxiliary cells are infective. They are rich in lipid deposits and have a more transitory storage function than vesicles. Extraradical hyphae also are not infective once mycorrhizal development has ceased (Biermann and Linderman, 1983; Pearson and Schweiger, 1994). This phenomenon is dramatically corroborated by the inability to reculture inocula of *Gigaspora* and *Scutellospora* accessions in INVAM from root fragments or hyphae after spores have died or are decomposed (Morton, unpublished).

Early events in mycorrhizal formation also may differ significantly between suborders, as evidenced by differential responses of two *Glomus* species and that of *Gigaspora margarita* in number and structure of appressoria formed on roots of myc⁻ mutants of *Medicago* (Bradbury et al., 1993). While these observations are limited in scope, they at least represent an explicit comparison of fungal genotypes in both suborders on a host genotype expressing considerable control over recognition and colonization processes.

Mode of spore formation also is unique to fungi in each suborder. In Glomineae, spores (*Glomus*) or "saccules" (*Acaulospora* and *Entrophospora*) develop from cylindrical to flared fertile hyphae. There is little developmental gap between these genera because several species (*A. gerdemannii, G. gerdemannii*) form spores both ways on the same mycelium (Morton et al., 1997). Spores of species in *Gigaspora* and *Scutellospora* (Gigasporineae) form on a morphologically distinct "sporogenous cell." Morphological organization and structure of mature spores are similar in appearance between suborders, which is why they share a common terminology (Morton, 1988). However, when these characters are interpreted according to their origin and position in differentiation sequences (Morton et al., 1995), they are ontogenetically (and hence phylogenetically) unique to members of each suborder.

Carbohydrate chemistry of fungal cell walls differs in each suborder as well. $\beta(1-3)$-Glucans are present in the hyphal walls of species in Glomineae, whereas they are absent in species of Gigasporineae (Gianinazzi-Pearson et al., 1993). Similar divergence has been found in other lineages of Zygomycota, suggesting that this phylum consists of a number of monophyletic groups. At the present time, it is not certain as to which of these groups represent close relatives to each suborder in Glomales.

Other endosymbionts of arbuscular fungi also appear to be unique to species of each suborder. Specific primers amplified rDNA of bacteria only of the genus *Burkholderia* in gigasporinean fungi and failed to amplify product from bacteria in spores of Glomineae species or of *Geosiphon pyriforme*, a putative ancestral relative (Bianciotto et al., 1996).

As mentioned previously, only one phylogeny conflicts with these congruent datasets that indicates dual rather than monophyletic origin of fungi in Glomales, and that is the one generated from 18S rDNA sequences. There are a number of theoretical and methodological reasons that may account for this conflict. Sanders et al. (1995) point out that the specific primers used in generating the 18S rDNA phylogeny may be amplifying copies that are not identical in different fungi. Even if there is no differential primer selectivity in amplification product, highly conserved rDNA regions may erroneously indicate monophyly among polyphyletic lineages, as appeared to be the case in the phylogenetic analysis of attine ant groups by different workers (Chapela et al., 1994; Hinkle et al., 1994). Another potential constraint is the narrow range of taxa sampled, which is exacerbated when the fungi being studied are long-lived, asexual, and contain numerous heterogeneous nuclei. A more fundamental consideration is that morphology and molecules of arbuscular fungi simply portray different phylogenetic histories (Doyle, 1992), as evidenced to some extent by conflicting groupings at the species level (Lloyd-MacGilp et al., 1996). Given the modular construction of arbuscular fungi, the different rate and magnitude of evolutionary change in structure (and possibly function) of these modules, and plasticity in regulation and behavior of fungal life history traits, incongruence between sequences in one gene fragment and all of the other evidence certainly is feasible. The evolution of two independent arbuscular symbioses is not surprising, given that other more recent mycorrhizal symbioses appear to have evolved more than once (LePage et al., 1997).

It is important to note at this point that the evolutionary events in origin of the AM symbiosis discussed at the beginning of this section apply only to the lineage classified as the suborder Glomineae. Origins of the suborder Gigasporineae lineage, in terms of both progenitor and age, cannot be determined from existing evidence. Molecular clock estimates suggest it is much later (Simon et al., 1993), but that conclusion also is based on a phylogeny that may be flawed. The implications for researchers are noteworthy given a propensity to apply experimental results to all of Glomales rather than to particular lineages of fungi. While there is every reason to believe that many symbiotic functions evolved convergently in both lineages, many more comparative studies are needed to identify them. Of equal importance, of course, is determining which processes and mechanisms are unique to each lineage.

3. COEVOLUTION AND SPECIATION

Pirozynski and Malloch (1975), Pirozynski (1981), Trappe (1987), and Morton (1990b) unanimously support the hypothesis that at least one primitive mycorrhizal association (most probably initiated by progenitors of *Glomus*, based on fossils and the *G. pyriforme* story) mediated radiation of plants with rootless axes in a terrestrial habitat and the subsequent rise and spread of angiosperms. Once established and canalized in the earliest symbiosis, the mycobiont became obligately associated with the phycobiont. As a result, arbuscular fungi evolved completely apart from other fungal groups for 400 million years as both a taxonomic and functional "cul-de-sac" (Malloch, 1987). The phytobiont also did not evolve alone from the outset and especially during the geological period when radiation of plant lineages was most "explosive" (Pirozynski, 1981; Pirozynski and Malloch, 1975). Both mycobiont and phytobiont are modular organisms so that each alone had a capacity for change and adaptation unheard of among unitary animal groups (Andrews, 1992). Together, however, the coevolutionary potential for new life cycle innovations unavailable to either of them separately was immense.

Once it is accepted that arbuscular fungi and their plant hosts evolved *together* from ancient ancestors, then the extraordinary compatibility between the two in modern interactions would be a predicted outcome as long as there is no detriment to the phytobiont as a higher unit of selection during the evolutionary process (see Buss, 1987, for its basis in hierarchical theory and Law and Lewis, 1983 for its relevance to the AM endosymbiosis). It then becomes apparent that coevolutionary accommodations of mycobiont and phytobiont may impact on speciation and population level adaptations.

Most hypotheses of coevolution have focused explicitly on speciation, in part because timing of arbuscule evolution and the explosion of plant diversification occurred so close together (Pirozynski, 1981) and in part because of the disproportionately large number of phytobiont species compared to that of the mycobiont (at least 1000-fold). Law and Lewis (1983) postulate that the low number of arbuscular fungal species is due to several factors, in particular asexual reproduction and the stabilizing influence of the endosymbiosis through time. Morton (1990b) attributes it to rapid speciation only during canalization of the symbiosis in ancient lineages followed by a long period of morphological stasis. Neither of these analyses considers the scope of possible variation within the somatic structures responsible for species level variation. Speciation in arbuscular fungi involves divergence only in component parts of single somatic cells that become spores (Morton et al., 1995). In plants, however, speciation encompasses multiple levels of cellular organization and multiple modes of reproduction (Grant, 1981). In this context, the number of plant species (over 250,000) is easily explained and the number of arbuscular fungal species described today (154) seems miraculously large (with at least twice that number expected as new species are discovered)! Thus, coevolutionary processes may be interpreted as expanding the evolutionary potential in *both* symbionts; its just not as overtly apparent in the mycobiont because of its simpler structure.

The role of the mycobiont in facilitating speciation in the phytobiont is intuitively obvious, given the correlation of origin of the arbuscular symbiosis and rapid radiation of terrestrial plant lineages. Research of contemporary processes indicates the benefits of an AM symbiosis on plant fecundity (Bryla and Koide, 1990), pollen production (Lau et al., 1995), seedling vigor (Stanley et al., 1993), phenology and P relationships (Merryweather and Fitter, 1995a), resistance to root pathogens (Newsham et al., 1994), and size/behavior of below-ground absorptive surfaces for "resource foraging" (Campbell et al., 1991). All of these differential interactions could individually or collectively expand or alter adaptive zones and promote speciation of the phytobiont. Pirozynski (1981) also recognized the potential of the AM symbiosis to promote new morphological innovations that accelerate speciation, especially major ones such as seed development. Less dramatic, but probably no less important on a geological scale, is the role of the mycobiont in mediating selection for heritable changes in hormone relationships (Beyrle, 1995), nutrient source-sink relationships (Eissenstat et al., 1993; Merryweather and Fitter, 1995b), root behavior and architecture (Hetrick et al., 1993), and basic dependency on the AM symbiosis (Hetrick et al., 1992; Merryweather and Fitter, 1996). Some evidence of such coevolutionary trends is evidenced in congruence of mycotrophy and other taxonomic traits in pteridophytes (Gemma et al., 1992).

The role of the phytobiont to advance speciation in the mycobiont has never been considered because the number of fungal species has been viewed as "low."

The number of species known or predicted from patterns of morphological diversity (Morton et al., 1995) can be accounted for just by stochastic mutation and genetic drift over 400 million years. Some evidence for this has been found in the discovery of an irreversible morphological mutant in successive pot cultures of an INVAM accession of *Scutellospora heterogama*, which would be diagnosed conclusively as a new species had it been found in any natural setting. Developmental comparisons with the wild-type parent indicate the mutant phenotype is the result of a single heterochronic change during spore differentiation (Heldreth and Morton, unpublished). Yet there is no measurable selective advantage to either wild-type or mutant, with no concomitant morphological change in other functional parts of the fungal organism (arbuscules, auxiliary cells, intra radical and extraradical hyphae) or differential behavior (growth rate, abundance of sporulation, etc.).

But the greater issue is the causal basis for the origin of such unique diversity in Glomales. It has no counterpart in any other group of the Kingdom Fungi, so it did not arise in fungal lineages that evolved in other niches. Obviously, the ancestral genotype is an important determinant, but the occurrence of convergent properties in separate lineages (Glomineae and Gigasporineae) that coevolved in an AM symbiosis suggests that the phycobiont played some role. Such a contribution is totally speculative at this time, but it conceivably could include any or all of the following mechanisms: (1) indirect regulation of sorting and/or partitioning of heterogeneous migratory nuclei, thus leading to unique gene combinations and phenotypes; (2) gene transfer (e.g., Nicety and Sugiyama, 1995) with a similar outcome; or (3) transfer of novel substrates to synthesize subcellular structures not obtainable by other means.

As asexual organisms, immediate fixation of whatever neutral or positive genotypic and phenotypic traits evolved over time assures that all progeny generations can be grouped diagnostically as a genealogically conserved morphospecies (Morton et al., 1992). It is less clear as to whether each of these morphospecies truly reflects the phylogeny of its inclusive clonal lineages. I offer three possibilities here. The most parsimonious scenario is one in which each species is a functional monophyletic group and coevolution of symbiotic partners occurred at a similar rate and magnitude of divergence as morphological evolution in spores. This pattern has been proven repeatedly to be false. Clonal lineages of a species can differ considerably in many symbiotic properties (Brundrett, 1991; Smith and Read, 1997; discussion below) and these phenotypes can be reasonably stable (Sylvia et al., 1993). These data indicate the genealogical patterns found in spores is not reflective of a "whole organism" phylogeny at the species level. The decoupling of species units from ecological units is evidenced further by the amicable coexistence of related and unrelated species in the same root system (Table 1). The other two possibilities are difficult to separate. Either clonal lineages coevolved with their plant hosts to produce many

Table 1 Taxonomic Structure of Arbuscular Fungal Communities at the Genus Level in 4-Month-Old Bait Pot Cultures (15-cm pots) on Sudangrass (*Sorghum sudanense*) of Inocula from Four Separate Geographic Locations

Fungal genus	Number of species per INVAM accession			
	KE104	MN409	VA105	VZ103
Acaulospora	2	2	1	2
Entrophospora	0	0	1	0
Glomus	4	2	1	3
Gigaspora	0	2	1	1
Scutellospora	1	2	2	2

Locations are: KE, Kenya; MN, Minnesota; VA, Virginia; VZ, Venezuela.

functionally diverse genotypes that arose from the same ancestor (which evolved a unique morphotype), or they coevolved with their plant hosts as distinct functional entities from two or more ancestors each of which convergently evolved the same morphotype. The persistence of ancestral genotypes along with their modern descendants (Morton, 1993) complicates resolution of a definitive pattern in part because it results in a narrowing of morphological gaps between species, genera, and even families within each suborder (but not between suborders).

Molecular patterns of divergence among arbuscular fungi, despite their resolving power, are hard to interpret because of the extensive coevolutionary trends over 400 million years impacting differentially on separate parts of the fungal organism or on segregation of heterogeneous copies, reversals, or other complicating genetic processes. Baum and Donoghue (1995) propose a genealogical species concept that groups members of a species by congruence among trees based on molecular character sets. Some potential for this approach is evident in isozyme profiles (Dodd et al., 1996) and polymerase chain reaction (PCR)–amplified RAPD bands (Lanfranco et al., 1995) that appear conserved enough to group clonal lineages into species. Other gene sequences appear to be more problematical. The multinucleate condition of fungal parts can create considerable "noise" if they are heterogeneous, as appears to be the case in the single spores from which DNA has been cloned (Lloyd-MacGilp et al., 1996). This may by partly why Simon (1996) finds divergence in conserved regions of the 18S rRNA gene between two clones of *Scutellospora heterogama* that are morphologically identical (Morton, unpublished), whereas Lloyd-MacGilp et al. (1996) can group clones of morphologically distinct morphospecies together using putatively more variable ITS regions.

4. COEVOLUTION AND POPULATION DYNAMICS

I reemphasize here that all of the evidence indicates that asexual species of fungi in Glomales are the product of genealogy and appear cohesive (whether monophyletic or polyphyletic) because of strong constraints on the direction and magnitude of morphological variation rather than any other process (Morton et al., 1992). The true functional entities are clonal lineages coevolving with each other (Table 1) and with the phycobiont lineages they inhabit. The only time the mycobiont functions independent of the phytobiont is when spores (all glomalean fungi) or hyphal fragments (Glomineae only) break dormancy and produce more hyphae that grow and branch through the soil matrix until contact with a root is established (Fig. 2a; Bonfante and Perotto, 1995). These processes are impacted directly by soil variables, especially those that are toxic to fungal propagules (e.g., Bartholomew-Estevan and Science, 1994; Laval et al., 1995; Tommerup, 1983). Anecdotal evidence from over 8000 cultures grown in INVAM between 1990 and 1996 suggests that in "typical" soils, two major factors determining mycorrhizal establishment by different fungi are disparity in temperature and pH in soils of greenhouse pot cultures vs. those at field sampling sites.

Once a mycorrhiza has formed, success of the symbiosis depends on compatibility between both symbionts, which is a function of the relative colonization ability of the various fungi that coinhabit a root, the status of the phycobiont genotype within its plant community, and the impact of environmental variables on the phycobiont directly and the mycobiont through the phycobiont "filter"

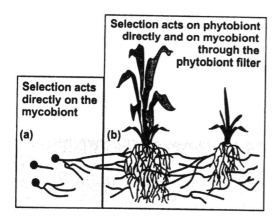

Figure 2 The focus of selection pressures on (a) the mycobiont directly and (b) the phytobiont directly and the mycobiont indirectly.

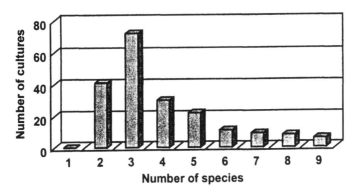

Figure 3 Number of sporulating arbuscular fungal species recovered from 15-cm-diameter pot cultures started with root-soil inoculum and grown for 4–5 months on sudangrass (*Sorghum sudanense*).

(Fig. 2b). A monospecific mycorrhiza is an extremely rare occurrence in nature or even in pot cultures of natural soils (Fig. 3; Table 1), so partitioning of niche occupation and functional contribution by each fungal genotype in a mycorrhizal community is hard to measure. Nonetheless, some evidence for differential preference by the phytobiont resulting in regulation of colonization or sporulation has been documented. McGonigle and Fitter (1988) were able to directly measure differential colonization by a "fine endophyte" species, *Glomus tenue*, since its mycorrhizal structures are morphologically unique from those of other AM fungi. Bever et al. (1995) also discovered host preference by members of a fungal community, but it was measured indirectly as a function of fecundity (sporulation). More indirectly, the phytobiont appears to have the physiological capability to selectively regulate levels of colonization (Pearson et al., 1993) and also efficiency of hyphal P transport (Ravnskov and Jakobsen, 1995) among cocolonizing fungal genotypes.

Irrespective of the complexity of interactions within a fungal community, a large body of evidence indicates the phytobiont is the ultimate arbitor of the success or failure of the symbiosis at the scale of the plant community and above. Moreover, coevolution of the mycobiont with most, if not all, ancestral plant lineages indicates that genetic and regulatory mechanisms are likely to differ greatly among plant lineages regardless of the community structure of the mycobiont. Probably the most conserved processes are those regulating recognition events and formation of arbuscule and hyphal interfaces in root cells. Blockage of mycorrhizal colonization from epidermal and hypodermal cell regions appears to be universal (Gianinazzi-Pearson et al., 1996). Another general phenomenon is a reduction in mycorrhizal development as soil (and plant) P levels increase

(Smith and Read, 1997). In this case, the causal mechanism diverges among plant lineages, from inhibition of entry point formation in leek (Amijee et al., 1989) to inhibition of infection unit growth and secondary colonization in cucumber (Bruce et al., 1994).

Mycorrhizal phenotypes of different host fungus combinations are known to vary considerably, which again can be attributed to divergence of coevolving mycobiont and phytobiont lineages. As an example, legume mutants which do not form bacterial nitrogen-fixing nodules differ considerably in their response to mycorrhizal development. Hyphal ingress into roots is blocked by both *Pisum* and *Medicago* mutants, but appressoria formation is not affected in the former and is significantly distorted in the latter (Bradbury et al., 1993). In contrast, nod⁻ mutants of *Glycine* do not inhibit mycorrhizal colonization (Wyss et al., 1990). Phytobiont species that have evolved into nonhosts also exhibit divergent mechanisms of resistance to the symbiosis that appear to involve different mechanisms from those elicited by plant pathogens (Giovannetti et al., 1995). The presence of physicochemical active defense mechanisms in incompatible (myc $^{-1}$) *Pisum* mutants suggests that plant-controlled defense genes are suppressed during mycorrhizal development. In only one report to date, a species in Chenopodiaceae produces localized necrosis in response to limited ingress by arbuscular fungi (Allen et al., 1989). In contrast, a number of *Brassica* species appear to effect some control over preinfection events in soil by failing to exude molecules that normally stimulate hyphal proliferation from spores and then show no indication of any active defense response at the root surface (Glenn et al., 1988). Inhibition of colonization and blockage of arbuscule formation in *Pisum* roots grafted to nonhost *Lupinus* indicates other mobile mechanisms of controls at different stages of mycorrhizal development (Gianinazzi, 1991). Such variation in mechanisms among lineages and at different stages of mycorrhizal development are an expected outcome of such long-term historical interactions.

Despite powerful phytobiont controls, the mycobiont still has considerable room to diversify in properties that do not impact negatively on long-term population dynamics of the phytobiont. While variation in fungal morphology in constrained by simplicity of design, hyphal architecture and behavior can change dramatically in response to phytobiont and environmental variables (Brundrett, 1991; Brundrett and Kendrick, 1990). Patterns and abundance of sporulation also are highly plastic, even under similar environmental conditions (Morton, unpublished). Contemporary differential responses among mycobiont genotypes have been measured in efficiency of P uptake independent of the degree of mycorrhizal development (Smith et al., 1994), efficiency of carbon and P exchange and translocation correlated with changes in fungal colonization (Pearson and Jakobsen, 1993a), and activation of root nutrient uptake (Pearson and Jakobsen, 1993b). Unfortunately, these data do not indicate whether the traits measured are heritable and thus phylogenetically significant. With the likelihood of heterogeneous nuclei in single cells (Lloyd-MacGilp et al., 1996) that have considerable mobility and are somatically transmitted, internal selection pressures could be intense and lead to diverse phenotypes that interact in complex ways with the phycobiont in response to local variables. Coevolution of *both* mycobiont and phytobiont is still occurring, albeit at a rate and magnitude that is poorly understood. To partition historical from local processes, more effort is needed to determine the conservativeness of processes being measured when host or fungus genotypes vary or when environmental conditions are altered.

5. CONCLUSIONS

The coevolution of arbuscular mycorrhizal fungi and most of today's plant lineages is considered in this chapter to have evolved twice, with most knowledge of ancestor properties and age confined to the lineage now classified as the suborder Glomineae (genera *Glomus, Sclerocystis, Acaulospora, Entrophospora*). Virtually nothing is known about evolutionary origins of the Gigasporineae lineage (genera *Gigaspora, Scutellospora*) since neither fossils nor extant fungi in other groups that might be related have been found. In both lineages, parallel function suggests similar coevolutionary trends. Speciation is considered to have been promoted in *both* symbionts following integration of the AM association, with the disproportionately greater number of species in the phytobiont being a function of morphological potential (complexity in design and life history traits) rather than any ecological causation. Ecology may be an important determinant of rate of speciation, but I propose that its central role has been in gradual coaccommodation and optimization of both symbionts as they changed and adapted in spatially and temporally complex communities. Universal mechanisms or models to explain mycorrhizal phenomena are expected to be rare because the symbiosis was established so early that the mycobiont and phytobiont evolved *together* and remained integrated *together* during their coevolution to the present day. The evolutionary stability of the AM symbiosis indicates a general positive outcome for the phytobiont, since it is the higher unit of selection (and hence regulator) of the symbiosis. However, the causal processes for mutualism are as diverse as the magnitude of genealogical divergence among phytobiont lineages, and extend far beyond the growth benefit so widely touted but so inconsistently expressed in nature. The mycobiont appears to be more homogeneous in structure and function, but accumulating evidence indicates considerable potential for divergence in behavior and physiology within the constraints imposed by the phytobiont. This divergence is a function of genealogy, coevolutionary pressures, and coadaptation.

ACKNOWLEDGMENTS

I thank co-workers Kelly Heldreth, Kris Nichol, and Sidney Stormier and a recent student visitor, Toby Kiers, for stimulating discussions that led to some of the ideas presented in this chapter. Insights also were generated from interactions with the COST-821 workgroup sponsored by the European Union. Research reported from our laboratory was funded by National Science Foundation Grant BIR-9015519.

REFERENCES

Allen, M. F., Allen E. B. and Friese, E. B. (1989). Responses of the non-mycotrophic plant *Salsola kali* to invasion by vesicular-arbuscular mycorrhizal fungi. *New Phytologist* 111, 45–49.

Amijee, F., Tinker, P. B. and Stribley, D. P. (1989). The development of endomycorrhizal root systems. VII. A detailed study of effects of soil phosphorus on colonization. *New Phytologist* 111, 435–446.

Andrews, J. H. (1992). Fungal life-history strategies. *In* G. C. Carroll and D. T. Wicklow (eds.), *The Fungal Community: Its Organization and Role in the Ecosystem*, Marcel Dekker, New York, pp. 119–145.

Balestrini, R., Hahn, M. G., Faccio, A., Mendgen, K. and Bonfante, P. (1996). Differential localization of carbohydrate epitopes in plant cell walls in the presence and absence of arbuscular mycorrhizal fungi. *Plant Physiology* 111, 203–213.

Bartolome-Esteban, H. and Schenck, N. C. (1994). Spore germination and hyphal growth of arbuscuclar mycorrhizal fungi in relation to soil aluminum saturation. *Mycologia* 86, 217–226.

Baum, D. A. and Donoghue, M. J. (1995). Choosing among alternative "phylogenetic" species concepts. *Systematic Botany* 20, 560–573.

Bever, J. D., Morton, J. B., Antonovics, J. and Schultz, P. A. (1995). Host specificity and diversity of arbuscular fungi in Glomales: an experimental approach using an old field soil. *Journal of Ecology* 84, 71–82.

Beyrle, H. (1995). The role of phytohormones in the function and biology of mycorrhizas. *In* A. Varma and B. Hock (eds.), *Mycorrhiza: Structure, Function, Molecular Biology and Biotechnology*, Springer-Verlag, Berlin, pp. 365–390.

Bianciotto, V., Bandi, C., Minerdi, D., Sironi, M., Tichy, H. V., and Bonfante, P. (1996). An obligately endosymbiotic mycorrhizal fungus itself harbors obligately intracellular bacteria. *Applied and Environmental Microbiology* 62, 3005–3010.

Biermann, B. and Linderman, R. G. (1983). Use of vesicular-arbuscular mycorrhizal roots, intra radical vesicles and extraradical vesicles as inoculum. *New Phytologist* 95, 97–105.

Bonfante, P. and Perotto, S. (1995). Tansley Review No. 82. Strategies of arbuscular mycorrhizal fungi when infecting host plants. *New Phytologist* 130, 3–21.

Bradbury, S. M., Peterson, R. L. and Bowley, S. R. (1993). Further evidence for a correlation between nodulation genotypes in alfalfa (*Medicago sativa* L.) and mycorrhiza formation. *New Phytologist* 124, 665–673.

Bruce, A., Smith, S. E. and Tester, M. (1994). The development of mycorrhizal infection in cucumber: effects of P supply on root growth, formation of entry points and growth of infection units. *New Phytologist* 127, 507–514.

Brundrett, M. (1991). Mycorrhizas in natural ecosystems. *Advances in Ecological Research* 21, 171–313.

Brundrett, M. and Kendrick, B. (1990). The roots and mycorrhizas of herbaceous woodland plants. II. Structural aspects of morphology. *New Phytologist* 114, 469–479.

Bryla, D. R. and Koide, R. T. (1990). Regulation of reproduction in wild and cultivated *Lycopersicon esculentum* Mill. by vesicular-arbuscular mycorrhizal infection. *Oecologia* 84, 74–81.

Buss, L. (1987). *The Evolution of Individuality*. Princeton University Press, Princeton, NJ.

Campbell, B. D., Grime, J. P. and Mackey, J. M. L. (1991). A trade-off between scale and precision in resource foraging. *Oecologia* 87, 532–538.

Chapela, I. H., Rehner, S. A., Schultz, T. R. and Mueller, U. G. (1994). Evolutionary history of the symbiosis between fungus-growing ants and their fungi. *Science* 266, 1691–1694.

Dodd, J. C., Rosendahl, S., Giovannetti, M., Broome, A., Lanfranco, L. and Walker, C. (1996). Inter- and intra specific variation within the morphologically-similar arbuscular mycorrhizal fungi *Glomus mosseae* and *Glomus coronatum*. *New Phytologist* 133, 113–122.

Doyle, J. J. (1992). Gene trees and species trees: molecular systematics as one-character taxonomy. *Systematic Botany* 17,144–163.

Eissenstat, D. M., Graham, J. H., Syvertsen, J. P. and Drouillard, D. L. (1993). Carbon economy of sour orange in relation to mycorrhizal colonization and phosphorus status. *Annals of Botany* 71, 1–10.

Franke, M. and Morton, J. (1994). Ontogenetic comparisons of arbuscular mycorrhizal fungi *Scutellospora heterogama* and *Scutellospora pellucida*: revision of taxonomic character concepts, species descriptions, and phylogenetic hypotheses. *Canadian Journal of Botany* 72, 122–134.

Gehrig, H., Schussler, A. and Kluge, M. (1996). *Geosiphon pyriforme*, a fungus forming endocytobiosis with *Nostoc* (cyanobacteria), is an ancestral member of the Glomales: evidence by SSU rRNA analysis. *Journal of Molecular Evolution* 43, 71–81.

Gemma, J. N., Koske, R. E. and Flynn, T. (1992). Mycorrhizae in Hawaiian pteridophytes: occurrence and evolutionary significance. *American Journal of Botany* 79, 843–852.

Gianinazzi, S. (1991). Vesicular-arbuscular (endo-)mycorrhizas: cellular, biochemical, and genetic aspects. *Agriculture Ecosystems and Environment* 35, 105–119.

Gianinazzi-Pearson, V., Dumas-Gaudot, E., Gollotte, A., Tahiri-Alaoui, A. and S. Gianinazzi, S. (1996). Cellular and molecular defense-related root responses to invasion by arbuscular mycorrhizal fungi. *New Phytologist* 133, 45–57.

Gianinazzi-Pearson, V., Lemaine, M. C., Arnould, C., Gollotte, A. and Morton, J. B. (1994). Localization of β(1-3)-glucans in spore and hyphal walls of fungi in the Glomales. *Mycologia* 86, 478–485.

Gianinazzi-Pearson, V., Gollotte, A., Lherminier, J., Tisserant, B., Franken, P., Dumas-Gaudot, E., Lemoine, M.C., van Tuinen, D. and S. Gianinazzi (1995). Cellular and molecular approaches in the characterization of symbiotic events in functional arbuscular mycorrhizal associations. *Canadian Journal of Botany* 73(Suppl. 1), 526–532.

Glenn, M. G., Chew, F. S. and Williams, P. H. (1988). Influence of glucosinolate content of *Brassica* (Cruciferae) roots on growth of vesicular-arbuscular mycorrhizal fungi. *New Phytologist* 110, 217–225.

Grant, V. (1981). *Plant Speciation*, 2nd ed. Columbia University Press, New York.

Hass, H., Taylor, T. N. and Remy, W. (1994). Fungi from the lower Devonian rhynie chert: mycoparasitism. *American Journal of Botany 81*, 29–37.

Hetrick, B. A. D., Wilson, G. W. T. and Cox, T. S. (1992). Mycorrhizal dependence of modern wheat varieities, landraces, and ancestors. *Canadian Journal of Botany* 70, 2032–2040.

Hetrick, B. A. D., Wilson, G. W. T. and Leslie, J. F. (1993). Root architecture of warm- and cool-season grasses: relationship to mycorrhizal dependence. *Canadian Journal of Botany* 69, 112–118.

Hinkle, G., Wetterer, J. K., Schultz, T. R. and Sogin, M. L. (1994). Phylogeny of the attine ant fungi based on analysis of small subunit ribosomal RNA gene sequences. *Science* 266, 1695–1697.

Jakobsen, I. and Rosendahl, L. (1990). Carbon flow into soil and external hyphae from roots of mycorrhizal cucumber plants. *New Phytologist* 115, 77–83.

Jasper, D. A., Abbott, L. K. and Robson, A. D. (1989). Soil disturbance reduces the infectivity of external hyphae of vesicular-arbuscular mycorrhizal fungi. *New Phytologist* 112, 93–99.

Kluge, M., Mollenhauer, D. and Mollenhauer, R. (1991). Photosynthetic carbon assimilation in *Geosiphon pyriforme* (Kützing) F. v. Wettstein, an endosymbiotic association of fungus and cyanobacterium. *Planta* 185, 311–315.

Lanfranco, L., Wyss, P., Marzachi, C. and Bonfante, P. (1995). Generation of RAPD-PCR primers for the identification of isolates of *Glomus mosseae*, an arbuscular mycorrhizal fungus. *Molecular Ecology* 4, 61–68.

Lau, T.-C., Lu, X., Koide, R. T. and Stephenson, A. G. (1995). Effects of soil fertility and mycorrhizal infection on pollen production and pollen grain size of *Cucurbita pepo* (Cucurbitaceae). *Plant Cell and Environment* 18, 169–177.

Laval, C., Weissenhorn, I., Glashoff, A. and Berthelin, J. (1995). Influence of heavy metals on germination of arbuscular-mycorrhizal fungal spores in soils. *Acta Botanica Gallica* 141, 523–528.

Law, R. and Lewis, D. H. (1983). Biotic environments and the maintenance of sex—some evidence from mutualistic symbioses. *Biological Journal of the Linnean Society* 20, 249–276.

LePage, B. A., Currah, R. S., Stockey, R. A. and Rothwell, G. W. (1997). Fossil ectomycorrhizae from the Middle Eocene. *American Journal of Botany* 84, 410–412.

Lloyd-MacGilp, S. A., Chambers, S. M., Dodd, J. C., Fitter, A. H., Walker, C. and Young, J. P. W. (1996). Diversity of the ribosomal internal transcribed spacers within and among isolates of *Glomus mosseae* and related fungi. *New Phytologist* 133, 103–112.

Malloch, D. M. (1987). The evolution of mycorrhizae. *Canadian Journal of Plant Pathology* 9, 398–402.

McGonigle, T. P. and Fitter, A. H. (1988). Ecological specificity of vesicular-arbuscular mycorrhizal associations. *Mycological Research* 94, 120–122.

Merryweather, J. and Fitter, A. (1995a). Phosphorus and carbon budgets: mycorrhizal contribution in *Hyacinthoides non-scripta* (L.) Chouard ex Rothm. Under natural conditions. *New Phytologist* 129, 619–627.

Merryweather, J. and Fitter, A. (1995b). Arbuscular mycorrhiza and phosphorus as controlling factors in the life history of *Hyacinthoides non-scripta* (L.) Chouard ex Rothm. *New Phytologist* 129, 629–636.

Merryweather, J. and Fitter, A. (1996). Phosphorus nutrition of an obligately mycorrhizal plant treated with the fungicide benomyl in the field. *New Phytologist* 132, 307–311.

Morton, J. B. (1988). Taxonomy of VA mycorrhizal fungi: classification, nomenclature, and identification. *Mycotaxon* 32, 267–324.

Morton, J. B. (1990a). Evolutionary relationships among arbuscular mycorrhizal fungi in the Endogonaceae. *Mycologia* 82, 192–207.

Morton J. B. (1990b). Species and clones of arbuscular mycorrhizal fungi (Glomales, Zygomycetes): their role in macro- and micro evolutionary processes. *Mycotaxon* 37, 493–515.

Morton, J. B. (1993). Problems and solutions for the integration of glomalean taxonomy, systematic biology, and the study of mycorrhizal phenomena. *Mycorrhiza* 2, 97–109.

Morton, J. B. and Benny, G. L. (1990). Revised classification of arbuscular mycorrhizal fungi (Zygomycetes): a new order, Glomales, two new suborders, Glomineae and Gigasporineae, and two new families, Acaulosporaceae and Gigasporaceae, with an emendation of Glomaceae. *Mycotaxon* 37, 471–491.

Morton, J. B., Bentivenga, S. P. and Bever, J. D. (1995). Discovery, measurement, and interpretation of diversity in symbiotic endomycorrhizal fungi (Glomales, Zygomycetes). *Canadian Journal of Botany* 73(Suppl. 1), 25–32.

Morton, J., Bentivenga, S. and Wheeler, W. (1993). Germ plasm in the International Collection of Arbuscular and Vesicular-arbuscular Mycorrhizal Fungi (INVAM) and procedures for culture development, documentation, and storage. *Mycotaxon* 48, 491–528.

Morton, J. B., Bever, J. D. and Pfleger, F. L. (1997). Taxonomy of *Acaulospora gerdemannii* and *Glomus leptotichum*, synanamorphs of one anamorphic fungus in Glomales. *Mycological Research* 101:625–631.

Morton, J. B., Franke, M. and Cloud, G. (1992). The nature of fungal species in Glomales and their role in endomycorrhizal associations. *In Mycorrhizas in Ecosystems*. D. J. Read, D. H. Lewis, A. Fitter and I. Alexander, (eds.), Arizona: University of Arizona Press, Tucson, pp. 65–73.

Newsham, K. K., Fitter, A. H. and Watkinson, A. R. (1994). Arbuscular mycorrhiza protect an annual grass from root pathogenic fungi in the field. *Journal of Ecology* 83, 991–1000.

Nicety, H. and Sugiyama, J. (1995). A common group I intron between a plant parasitic fungus and its host. *Molecular Biology and Evolution* 12, 883–886.

Pearson, J. N. and Jakobsen, I. (1993a). Symbiotic exchange of carbon and phosphorus between cucumber and three arbuscular mycorrhizal fungi. *New Phytologist* 124, 481–488.

Pearson, J. N. and Jakobsen, I. (1993b). The relative contribution of hyphae and roots to phosphorus uptake by arbuscular mycorrhizal plants, measured by dual labeling with ^{32}P and ^{33}P. *New Phytologist* 124, 489–494.

Pearson, J. N. and Schweiger, P. (1994). *Scutellospora calospora* (Nicol. & Gerd.) Walker & Sanders associated with subterranean clover produces non-infective hyphae during sporulation. *New Phytologist* 127, 697–701.

Pearson, J. N., Abbott, L. K. and Jasper, D. A. (1993). Mediation of competition between two colonizing VA mycorrhizal fungi by the host plant. *New Phytologist* 123, 93–98.

Pirozynski, K. A. (1981). Interactions between fungi and plants through the ages. *Canadian Journal of Botany* 59, 1824–1827.

Pirozynski, K. A. and Malloch, D. W. (1975). The origin of land plants: a matter of mycotropism. *BioSystems* 6, 153–164.

Ravnskov, S. and Jakobsen, I. (1995). Functional compatibility in arbuscular mycorrhizas measured as hyphal P transport to the plant. *New Phytologist* 129, 611–618.

Safir, G. R., ed. (1987). *Ecophysiology of VA Mycorrhizal Plants*. CRC Press, Boca Raton, FL.

Sanders, I. R., Alt, M., Groppe, K., Boller, T. and Wiemken, A. (1995). Identification of ribosomal DNA polymorphisms among and within spores of Glomales: application to studies on genetic diversity of arbuscular mycorrhizal fungal communities. *New Phytologist* 130, 419–427.

Schussler, A., Mollenhauer, D., Schnepf, E. and Kluge, M. (1994). *Geosiphon pyriforme*, an endosymbiotic association of fungus and cyanobacteria: the spore structure resembles that of arbuscular mycorrhizal (AM) fungi. *Botanica Acta* 107, 36–45.

Shachar-Hill, Y., Pfeffer, P. E., Douds, D., Osman, S. F., Doner, L. W. and Ratcliffe, R. G. (1995). Partitioning of intermediary carbon metabolism in VAM colonized leek. *Plant Physiology 108*, 7–15.

Sylvia, D. M., Wilson, D. O., Graham, J. M., Maddox, J. J., Millner, P., Morton, J. B., Skipper, H. D., Wright, S. F. and Jarfster, A. (1993). Evaluation of vesicular-arbuscular mycorrhizal fungi in diverse plants and soils. *Soil Biology and Biochemistry* 25, 705–713.

Simon, L. (1996). Phylogeny of the Glomales: deciphering the past to understand the present. *New Phytologist* 133, 95–101.

Simon, L., Bousquet, J., Lévesque, R. C. and Lalonde, M. (1993). Origin and diversification of endomycorrhizal fungi and coincidence with vascular land plants. *Nature* 363, 67–69.

Smith, S. E. and Read, D. J. (1997). *Mycorrhizal Symbiosis*. Academic Press, San Diego.

Smith, F. A. and Smith, S. E. (1996). Mutualism and parasitism: biodiversity in function and structure in the ''arbuscular'' (VA) mycorrhizal symbiosis. *Advances in Botanical Research* 22, 1–43.

Smith, S. E., Dickson, S., Morris, C. and Smith, F. A. (1994). Transfer of phosphate from fungus to plant in VA mycorrhizas: calculation of the area of symbiotic interface and of fluxes of P from two different fungi to *Allium porrum* L. *New Phytologist* 127, 93–99.

Stanley, M. R., Koide, R. T. and Shumway, D. L. (1993). Mycorrhizal symbiosis increases growth, reproduction and recruitment of *Abutilon theophrasti* Medic. in the field. *Oecologia* 94, 30–35.

Taylor, T. N., Remy, W., Hass, H. and Kerp, H. (1995). Fossil arbuscular mycorrhizae from the Early Devonian. *Mycologia* 87, 560–573.

Tommerup, I. C. (1983). Temperature relations of spore germination and hyphal growth of vesicular-arbuscular mycorrhizal fungi in soil. *Transactions of the British Mycological Society* 81, 381–388.

Trappe, J. M. (1987). Phylogenetic and ecological aspects of mycotrophy in the angiosperms from an evolutionary standpoint. *In* G. R. Safir (ed.), *Ecophysiology of VA Mycorrhizal Plants.* CRC Press, Boca Raton, FL, pp. 5–25.

Wyss, P., Mellor, R. B. and Wiemken, A. (1990). Vesicular-arbuscular mycorrhizas of wild-type soybean and non-nodulating mutants with *Glomus mosseae* contain symbiosis-specific polypeptides (mycorrhizins), immunologically cross-reactive with nodulins. *Planta* 182, 22–26.

7
Biodiversity and Evolution in Mycorrhizae of the Desert

Peter Herman
New Mexico State University, Las Cruces, New Mexico

1. INTRODUCTION

Interest in mycorrhizal associations has grown exponentially in the last 20 years. The 1990s have seen the publication of a number of general reference works on mycorrhizae (Allen, 1991; Smith and Read, 1997), methods for their study (Norris et al., 1992), and their roles in the environment (Read et al., 1992; Allen, 1992). In addition, one can no longer pick up an issue of the major plant, soil, or microbiological journals without finding articles on mycorrhizae. While this chapter will refer to some of the vast body of literature on arbuscular mycorrhizal fungi (AMF), it will focus on aspects of their role in plants from deserts or other arid and semiarid environments.

Desert and arid land plants face environmental stresses, which, while not unique to these environments, are often significant obstacles to survival and reproduction. In addition to the obvious low absolute water availability, rainfall patterns are highly unpredictable and any single year's pattern is likely to vary considerably from long-term averages (Zak et al., 1995). In general, mineral nutrients are also growth limiting. In the North American deserts, nitrogen is most frequently limiting, though low available phosphorus may also serve as a limiting nutrient (West and Klemmedson, 1978; Whitford, 1986). As with rainfall, nutrient distribution patterns can be highly variable temporally and spatially. The spatial distribution of nutrients appears correlated with vegetation type. Variability at fine scale is high throughout arid lands; however, at the 20- to 200-cm scale distributions are relatively even in arid grasslands and strongly patchy in shrub lands (Schlesinger et al., 1990).

Plants use a number of strategies in overcoming these stresses. The perennial plants generally have either deep tap roots to take advantage of water stored at depth in the soil profile or shallow fibrous roots to intercept available moisture before it reevaporates. Grasses such as black *gramma* (*Bouteloua eriopoda*) tend to use this later strategy while shrubs such as mesquite (*Prosopis glandulosa*) take advantage of the former. In addition, there is a seed bank of annuals that may be abundant in years with favorable precipitation timing and rare in adverse years.

Arbuscular mycorrhizal fungi have been shown to interact with their hosts in a variety of ways (Fig. 1). Studies have documented increased phosphorus, nitrogen, and water uptake in AMF-infected plants (Gerdemann, 1964; Daft and Nicolson, 1966; Baylis, 1967; Ames et al., 1983; Frey and Schüepp, 1992; Johansen et al., 1992, 1993; Nelson and Safir, 1982; Koide, 1985; Fitter, 1988). The presence of external mycelium of AMF plants modifies soil properties (Coleman et al., 1983). AMF provide protection against soil pathogens (Newsham et al., 1994) and may facilitate nutrient transfer between plants (Read et al., 1985; Newman, 1988). These phenomena have the potential to increase host plant stature, survivorship, fecundity, and fitness.

While not all of these benefits have been demonstrated unambiguously in

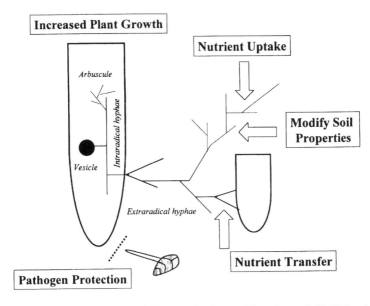

Figure 1   An overview of the organization and functions of AMF fungi in roots and soil.

desert plants, the stresses that are ameliorated by AMF associations are precisely those from which desert plants suffer and it is reasonable to infer that many accrue to desert plants in hostile environments.

2. COLONIZATION LEVELS IN ARID ZONE PLANTS

The fact that arid land plants are under stresses that are generally ameliorated by AMF would lead to the prediction that these plants are at least as heavily mycorrhizal as plants in general, if not more so. It is often difficult to assess levels of AMF occurrence because such a wide variety of descriptors are used in reporting AMF levels. They range from the number of plant families in a region with mycorrhizal associations or the number species in which AMF have been seen, to the percentage of plants of a species infected, or the proportion of roots of a particular species showing AMF infection. Clearly, the latter two measures give a more robust estimate of the importance of AMF to the plant community. The limited reports for arid and semiarid land plants would support the idea that AMF associations are widespread in arid environments. Dhillion and Zak (1993) report that 89% of 61 plant families surveyed from arid and semiarid lands were mycorrhizal. At least in some arid/semiarid environments, a remarkably high proportion of plants appear to be mycorrhizal. Rachel et al. (1989) reported that 95% of families and 97% of species surveyed in Warangal, Andhra Pradesh, India were endomycorrhizal. *Solanum surattense* was the only non-AMF-associated species among 31 species from 20 plant families. The proportions of roots infected ranged from 15.4% of segments scored for *Turnera subulata* to 94.7% for *Croton bonplandianus*.

In a relatively limited study, Wilson (1973) found that 11 of 13 common species in the Jornada Basin of New Mexico were mycorrhizal, although not all individuals within a species were mycorrhizal and the extent of colonization was variable among individuals. AMF appear to be common in extreme arid environments such as Arabian sand dunes where six species of dune stabilizing plants from five different plant families all showed colonization rates from 31–57% of root segments examined (Khaliel, 1989).

A number of studies have assessed colonization in selected taxa of arid or semiarid land plants. Mathew et al. (1991) looked at five species of spineless cacti in the genus *Opuntia* (Cactaceae) and one species of *Stapelia* (Asclepediceae) in Jodhpur, India. All six taxa were mycorrhizal with colonization rates ranging from 12% to 35.6% of root segments for the cactus species while *Stapelia gigantea* had 43% colonization. Neeraj et al. (1991) studied members of the Amaranthaceae from three different semiarid Indian sites. In all, 10 species were examined with colonization rates ranging from 2% to 65.5% of root segments. They found no correlation between soil moisture and frequency of colonization,

Table 1 AMF Colonization and Rhizobium Nodulation of Nonwoody Legumes from the Jornada Basin, New Mexico

Species	% Infected plants	% Infected segments	Range[a]	Nodulation[b]	Growth conditions
Astragalus mollissimus	93% (14/15)	59.03 ± 3.30	37–81	++	Growth chamber
Astragalus mollissimus	100% (23/23)	47.97 ± 3.29	11–79	++	Growth chamber
Astragalus mollissimus	100% (12/12)	34.71 ± 3.95	14–52	+	Field-collected
Astragalus tephrodes	100% (12/12)	56.16 ± 4.00	33–74	+++	Greenhouse
Cassia bauhinioides	100% (12/12)	76.55 ± 2.69	61–91	−	Greenhouse
Cassia bauhinioides	100% (12/12)	63.03 ± 2.39	45–74	−	Field-collected
Lupinus brevicaulis	0% (0/12)	0	nd	+++	Field-collected

* Percentage of segments colonized for the least and most colonized individual in the species; nd, not determined.
[b] The number of + indicate the degree of nodulation.
Source: Previously unpublished data from Y. Besmer's Examinsarbete, Swedish University of Agricultural Science and R.P. Herman.

whereas colonization was highest in the pH range 6.8–7.8. Four of the species were found at two of the three sites, allowing between site comparisons. In one case (*Celosia cristata*) colonization rates were similar at both sites (40.6 vs. 45.5%) while there were at least twofold differences in colonization rates between sites for the other three species (*Achyranthes aspera* 14.5 vs. 34.0%, *Celosia argentia* 26.3 vs. 60.0%, and *Amaranthus caudatus* 2.0 vs. 65.6%). There was no one obvious environmental variable that seemed to explain the variation in colonization between sites.

Selected nonwoody desert legumes originating in the Jornada Basin of New Mexico grown under a variety of conditions show considerable variability in the intensity of infection (Table 1). In the three species that were mycorrhizal (*Astragalus mollissimus*, *A. tephrodes*, and *Cassia bauhinioides*), virtually all individuals showed some colonization. On the other hand, no individuals of *Lupinus brevicaulis* showed any signs of AMF colonization, though an occasional segment was infected with septate mycelium. The proportion of segments colonized in mycorrhizal individuals showed significant variation. Furthermore, there were significant differences in colonization rates when the same plant species was field-collected as compared to grown under controlled conditions in field-collected soil.

3. DIVERSITY OF AM FUNGI IN ARID ZONE SOILS

Taxonomic diversity among AMF, assessed by morphology, is not linked to the diversity of higher plants (Allen et al., 1995), so there is no reason to expect that the relatively depauperate plant floras associated with deserts and arid or semiarid scrub lands would necessarily be linked to low AMF diversity. The diversity of AMF observed from any particular site is not only a function of the actual biodiversity of the site but also of the size of the area sampled and the length of time over which sampling occurred. The latter point is especially important in arid areas because weather conditions are often highly variable over time. In most studies, there is little effort to sample individuals spread out over one or more hectares. As a result, the reported AMF species diversity associated with an individual plant species usually represents AMF taxa sampled in only a small portion of a site on which the plant occurs. Studies of this type typically report one to three AMF species associated with a particular plant (Table 2).

Perhaps the most comprehensive study of the diversity of AMF associated with a single plant species was carried out with the arid scrub land species *Artemisia tridentata* and *Atriplex gardneri* by Allen et al. (1995). When sampled over a 1-ha area rather than in a restricted site, 12 species of AMF were associated with *Artemisia* and *Atriplex*. When *Artemisia* was surveyed at 68 sites over the entire western portion of North America, the number of associated AMF species

Table 2 Number of AMF Taxa Associated with Selected Plant Species

Site	No. plant species/locations observed	No. plant species with the indicated number of AMF taxa				Ref.
		1	2	3	>3	
Sand Dunes—Arabia	6	5 (83%)	1 (17%)			Khaliel 1989
Jodhpur India	7	0	4 (57%)	2 (29%)	1 (14%)	Mathew et al., 1991
Himalayan foothills	14	7 (50%)	4 (29%)	3 (21%)		Neeraj et al., 1991

increased to 48, including 20 previously undescribed. The number of AMF species associated with *Artemisia* rhizosphere soil varied from a high of 17 at one of the 68 sites to 1 AMF species at 9 sites. The largest number of sites had between 1 and 4 AMF species present. This study also suggested that the composition of AMF species at a particular site was controlled by site-edaphic characteristics and perhaps a longitudinal gradient. The relative diversity of AMF in arid western North America is low compared to that of some tropical systems. For example, Wilson et al. (1992) found a comparable number of vesicular arbuscular mycorrhizae (VAM) species (41) in only 10 ha of African moist tropical forests. On the other hand, low numbers of species per hectare have also been observed in tropics (Sieverding, 1990), making blanket comparisons difficult.

Just as area examined had a profound influence on the biodiversity of AMF associated with *Artemisia*, differences in sampling time also altered the observed AMF biodiversity. Eleven species of AMF were isolated from an *Artemisia/ Atriplex* steppe in southwestern Wyoming (USA) over a 3-year period (Allen et al., 1995). Of these, five (all species of *Glomus*) occurred at all three sampling periods, three were isolated in two of the three years, and three occurred only once.

It is becoming clear that species diversity as assessed by the morphological identification of spores associated with roots may represent only a small part of the diversity of AMF strains colonizing roots for two reasons. The first is that there may be significant ''cryptic'' genetic diversity present in morphologically identical spores. Sanders et al. (1995) reported that 10 morphologically identical *Glomus* spores were genetically different as determined by polymerase chain reaction/restriction fragment length polymorphism (PCR/RFLP) of the internally transcribed sequence (ITS) region of DNA from single spores. The second source of underestimated diversity is that root-associated AMF may not produce spores abundantly or in proportion to their presence in roots. This would obviously lead to an underestimation of some poorly sporulating or nonsporulating colonists. For example, English bluebells (*Hyacinthoides non-scripta*) roots analyzed by molecular techniques revealed the presence of three colonizing AMF species. On the other hand, only two species of spores were regularly recovered from rhizosphere soils (Clapp et al., 1995).

These two studies were carried out in temperate environments and cannot be directly applied to arid land associations. However, unless arid land AMF are under much heavier selection pressure than temperate zone species, it is reasonable to assume that cryptic diversity is also present in arid zone AMF. It is also likely that AMF in arid zone roots are at least as likely to encounter unfavorable sporulation conditions as their temperate counterparts. Allen et al. (1995) suggest that the differences they saw in AMF species in their multiyear study of the *Artemisia/Atriplex* steppe was due to failure of species present to sporulate. The year with the most diverse spore population was associated with greater than average rainfall, suggesting that more favorable conditions led to greater sporulation and thus increased observation of spores. Black gramma (*Bouteloua eriopoda*) plants in the Jornada Basin often show 90% or greater colonization with abundant vesicles but only one to three AMF spores per gram of soil (unpublished data), again suggesting that spore number may not be well correlated with AMF levels in desert soils.

4. FACTORS POTENTIALLY RESPONSIBLE FOR OBSERVED VARIABILITY IN COLONIZATION

The likelihood that an individual plant in an arid environment will form an AMF association can be highly variable. Species in Table 1 show 1.5- to 7.2-fold range between the highest and lowest number of segments colonized in an individual plant grown under identical conditions. There are not sufficient published data to determine if this variability is greater for arid zone plants than for AMF associations in other zones; however, differences in colonization rates could play an important role in the functioning of AMF plants in arid environments.

The likelihood that a compatible plant will form an AMF association and the degree to which it will due so, as measured by the proportion of root segments colonized, is a highly complex process. Paradoxically, it seems that under some conditions and in some plant types colonization rates correlate with apparent benefit to the host plant, whereas in other cases the process appears nearly stochastic (Fitter and Merryweather, 1992). Certainly, the process is controlled by some combination of environmental factors, AMF propagule load in the soil, initial colonization of the plant in question and its neighbors, as well as the genetic makeup of both the plant and the AMF partners.

Environmental factors that are known to affect colonization rates include phosphorus and nitrogen levels, light, and temperature. Because there can be significant heterogeneity in P and N levels in desert soils, microsite conditions could well influence the colonization rates of individuals of a mycorrhizal species. High phosphorus levels decrease AMF germ tube growth and/or hyphal elongation in the root cortex (Sanders, 1975; Bruce et al., 1994), which could influence both primary and secondary infections where phosphorus is not limiting. However, it is rare that phosphorus levels are high enough in typical arid environments for this to be a significant source of colonization suppression. In fact, available phosphorus may be sufficiently low in desert systems that locally increased levels could stimulate infection as has been observed in clover with low-level phosphorus additions in phosphorus-poor soils (Bolin et al., 1984).

Soil nitrogen status has also been reported to influence colonization, though the data are neither as extensive nor as consistent as the phosphorus data. The observed effects are not consistent across species of plants and AMF. Nitrogen additions have been shown both to decrease colonization with ammonium more effective than nitrate (Chambers et al., 1980) and to increase colonization (Aziz and Habte, 1989). Another factor further complicating the influence of mineral nutrient levels on colonization is a synergism between N and P. Sylvia and Neal (1990) have shown that nitrogen-deficient plants do not show the same phosphorus depression of colonization as do nitrogen-sufficient plants. To at least some degree, these nutrient effects may be due to the rate and extent of root mass growth, with rapidly growing roots outstripping the ability of AMF propagules to initiate colonizations (Bruce et al., 1994).

For arid zone plants, water regimes are highly variable and potentially influence AMF colonization rates. Two studies have looked at arid land plants and moisture and both found inconsistencies in response. Allen et al. (1989) found that in general both root length and AMF-colonized root length increased with increased water treatments; however, year-to-year variability was high and often greater than any treatment effects. In a study of the influence of wetting/drying cycles on colonization, the response was found to be highly variable. AMF propagules as either spores or in root segments showed either increases or decreases in infectivity dependent on species (Braunberger et al., 1996).

The extent of propagule load in the soil, both spores and external mycelium, is another factor that can influence the extent of infection. The buildup of propagules in an arid soil successional site has been nicely demonstrated by Allen and Allen (1992). After an initial lag phase of about 15 days where AMF spore populations were entirely due to incoming colonizers, spore numbers increased logarithmically with endogenously produced spores accounting for the bulk of the spores present. This increase in spores is the result of increases in hyphae in soil. The authors suggest that this hyphal development is not only a nutrient acquisition system but an organized method for searching for new roots to colonize. To further complicate the picture, infectivity of propagules in arid zone soils can vary significantly as a result of heating or wetting and drying cycles (McGee, 1989; Jasper et al., 1989a, b, 1993; Braunberger et al., 1994).

The final consideration in colonization in arid soils is the genetic makeup of the plant and AMF. Resistance to colonization in noncompatible plants is almost certainly a function of plant genotype. For example, species of *Lupinus* have been shown to resist colonization because their root exudates do not contain stimulants necessary for AMF propagule germination and early growth (Giovannetti et al., 1993) rather than active resistance mechanisms (Vierheilig et al., 1994). The potential interplay of plant and AMF genotypes in colonization is ripe for further study as we begin to increase our understanding of the genotypic diversity present in AMF. It is clear that the degree of host specificity in AMF colonization does not approach that seen in many pathogenic relationships; however, clear differences in colonization rates between different AMF types in the same plant have been demonstrated (McGonigle and Fitter, 1990). Most of the work on host specificity has relied on morphology to identify the AMF partner. It is entirely possible that a greater degree of genotypic specificity than previously suspected does exist in light of the genotypic diversity among identical morphotypes observed by Sanders et al. (1995).

5. PHOSPHORUS AND AMF INTERACTIONS IN ARID LAND PLANTS

The majority of benefits from AMF associations reported for arid land plants involve increased nutrient uptake, particularly of phosphorus. This is not surprising, as phosphorus uptake is perhaps the most commonly reported benefit of AMF in all systems. However, even here observations in arid zone plants are often highly variable or contradictory.

The general pattern of AMF influence has two aspects: an increase of tissue P in AMF-colonized plants and an increase in P flux into roots over non-AMF plants. The observation of increased P content in tissues dates back to early studies of AMF (Gerdemann, 1964; Daft and Nicolson, 1966; Baylis, 1967). Not long after, Sanders and Tinker (1971, 1973) demonstrated that AMF-colonized *Allium cepa* roots had higher P uptake on a unit area basis and calculated the contribution of AMF extraradical hyphae to that flux. This effect was also soon shown to be highly dependent on the particular AMF species involved (Sanders et al., 1977). The control of P uptake and its subsequent distribution in plants is complicated by plant phosphate demand, soil P availability, AMF nutrient demands, and the effects of AMF on root mass and root shoot ratio (Koide, 1993).

Desert and arid land sites do not uniformly follow this general pattern of increased P content/uptake. Some laboratory studies with desert species show elevated tissue P levels in AMF-colonized plants but this pattern has been difficult to demonstrate with any consistency in field studies, transplant studies, or in pot studies using field-collected material. In virtually all of these studies where one of the studied species shows an increase in P uptake, other species in the same study fail to do so. In other cases, no phosphorus effect at all can be shown.

Mycorrhizal western wheatgrass (*Agropyron smithii*) grown in soil columns with insoluble calcium phosphates as the major P source showed significantly higher tissue phosphorus levels than nonmycorrhizal plants (Knight et al., 1989). Root mass, though not shoot mass, was also increased. This study also measured CO_2 production by roots, with mycorrhizal plants showing higher levels than nonmycorrhizal plants. The authors suggest that the increased P levels in the plants resulted from the acid solubilization of rock phosphates. This conclusion is supported by their observation that total solution P was more highly associated with mycorrhizal plants and that CO_2 addition could mimic the AMF effect on P in solution.

Another soil column study with positive P effects of AMF measured plant growth and tissue P content in response to AMF inoculation or P addition in two tropical woody legumes, *Leucaena leucocephala* and *Acacia nilotica*, under ample water- and drought-stressed conditions (Michelsen and Rosendahl, 1990). Phosphorous uptake and shoot P content were higher in AMF-inoculated treatments for both species under both water regimes. Shoot growth in *L. leucocephala* was significantly greater in AMF-inoculated plants under both water treatments while *A. nilotica* showed increased growth relative to control only in the drought treatment. This suggests that some "luxury" accumulation of P may be taking place in well-watered *A. nilotica*.

Luxury P accumulation may also be an AMF effect in Indian ricegrass (*Oryzopsis hymenoides*) inoculated with native strains of *Glomus pallidum* and transferred to fumigated field plots in semiarid steppe. This bunch grass native to the study site showed no growth response to AMF inoculation but had significantly higher tissue P concentrations than non-AMF controls or plants inoculated with nonnative AMF (Trent et al., 1993). Interestingly, the introduced crested wheatgrass (*Agropyron desertorum*) did not show increases in P concentrations under similar conditions.

A number of studies have investigated symbionts in the genus *Hedysarum*, a mycorrhizal, nodulated legume that is native to arid zones. Carpenter and Allen (1988) measured P content of *Hedysarum boreale* leaflets over two seasons in plants germinated and inoculated in the greenhouse and transplanted to field plots. They were unable to detect any difference in tissue P on a per-gram-of-tissue or per-plant basis in plants inoculated with AMF alone. Roldan et al. (1992) tested five species of AMF in *H. confertum* and *H. spinosissimum*. In general, AMF inoculation did not improve P uptake; however, one AMF species, *Glomus macrocarpum*, slightly improved P uptake in both legume species.

Some studies have shown positive effects of AMF in P nutrition in some species while not in others from the same site. For example, of four grass species tested (*Aristida armata*, *Cenchrus ciliaris*, *Digiteria ammophilla*, and *Thyridolepis mitchelliana*), only *A. armata* showed an increased growth response to phosphorus in AMF-colonized individuals (Armstrong et al., 1992). Similarly, *Agave deserti* and *Ferocactus acanthoides* shoots contained significantly higher P concentrations in AMF-colonized treatments compared to non-AMF controls, while *Opuntia ficus-indica* did not (Cui and Nobel, 1992).

One of the most interesting recent findings relative to phosphorus uptake concerns the role of AMF in the acquisition of nutrients from patchy environments. This has special relevance in arid environments where nutrients like P are often patchy and in chemical forms not readily available to plants. *Agropyron desertorum* roots with or without AMF associations were allowed to explore soil volumes where the same quantity of P had been applied either uniformly or in restricted patches. AMF-associated plants showed increased P acquisition from both distributions compared to non-AMF plants when expressed on a shoot mass or root length basis. The improved performance was relatively greater in the patchy environment. Both mycorrhizal and nonmycorrhizal plants showed increased root densities in the patches. Interestingly, this root concentration in the nutrient patch was less pronounced in the mycorrhizal plants than in the nonmycorrhizal plants, suggesting that the extraradical hyphae were playing a significant role in the detection and acquisition of phosphorus (Cui and Caldwell, 1996a). This suggestion was further strengthened in experiments using two-compartment cylinders where the plant was separated from the evenly or patchily distributed nutrients by a fine-screen barrier that could be crossed by extra radical AMF mycelium but not by roots (Cui and Caldwell, 1996b). Mycorrhizal plants had significantly higher P concentrations than nonmycorrhizal plants in both distribution patterns. The extraradical hyphae in the root-free zone was equally efficient in delivering phosphorus to the plants in both nutrient distributions. Comparable AMF effects were not seen for either distribution pattern of a highly mobile nutrient, nitrate, when roots and extraradical hyphae or hyphae alone were exploring the soil volume (Cui and Caldwell, 1996a, b).

6. NITROGEN AND N/P SYNERGISTIC INTERACTIONS

Many studies have shown increased uptake of nitrogen by mycorrhizal compared to nonmycorrhizal plants. Based on laboratory studies, this effect in nature is assumed to come from exploitation of ammonium and/or organic nitrogen (Frey and Schüepp, 1992; Johansen et al., 1992, 1993). AMF appear to provide little benefit in the uptake of nitrate N (Tobar et al., 1994; Cui and Caldwell, 1996a, b). This observation has also been seen in field and field transplant studies with desert plants but, as with P, the effects are not uniform. For example, *Glomus pallidum* colonization increased shoot N concentration above fumigated control in both *Oryzopsis hymenoides* and *Agropyron desertorum* whereas *G. mosseae* had a positive effect in only *Oryzopsis* (Trent et al., 1993).

An area of some potential importance in nitrogen-poor arid soils is the role that AMF play in increasing nodulation and/or N fixation in desert legumes. This phenomenon has been well studied in a number of agronomic plants where, to a greater or lesser degree, the AMF effect can be mimicked by the addition of phosphorus fertilizers (Barea and Azcón-Aguilar, 1983; Barea et al., 1992; Bethlenfalvay, 1992). Studies with two different species of the arid zone herbaceous perennial legume *Hedysarum* demonstrate just how complex this AMF-N-P interaction is in the field. Carpenter and Allen (1988) found that *H. boreale* inoculated with AMF and *Rhizobium* had greater leaflet biomass, total above-ground biomass, above-ground N and P content, and survival than control or singly inoculated plants. There were two additional facets of the study worth noting here. First, AMF plants diminished soil N in their rooting zone to the same degree as non-AMF plants, indicating significant root uptake in AMF plants despite apparent increases in N fixation. Second, the doubly inoculated plants showed significant increases in tissue P that were not observed in plants inoculated with AMF alone. This synergistic effect could be the result of any of a series of nutrient supply-and-demand interactions, altered C balance, or factors resulting from altered root shoot ratios (Koide, 1993).

At about the same time, Barea et al. (1987) compared control, P-amended, and AMF-inoculated *H. coronarium* with regard to dry matter yield, N content, and N fixation vs. N uptake as sources of nitrogen over a series of four monthly harvests beginning when the plants were 3 months old. The stimulation in dry matter production and N fixation seen at first harvest in AMF plants was greater than control and similar to P-amended plants. In subsequent harvests, AMF plants showed greater dry matter production, N fixation, and N uptake from soil than either control or P-amended treatments. They suggest that the increase in fixation is mediated via a P-dependent mechanism while the increased uptake from soil reflects AMF stimulation of ammonium uptake.

7. AMF EFFECTS ON WATER AND OSMOTIC STRESSES

AMF colonization improvements in water relations and/or drought tolerance is perhaps mechanistically the most complex and controversial of AMF functions. Both the increase of bulk transport of liquid water from soil via AMF hyphae and the increase of water flux through plants by a number of mechanisms related to AMF colonization have been suggested, though it has been difficult to demonstrate bulk flow effects (Koide, 1993). It is clear that AMF colonization increases water flow through plants, both agronomic (Nelsen and Safir, 1982; Koide, 1985; Fitter, 1988) and arid land (Allen et al., 1981; Allen, 1982; Stahl and Smith, 1984; Cui and Nobel, 1992) as measured by increased transpiration. Cui and Nobel (1992) assessed both radial and axial hydraulic conductivity in *Agavi deserti* and found that AMF increased radial conductivity, the transport of water from the root surface to xylem, while not affecting axial, xylem vessel, conductivity. They suggest either increased membrane permeability as the result of improved P status or intraradical AMF hyphae providing a low-resistance pathway across the root cortex as possible mechanisms. The earlier studies also suggest that the AMF effect is an indirect one acting via P-mediated mechanisms on hydraulic resistance or stomatal resistance. As with other AMF effects, the degree to which transpiration is increased at any given soil water potential appears to depend on the particular AMF strain/plant pairing (Stahl and Smith, 1984).

AMF colonization has somewhat variable effects under drought or osmotic stress (Koide, 1993); however, some studies have shown improved AMF–plant resistance to these stresses (Fitter, 1988; Michelsen and Rosendahl, 1990; Dixon et al., 1993; Baker et al., 1995). The mechanisms for this improved stress toler-ance are not clear.

8. AMF AND PLANT FITNESS

The nearly universal nature of the plant–AMF symbiosis has led to the assump-tion that the relationship improves the fitness of both partners. It is understandable that little has been done to study fitness in the fungal partner since it does not reproduce sexually and, as an obligate biotroph, cannot reproduce even asexually without its plant partner. It is clear that many of the plant benefits associated with AMF colonization such as increased stature, improved nutrient or water relationships, or increased survivorship have the potential to increase fitness mea-sured in terms of fecundity or offspring superiority. Studies designed to explicitly test hypotheses related to AMF on fitness have been few, recent, and in nondesert plants. Nevertheless, they provide important insights for workers interested in arid zones. Stanley et al. (1993) measured a series of demographic parameters in *Abutilon theophrasti*, an early successional plant in the Malvaceae. They found that AMF colonization resulted in increases in vegetative growth, flower, fruit, and seed production and overall greater recruitment in the following year. In addition, seeds from AMF plants were larger and contained more phosphorus.

Several studies assessed the fecundity of field populations winter annual grass *Vulpia ciliata* in response to benomyl treatment to eliminate AMF (Carey et al., 1992; West et al., 1993a, b). In general, fecundity decreased in response to benomyl treatment. Dose-controlled double-fungicide experiments were car-ried out to separate the effects of soil pathogens and AMF (Newsham et al., 1994). They showed that most of the increase in fecundity in plants colonized by AMF resulted from AMF protection from pathogenic fungi rather than phos-phorus effects. This protection may be very significant in arid environments. In upland *Bouteloua eriopoda* grassland soils from New Mexico, spores and micro-sclerotia of potential pathogens are often more abundant than AMF spores per gram of soil, yet septate hyphae in *B. eriopoda* healthy roots are rare compared to AMF hyphae (the author's unpublished observations).

The other direct studies of AMF on fitness addresses offspring superiority. Koide and Lu (1992) showed that seeds of AMF-colonized *Avena fatua* produced more vigorous offspring than nonmycorrhizal plants and, while seed size was reduced in mycorrhizal plants, seed number and seed P content increased. This observation was repeated in *Abutilon* (Stanley et al., 1993). Koide and Lu (1995) investigated the hypothesis that the rapid root growth of seedlings produced from AMF plants and their increased root phosphatase activity allowed an increased rate of nutrient uptake and thus growth compared to seedlings from nonmycorrhi-zal plants. They found that short-term growth improvements were an intrinsic property of the seed produced by mycorrhizal plants since these seeds outperformed the control group in the absence of external nutrients. This seed superiority is potentially of significance to desert plants, particularly desert annuals, which depend on rapid growth from seeds during often short windows of opportunity.

9. CONCLUSIONS AND DIRECTIONS

AMF clearly play an important role in plants adapted to desert environments and contribute to nutrient and water relations in many cases. In some ways, the frustrating variability in responses to AMF opens fruitful avenues for further research. While most of the evidence suggests that AMF colonization has a rather broad host range, the differences in functional response in AMF–plant symbionts can be dramatic. What determines the suitability of matching between AMF species/strains and hosts as well as the extent to which hosts exert control over colonization by superior strains is an open question to which desert plant AMF pairs can contribute. This is a question that can be addressed at a variety of levels ranging from the cell and molecular, through the physiology of the symbiont pair, to the ecology of the AMF–plant unit. These questions have both basic biological interest and major practical importance in expanding the use of AMF inocula to facilitate landscape restoration efforts.

Biodiversity of AMF themselves in arid ecosystems and the role that they play in the biodiversity of higher plant communities is an area that has recently become amenable to study. The emergence of selective molecular tools has opened the possibility of looking at the genetic structure of AMF communities in soils and plants, something impossible even in the early 1990s. Hopefully, these tools will allow us new ways to look at the extent and diversity of the mycelial net in soils and help us to understand the role of AMF in the establishment, persistence, and patterning of vegetation in arid ecosystems. Finally, the emerging direct investigations of relative fitness contributions of AMF–plant symbionts have the potential to open new ways of thinking about the biology of desert plants. To fail to include AMF in our thinking about the plant and soil biology of arid ecosystems is to ignore a major player in these systems.

ACKNOWLEDGMENTS

I thank E. B. Allen for introducing me to AM fungi and their ecological importance many years ago. Anna Mårtensson, host on my recent sabbatical, and many colleagues at the Swedish University of Agricultural Science renewed that interest. Finally visiting students Ylva Lekberg Besmer, Adam Langley, and Heather Ambro contributed an infusion of new energy to my ongoing studies in desert mycorrhizae. NSF Grants INT 9512776 and DEB-94-11971 also contributed to these studies.

REFERENCES

Allen, E. B., Allen, M. F., Helm, D. J., Trappe, J. M., Molina, R. and Rincon, E. (1995). Patterns and regulation of mycorrhizal plant and fungal diversity. *Plant and Soil* 170, 47–62.

Allen, M. F. and Allen, E. B. (1992). Development of mycorrhizal patches in a successional arid ecosystem. *In* D. J. Read, D. H. Lewis, A. H. Fitter and I. J. Alexander (eds.), *Mycorrhizas in Ecosystems* CAB International, Wallingford, UK. pp. 171–176.

Allen, M. F. (1982). Influence of vesicular-arbuscular mycorrhizae on water movement through *Bouteloua gracilis* (HBK). Lag ex Steud. *New Phytologist* 91, 191–196.

Allen, M. F. (1991). *The Ecology of Mycorrhizae.* Cambridge University Press, Cambridge, UK.

Allen, M. F. (Ed.). (1992). *Mycorrhizal Functioning.* Chapman and Hall, New York.

Allen, M. F., Richards, J. H. and Busso, C. A. (1989). Influence of clipping and soil water status on vesicular arbuscular mycorrhizae of two semi-arid tussoc grasses. *Biology Fertility and Soil* 8, 285–289.

Allen, M. F., Smith, W. K., Moore, T. S. and Christensen, M. (1981). Comparative water relations and photosynthesis of mycorrhizal and non-mycorrhizal *Bouteloua gracilis* HBK Lag ex Steud. *New Phytologist* 88, 683–693.

Ames, R. N., Reid, C. P. P., Porter, L. and Cambardella, C. (1983). Hyphal uptake and transport of nitrogen from two ^{15}N-labelled sources by *Glomus mosseae,* a vesicular-arbuscular mycorrhizal fungus. *New Phytologist* 95, 381-396.

Armstrong, R. D., Helyar, K. R. and Christie, E. K. (1992). Vesicular arbuscular mycorrhizae in semi-arid pastures of south-west Queensland and their effect on growth responses to phosphorus fertilizers by grasses. *Australian Journal of Agriculture Research* 43, 1143–1155.

Aziz, T., and Habte, M. (1989). Influence of inorganic N on mycorrhizal activity, nodulation, growth of *Leucaena leucocephala* in oxisol subject to simulated erosion. *Communications in Soil Science and Plant Analysis* 20, 239–251.

Baker, A., Sprent, J. I. and Wilson, J. (1995). Effects of sodium chloride and mycorrhizal infection on the growth and nitrogen fixation of *Prosopis juliflora. Symbiosis* 19, 39–51.

Barea, J. M. and Azcón-Aguilar, C. (1983). Mycorrhizas and their significance in modulating nitrogen-fixing plants. *Advances in Agronomy* 36, 1–54.

Barea, J. M., Azcón, R. and Azcón-Aguilar, C. (1992). Vesicular-arbuscular mycorrhizal fungi in nitrogen-fixing systems. *Methods in Microbiology* 24, 391–416.

Barea, J. M., Azcón-Aguilar, C. and Azcón, R. (1987). Vesicular-arbuscular mycorrhizae improve both symbiotic N$_2$ fixation and N uptake from soil as assessed with a ^{15}N technique under field conditions. *New Phytologist* 106, 717–725.

Baylis, G. T. S. (1967). Experiments on the ecological significance of phycomycetous mycorrhizas. *New Phytologist* 66, 231–243.

Bethlenfalvay, G. J. (1992). Vesicular-arbuscular mycorrhizal fungi in nitrogen-fixing legumes: problems and prospects. *Methods in Microbiology* 24, 375–389.

Bolan, N. S., Robson, A. D. and Barrow, N. J. (1984a). Increasing phosphorus supply can increase the infection of plant roots by vesicular-arbuscular mycorrhizal fungi. *Soil Biology and Biochemistry* 16, 419–420.

Braunberger, P. G., Abbot, L. K. and Robson, A. D. (1996). Infectivity of arbuscular mycorrhizal fungi after wetting and drying. *New Phytologist* 134, 673–684.

Braunberger, P. G., Abbott, L. K. and Robson, A. D. (1994). The effect of rain in the dry season on the formation of vesicular-arbuscular mycorrhizas in the growing season of annual clover-based pastures. *New Phytologist* 127, 107–114.

Bruce, A., Smith, S. E. and Tester, M. (1994). The development of mycorrhizal infection in cucumber: effects of P supply on root growth, formation of entry points and growth of infection units. *New Phytologist* 127, 507–514.

Carey, P. D., Fitter, A. H. and Watkinson, A. R. (1992). A field study using the fungicide benomyl to investigate the effect of mycorrhizal fungi on plant fitness. *Oecologia* 90, 550–555.

Carpenter, A. T. and Allen, M. F. (1988). Responses of *Hedysarum boreale* Nutt. to mycorrhizas: plant and soil nutrient changes in a disturbed shrub-steppe. *New Phytologist* 109, 125–132.

Chambers, C. A., Smith, S. E. and Smith, F. A. (1980). Effects of ammonium and nitrate ions on mycorrhizal infection, modulation and growth of *Trifolium subterraneum*. *New Phytologist* 85, 47–62.

Clapp, J. P., Young, J. P. W., Merryweather, J. and Fitter, A. H. (1995). Diversity of fungal symbionts in arbuscular mycorrhizas from a natural community. *New Phytologist* 130, 259–265.

Coleman, D. C., Reid, C. P. and Cole, C. P. (1983). Biological strategies in nutrient cycling in soil systems. *Advances in Ecological Research* 13, 1–55.

Cui, M. and Caldwell, M. M. (1996a). Facilitation of plant phosphate acquisition by arbuscular mycorrhizas from enriched soil patches. I. Roots and hyphae exploiting the same soil. *New Phytologist* 133, 453–460.

Cui, M. and Caldwell, M. M. (1996b). Facilitation of plant phosphate acquisition by arbuscular mycorrhizas from enriched soil patches. II Hyphae exploiting root-free soil. *New Phytologist* 133, 461–467.

Cui, M., and Nobel, P. S. (1992). Nutrient status, water uptake and gas exchange for three desert succulents infected with mycorrhizal fungi. *New Phytologist* 122, 643–649.

Daft, M. J. and Nicolson, T. H. (1966). Effect of *Endogone* mycorrhiza on plant growth. *New Phytologist* 65, 343–350.

Dhillion, S. S. and Zak, J. C. (1993). Microbial dynamics in arid ecosystems: desertification and the potential role of mycorrhizas. *Revista Chilena de Historia Natural* 66, 253–270.

Dixon, R. K., Garg, V. K. and Roa, M. V. (1993). Inoculation of *Luecaena* and *Prosopis* seedlings with *Glomus* and *Rhizobium* species in saline soil: rhizosphere relations and seedling growth. *Arid Soil Research Rehabilatation* 7, 133–144.

Fitter, A. H. (1988). Water relations of red clover, *Trifolium pratense* L., as affected by VA mycorrhizal infection and phosphorus supply before and during drought. *Journal of Experimental Botany* 39, 595-604.

Fitter, A. H. and Merryweather, J. W. (1992). Why are some plants more mycorrhizal than others? An ecological enquiry. *In* D. J. Read, D. H. Lewis, A. H. Fitter and I. J. Alexander (eds.), *Mycorrhizas in Ecosystems.* CAB International, Wallingford, UK, pp. 26–36.

Frey, B. and Schüepp, H. (1992). Transfer of symbiotically fixed nitrogen from berseem (*Trifolium alexandrium* L.). to maize via vesicular-arbuscular mycorrhizal hyphae. *New Phytologist* 122, 447–454.

Gerdemann, J. W. (1964). The effect of mycorrhizas on the growth of maize. *Mycologia* 56, 342–349.

Giovannetti, M., Sbrana, C., Avio, L., Citemesi, A. S. and Logi, C. (1993a). Differential hyphal morphogenesis in arbuscular mycorrhizal fungi during pre-infection stages. *New Phytologist* 125, 587–593.

Jasper, D. A., Abbott, L. K. and Robson, A. D. (1989a). Soil disturbance reduces the infectivity of external hyphae of vesicular-arbuscular mycorrhizal fungi. *New Phytologist* 112, 93–99.

Jasper, D. A., Abbott, L. K. and Robson, A. D. (1989b). Hyphae of a vesicular-arbuscular mycorrhizal fungus maintain infectivity in dry soil except when the soil is disturbed. *New Phytologist* 112, 101–107.

Jasper, D. A., Abbott, L. K. and Robson, A. D. (1993). The survival of infective hyphae of vesicular-arbuscular mycorrhizal fungi in dry soil: an interaction with sporulation. *New Phytologist* 124, 473–479.

Johansen, A., Jakobsen, I. and Jensen, E. S. (1992). Hyphal transport of ^{15}N-labelled nitrogen by a vesicular-arbuscular mycorrhizal fungus and its effect on depletion of inorganic soil N. *New Phytologist* 122, 281–288.

Johansen, A., Jakobsen, I. and Jensen, E. S. (1993). Hyphal transport by a vesicular-arbuscular mycorrhizal fungus associated with *Trifolium subterraneum* L. 3. Hyphal transport of ^{32}P and [15]N. *New Phytologist* 124, 61–68.

Khaliel, A. S. (1989). Mycorrhizal status of some desert plants and correlation with edaphic factors. *Transactions of the Mycological Society of Japan* 30, 231–238.

Knight, W. G., Allen, M. F. Jurniak, J. J. and Dudley, L. M. (1989). Elevated carbon dioxide and solution phosphorus is soil with vesicular arbuscular mycorrhizal western wheatgrass. *Soil Science Society of America Journal* 53, 1075–1082.

Koide, R. (1985). The effect of VA mycorrhizal infection and phosphorus status on sunflower hydraulic and stomata properties. *Journal of Experimental Botany* 36, 1087–1098.

Koide, R. (1993). Physiology of the mycorrhizal plant. *Advances in Plant Pathology* 9, 33–54.

Koide, R. T. and Lu, X. (1992). Mycorrhizal infection of wild oats: maternal effects on offspring growth and reproduction. *Oecologia* 90, 218–226.

Koide, R. T and Lu, X. (1995). On the cause of offspring superiority conferred by mycorrhizal infection of *Abutilon theophrasti*. *New Phytologist* 131, 435-441.

Mathew, J., Shankar, A., Neeraj, A., and Varma, A (1991). Glomalaceous fungi associated with spineless cactus : A fodder supplement in deserts. *Transaction of the Mycological Society of Japan* 32, 225–234.

McGee, P.A. (1989). Variation in propagule numbers of vesicular-arbuscular mycorrhizal fungi in semi-arid soil. *Mycological Research* 92, 28–33.

McGonigle, T. P. and Fitter, A. H. (1990). Ecological specificity of vesicular arbuscular mycorrhizal associations. *Mycological Research* 94, 120–122.

Michelsen, A., and Rosendahl, S. (1990). The effect of VA mycorrhizal fungi, phosphorus and drought stress on the growth of *Acacia nilotica* and *Leucaena leucocephala* seedlings. *Plant and Soil* 124, 7–13.

Neeraj, Shankar, A., Mathew, J. and Varma, A. (1991). Occurrence of vesicular-arbuscular mycorrhizae with Amaranthaceae in soils of the Indian semi-arid region. *Biology and Fertility of Soils* 11, 140–144.

Nelsen, C. E. and Safir, G. R. (1982). The water relations of well watered, mycorrhizal and non-mycorrhizal onion plants. *Journal of the American Society of Horticulture Science* 107, 271–274.

Newman, E. I. (1988). Mycorrhizal links between plants: their functioning and ecological significance. *Advances in Ecological Research* 18, 243–270.

Newsham, K. K., Fitter, A. H. and Watkinson, A. R. (1994). Root pathogenic and arbuscular mycorrhizal fungi determine fecundity of asymptomatic plants in the field. *Journal of Ecology* 82, 805–814.

Norris, J. R., Read, D. J. and Varma, A. K., eds. (1992). *Methods in Microbiology*, Vol. 24. Academic Press, London.

Rachel, E. K., Reddy, S. R. and Reddy, S. M. (1989). VA mycorrhizal colonization of different angiospermic plant species in the semi-arid soils of Andhra Pradesh (India). *Acta Botanica Indica* 17, 225–228

Read, D. J., Francis, R. and Finlay, R. D. (1985). Mycorrhizal mycelia and nutrient cycling in plant communities. *In* A. H. Fitter, D. Atkinson, D. J. Read and M. B. Usher (eds.), *Ecological Interactions in Soil.* Blackwell Scientific, Oxford, UK, pp. 193–217.

Read, D. J., Lewis, D. H., Fitter, A. H. and Alexander, I. J. eds. (1992). *Mycorrhizas in Ecosystems.* CAB International, Wallingford, UK.

Roldan, A., Diaz, G. and Albaladejo, J. (1992). Effect of VAM-fungal inoculation on growth and phosphorus uptake of two *Hedysarum* species in a xeric torriorthent soil from southeast Spain. *Arid Soil Research and Rehabilitation* 6, 33–39.

Sanders, E. E. (1975). The effect of foliar-applied phosphate on the mycorrhizal infections of onion roots. *In* F. E. Sanders, B. Mosse and P. B. Tinker (eds.), *Endomycorrhizas.* Academic Press, London, pp. 261–276.

Sanders, F. E. and Tinker, P. B. (1971). Mechanism of absorption of phosphate from soil by *Endogone* mycorrhizas. *Nature* 233, 278–279.

Sanders, F. E. and Tinker, P. B. (1973). Phosphate flow into mycorrhizal roots. *Pesticide Science* 4, 385–395.

Sanders, F. E., Tinker, P. B., Black, R. L. B. and Paimerley, S. M. (1977). The development of endomycorrhizal root systems. 1. Spread of infection and growth promoting effects with four species of vesicular-arbuscular mycorrhizas. *New Phytologist* 78, 257-268.

Sanders, I. R., Alt, M., Groppe, K., Boller, T. and Wiemken, A. (1995). Identification of ribosomal DNA polymorphisms among and within spores of the Glomales: application to studies on the genetic diversity of arbuscular mycorrhizal fungal communities. *New Phytologist* 130, 419–427.

Schlesinger, W. H., Reynolds, J. F., Cunningham, G. L., Huenneke, L. F., Jarrell, W. M., Virginia, R. A. and Whitford, W. W. (1990). Biological feedbacks in global desertification. *Science* 247, 1043–1048.

Sieverding, E. (1990). Ecology of VAM fungi in tropical agroecosystems. *Agriculture Ecosystem and Environment* 29, 369–390.

Smith, S. E. and Read, D. J. (1997). *Mycorrhizal Symbiosis*, 2nd ed. Academic Press, San Diego.

Stahl, P. D., and Smith, W. K. (1984). Effects of different geographical isolates of *Glomus* on the water relations of *Agropogon smithii*. *Mycologia* 76, 261–267.

Stanley, M. R., Koide, R. T. and Shumway, D. L. (1993). Mycorrhizal symbiosis increases growth, reproduction and recruitment of *Abutilon theophrasti* Medic in the field. *Oecologia* 94, 30–35.

Sylvia, D. M. and Neal, L. H. (1990). Nitrogen affects the phosphorus response of VA mycorrhiza. *New Phytologist* 115, 303–310.

Tobar, R., Azcón, R. and Barea, J.-M. (1994). Improved nitrogen uptake and transport from 15N-labelled nitrate by external hyphae of arbuscular mycorrhiza under water-stressed conditions. *New Phytologist* 126, 119–122.

Trent, J. D. Svejcar, A. J. and Bethlenfalvay, G. J. (1993). Growth and nutrition of combinations of native and introduced plants and mycorrhizal fungi in a semi-arid range. *Agriculture Ecosystem and Environment* 45, 13–23.

Vierheilig, H., Alt, M., Mohr, U., Boller, T. and Wiemken, A. (1994). Ethylene biosynthesis and activities of chitinase and β-1-3 glucanase in the roots of host and non-host plants of vesicular-arbuscular mycorrhizal fungi after inoculation with *Glomus mosseae*. *Journal of Plant Physiology* 143, 337–343.

West, H. M., Fitter, A. H. and Watkinson, A. R. (1993a). Response of *Vulpia ciliata ssp. ambigua* to removal of mycorrhizal infection and to phosphate application under field conditions. *Journal of Ecology* 81, 351–358.

West, H. M., Fitter, A. H. and Watkinson, A. R. (1993b). The influence of three biocides on the fungal associates of the roots of *Vulpia ciliata ssp. ambigua* under natural conditions. *Journal of Ecology* 81, 345–350.

West, N. E. and Klemmedson, J. O. (1978). Structural distribution of nitrogen in desert ecosystems. *In* N. E. West & Skujins J. J. (ed.), *Nitrogen in Desert Ecosystems*. Dowden, Hutchinson and Ross, Stroudsburg, PA, pp. 1–19.

Whitford, W. G. (1986). Decomposition and nutrient cycling in deserts. *In* W. G. Whitford (eds.), *Pattern and Process in Desert Ecosystems*. University of New Mexico Press, Albuquerque, pp. 93–118.

Wilson, J., Ingleby, K., Mason, P. A., Ibrahim, K. and Lawson, G. J. (1992). Long term changes in vesicular-arbuscular mycorrhizal spore populations in *Terminalia* plantations in Côte d'Ivoire. *In* D. J. Read, D. H. Lewis, A. H. Fitter and I. J. Alexander (eds.), *Mycorrhizas in Ecosystems*. CAB International, Wallingford, UK, pp. 268–275.

Wilson, S. A. (1973). Endomycorrhizae on desert plants with primary emphasis on *Prosopis juliflora* [M. S.Thesis]. New Mexico State University, Las Cruces.

Zak, J. C., Sinsabaugh, R. and MacKay, W. P. (1995). Windows of opportunity in desert ecosystems: their implications to fungal development. *Canadian Journal of Botany* 73 (Suppl. 1), S1407–1414.

8

Mycorrhizal Endosymbiosis

Marion M. Kyde
The Tulgey Wood, Ottsville, Pennsylvania

Ann B. Gould
Cook College, Rutgers University, New Brunswick, New Jersey

1. INTRODUCTION

Certain specialized members of the Zygomycetes, Ascomycetes, and Basidiomycetes form symbioses with species in most plant families, including many angiosperms, gymnosperms, Pteridophytes, and Thallophytes (Harley and Smith, 1983; Harley and Harley, 1987; Morton, 1988). In 1885, A. B. Frank (1885), while investigating the possibility of introducing truffle cultivation to Russia, first described in detail these associations, which are characterized by recognizable structures, in beech, oak, and conifer roots. Frank coined the term ''mykorrhizen,'' or ''fungus root'' for these structures, and later characterized them as ''ectoptropisch'' (Frank, 1885) when characterized by a conspicuous sheath of fungal tissue around the host root, or ''endotropisch'' where the external sheath was lacking (Harley and Smith, 1983; Harley, 1985; Morton, 1988; Schenck, 1982). Even earlier, the Polish botanist Kamienski (1882) described a mutualistic relationship between an unknown fungus and an achlorophyllous plant, *Monotropa hypopitys* L. Kamienski commented on the similarity between the ''deformed'' roots of *Monotropa* and those of nearby trees. Unlike Frank, however, who specifically pointed out the ubiquity and necessity of the mutualism between fungus and tree (Frank, 1885), Kamienski assumed that the relationship was a parasitic one.

Frank's ''endotropisch'' fungi had received some prior attention from other researchers: in 1842, Nageli (1842, as cited in Trappe and Berch, 1985) described hyphae and what are now known as arbuscules from the roots of iris; Mollberg (1884, as cited in Trappe and Berch, 1985) noted the presence of mycorrhizal-like fungi in orchid roots; and in separate studies Treub (1884, as cited in Trappe and Berch, 1985) and Bruchman (1885, as cited in Trappe and Berch, 1985) described apparent mycorrhizal fungi in lycopods and ferns, and commented that their presence caused little or no harm to the hosts. In 1889, a student of Frank's, A. Schlicht, presented the first definitive morphological descriptions and ecological observations of the "endotropisch" (now known as the Glomalean) mycorrhizal relationship (Schlicht, 1889, as cited in Mosse,1985). He stated that these fungi always entered from the soil; caused no change in root anatomy (affected roots remained vigorous and healthy); were rarely septate and contained oil droplets; produced cauliflower-like shapes inside root cells; were confined to the primary cortex and did not colonize the main root; rarely colonized water plants but might be found in plants occasionally flooded; and represented several species. As early as 1900 it was known that Glomalean mycorrhizas were the most common type of mycorrhizal association (Mosse, 1985).

By the turn of the 20th century, the morphology of vesicles and arbuscules had been described, and potassium hydroxide was used as a staining technique for colonized roots (Mosse, 1985). Most studies in the first half of the 20th century focused on the occurrence of Glomalean relationships with different families and species, growth comparisons between mycorrhizal and nonmycorrhizal plants, and attempts to isolate the fungi in culture, which were unsuccessful. However, early researchers were often working with the entire range of mycorrhizal types and looking for universal explanations to cover all of the observed phenomena (Mosse, 1985). Whereas it was evident that ectomycorrhizal and orchidaceous associations were beneficial, the value of the Glomalean association was less well accepted; indeed, it was generally believed that the fungi involved were controlled parasites (Mosse, 1985) and that their taxonomic status was uncertain. For those reasons, and the fact that Glomalean fungi could not be cultured, interest in them languished, and in the early 1950s there were fewer than 10 researchers working with this group of mycorrhizal fungi (Schenck, 1985).

Two papers published during the 1950s produced the keys that unlocked the door to mycorrhizal research. The first was Barbara Mosse's paper in 1953 in which she demonstrated the establishment of a pure pot culture of mycorrhizae on strawberries using spores of a fungus she found attached to strawberry roots in the field (Mosse, 1953). In the second seminal paper, Gerdemann (1955) linked spores isolated from agricultural soils to mycorrhizal colonization and arbuscule formation in four crop species in pot culture. Gerdemann's use of a method borrowed from nematologists, known as wet sieving and decanting, provided the key to obtaining phycomycetous (*sic*, now Glomalean) spores in quantity from soils to use in laboratory experiments. With the publication of these two influential papers, endomycorrhizal researchers began to proliferate, and their ranks swelled nearly 50-fold from the 1950s to the present.

2. TYPES OF MYCORRHIZAE

Allen (1991) provides a clear and succinct working definition for mycorrhizae: "a mutualistic symbiosis between plant and fungus localized in a root or root-like structure in which energy moves primarily from plant to fungus and inorganic resources move from fungus to plant." Within this definition, there are four major kinds of mycorrhizas: endo-(Glomalean), ecto-, ericoid, and orchidaceous mycorrhizae, with associated subtypes (Harley and Smith, 1983).

Endo-(Glomalean) mycorrhizae [also previously referred to as endo-, arbuscular, vesicular, or vesicular-arbuscular mycorrhizae (VAM)] are formed between aseptate phycomycetes classified in the Glomales (Morton and Benny, 1990) and plants within the Bryophytes, Pteridophytes, gymnosperms, and angiosperms. Exceptions include the Pinaceae, Betulaceae, Fagaceae, Salicaceae, most Cruciferae, Orchidaceae, and Ericales (Berch et al., 1985). In these mycorrhizae, fungal hyphae grow between and within the cells of the plant root cortex, and may form branched arbuscules, round to oval vesicles, and peletons or hyphal coils (Harley and Smith, 1983; Kinden, and Brown, 1975). The arbuscules are considered to be the organs of nutrient exchange between phycobiont and autobiont (Gianinazzi et al., 1979). Although transitory in nature, their presence alone is enough to distinguish this kind of mycorrhiza (Morton, 1988). Intraradial vesicles may be variable in size, shape, or frequency (Morton, 1985) or lacking, replaced instead by external auxiliary cells (originally called "vesicles") (Gerdemann and Trappe, 1974). Of all the mycorrhizal types, Glomalean fungi associate with the largest number of species across the plant kingdom (Gerdemann, 1968).

3. ECTOMYCORRHIZAE

Ectomycorrhizae are formed between many basidiomycete genera, a few hypogeous ascomycetes, and the plant families Betulaceae, Salicaceae, Fagaceae, Pinaceae and Myrtaceae (Berch et al., 1985). Ectomycorrhizae are characterized by the formation of two structures: the mantle and the Hartig net. The mantle (a sheath) of fungal tissue closely encircles the rootlet, is thick (20–100 μm), and may have two distinct layers (Harley and Harley, 1987). An intercellular network of hyphae that surrounds but does not penetrate the cells of the root cortex (Harley and Smith, 1984) characterizes the Hartig net.

Thirty-three families of basidio- and ascomycetous fungi are known to contain ectomycorrhizal members (Miller, 1984). In addition, two species in the Zygomycetes, *Endogone lactiflua* Berch, Berk. and Broome, and *E. flamnicorona* Trappe and Gerd. form ectomycorrhizal relationships (Gerdemann and Trappe,

1974; Miller, 1984). In these associations the sheath is very thin and the Hartig net is well developed (Harley and Smith, 1983).

4. ERICOID

Mycorrhizas of Ericales include the ericoid and arbutoid types (Read, 1984a). Ericoid mycorrhizas are found in the Ericaceae, Epacridaceae, and Empetraceae, the roots of which are small in diameter and hair-like (Harley and Harley, 1987). In an ericoid relationship, the fungus forms a loose sheath of tissue from which hyphae penetrate the cells behind the root apex to form tight, intracellular hyphal coils (Harley and Harley, 1987). Among the fungi thus far proven to form ericoid relationships are *Pezizella ericae* Read, an ascomycete (Read, 1974), *and Oidiodendron griseum* Robak, an ascomycete anamorph, isolated from both *Vaccinium* and *Gaultheria* species (Xiao and Berch, 1992). Other fungi from basidiomycete (particularly *Clavaria* ssp.) (Harley and Smith, 1983) and ascomycete taxa have been implicated (Harley and Harley, 1987).

Arbutoid mycorrhizae, found mainly on *Arbutus* and *Arctostaphylos* species, combine characteristics of both ecto- and endomycorrhizae. They possess a well-developed fungal sheath, an extensive Hartig net, and copious intracellular penetration of root cells as well (Harley and Smith, 1983; Harley and Harley, 1987). The fungal partners are basidiomycetes, including species of *Cortinarius*, *Laccaria*, *Thelephora*, and *Pisolithus*, all of which also form typical ectomycorrhizae with forest trees (Zak, 1974, 1976a, b).

A special case exists for mycorrhizae on plants in the Monotropaceae. Achlorophyllous *Monotropa* spp. are always associated with the roots of autotrophic plants through connections made by fungal rhizomorphs (Duddridge and Read, 1982). Although the mycorrhizae on *Monotropa* roots appear superficially similar to both ecto- and arbutoid mycorrhizae, they are characterized by the formation of fungal pegs, which never actually penetrate the cortical cells but serve as sites of nutrient transfer (Duddridge and Read, 1982). Duddridge and Read considered this relationship to be sufficiently different from the others to refer to it as "monotropoid." At least one discomycete, *Hymenocyphus ericae* (Read) Korf and Kernan, forms monotropoid mycorrhizae (Kernan and Finocchio, 1983). *M. uniflora* L. may be associated with the basidiomycete *Armillaria mellea* (Vahl: Fr.) P. Kumm. in a similar manner (Campbell, 1971).

5. ORCHIDACEOUS MYCORRHIZAE

Orchidaceous mycorrhizae form between members of the Orchidaceae and certain basidiomycetes which have the ability to break down cellulose and sometimes lignin (Harley and Smith, 1983). In nature, all orchids pass through a period of development in which they are dependent on external sources of carbon. In a surprising reversal of roles, the fungi in orchid mycorrhizae provide this needed carbon to the orchid protocorm (Harley and Harley, 1987). Fungi known to form orchidaceous mycorrhizae include members of the form genus *Rhizoctonia* (Currah et al.,1988), perfect stages of which include *Thanetophorus, Sebacina, Ceratobasidium, Upsilonidium*, and *Tulasnella* (Warcup, 1981). Other fungi known to form orchid mycorrhizae include the very common saprophytes, *Coriolus versicolor* (= *Trametes versicolor* L.:Fr.) Pilat, *Fomes* spp, and *Armillaria mellea*, a serious tree pathogen (Harley and Smith, 1983).

Reports of other types of mycorrhizae on trees, variously called "ectendomycorrhizas," "pseudomycorrhizas" and "E-strain mycorrhizas" exist in the literature. These mycorrhizae exhibit characteristics of both ecto- and endomycorrhizae and seem related in several ways to ericoid, arbutoid, and orchidaceous mycorrhizae (Harley and Smith, 1983). In many cases, the same fungus will form ectomycorrhizae on spruce and ectendomycorrhizae on pine. In addition, several ectomycorrhizal basidiomycetes have been shown to form ectendomycorrhizae on *Arbutus* and *Arctostaphylos* species (Zak, 1974, 1976a, b). Ascomycete genera are also able to form ectendomycorrhizae; in 1984 a new species of operculate discomycete was obtained from pot cultures of red pine (*Pinus resinosa* Ait) inoculated with asexual spores of an E-strain fungus. This fungus not only was able to reproduce identical chlamydospores in culture, but could also form ectendomycorrhizas with red pine (Yang and Wilcox, 1984). In 1985, three additional previously unknown hyphomycetes were isolated and identified from pseudo- and ectendomycorrhizae in conifers (Wang and Wilcox, 1985). Ongoing research by Read and others will probably clear up some of this confusion.

6. MORPHOLOGY OF GLOMALEAN MYCORRHIZAL FUNGI

Glomalean fungi pass through several structural phases during their life cycles. Within the root they form unbranched intracellular hyphae and hyphal coils, intercellular hyphae, highly dichotomously branched arbuscules, and may also form vesicles and internal spores (Bonfante-Fasolo, 1984). Externally they form long aseptate hyphae, appressoria, spores, and auxiliary cells (called vesicles in the older literature). Although colonization by Glomalean fungi usually results in very little or no modification to the external morphology of host roots (Gerdemann, 1968), in some plant species mycorrhizal roots may have a yellow color (Gerdemann, 1961), or may be thicker (Mosse, 1959), more brittle, contorted, branched, or have fewer root hairs (Gerdemann, 1968).

6.1. External Hyphae

The external hyphae of Glomalean fungi surround root hairs with extensive loose networks that may penetrate into the soil as far as 1 cm (Mosse, 1959), and may be 80–134 times longer than the length of colonized root (Bonfante-Fasolo, 1984). External hyphae also grow tightly oppressed to the root epidermis (Gerdemann, 1968) but do not form a solid mantle around the root, as do ectomycorrhizal fungi. External hyphae are dimorphic: larger strands are coarse, irregular, and thick-walled, whereas secondary lateral branches are smaller in diameter and thin-walled (Brown and King, 1982; Gerdemann, 1968). The larger coarser strands, or runner hyphae, seek out new infection sites, either on the same root or on roots of other plants, by forming hyphal bridges (Freise and Allen, 1991). Branching angles of runner hyphae are about 45 degrees (Baco et al., 1998). These secondary absorptive hyphae may bear numerous short arbuscule-like branches and form anastomoses (Chabot et al., 1992; Friese and Allen, 1991). Branching patterns are always dichotomous and may include eight orders of branching (Friese and Allen, 1991). In the Graminae, at least, the lateral branches do not persist for very long and become septate as they die (Nicholson, 1959). Freise and Allen found the lateral branches to die back within 5–7 days; as hyphal networks die back, others form along the root (Friese and Allen, 1991). External hyphae bear spores, either singly or in loose sporocarps, and in the Gigasporineae Morton and Benny produce auxiliary cells as well (Morton and Benny, 1990). Recently developed monoxenic culture techniques for Glomalean fungi are providing even more detailed information about their external architecture. Spore production, for example, does not occur equally on all hyphal branches of a Glomalean rhizosphere colony but appears to be associated with special dichotomous highly branched "horsetail-like" structures that form following extensive substrate colonization (Baco et al, 1998).

6.2. Appressoria

Upon contact with a root surface, external hyphae form appressoria behind the meristematic region at the root tip (Bonfante-Fasolo, 1984). The number of entry points may vary; in one report the number ranged between 2.6 and 21.1 per mm of root length (Mosse, 1959). On the other hand, Nicholson in his 1959 study of mycorrhizae in the Graminae found as few as six entry points per cm.

6.3. Auxiliary Cells

Auxiliary cells are formed only by the Gigasporaceae (Morton 1990), which includes the modern genera *Gigaspora* and *Scutellospora*. As described by Gerdemann in 1955 (Gerdemann, 1955), auxilliary cells appear singly or in clusters on coiled hyphae in the soil after a plant root has been colonized. They are thin-walled, aseptate, contain oil droplets (Nicholson and Gerdemann, 1968), and are shown to contain high concentrations of phosphorus and calcium with lesser amounts of several other elements (Jabaji-Hare et al.,1986). Although they have been postulated to serve as temporary storage organs for nutrients (Gerdemann and Trappe, 1974; Nicholson and Gerdemann, 1968), Morton and Benny (1990) state that there is no evidence that they are substitutes for the vesicles found in other families, but resemble more closely relict reproductive spores. Auxiliary cells break down before reproductive spores are mature (Nicholson and Gerdemann, 1968).

6.4. Spores

The size, color, shape, wall structure, subtending hyphal morphology, staining reaction, and germination characteristics of Glomalean spores are the characters used to speciate this group of mycorrhizal fungi (Morton, 1988, 1989). Spores may be produced inside or outside the roots; they may be solitary, in loose aggregations or in more or less organized sporocarps (Morton, 1988). Spore color may range from hyaline or white through yellow and shaded brown to black; color is variable within a species and may depend on the spore age and condition (Morton, 1988). Although *Scutellospora* and *Gigaspora* spores are usually larger than those of the other genera, there is a great deal of overlap in size, both within and between genera and species. Spore size, therefore, is not considered to be a very useful taxonomic character (Morton, 1988). Diagrams of spore diameter frequency, such as those published by Berch and Koske (1986), Koske and Gemma (1986), and Bentivenga and Morton (1995), when developed for each known and new species, are better indicators of species than the size ranges of spores alone.

The shape of Glomalean spores may range from globose and subglobose (most common) to ellipsoid, pyriform, or to the more unusual shapes such as clavate, reniform, or oblong (Morton, 1988). Except for the sporocarpic species, unusual spore shapes are rare. Of all spore characteristics, however, spore wall structure is the main character used to speciate.

6.5. Glomalean Fungi

Differences in wall type, ornamentation, number, and staining reactions, especially in Melzer's reagent, are the details that make taxonomic diagnoses possible (Morton, 1988). In addition, *Glomus* spores form on cylindrical to flared sporogenous hyphae, which may be single, or loosely to tightly aggregated; *Acaulospora* and *Entrophospora* spores form laterally from a hypha terminating in a sporiferous saccule, while spores of *Gigaspora* and *Scutellospora* expand from a bulbous sporogenous cell (Morton and Bentivenga, 1994).

6.6. Internal Hyphae

Non-septate internal hyphae grow inter- and/or intracellularly in the root cortex of the phytobiont (Brown and King, 1982; Gerdemann, 1968). They are variable and irregular in shape; complex coils (peletons), anastomoses, and loops are common (Gerdemann,1968). Though coiled hyphae are produced only occasionally in *Gigaspora* species (Bentivenga and Morton, 1995), *Glomus mosseae* (Nicolson & Gerdemann) Gerdemann & Trappe (Brown and King, 1982) and *Endogone fascicula* Thaxter [now *Glomus facsiculatum* (Thaxter) Gerdemann & Trappe emend. Walker and Koske] endomycorrhizae are composed largely of coils, especially in tulip tree roots (Gerdemann, 1965). Some data suggest that the host may determine the spacial distribution of internal hyphae; for example, *Glomus diaphanum* Morton and Walker produced long parallel strands in red clover, but mostly coils in corn, sudangrass, and fescue (Morton, 1985).

6.7. Arbuscules

In his 1988 taxonomy, Morton states that "arbuscular development alone . . . is enough of a distinction to classify genera of VA mycorrhizal fungi, apart from Endogone" (Morton, 1988). Arbuscules are dichotomously branched structures formed within cortex cells by hyphae from adjoining cells (Brown and King, 1982; Cox and Sanders, 1974; Gerdemann, 1968) and are the structures primarily responsible for bidirectional transport of nutrients between symbionts (Brown and King, 1982; Gianinazzi et al., 1979). As hyphae grow through the roots, arbuscules are progressively produced within cells of the inner cortex and collapse into dense masses within a few days after formation (Brown and King, 1982). The average lifetime for an arbuscule is 4–10 days (Sanders et al., 1977).

6.8. Vesicles

Vesicles produced by Glomalean fungi may be inter- or intracellular, and terminal or intercalary (Brown and King, 1982). They are ovate to spherical or irregular and contain droplets of oil (Brown and King, 1982; Gerdemann, 1968). They develop after the formation of arbuscules (Morton, 1985) and are thought to be storage organelles (Brown and King, 1982; Gerdemann, 1968) Some vesicles develop thick walls and appear identical with the spores for that species (Gerdemann, 1968). They may even be used successfully as inoculum. Taxa in *Gigaspora* and *Scutellospora* do not form vesicles (Morton, 1988).

7. TAXONOMY

Endomycorrhizal fungi are Zygomycetes that were placed into their own order, the Glomales, in 1990 (Morton and Benny, 1990). They have presented taxonomic problems almost since they were first discovered (Gerdemann and Trappe, 1974). The first person to name an arbuscular mycorrhizal fungus was Dangeard in 1900, who described a typical mycorrhiza on poplar and called it *Rhizophagus populinus* (Gerdemann, 1969). He had no idea of its true function. In 1922, Roland Thaxter monographed the Endogonaceae and placed mycorrhizal fungi with zygospores, chlamydospores, or sporangiospores in sporocarps into one of four different genera (*Endogone, Sphaerocreas, Sclerocystis,* and *Glaziella*) (Thaxter, 1922). One year later, Peyronel traced the hyphae of an endophyte to endogonaceous sporocarps in the soil (Gerdemann, 1969; Harley and Smith, 1983; Trappe and Schenck, 1982). He later identified the sporocarpic fungus as *Endogone vesiculifera* Thaxter [now *Glomus vesiculiferum* (Thaxter) Gerdemann & Trappe] and *E. fuegiana* Spegazzini [now *Glomus fuegianum* (Spegazzini) Trappe & Gerdemann] (Gerdemann, 1968); these species fit into the family as described by Thaxter. In 1939, Butler concluded that *Rhizophagus* species such as those described by Dangeard were imperfect members of the Endogonaceae because of the close resemblance of vesicles on the hyphae to *Endogone* chlamydospores (Gerdemann, 1961; Harley and Smith, 1983).

Following publication of the two aforementioned pivotal papers by Mosse (1953), who demonstrated that species in the Endogonaceae can and do form Glomalean mycorrhizal relationships, and Nicholson and Gerdemann (1958) who described an easy isolation technique for spores from soil, more and more endomycorrhizal species were described. Since many of them did not produce sporocarps and thus could not be placed in Thaxter's *Endogone*, Nicholson and Gerdemann emended the genus to include nonsporocarpic types (Nicholson and Gerdemann, 1968). This, they felt, was justified by the fact that at least two species, *E. fasiculata* Thaxter and *E. microcarpa* (Tul. & Tul.) Tul.& Tul. produced spores both within sporocarps (endocarpic) and singly on isolated hyphae (ectocarpic). They described four new endogonaceous species which included *E. mosseae* Nicolson & Gerdemann (to honor B. Mosse's work), *E. heterogamma* Nicolson & Gerdemann, *E. gigantea* Nicolson & Gerdemann, and *E. calospora* Nicolson & Gerdemann. These latter three species were very different from previously described zygosporic *Endogone* in two ways: (1) the very large "zygospores" were produced on single suspensors instead of two; and (2) they produced vesicles on coiled hyphae singly or in clusters.

In 1974 Gerdemann and Trappe published the first formal taxonomic work (Bentivenga and Morton, 1994) of vesicular arbuscular (or Glomalean) mycorrhizal fungi (Gerdemann and Trappe, 1974). A total of 44 species was included. The authors divided the heterogeneous *Endogone* into four separate genera: *En-*

dogone (11 species, sporocarpic, true zygospores formed above the union of two gametangia—possibly ectomycorrhizal); *Glomus* (19 species, sporocarpic or not, chlamydospores formed terminally on undifferentiated hyphae, endomycorrhizal with internal vesicles and arbuscules); *Gigaspora* (5 species, nonsporocarpic, azygospores borne on a large bulbous suspensor, germination tubes produced directly through wall near base of spore, thin-walled vesicles formed in soil on coiled hyphae, endomycorrhizal; and *Modicella* (sporocarpic with sporangia, mycorrhizal relationships unknown). Gerdemann and Trappe also included three other genera in this group: *Sclerocystis* (chlamydospores arranged around a central plexus); *Glaziella* (extralimital), and a new mycorrhizal genus, *Acaulospora* (azygospores produced laterally on the stalk of a terminal vesicle), which contained two previously undescribed species.

Within the next two years, a third *Acaulospora* species (*A. trappei* Ames & Linderman) was named by Ames and Linderman (1976) in honor of James Trappe. Following this, new Glomalean species were described in increasing numbers from all over the world. In 1977, the first *Acaulospora* species to stain red in Melzer's reagent was described (Trappe, 1977); at that time this reaction was unique to all known Endogonaceae.

In 1979, the genus *Entrophospora* was added to the family because of the unique method of spore formation inside a vesicular stalk. The type species for the genus had been previously described as a *Glomus* (Hall, 1977) and was renamed *Entrophospora infrequens* (Hall) Ames and Schneider comb nov. That same year, *G. gerdemanii* Rose, Daniels & Trappe was described and named in honor of Gerdemann (Rose et al., 1979). This species appeared to link the chlamydosporic and azygosporic members of the Endogonaceae because of its very complex wall structure of five walls, two of which were sloughing.

Walker (1979) tentatively added a new genus, *Complexipes moniliformis* Walker, to the family. This genus, separated from other species by its moniliform subtending hypha and highly ornamented outer coat, warranted placement into the Endogonaceae by the resemblance of its spores and arbuscule-like structures to other family members. Unlike the other endogonaceous species, however, *Complexipes* was sparsely septate and formed ectendomycorrhizae with pines.

In the same year, to accommodate the new taxa formally and informally described following Gerdemann and Trappe's 1974 monograph, Hall and Fish (1979) published a key to the Endogonaceae. Ninety-four species, named and unnamed, were included in the key, including 21 species from Florida enumerated that year by Nicolson and Schenck (1979). Two years later, Walker and Trappe (1981) produced a useful key to the *Acaulospora* and introduced a new species, *A. spinosa* Walker & Trappe, which brought the total for that genus to eight. Schenck and Smith (1982) widened the known distribution of three *Acaulospora* species and described six new *Glomus* and one *Gigaspora* species from Florida; they also also published a synoptic key to all 34 known endomycorrhizal fungi from Florida. In 1982, Trappe (1982) removed *Modicella* and *Complexipes* from the Endogonaceae, the former because it forms only sporangia and is saprobic, and the latter because it was proven to be a discomycete (Danielson, 1982). To help workers just entering the field, he provided keys to eight *Acaulospora*, 17 *Gigaspora*, 40 *Glomus,* and all the known species of *Entrophospora, Glaziella,* and *Sclerocystis.*

As more species were described, the speciation of members within a genus relied more heavily on spore wall characteristics. In an attempt to clarify wall descriptions, Walker (1983) provided a consistent terminology and a visual short-hand (murograph) for future species descriptions. He identified four wall types that could be easily observed by light microscopy in well-preserved spores: unit, laminated, evanescent, and membranous. Although such murographs began to appear in publications of new Endogonaceae (Koske, 1985; Koske and Walker, 1984; Koske and Walker, 1985; Morton and Walker, 1984; Rothwell and Victor, 1984; Schenck et al., 1984; Smith and Schenck, 1985), their use was not consistent (Berch and Trappe, 1985; Trappe et al.,1984).

In 1984, *A. appendicula* Spain, Sieverding et Schenck was described (Schenck et al., 1984). This new species produced both azygospores and *Glomus*-like chlamydospores and appeared to provide a link between *Glomus* and *Acaulospora* genera. In this same publication, Schenck and co-workers published a dichotomous key to the genus, including 13 published species. Berch (1985) emended the genus to include sporocarpic species when she described from Arizona an *Acaulospora* that formed sporocarps. Prior to this paper only *Glomus* species were known to be sporocarpic. In the first published book devoted solely to vesicular-arbuscular (Glomalean) fungi, Hall (1984) provided useful keys to sporocarpic and nonsporocarpic species and genera published up to 1982. Although he made no revisions to the Gerdemann and Trappe system of classification, he clarified the status of many informally described species.

The year 1986 was very productive for researchers in the field of endomycorrhizal fungi. Walker (1986) added a fifth wall type, "coriaceous" (thick, tough, and flexible), to his murographic *Gigaspora*. Other new species described in 1986 expanded the morphological characteristics of vesicular-arbuscular mycorrhizal spores: complete endospore formation in *Glomus* (Miller and Walker, 1986); multiple spore configurations (Schenck et al., 1986); spore formation within dead spores of other Glomalean fungi (Koske and Gemma, 1986); unusual peridium formation (Koske and Walker, 1986a); ornamentation of outer (Koske et al., 1986) and/or inner walls (Morton, 1986); and the new wall types, "expanding" (Berch and Koske, 1986) and "amorphous" (Morton, 1986). Based on differences in spore wall structure, germination characteristics, and ornamentation of auxiliary cells, Walker and Sanders (1986) separated *Scutellospora* from *Gigaspora*. Species that germinated directly through the spore wall and possessed echinulate auxiliary cells were retained in *Gigaspora* (six species), while those with distinct germination shields on interior walls, one or more flexible inner walls, and knobbed or broadly papillate auxiliary cells were placed into the new genus. The genus *Glaziella* was also removed from the Endogonaceae (Gibson et al., 1986) because its spores, previously assumed to be chlamydospores, proved to be ascospores, its hyphae were regularly septate, and the septa were associated with typical ascomycete Woronin bodies.

In 1988 Morton listed 126 species of vesicular-arbuscular (Glomalean) fungi (22 *Acaulospora*, 3 *Entrophospora*, 6 *Gigaspora*, 67 *Glomus*, 9 *Sclerocystis*, and 19 *Scutellospora* species) evaluated then current criteria for classification and identification, and proposed techniques and approaches needed to clarify the taxonomy of the group (Morton, 1988). Shortly thereafter, Pirozynski and Dalpe (1989) separated *Glomus* and *Sclerocystis* from the Endogonaceae and placed these genera into a new family, the Glomaceae, based on paleontological evidence of intraradial fossilized spores resembling extant *Glomus* spores. This left the classification of endomycorrhizal fungi in a very awkward arrangement with two clearly "vesicular-arbuscular" mycorrhizal genera in one family and other clearly "vesicular-arbuscular" mycorrhizal genera lumped with *Endogone* in another.

In the most recent attempt to organize and understand these fungi, Morton and Benny (1990) published the revised classification scheme that is used today. Since, prior to this, taxonomic relationships were not well enough understood to erect a more natural or phylogenetically based taxonomy, all classification schemes were of necessity somewhat artificial. The scheme of Morton and Benny was meant to address two major guidelines for natural classification: (1) the use of identifying characters with the most physiological importance to the organism; and (2) the demonstration of evolutionary relationships. The authors placed *Endogone* into an order of its own and erected a new order, the Glomales, for endomycorrhizal fungi that formed arbuscules in obligate mutualistic associations with living plants. Two new suborders, Gigasporineae and Glomineae, separated those species that formed auxiliary cells in the soil from those that did not. Into the Glomineae they placed two families separated mainly by their spore formation and wall structure: an emended Glomaceae (containing *Glomus* and *Sclerocystis)* and Acaulosporaceae (containing *Acaulospora* and *Entrophospora*). Into the Gigasporineae they placed the family, Gigasporaceae, with its two genera, *Gigaspora* and *Scutellispora* [*sic*], following in their spelling the recommendation made by Almeida (1989). In 1991, Walker succinctly defended the spelling of *Scutellospora* and emended Morton and Benny's Gigasporaceae to reflect the original generic appellation (Walker, 1991). Compared to *Scutellospora*, *Gigaspora* lacks inner flexible walls and germination shields, and possesses echinulate instead of papillate, knobby, or smooth auxiliary cells. Morton and Benny's classification contains 149 species in one of six genera (of which genera only three were represented in the 1974 monograph of Gerdemann and Trappe).

Other changes in Glomalean taxonomy have since reflected changes in genus and species concepts, but, in general, the emphasis has shifted from alpha taxonomy to systematic analysis. Almeida and Schenck (1990) revised the genus *Sclerocystis*, leaving it with but one species, *S. coremioides* Berk. & Broome, and made synonyms or transferred all others to *Glomus*. Their revision was based on four unique features in the mode of sporocarp formation and spore development in *S. coremioides*. Wu (1993) rejected the concept of *Sclerocystis* as a monotypic genus, citing numerous examples from Taiwan, as well as micrographic evidence that the four features of Almeida and Schenck were not unique to *S. coremioides*. He retained six genera in *Sclerocystis*, and suggested three dimorphic North American *Glomus* species as transitional taxa linking the two genera (Wu, 1993). Inasmuch as several researchers are now focusing on developmental processes and evolutionary relationships rather than static phenotypic differences (Franke and Morton, 1994; Maia et al., 1994; Morton, 1990, 1995; Wu and Sylvia, 1993), uncertainties in the relationship of Glomalean fungi to one another such as this may be resolvable. Using just such techniques, Bentivenga and Morton (1995) recently monographed the *Gigaspora*. The genus and five retained species were redescribed using both morphological and developmental characters to delineate the taxa.

New species of Glomales continue to be discovered (Blaszkowski, 1995; Dalpe et al., 1992; Koske and Gemma, 1995). It seems likely that the taxonomy of the group will continue to be revised and refined in the future because, as Morton (1993) states, "so few areas of the world have been extensively sampled for indigenous species." A species concept based on sexual reproduction is not applicable to Glomalean fungi because sexual reproduction is nonexistent to rare within the group. (Morton, 1993). According to Morton, species definitions in this group can be based only on conserved characters passed on from a unique ancestor to all of its descendants, a situation unlike any other in the fungal kingdom. The current working definition of a glomalean species is "the smallest assemblage of reproductively isolated individuals or populations diagnosed by epigenetic, morphological, or organizational properties of fungal spores that specify a unique genealogical origin and continuity of descent according to the criteria of monophyly" (Morton et al., 1992).

8. ECOLOGY OF GLOMALEAN FUNGAL RELATIONSHIPS

8.1. Origin of Symbiosis

Mutualistic relationships between biological organisms are both important and widespread (Boucher, 1985), but none is more important or ubiquitous than the association between Glomalean mycorrhizal fungi and terrestrial plants (Pirozynski and Dalpe, 1989). These associations are proposed to have made possible the move from sea to land by the first primitive plants (Pirozynski and Malloch, 1975; Pirozynski, 1981). Fossil VAM (vesicular-arbuscular mycorrhizal) fungi appear unquestionably in tracheophytes from the Early Devonian (410–360 myBP) (Pirozynski and Dalpe, 1989; Taylor et al., 1995), placing the origin of the putative symbiosis over 400 million years ago in the Late Silurian (Pirozynski and Malloch, 1975; Stubblefield et al., 1987). Kidston and Lang (1996), however, argue that these early records show degraded plant material where the fungi are present, but intact plant material where fungi are absent, pointing to their role as saprophytic rather than symbiotic.

Pirozynski and Dalpe (1989) erected the Glomaceae to encompass these ancient fossil fungi and extant *Glomus* species, demonstrating their close taxonomic affinities. Fossil representatives of both the Glomineae and Gigasporineae appear in *Antarcticycas* fossils from the Middle Triassic cherts of Antarctica (Phillips and Taylor, 1996) Both suborders were well established in the Triassic period (250 million years BP), which is the period indicated by ribosomal DNA data for the split between suborders (Simon et al., 1993).

8.2. Ecosystem Role

Mycorrhizae exist in almost every ecosystem from aquatic to desert (Allen, 1983; Anderson et al., 1984; Bagyaray et al., 1979; Bethlenfalvay, 1984; Miller, 1979; Pendleton and Smith, 1983; Reeves et al., 1977; Trappe, 1987), from sea level to snow line (Allen and Allen, 1990; Allen et al., 1987; Gemma and Koske, 1988; Gemma et al., 1989; Janos 1987; Koske and Halverson, 1981; Malloch and Malloch, 1981, 1982; Nicolson, 1960; Read and Haselwandter, 1981; St. John, 1980), in grasslands and forests (Berliner and Torrey, 1989; Cook et al., 1988; Ebbers et al., 1987; Hetrick and Bloom , 1983; Hopkins, 1987; Liberta and Anderson 1986; Miller, 1987; Moutoglis and Widden, 1996; Mulahey and Speed 1990), and even on midoceanic islands (Gemma and Koske, 1990; Koske and Gemma, 1990). Of the hosts that form mycorrhizal relationships in different ecosystems, the type of mycorrhizae present varies from 100% endomycorrhizal in semiarid sagebrush communities of Colorado (Reeves et al., 1979) and coastal fixed dune communities in Baja California (Siguenza et al., 1996) to 90% endomycorrhizal in a short grass prairie community (St. John and Coleman, 1983), 80% 4% 3% (endo- ecto- orchidaceous mycorrhizal) in a sugar maple forest (Brundrett and Kendrick, 1988) and 71% 22% endo-ectomycorrhizal) in a temperate mixed deciduous forest in Massachusetts (Berliner and Torrey, 1989). Reports of lower percentages (less than 60%) of endomycorrhizal host plants from the late 1920s (McDougal and Liebtag, 1928) in a similar temperate forest may have been due to the relatively small number of observations or to the less sophisticated diagnostic techniques available prior to the middle of this century.

Habitats without mycorrhizal fungi and mycorrhizae do exist; these include severely disturbed areas, such as the slopes of Mt. St. Helens following the volcanic eruption (Allen, 1987), strip mines (Schramm, 1966), severely eroded sites (Powell, 1980), and tidal sands on coasts and barrier beaches (Siguenza et al., 1996). With these exceptions, the high percentages of mycorrhizal plants in normal ecosystems make it clear that although the formation of mycorrhizae may be costly to a plant in terms of fixed carbon allocation, the value of the symbiosis is worth the cost. Fitter (1991) calls inconceivable the idea that such a longstanding association could be merely neutral and hypothesizes that the benefits to the plant partner accrue when the demand for phosphorus exceeds the capacity of the root system alone to provide it. Allen and Allen (1986) suggest that the symbiosis permits plants to survive periods of acute stress, which can be agents of natural selection.

The role that mycorrhizae play in an ecosystem is based on the nature of this symbiotic association. Symbiosis, or the state of living together, as originally described by de Bary may range from mutualism through commensualism to neutralism or even parasitism and antagonism (Allen, 1991). Though the line between benefit and detriment is not always clear, the mycorrhizal symbiosis is usually positive for both partners (Harley and Smith, 1983). Fitter (1977) demonstrated, however, that mycorrhizae may be selectively detrimental where interspecific competition exists. The difference between the mycorrhizal condition and disease is that the former is the normal state of both partners in their ecosystem (Harley and Smith, 1983) and forms essential links between the inner host plant and the outer environment or ecosystem (Allen, 1991). For example, nutrient-rich patches in the soil, otherwise unavailable to plants, may be linked by mycorrhizal hyphae to their plant partners (St. John and Coleman, 1983). Glomalean hyphae may even form links between neighboring mycotrophic plants in an ecosystem (Chiarello et al., 1982; Read, 1984), permitting more than one plant to benefit from the same nutrient pool. According to Read et al. (1985), seedlings may be able to tap into the carbon transfer between plants to help establish themselves. Read and his associates had earlier shown that seedling infection by Glomalean fungi in grassland communities occurred at a very early developmental stage. They postulated that under conditions of limited water and nutrients, such infection was probably essential for survival. (Read et al., 1976).

A single plant partner may form simultaneous mycorrhizae with different endomycorrhizal fungi (Koske and Halverson, 1981; Molina et al., 1978), which in turn may be symbiotic with other hosts (Heap and Newman, 1980), forming what Harley and Smith (1983) call a possible ''social complex of organisms.'' A new technique developed by Clapp et al. (1995) that uses selective enrichment of amplified DNA to identify to species the AM fungi in roots taken directly from natural communities may be able to prove the existence of these social complexes. Through the use of this technique, for example, Clapp's group demonstrated the simultaneous presence of three different Glomalean genera in endomycorrhizae of bluebells (*Hyacinthoides non-scripta*). Two of these had been identified previously by spore surveys, but the cryptic presence of the third had not been suspected.

Within an ecosystem the microhabitats or niches of plants and their fungal associates are quite different with respect to all parameters (e.g., light, temperature, and moisture). Mycorrhizae are able to alter the "operational niche," the actual niche, and the reproductive ability of the plant partner (MacMahon et al., 1981). All of these changes affect the plant's fitness or ability to survive and reproduce successfully. According to Allen (1991), mycorrhizae may alter the host niche by making available resources that were previously unavailable and by increasing the input from currently available sources. Harley and Smith (1983) cite numerous studies that relate the direct transport of labeled nutrients such as phosphorus, calcium, zinc, and sulfur from soil to host by Glomalean fungi. Evidence is also available that depicts the transport of water from soil to host through hyphae (Allen, 1982; Allen and Allen, 1986; Age et al., 1986). Transport of carbon occurs from source plants to seedlings, especially if the seedlings are shaded; that this occurs by direct linkages or hyphal bridges between roots rather than from reabsorption of leaked nutrients has been demonstrated by autoradiographs (Francis and Read, 1984).

Mycorrhizal roots take up more soil nutrients, at least phosphate, from lower concentration pools than do nonmycorrhizal roots (Cress et al., 1979). Mycorrhizal plants take up more water during periods of drought than nonmycorrhizal plants (Allen et al., 1981; Allen and Boosalis, 1983) and have been shown to delay flowering (Allen and Allen, 1986) and to retain leaves longer, thus fixing a greater amount of carbon (Kormanik, 1985). Different Glomalean species have distinct external hyphal characteristics that may permit differences in resource acquisition (Abbott and Robson, 1985). Since more than one Glomalean fungus can occupy a root system and, according to Read et al. (1985), root systems may be connected by a mycelial matrix, mycorrhizal plants may benefit from the differential resource acquisition of several Glomalean fungi simultaneously. There is in addition some evidence of seasonal variation in activity among Glomalean fungi (Allen et al.,1984a; Jakobson and Neilson, 1983), which can thus spread the benefits of the mycorrhizal association over the entire growing season.

In these various ways, mycorrhizal fungi contribute to the survival and fitness of plants in an ecosystem, particularly under conditions of stress (i.e., drought, resource depletion, and seedling establishment). Although many studies have looked at the effects of Glomalean fungi on plant growth, after perusing the ecological literature it is hard not to conclude that the real benefit to plants from the mycorrhizal association is the enhanced ability to survive and reproduce, to withstand the "ecological crunches" postulated by Allen and Allen (1986).

8.3. Dispersal and Establishment in New Habitats

Glomalean fungi can persist as propagules for varying lengths of time in the absence of adequate nutrition or suitable host plant (Allen et al., 1984b; Carpenter et al., 1987). For example, Allen and Boosalis (1983) reported that *G. fascicula-tum* (Thaxter) Gerdemann & Trappe emend Walker and Koske on native perennial grasses was able to survive a year with no photosynthetic input. The fungus could not, however, persist two years with a fallow year between annual host species, and was replaced instead by larger spored invading *Glomus* species. In another study hyphae within stored mycorrhizal roots survived six years in dried soil (even though viability declined as much as 80%) and low percentages (4 to 22%) of viable spores of *Glomus*, *Acaulospora*, and *Gigaspora* species survived in air dried soil for periods of up to 18 years (Tommerup, 1985). Christiansen and Allen (1980) found active inoculum in topsoil from a uranium mine which had been stored for 12 years, and Gould et al., (1996) detected mycorrhizal spores, albeit in low populations, in topsoil of newly reclaimed mine spoils which had been buried for more than thirty years. Koske et al. (1996) reported that *Gi. gigantea* (Nicol. & Gerd.) Gerdemann & Trappe spores isolated from sand dune soil and immersed in seawater maintained viability for 20 days with only a slight linear decrease in germination. The authors concluded that this represents ample time for dispersal to other beach sites via long shore currents.

This longevity of viable spores, however, is very important to organisms that are obligately symbiotic. According to Read et al. (1976), most mycorrhizal fungi spread by root-to-root contact; hyphae grow along with the root and initiate new infections as other roots are encountered. Greenhouse experiments have estimated annual growth rates of up to 30 cm (Powell, 1979) and in the field Warner and Mosse (1980) reported rates of spread of a few decimeters per year. These rates and distances are not sufficient, however, to permit the colonization of large-scale disturbances, natural (e.g., landslides or volcanic eruptions) or man-made (e.g., extractive mining or agriculture). Therefore, some other mechanism must account for the dispersal of fungal symbionts to potential new habitats. Although colonized root pieces may be active propagules of Glomalean fungi, Gould and Liberta (1981) found the viability of this kind of propagule to decline after 3 years, which leaves spore dispersal as the more probable long-term long-distance mechanism for the colonization of new habitats by Glomalean fungi.

Since Glomalean spore formation is underground, dispersal by air currents is not a primary factor in the spread of these fungi as it is for ectomycorrhizal fungi. Of greater importance is the movement of spore-containing soil by burrowing animals. The dispersal of viable spores to the surface by gophers following the Mt. St. Helens eruption permitted the rapid colonization of facultatively mycorrhizal plants (Allen and MacMahon, 1988); this movement was shown to be sufficient to restore mycorrhizae to the entire eruption zone. In other studies,

Glomalean spores were disseminated when passed undamaged through the digestive tracts of small mammals (Gerdemann and Trappe, 1974; Trappe and Maser, 1976; Warner et al., 1987); spores such as these germinated in the presence of a suitable host (McIlveen and Cole, 1976) or in distilled water (Warner et al., 1987). Grasshoppers transport Glomalean spores on their bodies (Warner et al., 1987) and viable spore transport has also been shown for such diverse animals as earthworms (several species), ants, wasps, and two species of mud nest-building birds (Reddell and Spain, 1991; Warner et al., 1987). Though the burrowing activity of earthworms and ants can move Glomalean spores only a few meters per year (Reddell and Spain, 1990), this is considerably greater than the rate of hyphal spread, and removal to the soil surface does permit their subsequent dispersal by wind or water erosion.

Two other dispersal vectors are important for Glomalean fungi: water and wind. Water erosion on the tephra of Mt. St. Helens exposed and washed away Glomalean spores, effectively moving them to lower lying soil deposits in which succession could proceed (Allen et al., 1984b; Carpenter et al., 1987). Salt water in various concentrations and seawater do not significantly reduce the viability of Glomalean spores (Hirrel, 1981; Koske and Gemma, 1990) and fungal propagules can survive in roots and vegetative fragments of plants for some time (Koske et al., 1975; Taber and Trappe, 1982). With this background, Koske and Gemma (1990) presented evidence for codispersal of Glomalean fungi and mycorrhizal plant fragments by oceanic transport and wave action from coastal mainlands to the Hawaiian Islands. This mechanism may account for the high frequency of mycotrophy in many tropical Pacific island strand communities.

Although dispersal by animal vectors seems to be more common in mesic habitats, erosion by wind may be more important in arid climes. Microspheres, equal in size and weight to spores, were used to identify wind dispersal patterns in a disturbed arid ecosystem and traveled as much as 2 km from the release site (Warner et al., 1987). In addition, capsules and seeds of common top soil weeds, notably purslane (*Portulaca oleracea*), may contain endomycorrhizal spores (Taber, 1982); winds at the soil surface presumably carry mycorrhizal spores into decumbent seed capsules of the plants. Inasmuch as some purslane seeds can remain viable for 40 years (Taber, 1982), this may be an additional dispersal mechanism for Glomalean spores.

It is not enough for spores to disperse, however; as obligate symbionts they must find other suitable hosts quickly if they are to germinate and grow. As Allen (1991) remarks, the ability to find a new host quickly could be crucial to survival strategy. Specificity of host would be selected against. The nonspecific nature of host selection for these fungi is obvious upon examination of the INVAM germ plasm collection data (Morton et al.,1993); *Acaulospora scrobiculata* Trappe, for example, is known to be symbiotic with such diverse hosts as cassava (*Manihot esculenta*), sugarcane (*Saccharum officinarum*), goldenrod (*Solidago* ssp.), beachgrass (*Ammophila* ssp.), apple (*Malus domestica*), and foxgrape (*Vitis lambrusca*). Likewise, *Scutellospora pellucida* (Nicol. & Schenck) Walker & Sanders has been found with wild grape, various composites, legumes, fescues, multiflora rose, peach trees, broomsedge (*Andropogon virginicus*), wild oat grass (*Danthonia spicata*), and bird's foot trefoil (*Lotus corniculatus*) (Morton et al., 1993).

8.4. Interactions with the Community

MacMahon et al. (1981) provide a useful definition to use when considering mycorrhizae/mycorrhizal fungi as members of a community. A community comprises "the organisms which affect, directly or indirectly, the expected reproductive success of a reference organism." Mycorrhizae affect all those organisms which use the same resources as either symbiont and those which use either symbiont as a resource (Allen, 1991).

8.4.1. Interactions with Other Arbuscular Mycorrhizal Fungi

Species of Glomalean fungi, under competition with other such fungi, differ in their rate of spread and ability to remain colonized within roots (Powell, 1979). Ross and Ruttencutter (1977) described the competitive relationship between *Gigaspora macrocarpun* Tul. & Tul. and *Gi. gigantea* (Nicol. & Gerd.) Gerdemann & Trappe in peanut and soybean. Although percent root colonization was greater when roots were inoculated with both fungi, the percentage of roots colonized by *G. macrocarpum* was suppressed compared to that colonized by *G. macrocarpum* inoculated alone. Gemma et al. (1989) observed that high populations of one Glomalean species in sand dune soil were strongly associated with low populations of the other species recovered from the soil, suggesting antagonism and competition between species. In a sagebrush (*Artemesia tridentata*) community of only 8 plant species, 15 Glomalean species in 5 genera were recovered from samples taken along a 100-m transect (Allen, 1991). Seven species in 3 genera were associated with a single plant species (*Uniola paniculata*) on foredunes in Florida (Sylvia, 1986). Wilson (1984) proposed that successful competition could be explained by differences in inoculum potentials.

8.4.2. Interactions with Other Soil Organisms

Glomalean fungi also interact with other soil organisms including fungi, bacteria, and invertebrates. Bacteria may destroy spores by direct penetration or through production of cell wall-degrading enzymes on the spore surface; in addition, they may induce autolysis or deplete the nutrient zone around spores, causing nutrients to leach from the spores themselves (Hetrick,1984). Although their impact on Glomalean spore populations is not well understood (Schenck and Nicolson,

1977), mites, collembolans, nematodes, and amebae inhibit mycorrhizal activity by grazing on the mycelium as well as spores (Ingham et al., 1986; McGonigle and Fitter, 1988; Moore et al., 1986; Rabitin and Stinner, 1985). McGonigle and Fitter (1988) suggest that invertebrate grazing may account for the lack of response by plants to mycorrhizae in many field studies. However, Klironomos and Kendrick (1996) feel that grazing by mites and springtails may not be of much importance to the Glomalean fungi. When six mite and collembolem species were presented with a choice between conidial fungi and *G. macrocarpum*, the soil fauna strongly preferred the conidial fungi. Even sterile control roots were favored over the *Glomus* by all but one of the experimental animals. When the mycorrhizal fungus was grazed, only hyphae narrower than 5 μm were used as food.

Bird et al. (1974) found that mycorrhizal activity increased when populations of root nematodes were controlled by treating soils with nematicides; however, Stanton et al. (1981) reported no significant changes in mycorrhizal populations following nematicide treatments. Later research by Allen et al. (1987) supported Bird's observations. They found that mycorrhizal activity in a semiarid climate was reduced during wet years (which should have been good years for growth) where nematode activity was high. Decreased mycorrhizal activity was correlated with decreased productivity and survival of the plant community.

A number of fungal hyperparasites of Glomalean spores are reported in the literature (Daniels and Mange, 1980; Koske, 1981; Ross and Ruttencutter, 1977; Schenck and Nicolson, 1977). Ross and Ruttencutter (1977) showed that spore production by *Glomus* and *Gigaspora* species was reduced due to parasitism by soil fungi, and Daniels and Mange (1980) observed that light-colored spores are more susceptible to such attack. According to Hetrick (1984), some of these hyperparasites parasitize hyphae as well as spores.

In some studies, colonization by Glomalean fungi reduced disease severity in hosts affected by *Phytophthora parasitica* (Davies and Mange, 1981) and other root pathogens (Graham and Mange, 1982). The mechanism proposed is indirect. Unlike ectomycorrhizal fungi that reduce pathogen activity by enclosing the root with a mantle and/or producing antibiotic compounds (Marx, 1969), Glomalean fungi improve root phosphate levels, which reduces root exudates and makes roots less attractive to many soil pathogens.

To the benefit of their plant partners, mycorrhizae can interact synergistically with other symbiotic microorganisms. Compared to inoculation of either microorganism alone, Glomalean fungi combined with *Frankia* in snowbrush (*Leanothus velutinus*) (Rose and Youngberg, 1981) and *Rhizobium* in sweet vetch (*Hedysarum borealis*) (Carpenter and Allen, 1988) enhanced plant growth and, in the case of sweet vetch, flowering and survival as well. As a result of combined inoculation of tomato with *Azotobacter chroococcum* and *G. fasciculatum,* host growth, *Azotobacter* populations in the rhizosphere, and percent colonization and spore production by the fungus were all enhanced (Bagyaraj and Mange, 1978). Bagyaraj (1984) cites 20 or more studies in which mycorrhizae added to the effect of *Rhizobium* on plant growth through increased nodulation, nitrogen fixation, nitrogen and phosphorus concentrations, number of leaves, plant height, above- and below-ground biomass, and crop yield. He speculates that whereas these effects may be largely due to improved phosphorus nutrition, mycorrhizae may be influencing other processes, e.g., copper and zinc uptake, hormone production, and photosynthesis, in a beneficial way.

8.5. Interactions with Animals

Mycorrhizae interact with animals other than soil inhabitants. They are in direct competition for the fixed carbon of the plant community with herbivores, and indirect competition with the predators of those herbivores. Stanton et al. (1981) estimated that animals grazing above ground consume 10% of the primary plant material produced in a shortgrass prairie, whereas animals grazing below ground consume up to 25%. Estimates of the amount of fixed carbon used by mycorrhizal fungi vary greatly according to St. John and Coleman (1983), but the actual amounts may approach 40–60% over the lifetime of the plant. Data are inconclusive as to the relative winners and losers when these needs conflict. Reece and Bonham (1978) found that heavy grazing increased mycorrhizal activity in a shortgrass prairie; conversely, Bethlenfalvay and Dakessian (1984) noted that heavy grazing reduced mycorrhizal activity and altered the fungal flora in a semi-arid rangeland. Wallace (1987) suggested that it was not the grazing but the trampling by large animals that reduced mycorrhizal activity.

8.6. Animals Also Interact Favorably with Mycorrhizal Fungi

As mentioned previously, small rodents, ants, gophers, wasps, earthworms, and millipedes can disperse and transport Glomalean inoculum. Harvester ants line their tunnels with roots colonized by Glomalean fungi (Freise and Allen, 1988), thus creating oases of concentrated inoculum in a shrub-grassland community, which is of great benefit to nearby plants. Classic interdependence of community members was demonstrated by Allen's 1987 study of the reestablishment of mycorrhizae on Mt. St. Helens. Wapiti feeding on grasses in undisturbed montane meadows ingested roots and deposited them along with plant propagules on the bare volcanic debris. Plants sprouting from these droppings were mycorrhizal, while nearby seedlings (presumably windborne) were not (Allen, 1987).

8.7. Interactions with the Plant Community

Mycorrhizae affect both the patterns of community composition and the mechanisms of interaction between plant members of a community. Although nutrient transfer between plants has been demonstrated by Read and his associates (1985), its importance to plant communities is not well understood (Miller and Allen, 1992; Molina et al.,1992). Chiariello et al. (1982) have suggested that if nutrient transfer between plants is significant in quantity, some of the theories concerning community functioning, in particular competition theory, will have to be reexamined.

Two traditional assumptions are used to explain coexistence between members in communities composed of multiple species: (1) by minimizing interactions, niche differentiation minimizes competition; and (2) potentially negative interactions in a community are converted by synergism to beneficial interactions (Janos, 1985). These traditional assumptions are valid only if the fungal network within a community is thought of as a static conduit and plants are emphasized as resource sinks (Miller and Allen, 1992). Janos (1985) claims, however, that the mycorrhizal relationship greatly influences community composition by affecting the competitive abilities of plants. He hypothesizes that nonmycorrhizal species can outcompete mycorrhizal species only if (1) the onset of rapid growth by the mycotrophic species is delayed due to slow or delayed colonization; (2) the nonmycotrophs secrete alleopathic chemicals against the mycotrophic species; and (3) the habitat-influenced benefit/cost ratio of mineral uptake adaptions by the nonmycotrophs exceeds the benefit/cost ratio of mycorrhizae to the mycotrophic species.

Janos further suggests that, in the presence of mycorrhizal fungi, nonmycorrhizal species, which do well at intermediate or high levels of soil fertility, are unable to compete with obligate mycotrophs in low-fertility soils. If, however, neither of the two competing species rejects mycorrhizal infection, colonization levels in one or both may be considerably greater than when growing alone. A corollary to this theory suggests that the energy cost of mycorrhizae (to the plant) retards the growth of species less dependent on the relationship; if soil fertility is not high enough for the less dependent competitor to reject infection, it will be out competed by the species that is more dependent on mycorrhizal infection.

Evidence that supports Janos's hypothesis is accumulating in the literature. Crowell and Boerner (1988) have demonstrated that in poor nutrient soil ragweed (*Ambrosia artemesiifolia*) competes strongly with the nonmycotrophic black mustard (*Brassica nigra*) in the presence, but not the absence, of Glomalean fungi. Allen and Allen (1984) showed that inoculation of two grasses with Glomalean fungi could overcome negative competition by the nonmycotrophic Russian thistle (*Salsola kali*). Of four species from genera considered weakly mycotrophic, the one most responsive to inoculation with Glomalean fungi was also the one that persisted longest in old field secondary succession (Boerner, 1992b). In a related paper, Boerner (1992a) presented a table of data derived from 26 different studies; these data showed that annuals were in general less responsive to Glomalean fungi than herbaceous perennials which were, in turn, less responsive than woody perennials. All were more responsive at lower phosphorus levels (which may be interpreted as stress), and perennial grasses were as responsive to inoculation with Glomalean fungi at higher phosphorus levels as annual grasses at lower levels. Since levels of available phosphorus decrease over time during secondary succession (Odum, 1969), responsiveness to Glomalean fungi must be considered an important factor that drives secondary succession (Boerner, 1992a). Boerner also showed that mycotrophic species could utilize pools of phosphorus unavailable to nonmycorrhizal plants and could reduce this resource to levels that would not support continued growth of the nonmycotrophs (Boerner, 1992c). Since different Glomalean species are differentially efficient at obtaining soil phosphorus (Reeves, 1985), the dynamics of succession may even depend on which Glomalean fungi are acquired by competing mycotrophs.

Variation in spore population density is another factor that may contribute to diversity in plant community structure. In one study, for example, spore populations varied from 0 to 100 spores per gram within a few centimeters (Allen and MacMahon, 1985). Differences in inoculum density, which permit mycotrophs to occupy certain patches of ground and leave other patches to nonmycotrophs, may be particularly important in early successional seres. Pendleton and Smith (1983) found a wide variation in the density of mycorrhizae in the pioneer vegetation of disturbed sites in Utah. Allen and Allen (1984) suggest that inoculum density drives the rate of succession; different species and even different ecotypes of Glomalean fungi may result in specialization of mycorrhizal roles.

For example, although *Gigaspora margarita* Becker & Hall and *Glomus fasciculatum* probably have the widest worldwide distribution of any Glomalean fungi (Walker, 1985), they may have different ecotypes that alter water relations, even within the same host species (Stahl and Smith, 1984). One Glomalean fungus may reduce drought tolerance whereas another may improve it (Allen and Boosalis, 1983). Hyphal length per unit root length may differ among fungal isolates and may affect nutrient relationships within a particular host (Abbott and Robson, 1985; Graham et al., 1982). Koske (1987) documented changes in the composition of Glomalean species associated with three herbaceous dune plants over a temperature gradient and noted that certain fungi were favored by cooler soil temperatures and others by warmer. Although *Gi. gigantea* dominated both cooler and warmer coastal dune communities, it was more dominant in the northern half of Koske's transect (87% of root samples) than in the southern (74%). Species that preferred warmer temperatures included *S. fulgida* Koske et Walker, *S. dipapillosa* (Walker & Koske) Walker & Sanders, and *Gi. albida* Schenck & Smith (found only in the southern zone); species that preferred cooler soil temperatures included *G. gigantea*, *S. persica* (Koske & Walker) Walker & Sanders, and *S. calospora* (Nicol. & Gerd.) Gerdemann & Trappe. Although none of the species recovered from the cooler climes was found in Florida sand dunes (Sylvia, 1986), four of Koske's species that favored warmer temperatures were present, i.e., *A. scrobiculata* Trappe, *S. verrucosa* (Koske & Walker) Walker & Sanders, *S. weresubiae* Koske & Walker, and *G. aggregatum* Schenck & Smith emend. Koske. Ebbers et al. (1987) found that both fungal species and distribution were associated with soil moisture and nutrient gradients. In western sagebrush grasslands, both spore population density and diversity were inversely correlated with increasing aridity (Stahl and Christiansen, 1983).

Topography and frequency of fire were mainly responsible for Glomalean species distribution gradients in a Kansas tall-grass prairie (Gibson and Hetrick, 1988). Conversely, disturbance by fire did not cause any significant changes in the Glomalean community in disturbed sites in Utah (Pendleton and Smith, 1983). Although spore abundance decreased with increasing soil moisture and nutrient levels, availability of calcium was the best predictor of spore abundance across a complex gradient from dry mesic to emergent aquatic vegetation (Anderson et al., 1984). The composition of the fungal community also varied across this gradient; *G. gigantea* predominated at the wetter end and *G. caledonium* (Nicol. & Gerd.) Trappe & Gerdemann restricted itself to the drier.

Variation in both inoculum (i.e., spore population) density and species composition may occur not only horizontally but also vertically within the soil profile. In one tall grass prairie study, Glomalean species composition was richest in the top 60 cm (the A horizon) of the soil profile, limited to two species in the B horizon, and composed of *G. fasciculatum* below 100 cm (Zajicek et al., 1986). In a dry land *Artemesia* community four Glomalean species were associated with *Artemesia tridentata* in the top 20 cm of soil, but only *G. microcarpum* Tul. & Tul. persisted as deep as 60 cm (Allen, 1991).

9. IMPORTANCE OF CONTINUED RESEARCH

Nearly all land plants show some reaction to mycorrhizal fungi (Allen, 1992). In temperate, subtemperate, and tropical plant communities around the world, most plant species are symbiotically associated with Glomalean mycorrhizal fungi (Trappe, 1987). Infection of germinating seedlings occurs rapidly in such communities (Read and Berch, 1988), and the ever-present population of Glomalean symbionts with low host specificity integrates and interconnects community plants (Francis and Read, 1994; Molina et al., 1992). Molina et al. recognize six phenomena that determine the occurrence and function of these interconnections: (1) dependency vs. independency; (2) facultative vs. obligate symbionts; (3) fidelity to a class of mycorrhizas; (4) host range of mycorrhizal fungi; (5) host receptivity; and (6) ecological specificity (Molina et al., 1992). These phenomena affect plant community development in many ways, most of which are inadequately understood. For example, Molina et al. indicate that, although Glomalean fungi are generally considered non-host-specific, ecological specificity and functional compatibility have barely been examined in natural ecosystems, and it is within this setting that the impact of Glomalean fungi needs to be understood (Molina et al., 1992). Additionally the dynamics of root colonization within and between seasons, and the role played by diversity within Glomalean populations, have not been explored (Abbott and Gazey, 1994). Mycorrhizal fungal diversity does not echo patterns of plant diversity, nor do different ecotypes of the same Glomalean species produce identical responses in the same plant species (Allen et al., 1995). Conversely, plant response varies seasonally to different Glomalean species in the same ecosystem as well (Allen, 1992). Clearly, despite the extensive and growing body of literature describing the various aspects of Glomalean biology, many questions remain unanswered concerning these important symbionts. Indeed, Allen et al. (1995) suggest that "from the plant's point of view, the fungi are likely dissimilar in ways we have not even begun to examine." In view of increasing habitat destruction and extirpation of plant (and possibly fungal) species worldwide, finding the answers to these questions is becoming more urgent.

REFERENCES

Abbott, L. K. and Gazey, G. (1994). An ecological view of the formation of VA mycorrhizas. *Plant and Soil* 159, 69–78.

Abbott, L. K. and Robson, A. D. (1985). Formation of external hyphae in soil by four species of vesicular-arbuscular mycorrhizal fungi. *New Phytologist.* 99, 245–255.

Age, R. M., Schekel, K. A. and Wample, R. L. (1986). Greater leaf conductance of well-watered VA mycorrhizal plants is not related to phosphorus nutrition. *New Phytologist* 103, 107–116.

Allen, E. B. and Allen, M. F. (1984). Competition between plants of different successional stages: mycorrhizae as regulators. *Canadian Journal of Botany* 62, 2625–2629.

Allen, E. B. and Allen, M. F. (1986). Water relations of xeric grasses in the field: interactions of mycorrhizae and competition. *New Phytologist* 104, 559–571.

Allen, E. B. and Allen, M. F. (1990). The mediation of competition by mycorrhizae in successional and patchy environments. *In* J. B. Grace and G. D. Tilman (eds.), *Perspectives on Plant Competition.* Academic Press, New York, pp. 367–389.

Allen, E. B., Allen, M. F., Helm, D. J., Trappe, J. M., Molina, R. and Rincon, E. (1995). Patterns and regulation of mycorrhizal plant and fungal diversity. *Plant and Soil* 170, 47–62.

Allen, M. F. (1982). Influence of vesicular-arbuscular mycorrhizae on water movement through *Bouteloua gracilis* (H.B.K.)Lag ex Steud. *New Phytologist* 91, 191–196.

Alleń, M. F. (1983). Formation of vesicular-arbuscular mycorrhizae in *Atriplex gardneri* (Chenopodiaceae): seasonal response in a cold desert. *Mycologia* 75, 773–776.

Allen, M. F. (1987). Re-establishment of mycorrhizae on Mt. St. Helens: migration vectors. *Transactions of the British Mycological Society* 88, 413–417.

Allen, M. F. (1991). *The Ecology of Mycorrhizae.* Cambridge University Press, Cambridge, UK.

Allen, M. F. (1992). *Mycorrhizal Functioning.* Chapman and Hall, New York.

Allen, M. F., Allen, E. B., and Stahl, P. D. (1984a). Differential niche response of *Bouteloua gracilis* and *Pascopyrum smithii* to VA mycorrhizae. *Bulletin Torrey Botanical Club* 111, 316–325.

Allen, M. F., Allen, E. B. and West, N. E. (1987). Influence of parasitic and mutualistic fungi on *Artemisia tridentata* during high precipitation years. *Bulletin Torrey Botanical Club* 114, 272–279.

Allen, M. F. and Boosalis, M. G. (1983). Effects of two species of VA mycorrhizal fungi on drought tolerance of winter wheat. *New Phytologist* 93, 67–76.

Allen, M. F. and MacMahon, J. A. (1985). Importance of disturbances on cold desert fungi: comparative microscale dispersion patterns. *Pedobiologia* 28, 215–224.

Allen, M. F. and MacMahon, J. A. (1988). Direct VA mycorrhizal inoculation of colonizing plants by pocket gophers (*Thomomys talpoides*) on Mount St. Helens. *Mycologia* 80, 754–756.

Allen, M. F., MacMahon, J. A. and Andersen, D. C. (1984b). Reestablishment of Endogonaceae on Mount St. Helens: survival of residuals. *Mycologia* 76, 1031–1038.

Allen, M. F., Smith, W. K., Moore, T. S., Jr. and Christiansen, M. (1981). Comparative water relations and photosynthesis of mycorrhizal and nonmycorrhizal *Bouteloua gracilis* (J.B.K.) Lag ex Steud. *New Phytologist* 88, 683–693.

Almeida, R. T. (1989). Scientific names in the Endogonales, Zygomycotina. *Mycotaxon* 36, 147–159.

Almeida, R. T. and Schenck, N. C. (1990). A revision of the genus *Sclerocystis* (Glomaceae, Glomales) *Mycologia* 82, 703–714.

Ames, R. N. and Linderman, R. G. (1976). *Acaulospora trappei* sp.nov. *Mycotaxon* 3, 565–569.

Ames, R. N., and Schneider, R. W. (1979). *Entrophospora*, a new genus in the Endogonaceae. *Mycotaxon* 8, 347–352.

Anderson, R. C., Liberta, A. E. and Dickman, L. A. (1984). Interaction of vascular plants and vesicular-arbuscular mycorrhizal fungi across a soil moisture–nutrient gradient. *Oecologia* 64, 111–117.

Baco, B., Azcon-Aguilar, C. and Piche, Y. (1998). Architecture and developmental dynamics of the external mycelium of the arbuscular mycorrhizal fungus *Glomus intraradices* grown under monoxenic conditions. *Mycologia* 90. 52–62.

Bagyaraj, D. J. (1984). Biological interactions with VA mycorrhizal fungi. *In* C. L. Powell and D. J. Bagyaraj (eds.), *VA Mycorrhiza.* CRC Press, Boca Raton, FL, pp. 131–153.

Bagyaraj, D. L. and Mange, J. A. (1978). Interaction between a VA mycorrhiza and *Azotobacter* and their effects on the rhizosphere microflora and plant growth. *New Phytologist* 80, 567–573.

Bagyaraj, D. L., Manjunath, A. and Patil, R. B. (1979). Occurrence of vesicular-arbuscular mycorrhizas in some tropical plants. *Transactions of the British Mycological Society* 72, 164–167.

Bentivenga, S. P. and Morton, J. B. (1994). Systematics of Glomalean endomycorrhizal fungi: current views and future directions. *In* F. L. Pfleger and R. G. Linderman (eds.), *Mycorrhizae and Plant Health*. APS Press, St. Paul, MN. pp. 283–308.

Bentivenga, S. P. and Morton, J. B. (1995). A monograph of the genus *Gigaspora*, incorporating developmental patterns of morphological characters. *Mycologia* 87:719–731.

Berch, S. M. (1985). *Acaulospora sporocarpia*, a new sporocarpic species, and emendation of the genus *Acaulospora*, Endogonaceae, Zygomycotina. *Mycotaxon* 23, 409–418.

Berch, S. M. and Koske, R E. (1986). *Glomus pansihalos*, a new species in the Endogonaceae, Zygomycetes. *Mycologia* 78, 822–836.

Berch, S. M. and Trappe, J. M. (1985). A new species of Endogonaceae, *Glomus hoi*. *Mycologia* 77, 654–657.

Berch, S. M., Miller, O. K. and Thiers, H. D. (1985). Evolution of mycorrhizae. *In Proceedings of the Sixth NACOM*. For. Res. Lab., Oregon State Univ. Corvallis, OR, pp. 189–192.

Berliner, R. and Torrey, J. G. (1989). Studies on mycorrhizal associations in Harvard Forest, Massasschusetts. *Canadian Journal of Botany* 67, 2245–2251.

Bethlenfalvay, G. J. and Dakessian, S. (1984). Grazing effects on mycorrhizal colonization and floristic composition of the vegetation on a semi-arid range in northern Nevada. *Journal of Range Management*. 37, 312–316.

Bethlenfalvay, G. J. Dakessian, S. and Pacovsky, R. S. (1984). Mycorrhizae in a southern California desert: ecological implications. *Canadian Journal of Botany* 62, 519–524.

Bird, G. W., Rich, J. R. and Glover, S. U. (1974). Increased endomycorrhizae of cotton roots in soil treated with nematicides. *Phytopathology* 64, 48–51.

Blaszkowski, J. (1995). *Glomus corymbiforme*, a new species in Glomales from Poland. *Mycologia* 87, 732–737.

Boerner, R. E. J. (1986). Seasonal nutrient dynamics, nutrient resorption and mycorrhizal infection intensitu of two perennial forest herbs. *American Journal of Botany* 3, 1249–1257.

Boerner, R. E. J. (1992a). Plant life span and response to inoculation with vesicular-arbuscular mycorrhizal fungi. I. Annual versus perennial grasses. *Mycorrhiza* 1, 153–161.

Boerner, R. E. J. (1992b). Plant life span and response to inoculation with vesicular-arbuscular mycorrhizal fungi. II. Species from weakly mycotrophic genera. *Mycorrhiza* 1, 163–167.

Boerner, R. E. J. (1992c). Plant life span and response to inoculation with vesicular-arbuscular mycorrhizal fungi. III. Responsiveness and residual soil P levels. *Mycorrhiza* 1, 169–174.

Bonfante-Fasolo, P. (1984). Anatomy and morphology of VA mycorrhizae. *In* C. L. Powell and D. J. Bagyaraj (eds.), *VA Mycorrhiza*. CRC Press, Boca Raton, FL, pp. 5–33.

Boucher, D. H. (1985). *The Biology of Mutualism, Ecology and Evolution*. Oxford University Press, New York.

Brown, M. F. and King, E. J. (1982). Morphology and histology of vesicular-arbuscular mycorrhizae. A. Anatomy and cytology. *In* N. C. Schenk (ed.), *Methods and Principles of Mycorrhizal Research*. Am. Phytopathol. Soc., St. Paul, MN. pp. 15–21.

Brundrett, M. C. and Kendrick, B. (1988). The mycorrhizal status, root anatomy, and phenology of plants in a sugar maple forest. *Canadian Journal of Botany*. 66, 1153–1173.

Campbell, E. O. (1971). Notes on the fungal associations of two *Monotropa* species in Michigan. *Michigan Botanist*. 10, 63–67.

Carpenter, A. T. and Allen, M. F. (1988). Responses of *Hedysarum boreale* to mycorrhizas and *Rhizobium*: plant and soil nutrient changes. *New Phytologist* 109, 125–132.

Carpenter, S. E., Trappe, J. M. and Ammirati, J., Jr. (1987). Observations of fungal succession in the Mount St. Helens devastation zone, 1980–1983. *Canadian Journal of Botany* 65, 716–728.

Chabot, S., Becard, G. and Piche, Y. (1992). Life cycle of *Glomus intraradix* in root organ culture. *Mycologia* 84, 315–321.

Chiarello, N., Hichman, J. C. and Mooney, H. A (1982). Endomycorrhizal role for interspecific transfer of phosphorus in a community of annual plants. *Science* 217, 941–943.

Christiansen, M. and Allen, M. F. (1980). Effect of VA mycorrhizae on water stress tolerance and hormone balance in native western plant species. 1979 Final Report to RMIEE, Laramie, WY.

Clapp, J. P., Young, J. P. W., Merryweather, J. W., and Fitter, A. H. (1995). Diversity of fungal symbionts in arbuscular mycorrhizae from a natural community. *New Phytologist* 130, 259–265.

Cook, B. D., Jastrow, J. D. and Miller, R. M. (1988). Root and mycorrhizal endophyte development in a chronosequence of restored tall grass prairie. *New Phytologist* 11, 355–362.

Cox, G., and Sanders, F. (1974). Ultrastructure of the host-fungus interface in a vesicular-arbuscular mycorrhiza. *New Phytologist* 73, 901–912.

Cress, W. A., Throneberry, G. D. and Lindsey, D. L . (1979). Kinetics of phosphorus absorption by mycorrhizal and non-mycorrhizal tomato roots. *Plant Physiology* 64, 484–487.

Crowell, H. F. and Boerner, R. E. J. (1988). Influence of mycorrhizae and phosphorus on below ground competition between two old-field annuals. Environmental and *Experimental Botany* 28, 381–392.

Currah, R. S., Hambleton, S. and Smrecia, A. (1988). Mycorrhizae and mycorrhizal fungi of *Calypso bulbosa*. *American Journal of Botany*. 75, 739–752.

Dalpe, Y., Koske, R. E. and Tews, L. L. (1992). *Glomus lamellosum* sp nov., a new Glomaceae associated with beach grass. *Mycotaxon* 43, 289–293.

Daniels, B. A. and Mange, J. A. (1980). Hyperparasitism of vesicular-arbuscular mycorrhizal fungi. *Phytopathpology* 70, 584–588.

Danielson, R. M. (1982). Taxonomic affinities and criteria for identification of the common ectendomycorrhizal symbiont of pines. *Canadian Journal of Botany* 60, 7–18.

Davies, R. M. and Mange, J. A. (1981). *Phytophthora parasitica* inoculation and intensity of vesicular-arbuscular mycorrhizae in citrus. *New Phytologist* 87, 705–715.

Duddridge, J. A. and Read, D. J. (1982). An ultrastructural analysis of the development of mycorrhizae in *Monotropa hypopitys* L. *New Phytologist* 92, 203–214.

Ebbers, B. C., Anderson, R. C. and Liberta, A. E. (1987). Aspects of the mycorrhizal ecology of prairie dropseed, *Sporobolus heterolepis* (Poaceae). *American Journal of Botany* 74, 564–573.

Fitter, A. H. (1977). Influence of mycorrhizal infection on competition for phosphorus and potassium by two grasses. *New Phytologist* 79, 19–125.

Fitter, A. H. (1991). Costs and benefits of mycorrhizas. Implications for functioning under natural conditions. *Experientia* 47, 350–355.

Francis, R. and Read, D .J. (1984). Direct transfer of carbon between plants connected by vesicular-arbuscular mycorrhizal mycelium. *Nature* 233, 133.

Francis, R. and Read, D. J. (1994). The contributions of mycorrhizal fungi to the determination of plant community structure. *Plant and Soil* 159, 11–25.

Frank, B. (1885). On the root-symbiosis-depending nutrition through hypogeous fungi on certain trees. Translated by J. M. Trappe from Berichte der deutschen botanishen Gesellschaft 3:128–145. *In* R. Molina (ed.), *Proc. Sixth NACOM.* For. Res. Lab., Oregon State Univ., Corvallis, OR, pp. 18–25.

Franke, M. and Morton, J. B. (1994). Ontogenetic comparisons of the endomycorrhizal fungi *Scutellospora heterogamma* and *S. pellucida:* revision of taxonomic character concepts, species descriptions and phylogenetic hypotheses. *Canadian Journal of Botany* 72, 122–134.

Freise, C. F. and Allen, M. F. (1988). The interaction of harvester ant activity and VA mycorrhizal fungi. *Proceedings of the Royal Society of Edinborough* 94B, 176.

Freise, C. F. and Allen, M. F. (1991). The spread of mycorrhizal fungal hyphae in the soil: inoculum types and external hyphal architecture. *Mycologia* 83, 409–418.

Gemma, J. N. and Koske, R. E. (1988). Seasonal variation in spore abundance and dormancy of *Gigaspora gigantea* and in the mycorrhizal potential of a sand dune soil. *Mycologia* 80, 211–216.

Gemma, J. N. and Koske, R. E. (1990). Mycorrhizae in recent volcanic substrates in Hawaii. *American Journal of Botany* 77, 1193–1200.

Gemma, J. N., Carriero, M. and Koske, R. E. (1989). Seasonal dynamics of selected species of VA mycorrhizal fungi in a sand dune. *Mycological Research 92, 317–321.*

Gerdemann, J. W. (1955). Relation of a large soil-bourne spore to phycomycetous mycorrhizal infections. *Mycologia* 47, 619–632.

Gerdemann, J. W. (1965). VA mycorrhizae formed on maize and tulip tree by *Endogone fasciculata. Mycologia* 57, 562–575.

Gerdemann, J. W. (1968). Vesicular-arbuscular mycorrhizas and plant growth. *Annual Review of Phytopathology* 6, 397–419.

Gerdemann, J. W. (1969). Fungi that form the vesicular-arbuscular type of endomycorrhiza. *In* E. Hacskaylo (ed.), *Mycorrhizae. Proc. First NACOM.* Misc. pub No. 1189. USDA Forest Service. U.S. Government Printing Service, Washington, DC, pp. 9–18.

Gerdemann, J. W. and Trappe, J. M. (1974). The Endogonaceae in the Pacific Northwest. Mycol. Mem. No. 5, 1–76. New York Botanical Garden. Bronx, NY, pp. 1–76.

Gianinazzi, S., Gianinazzi-Pearsons, V. and Dexheimer, J. (1979). Enzymatic studies on the metabolism of vesicular-arbuscular mycorrhizae. III. Ultrastructural location of acid and alkaline phosphatase in onion roots affected with *Glomus mosseae* Nicol. and Gerd. *New Phytologist* 82, 127–132.

Gibson, D. J. and Daniels Hetrick, B. A. (1988). Topographic and fire effects on the composition and abundance of VA mycorrhizal fungi in tall grass prairie. *Mycologia* 80, 433–441.

Gibson, J. L., Kimbrough, J. W. and Benny, G. L. (1986). Ultrastructural observations on Endogonaceae (Zygomycetes). II. Glaziellales ord nov and Glaziellaceae fam nov. New taxa based upon light and electron microscopic observations of *Glaziella aurantica*. *Mycologia* 76, 941–954.

Gould, A. B., Hendrix, J. W. and Ferriss, R. S. (1996). Relationship of mycorrhizal activity to time following reclamation of surface mine land in western Kentucky. I. Propagule and spore population densities. *Canadian Journal of Botany* 74, 247–261.

Gould, A. B. and Liberta, A. E. (1981). Effects of topsoil storage during surface mining on the viability of vesicular-arbuscular mycorrhiza. *Mycologia* 73, 914–922.

Graham, J. H. and Mange, J. A. (1982). Influence of vesicular-arbuscular mycorrhizae and soil phosphorus on take-all disease of wheat. *Phytopathology* 72, 95–98.

Graham, J. H., Linderman, R. C. and Mange, J. A. (1982). Development of external hyphae by different isolates of mycorrhizal *Glomus* in relation to root colonization and growth of Troyer citrange. *New Phytologist* 91, 183–189.

Hall, L. R. (1977). Species and mycorrhizal infection of New Zealand Endogonaceae. *Transactions of the British Mycologial Society*. 68, 341–356.

Hall, L. R. (1984). Taxonomy of VA mycorrhizal fungi. *In* C. L. Powell and P. J. Bagyaraj (eds.), *VA Mycorrhiza*. CRC Press, Boca Raton, FL, pp. 57–94.

Hall, L. R. and Fish, B. J. (1979). A key to the Endogonaceae. *Transactions of the British Mycological Society* 73, 261–270.

Harley, J. L. (1985). Mycorrhizae: the first 65 years; from the time of Frank till 1950. *In* R. Molina (ed.), *Proc. Sixth NACOM*. For. Res. Lab., Oregon State Univ., Corvallis, OR, pp. 26–33.

Harley, J. L. and Smith, S. E. (1983). *Mycorrhizal Symbiosis*. Academic Press, London, UK.

Harley, J. L. and Harley, E. L. (1987). A check list of mycorrhizae in the British flora. *New Phytologist* 105, 1–102.

Heap, A . J. and Newman, E. I. (1980). Links between roots by hyphae of vesicular arbuscular mycorrhizas. *New Phytologist* 85, 169–171.

Hetrick, B. A. D. (1984). Ecology of VA mycorrhizal fungi. *In* C. L. Powell and D. J. Bagyaraj (eds.), *VA Mycorrhizae*. CRC Press, Boca Raton, FL, pp. 35–55.

Hetrick, B. A. D. and Bloom, J. (1983). Vesicular-arbuscular mycorrhizal fungi associated with native tall grass prairie and cultivated winter wheat. *Canadian Journal of Botany* 61, 2140–2146.

Hirrell, M. C. (1981). The effect of sodium and chloride salts on the germination of *Gigaspora margarita*. *Mycologia* 73, 610–617.

Hopkins, N. A. (1987). Mycorrhizae in a California serpentine grassland community. *Canadian Journal of Botany* 65, 484–487.

Ingham, E. R., Trofymow, J. A., Ames, R. N., Hunt, H. W., Morley, C. R., Moore, J. C. and Coleman, D. C. (1986). Trophic interactions and nitrogen cycling in a semi-arid grassland soil. II. System responses to removal of different groups of soil microbes or fauna. *Journal of Applied Ecology* 23, 615–630.

Jabaji-Hare, S. H., Piche, Y. and Fortin, J. A. (1986). Isolation and structural characterization of soil bourne auxilliary cells of *Gigaspora margarita* Becker and Hall, a vesic-ular-arbuscular mycorrhizal fungus. *New Phytologist* 103, 777–784.

Jakobson, I. and Neilsen, N. E. (1983). Vesicular-arbuscular infection in cereals and peas at various times and soil depths. *New Phytologist* 93, 401–413.

Janos, D. P. (1985). Mycorrhizal fungi: agents or symptoms of tropical community compo-sition? *In* R. Molina (ed.), *Proc. Sixth NACOM.* For. Res. Lab., Oregon State Univ., Corvallis, OR, pp. 98–103.

Janos, D. P. (1987). VA mycorrhizas in humid tropical ecosystems. *In* G. R. Safir (ed.), *Ecophysiology of VA Mycorrhizal Plants.* CRC Press, Boca Raton, FL, pp. 107–134.

Kamienski, F. (1882). The vegetative organs of *Monotropa hypopitys* L. Translated by S. M. Berch from Memoires-Societe Nationale des Science Naturelle et Mathema-tiques de Cherbourg 24: 5–40. *In* R. Molina (ed.), *Proc. Sixth NACOM.* For. Res. Lab., Oregon State Univ., Corvallis, OR, pp. 12–17.

Kernan, M. J. and Finocchio, A. F. (1983). A new discomycete associated with the roots of *Monotropa uniflora* (Ericaceae). *Mycologia* 75, 916–920.

Kidston, R. and Lang, W. H. (1996). On old red sandstone plants showing structure from the Rhynie Chart Bed, Aberdeen shire. *Transactions of the Royal Society of Edin-burgh* 87, 423–460.

Kinden, D. A. and Brown, M. F. (1975). Electron microscopy of vesicular-arbuscular mycorrhizae of yellow poplar. I. Characterization of endophytic structures by scan-ning electron microscopy. *Canadian Journal of Microbiology.* 21, 989–993.

Klironomos, J. N. and Kendrick, B. (1996). Palatability of micro fungi to soil arthropods in relation to the functioning of endomycorrhizal associations. *Biology and Fertility of Soils* 21, 43–52.

Kormanik, P. P. (1985). Effects of phosphorus and vesicular-arbuscular mycorrhizae on growth and leaf retention of black walnut, *Juglans nigra*, seedlings. *Canadian Jour-nal of Forest Research* 15, 688–693.

Koske, R .E. (1981). *Labyrinthula* inside the spores of a vesicular-arbuscular mycorrhizal fungus. *Mycologia* 73, 1175–1180.

Koske, R. E. (1985). *Glomus aggregatum* emended: a distinct taxon in the *Glomus fascicu-latum* complex. *Mycologia* 77, 619–630.

Koske, R. E. and Gemma, J. N. (1986). *Glomus microaggragatum*: a new species in the Endogonaceae. *Mycotaxon* 26, 125–132.

Koske, R. E. and Gemma, J. N. (1990). VA mycorrhizae in strand vegetation of Hawaii–evidence for long distance codispersal of plants and fungi. *American Journal of Botany* 77, 466–474.

Koske, R. E. and Gemma, J. N. (1995). *Scutellospora hawaiiensis:* a new species of arbus-cular mycorrhizal fungus from Hawaii. *Mycologia* 87, 678–683.

Koske, R. E. and Halverson, W. L. (1981). Ecological studies of vesicular-arbuscular mycorrhizae in a barrier sand dune. *Canadian Journal of Botany* 59, 1413–1422.

Koske, R. E. and Walker, C. (1984). *Gigaspora erythropa*, a new species forming arbuscu-lar mycorrhizae. *Mycologia* 76, 250–255.

Koske, R. E. and Walker, C. (1985). Species of *Gigaspora* (Endogonaceae) with rough-ened outer walls. *Mycologia* 77 702–720.

Koske, R. E. and Walker, C. (1986a). *Glomus globiferum*, a new species of Endogonaceae with a hyphal perideum. *Mycotaxon* 26, 133–142.

Koske, R., Bonin, C., Kelly, J. and Martinez, C. (1996). Effects of sea water on spore germination of a sand-dune-inhabiting arbuscular mycorrhizal fungus. *Mycologia* 88, 947–950.

Koske, R. E., Freise, C., Walker, C. and Dalpe, Y . (1986b). *Glomus pustulatum:* a new species in the Endogonaceae. *Mycotaxon* 26, 143–149.

Koske, R. E., Sutton, J.C. and Sheppard, B. R. (1975). Ecology of *Endogone* in Lake Huron sand dunes. *Canadian Journal of Botany* 53, 87–93.

Liberta, A. E. and Anderson, R. C. (1986). Comparison of vesicular-arbuscular mycorrhiza species composition, spore abundance and inoculum potential in an Illinois prairie and adjacent agricultural sites. *Bulletin of the Torrey Botanical Club* 113, 178–182.

MacMahon, J. A., Schimpf, D J., Anderson, D. C., Smith, K. G. and Bayn, R. L., Jr. (1981). An organism centered approach to some community and ecosystem concepts. *Journal of Theoretical Biology* 88, 287–307.

Maia, L. C., Kimbrough, J. W. and Benny, G. L. (1994). Ultrastructure of spore germination in *Gigaspora albida* (Glomales). *Mycologia* 86, 343–349.

Malloch, D. and Malloch, B. (1981). The mycorrhizal status of boreal plants: species from northeastern Ontario. *Canadian Journal of Botany* 59, 2167–2172.

Malloch, D. and Malloch, B. (1982). The mycorrhizal status of boreal plants: additional species from northeastern Ontario. *Canadian Journal of Botany* 60, 1035–1040.

Marx, D. H. (1969). Antagonism of mycorrhizal fungi to root pathogenic fungi and bacteria. *Phytopathology* 56, 53–63.

McDougall, W. B. and Liebtag, C. (1928). Symbiosis in a deciduous forest. III. Mycorrhizal relations. *Botanical Gazette* 86, 226–234.

McGonigle, T. P. and Fitter, A. H. (1988). Ecological consequences of arthropod grazing on VA mycorrhizal fungi. *Proceedings of the Royal Society Edinburgh* B94, 25–32.

McIllveen, W. D., and Cole, H., Jr. (1976). Spore dispersal of Endogonaceae by worms, ants, wasps and birds. *Canadian Journal of Botany.* 54, 1486–1489.

Miller, D. D. and Walker, C. (1986). *Glomus maculosum*, sp nov (Endogonaceae): an endomycorrhizal fungus. *Mycotaxon* 25, 217–227.

Miller, O. K. (1984). Taxonomy of ecto- and ectendomycorrhizal fungi. *In* N. C. Schenck (eds.), *Methods and Principles of Mycorrhizal Research*. American Phytopathological Society, St. Paul, MN. pp 91–101.

Miller, R. M. (1979). Some occurrences of vesicular-arbuscular mycorrhizae in natural and disturbed ecosystems of the Red Desert. *Canadian Journal of Botany* 57, 619–623.

Miller, R. M. (1987). The ecology of vesicular-arbuscular mycorrhizae in grass- and shrublands. *In* G. R. Safir (ed.), *Ecophysiology of VA Mycorrhizal Plants*. CRC Press, Boca Raton, FL pp. 135–170.

Miller, S. L. and Allen, E. B. (1992). Mycorrhizal nutrient translocation and interactions between plants. *In* M. F. Allen (ed.), *Mycorrhizal Functioning*. Chapman and Hall, New York, pp. 301–332.

Molina, R., Massicote, H. and Trappe, J. M. (1992). Specificity phenomena in mycorrhizal symbioses: community-ecological consequences and practical implications. *In* M. F. Allen (eds.), *Mycorrhizal Functioning*. Chapman and Hall, New York, pp. 357–423.

Molina, R. J., Trappe, J. M. and Strickler, G. S. (1978). Mycorrhizal fungi associated with *Festuca* in western United States and Canada. *Canadian Journal of Botany* 56, 1691–1695.

Moore, J. C., St John, T. V. and Coleman, D. C. (1986). Ingestion of vesicular-arbuscular mycorrhizal hyphae and spores by soil microarthropods. *Ecology* 66, 979–981.

Morton, J. B. (1985). Variation in mycorrhizal and spore morphology of *Glomus occultum* and *Glomus diaphanum* as influenced by plant host and soil environment. *Mycologia* 77, 192–204.

Morton, J. B. (1986). Three new species of *Acaulospora* (Endogonaceae) from high aluminum low pH soils in West Virginia. *Mycologia* 78, 641–648.

Morton, J. B. (1988). Taxonomy of VA mycorrhizal fungi: classification, nomenclature and identification. *Mycotaxon* 32, 261–324.

Morton, J. B. (1989). Mycorrhizal Fungi Slide Set: Morphological Characters Important in Identifying Endomycorrhizal Fungi in the Zygomycetes. Audio-Visual Publication No. 2. West Virginia Univ. Agri. and For. Expt. Stn., Morgantown, WV.

Morton, J. B. (1990a). Species and clones of arbuscular mycorrhizal fungi (Glomales, Zygomycetes): their role in macro- and microevolutionary processes. *Mycotaxon* 37, 493–515.

Morton, J. B. (1990b). Evolutionary relationships among arbuscular mycorrhizal fungi in the Endogonaceae. *Mycologia* 82, 192–207.

Morton, J. B. (1993). Problems and solutions for the integration of glomalean taxonomy, systematic biology and the study of endomycorrhizal phenomena. *Mycorrhiza* 2, 97–109.

Morton, J. B. (1995). Taxonomic and phylogenetic divergence among *Scutellospora* species based on comparative developmental sequences. *Mycologia* 87, 127–137.

Morton, J. B. and Benny, G. L. (1990). Revised classification of arbuscular mycorrhizal fungi (Zygomycetes): a new order, Glomales, two new suborders, Glomineae and Gigasporineae and two new families, Acaulosporaceae and Gigasporaceae with an emendation of Glomaceae. *Mycotaxon* 37, 471–491.

Morton, J. B. and Bentivenga, S. P. (1993). Glomales Identification Workshop. Ninth NACOM. University of Guelph, Ontario, Canada.

Morton, J. B. and Bentivenga, S. P. (1994). Levels of diversity in endomycorrhizal fungi (Glomales, Zygomycetes) and their role in defining taxonomic and non-taxonomic groups. *Plant and Soil* 159, 47–59.

Morton, J. B. and Walker, C. (1984). *Glomus diaphanum*: a new species in the Endogonaceae common in West Virginia. *Mycotaxon* 21, 431–440.

Morton, J. B., Bentivenga, S. P. and Wheeler, W. W. (1993). Germ plasm in the international collection of arbuscular and vesicular-arbuscular mycorrhizal fungi— INVAM. *Mycotaxon* 48, 491–528.

Morton, J. B., Franke, M. and Cloud, G. (1992). The nature of fungal species in Glomales (Zygomycetes). *In* D. J. Read, D. H. Lewis, A. Fitter and I. Alexander (eds.), *Mycorrhizas in Ecosystems*. CAB Int., Univ. of Ariz. Press, Tucson, pp. 65–73.

Mosse, B. (1953). Fructifications associated with mycorrhizal strawberry roots. *Nature* 171, 974.

Mosse, B. (1959). Observations on the extramatrical mycelium of a vesicular-arbuscular endophyte. *Transaction of the British Mycological Society*. 42, 439–448.

Mosse, B. (1985). Endotropic mycorrhiza (1885–1950): the dawn and middle ages. *In* R. Molina (ed.), *Proc. Sixth NACOM*. Forest Research Laboratory, Oregon State Univ., Corvallis, OR, pp. 48–55.

Moutoglis, P. and Widden, P. (1996). Vescicular-arbuscular mycorrhizal spore populations in sugar maple (*Acer saccharum* marsh. L.) forests. *Mycorrhiza* 6, 91–97.

Mullahey, J. J. and Speed, C. S. (1990). The occurrence of vesicular-arbuscular mycorrhizae on Florida range grasses. *Soil Crop Science Society, Florida Proceedings* 50, 44–47.

Nicholson, T. H. (1960). Mycorrhizae in the Graminaea. II. Development on different habitats particularly sand dunes. *Transactions of the British Mycological Society*. 43, 132–145.

Nicholson, T. H. and Gerdemann, J. W. (1968). Mycorrhizal *Endogone* species. *Mycologia* 60, 313–325.

Nicholson, T. H. and Schenck, N. C. (1979). Endogonaceous mycorrhizal endophytes in Florida. *Mycologia* 71, 178–198.

Odum, E. P. (1969). The strategy of ecosystem development. *Science* 164, 262–270.

Pendleton, R. L. and Smith, B. N. (1983). Vesicular-arbuscular mycorrhizae of weedy and colonizer plant species at disturbed sites in Utah. *Oecologia* 59, 296–301.

Phillips, C. J. and Taylor, T. N. (1996). Mixed arbuscular mycorrhizae from the Triassic of Antarctica. *Mycologia* 88, 707–714.

Pirozynski, K. A. (1981). Interactions between fungi and plants through the ages. *Canadian Journal of Botany* 59, 1824–1827.

Pirozynski, K. A. and Malloch, D. (1975). The origin of land plants, a matter of mycotropism. *BioSystems* 6, 153–164.

Pirozynski, K. A. and Dalpe, Y. (1989). Geological history of the Glomaceae with particular reference to mycorrhizal symbiosis. *Symbiosis* 7, 1–36.

Powell, C. L. (1979). Spread of mycorrhizal fungi through soil. *New Zealand Journal of Agricultural Research* 22, 335–339.

Powell, C. L. (1980). Mycorrhizal infectivity of eroded soils. *Soil Biology and Biochemistry* 12, 247–257.

Rabatin, S. C. and Stinner, B. R. (1985). Arthropods as consumers of vesicular-arbuscular mycorrhizal fungi. *Mycologia* 77, 320–322.

Read, D. J. (1984a). Progress, problems and prospects in research on Ericaceous and Orchidaceous mycorrhizas. *In* R. Molina (eds.), Proc. Sixth NACOM. For. Res. Lab. Oregon State Univ., Corvallis, OR, pp. 202–206.

Read, D .J. (1984b). The structure and function of the vegetative mycelium of mycorrhizal roots. *In* D. H. Jennings and A. D. M. Rayner (eds.), *The Ecology and Physiology of the Fungal Mycelium*. Cambridge University Press, Cambridge, UK, pp. 215–240.

Read, D. J. and Berch, C. P. D. (1988). The effects and implications of disturbance of mycorrhizal mycelial systems. *Proceedings of the Royal Society of Edinburgh* 94B, 13–24.

Read, D. J. and Haselwandter, K. (1981). Observations on the mycorrhizal status of some alpine plant communities. *New Phytologist* 88, 341–352.

Read, D. J., Francis, R. and Finlay, R. D. (1985). Mycorrhizal mycelia and nutrient cycling in plant communities. *In* A. H. fitter (ed.), *Ecological Interactions in Soil.* Blackwell Scientific, Oxford, UK, pp. 193–217.

Read, D. J., Koncheki, H. K. and Hodgson, J. (1976). Vesicular-arbuscular mycorrhiza in natural vegetation systems. I. The occurrence of infection. *New Phytologist* 77, 641–653.

Reddell, P. and Spain, A. V. (1991). Earthworms as vectors of viable propagules of mycorrhizal fungi. *Soil Biology and Biochemistry* 23, 767–774.

Reece, P. E. and Bonham, C. D. (1978). Frequency of endomycorrhizal infection in-grazed and ungrazed bluegramma plants. *Journal of Range Management* 31, 149–151.

Reeves, F. B. (1985). Survival of VA mycorrhizal fungi: interactions of secondary succession, mycorrhizal dependency in plants and resource competition. *In* R. Molina (ed.), *Proc. Sixth NACOM.* Forage Research Laboratory, Oregon State Univ., Corvallis, OR, pp. 110–113.

Reeves, F. B., Wagner, D. W., Moorman, T. and Keil, J. (1979). The role of endomycorrhizae in revegetation practices in the semi-arid west. I. A comparison of incidence of mycorrhizae in severely disturbed vs. natural environments. *American Journal of Botany*, 66, 1–13.

Rose, S. L., Daniels, B. A. and Trappe, J M. (1979). *Glomus gerdemanii* sp nov. *Mycotaxon* 8, 297–301.

Rose, S. L. and Youngberg, C. T. (1981). Tripartite associations in snowbrush (*Leanothus velutinus*): effect of vesicular-arbuscular mycorrhizae on growth, nodulation and nitrogen fixation. *Canadian Journal of Botany.* 59, 34–39.

Ross, J. P. and Ruttencutter, R. (1977). Population dynamics of two vesicular-arbuscular endomycorrhizal fungi and the role of hyperparisitic fungi. *Phytopathology* 67, 490–496.

Rothwell, F. M. and Victor, B. J. (1984). A new species of Endogonaceae: *Glomus botryoides. Mycotaxon* 20, 163–167.

Sanders, F. E., Tinker, P. B., Black, R. L. B. and Palmerly, S. M. (1977). The development of endomycorrhizal root systems. I. Speed of infection and growth-promoting effects with four species of vesicular-arbuscular endophyte. *New Phytologist* 78, 257–268.

Schenck, N. C. (1982). Introduction. *In* N. C. Schenck (ed.), *Methods and Principles of Mycorrhizal Research.* American Phytopathological Society, St. Paul, MN.

Schenck, N. C. (1985). VA mycorrhizal fungi 1950 to the present: the era of enlightenment. *In* R. Molina (ed.), *Proc. Sixth NACOM.* Forestry Research Laboratory. Oregon State Univ., Corvallis, OR, pp. 56–60.

Schenck, N. C. and Nicolson, T. H. (1977). A zoosporic fungus occurring on species of *Gigaospora margarita* and other vesicular-arbuscular mycorrhizal fungi. *Mycologia* 69, 1049–1053.

Schenck, N. C. and Smith, G. S. (1982). Additional new and unreported species of mycorrhizal fungi (Endogonaceae) from Florida. *Mycologia* 74, 77–92.

Schenck, N. C., Spain, J. L., Sieverding, E. and Howeler, R. H. (1984). Several new and unreported vascicular-arbuscular mycorrhizal fungi (Endogonaceae) from Colombia. *Mycologia* 76, 685–699.

Schenck, N. C., Spain, J. L. and Sieverding, E. (1986). A new sporocarpic species of *Acaulospora* (Endogonaceae). *Mycotaxon* 25, 111–117.

Schramm, J. R. (1966). Plant colonizing studies on black wastes from anthracite mining in Pennsylvania. *Transactions of the American Philosophical Society* 47, 1–331.

Siguenza, C., Espejel, J. and Allen, E. B. (1996). Seasonality of mycorrhizae in coastal sand dunes of Baja California. *Mycorrhiza* 6, 151–157.

Simon, L., Bausquret, J., Levesque, R. C. and Lalonde, M. (1993). Origin and diversification of endomycorrhizal fungi and coincidence with vascular land plants. *Nature* 363, 67–69.

Smith, G. S. and Schenck, N. C. (1985). Two new dimorphic species in the Endogonaceae: *Glomus ambisporum* and *G. heterosporum. Mycologia* 77, 566–574.

St. John, T. V. (1980). A survey of mycorrhizal infection in an Amazonian rain forest. *Acta Amazonica* 10, 527–533.

St. John, T. V. and Coleman, D. C. (1983). The role of mycorrhizae in plant ecology. *Canadian Journal of Botany* 61, 1005–1014.

Stahl, P. D. and Christensen, M. (1983). Mycorrhizal fungi associated with *Bouteloua* and *Agropyron* in Wyoming sagebrush-grass lands. *Mycologia* 74, 877–885.

Stahl, P. D. and Smith, W. K. (1984). Effects of different geographic isolates of *Glomus* on the water relations of *Agropyron smithii. Mycologia* 76, 261–267.

Stanton, N. L., Allen, M. F. and Campion, M. (1981). The effect of the pesticide carbofuran on soil organisms and root and shoot production in shortgrass prairie. *J. Applied Ecology.* 18, 434–437.

Stubblefield, S. P., Taylor, T. N. and Trappe, J. M. (1987). Fossil mycorrhizae: a case for symbiosis. *Science* 237, 59–60.

Sylvia, D M. (1986). Spacial and temporal distribution of vescicular-arbuscular mycorrhizae associated with *Uniola paniculata* in Florida foredunes. *Mycologia* 78, 728–734.

Taber, R. A. (1982). Occurrence of *Glomus* spores in weed seeds in soil. *Mycologia* 74, 515–520.

Taber, R. A. and Trappe, J. M. (1982). Vesicular-arbuscular mycorrhizal fungi in rhizomes, scale-like leaves, roots and xylem of ginger. *Mycologia* 74, 154–161.

Taylor, T. N., Remy, W., Hass, H. and Kerp, H. (1995). Fossil arbuscular mycorrhizae from the early Devonian. *Mycologia* 87, 560–573.

Thaxter, R. (1922). A revision of the Endogonaceae. *Proceedings of the American Academy of Arts Science.* 57, 291–351.

Tommerup, I. C. (1985). Strategies for long- term preservation of VA mycorrhizal fungi. *In* R. Molina (ed.), *Proc. Sixth NACOM.* For. Res. Lab., Oregon State Univ., Corvallis, OR, pp. 87–88.

Trappe, J. M. (1977). Three new Endogonaceae: *Glomus constrictus, Sclerocystis clavispora,*and *Acaulospora scrobiculata. Mycotaxon* 6, 359–366.

Trappe, J. M. (1982). Synoptic keys to the genera and species of zygomycetous mycorrhizal fungi. *Phytopathology* 72, 1102–1108.

Trappe, J. M. (1987). Phylogenetic and ecologic aspects of mycotrophy in the angiosperms from an evolutionary aspect. *In* G. R. Safir. (ed.), *Ecophysiology of V.A. Mycorrhizal Plants.* CRC Press, Boca Raton, FL, pp.5–25.

Trappe, J. M. and Berch, S. M. (1985). The prehistory of mycorrhizae: A.B. Frank's predecessors. *In* R. Molina (ed.), *Proc. Sixth NACOM.* For. Res. Lab., Oregon State Univ., Corvallis, OR, pp. 2–11.

Trappe, J. M. and Maser, C. (1976). Germination of spores of *Glomus macrocarpus* (Endogonaceae) after passage through a rodent digestive tract. *Mycologia* 68, 433–436.

Trappe, J. M. and Schenck, N. C. (1982). Taxonomy of the fungi forming endomycorrhizae. *In* N. C. Schenck (ed.), *Methods and Principles of Mycorrhizal Research.* Am. Phytopathol. Soc., St. Paul, MN, pp. 1–9.

Trappe, J. M., Bloss, H. E. and Mange, J. A. (1984). *Glomus deserticola* sp nov. *Mycotaxon* 20, 123–127.

Walker, C. (1979). *Complexipes moniliformis*: a new genus and species tentatively placed in the Endogonaceae. *Mycotaxon* 10, 99–104.

Walker, C. (1983). Taxonomic concepts in the Endogonaceae: spore wall characteristics in species descriptions. *Mycotaxon* 18, 443–455.

Walker, C. (1985). Taxonomy of the Endogonaceae. *In* R. Molina (ed.), *Proc. Sixth NACOM.* For. Res. Lab., Oregon State Univ., Corvallis, OR, pp. 193–199.

Walker, C. (1986). Taxonomic concepts in the Endogonaceae. II. A fifth morphological wall type in Endogonaceous spores. *Mycotaxon* 25, 95–99.

Walker, C. (1991). *Scutellospora* is *Scutellospora*. *Mycotaxon* 40, 141–143.

Walker, C. and Sanders, F. E. (1986). Taxonomic concepts in the Endogonaceae: III. The separation of *Scutellospora* gen nov from *Gigaspora* Gerd. and Trappe. *Mycotaxon* 27, 169–182.

Walker, C. and Trappe, J. M. (1981). *Acaulospora spinosa* sp nov with a key to the species of *Acaulospora*. *Mycotaxon* 12, 515–521.

Wallace, L. L. (1987). Mycorrhizas in grasslands: interactions of ungulates, fungi and drought. *New Phytologist* 105, 619–632.

Wang, C. J. K. and Wilcox, H. E. (1985). New species of ectendomycorrhizal and pseudomycorrhizal fungi: *Phialophora finlandia, Chloridium paucisporum,* and *Phialocephala fortinii. Mycologia* 77, 951–958.

Warcup, J. H. (1981). The mycorrhizal relationship of Australian orchids. *New Phytology* 87, 371–387. Warner, N. J., Allen, M. F. and MacMahon, J. A. (1987). Dispersal agents of vesicular-arbuscular mycorrhizal fungi in a disturbed arid ecosystem. *Mycologia* 9, 721–730.

Warner, A. and Mosse, B. (1980). Independent spread of vesicular-arbuscular mycorrhizal fungi in soil. *Transactions of the British Mycological Society* 74, 407–410.

Wilson, J. M. (1984). Competition for infection between vesicular-arbuscular mycorrhizal fungi. *New Phytologist* 97, 427–435.

Wu, C. G. (1993). Glomales of Taiwan: a comparative study of spore ontogeny in *Sclerocystis* (Glomaceae, Glomales). *Mycotaxon* 47, 25–39.

Wu, C. G. and Sylvia, D. M. (1993). Spore ontogeny of *Glomus globiferum. Mycologia* 85, 317–322.

Xiao, G. and Berch, S. M. (1992). Ericoid mycorrhizal fungi of *Gaultheria shallon. Mycologia* 84, 470–471.

Yang, C. S. and Wilcox, H. E. (1984). An E-strain ectendomycorrhiza formed by a new species *Tricharina mikolae*. *Mycologia* 76: 675–684.

Zajicek, J. M., Daniels Hetrick, B. A. and Owensby, G. E. (1986). The influence of soil depth on mycorrhizal colonization of forbs in the tall grass prairie. *Mycologia* 78, 316–320.

Zak, B. (1974). Ectendomycorrhizas of Pacific madrone (*Arbutus menziesii*) *Transactions of the British Mycological Society*. 62, 202–204.

Zak, B. (1976a). Pure culture synthesis of bearberry mycorrhizae. *Canadian Journal of Botany* 54, 1297–1305.

Zak, B. (1976b). Pure culture synthesis of Pacific madrone endomycorrhizae. *Mycologia* 68, 362–369.

9

Bacterial Endophytes and Their Effects on Plants and Uses in Agriculture

D. Y. Kobayashi and J. D. Palumbo
Cook College, Rutgers University, New Brunswick, New Jersey

1. INTRODUCTION

Although the term *endophyte* is most commonly associated with fungal organisms, there is sufficient literature pertaining to bacteria as endophytes, some of which are regarded to have a beneficial effect, while others are regarded to have a neutral or detrimental effect on plants. The recent increase in publications on bacterial endophytes reflects an interest in their potential benefit to agriculture. This chapter will focus on major themes and issues surrounding bacterial endophytes and their potential use in agriculture.

Bacteria have been proposed to exist inside plants without causing disease symptoms for over 50 years (Tervet and Hollis, 1948; Hollis, 1951). Various reports indicate that endophytic bacteria exist in a variety of tissue types within numerous plant species (Table 1), suggesting a ubiquitous existence in most if not all higher plants. Mundt and Hinkle (1976) demonstrated the diversity of host plants harboring endophytes by isolating bacteria from seeds and ovules of 27 different plant species. The diversity of plant species has expanded since the study by Mundt and Hinkle. Endophytic bacteria have been isolated from both monocotyledonous and dicotyledonous plants that range from woody tree species such as oak (Brooks et al., 1994) and pear (Whitesides and Spotts, 1991), to herbaceous crop plants such as sugar beets (Jacobs et al., 1985) and maize (Fischer et al., 1992; Lalande et al, 1989; McInroy and Kloepper, 1995b).

Diversity associated with bacterial endophytes exists not only in the plant

Table 1 Bacterial Species Isolated as Endophytes and the Host and Tissues from Which They Were Isolated[a]

Species[b]	Host[c]	Tissue	Ref.
Gram-positive			
Arthrobacter			
ureafaciens	Aspen; potato	Woody; tuber	Bacon and Mead, 1971; Hollis, 1951
spp.	Corn; cotton; lemon;	Root and stem; stem; root	McInroy and Kloepper, 1995b; McIn-roy and Kloepper, 1995b; Gardner et al., 1982
Aureobacterium			
saperdae	Cotton	Seed, root, stem, flower, and boll	Misaghi and Donndelinger, 1990
spp.	Corn; cotton	Stem; root and stem	McInroy and Kloepper, 1995b; McIn-roy and Kloepper, 1995b
Bacillus			
alclophilus	Potato	Tuber	Sturz and Matheson, 1996
alvei	Oak	Woody	Brooks et al., 1994
amyloliquefaciens	Potato	Tuber	Sturz and Matheson, 1996
brevis	?; cotton; potato	Seed/ovule; seed, root, stem, flower and boll; tuber	Mundt and Hinkle, 1976; Misaghi and Donndelinger, 1990; Sturz and Matheson, 1996
carotarum	Maple	Woody	Hall et al., 1986
cereus	?; cauliflower; *Sinapis*	Seed/ovule; seed; seed	Mundt and Hinkle, 1976; Pleban et al., 1995; Pleban et al., 1995
circulans	?	Seed/ovule	Mundt and Hinkle, 1976
coagulans	Maple	Woody	Hall et al., 1986
fastidiosus	Grape	Stem	Bell et al., 1995
insolitus	Grape	Stem	Bell et al., 1995
megatherium	?; clover; corn; cotton; po-tato; potato	Seed/ovule; root; root and stem; root and stem; tuber; tuber	Mundt and Hinkle, 1976; Philipson and Blair, 1957; McInroy and Kloep-per, 1995b; McInroy and Kloepper, 1995b; Hollis, 1951; Sturz and Matheson, 1996

	Host	Tissue	Reference
pasteurii	Potato	Tuber	Sturz and Matheson, 1996
pumilus	?; corn; cotton; cotton; oak; sunflower	Seed/ovule; root and stem; seed, root, stem, flower and boll; root and stem; woody; seed	Mundt and Hinkle, 1976; McInroy and Kloepper, 1995b; Misaghi and Donndelinger, 1990; McInroy and Kloepper, 1995b; Brooks et al., 1994; Pleban et al., 1995
sphaericus	Potato	Tuber	Sturz and Matheson, 1996
subtilis	?; corn; cotton; maple; onion; sugar beet	Seed/ovule; root and stem; root and stem; woody; flower; root	Mundt and Hinkle, 1976; McInroy and Kloepper, 1995b; McInroy and Kloepper, 1995b; Hall et al. 1986; Pleban et al., 1995; Jacobs et al., 1985
thuringiensis	Corn; cotton	Root and stem; root	McInroy and Kloepper, 1995b; McInroy and Kloepper, 1995b
spp.	Aspen; corn; corn; cotton; cotton; lemon; potato; potato	Woody; root and stem; root; seed, root, stem, flower and boll; root; root and stem; stem and tuber; tuber	Knutson, 1973; McInroy and Kloepper, 1995b; Lalande et al., 1989; Misaghi and Donndelinger, 1990; McInroy and Kloepper, 1995b; Gardner et al., 1982; De Boer and Copeman, 1974; Sturtz, 1995
Brevibacterium			
ammoniagenes	Alder; aspen	Woody; woody	Bacon and Mead, 1971; Bacon and Mead, 1971
insectiphilium	Aspen	Woody	Bacon and Mead, 1971
linens	?; alder; aspen; pine	Seed/ovule; woody; woody; woody	Mundt and Hinkle, 1976; Bacon and Mead, 1971; Bacon and Mead, 1971; Bacon and Mead, 1971
Cellulomonas			
acidula	Aspen	Woody	Bacon and Mead, 1971
spp.	Cotton	Root and stem	McInroy and Kloepper, 1995b

Table 1 Continued

Species[b]	Host[c]	Tissue	Ref.
Clavibacter			
michiganensis	?; corn; cotton	Seed/ovule; root and stem; root and stem	Mundt and Hinkle, 1976; McInroy and Kloepper, 1995b; McInroy and Kloepper, 1995b
spp.	Cotton	Seed, root, stem, flower and boll	Misaghi and Donndelinger, 1990
Clostridium			
carbonei	Pinto bean	Stem	Thomas and Graham, 1952
Corynebacterium			
fimi	Pinto bean	Stem	Thomas and Graham, 1952
helvolum	Pinto bean	Stem	Thomas and Graham, 1952
humiferum	Alder; aspen	Woody; woody	Bacon and Mead, 1971; Bacon and Mead, 1971
hypertrophicans	?	Seed/ovule	Mundt and Hinkle, 1976
spp.	?; corn; lemon; sugar beet; tomato	Seed/ovule; root; root; root; fruit	Mundt and Hinkle, 1976; Lalande et al., 1989; Gardner et al., 1982; Jacobs et al., 1985; Samish et al., 1963
Curtobacterium			
citreum	Potato	Tuber	Sturz and Matheson, 1996
faccumfaciens	?; grape; pinto bean	Seed/ovule; stem; stem	Mundt and Hinkle, 1976; Bell et al., 1995; Thomas and Graham, 1952
luteum	Potato	Tuber	Sturz and Matheson, 1996
pusillum	Grape	Stem	Bell et al., 1995
spp.	Corn; cotton; yam	Root and stem; root and stem; shoot	McInroy and Kloepper, 1995b; McInroy and Kloepper, 1995b; Tör et al., 1992

Organism	Host	Tissue	Reference
Lactobacillus sp.	Sugar beet	Root	Jacobs et al., 1985
Leuconostoc mesenteroides	?	Seed/ovule	Mundt and Hinkle, 1976
Microbacterium spp.	Corn; cotton	Root and stem; root and stem	McInroy and Kloepper, 1995b; McInroy and Kloepper, 1995b
Micrococcus luteus	?	Seed/ovule	Mundt and Hinkle, 1976
spp.	?; corn; cotton; potato	Seed/ovule; root and stem; root and stem; tuber	Mundt and Hinkle, 1976; McInroy and Kloepper, 1995b; McInroy and Kloepper, 1995b; De Boer and Copeman, 1974
Nocardia salmonicolor	?	Seed/ovule	Mundt and Hinkle, 1976
Rhodococcus luteus	Grape	Stem	Bell et al., 1995
Staphylococcus spp.	Corn; cotton; grape	Root and stem; root and stem; stem	McInroy and Kloepper, 1995b; McInroy and Kloepper, 1995b; Bell et al., 1995
Streptomyces albolongus	?	Seed/ovule	Mundt and Hinkle, 1976
albovinaceus	Elm	Woody	O'Brien et al., 1984
faecium	?	Seed/ovule	Mundt and Hinkle, 1976
griseus	Elm	Woody	O'Brien et al., 1984
Actinomycetes	Lemon; potato; squash	Root; tuber; fruit	Gardner et al., 1982; Sturtz, 1995; Sharrock et al., 1991
Gram-negative			
Achromobacter parvulus	Alder; aspen; pine	Woody; woody; woody	Bacon and Mead, 1971; Bacon and Mead, 1971; Bacon and Mead, 1971

Table 1　Continued

Species[b]	Host[c]	Tissue	Ref.
Achromobacter spp.	?; Lemon	Seed/ovule; root	Mundt and Hinkle, 1976; Gardner et al., 1982
Acinetobacter			
baumannii	Cotton	Stem	McInroy and Kloepper, 1995b
calcoaceticus	?	Seed/ovule	Mundt and Hinkle, 1976
lwoffii	Lemon	Root	Gardner et al., 1982
spp.	Potato	Tuber	Sturtz, 1995
Aerobacter			
cloaceae	Clover; potato; tomato	Root; tuber; fruit	Philipson and Blair, 1957; Hollis, 1952; Samish et al., 1963
Agrobacterium			
radiobacter	Corn; cotton	Root and stem; root and stem	McInroy and Kloepper, 1995b; McInroy and Kloepper, 1995b
spp.	Potato; potato	Tuber; tuber	De Boer and Copeman, 1974; Sturtz and Matheson, 1996
Alcaligenes			
faecalis	?	Seed/ovule	Mundt and Hinkle, 1976
spp.	Cotton; lemon	Root; root	McInroy and Kloepper, 1995b; Gardner et al., 1982
Burkholderia (*Pseudomonas*)			
cepacia	Corn; cotton; lemon	Root and stem; root and stem; root	McInroy and Kloepper, 1995b; McInroy and Kloepper, 1995b; Gardner et al., 1982
gladioli	Corn; cotton	Root and stem; root	McInroy and Kloepper, 1995b; McInroy and Kloepper, 1995b
picketii	Corn; cotton	Root and stem; root and stem	McInroy and Kloepper, 1995b; McInroy and Kloepper, 1995b

Citrobacter			
freundii	Lemon	Root	Gardner et al., 1982
koseri	Corn	Root and stem	McInroy and Kloepper, 1995b
sp.	Cucumber	Fruit	Meleney and Stanghellini, 1974
Comamonas			
terrigena	Grape	Stem	Bell et al., 1995
testosteroni	Cotton	Root	McInroy and Kloepper, 1995b
Cytophaga			
hutchinsonii	?	Seed/ovule	Mundt and Hinkle, 1976
rubra	?	Seed/ovule	Mundt and Hinkle, 1976
Enterobacter			
aerogenes	?; lemon	Seed/ovule; root	Mundt and Hinkle, 1976; Gardner et al., 1982
agglomerans	Corn; grape; lemon	Stem; stem; root	Fisher et al., 1992; Bell et al., 1995; Gardner et al., 1982
asburiae	Corn; cotton	Root and stem; root and stem	McInroy and Kloepper, 1995b; McInroy and Kloepper, 1995b
cloacae	Corn; corn; cotton; cucumber; grape; lemon	Root; root and stem; root; fruit; stem; root	Hinton and Bacon, 1995; McInroy and Kloepper, 1995b; McInroy and Kloepper, 1995b; Meleney and Stanghellini, 1974; Bell et al., 1995; Gardner et al., 1982
sakazakii	Lemon	Root	Gardner et al., 1982
spp.	Corn; cotton; grape	Root and stem; root and stem; stem	McInroy and Kloepper, 1995b; McInroy and Kloepper, 1995b; Bell et al., 1995
Erwinia			
amylovora	?; aspen; pine	Seed/ovule; woody; woody	Mundt and Hinkle, 1976; Bacon and Mead, 1971; Bacon and Mead, 1971
carotovora	?; cotton	Seed/ovule; root and stem	Mundt and Hinkle, 1976; McInroy and Kloepper, 1995b

Table 1 Continued

Species[b]	Host[c]	Tissue	Ref.
herbicola	?; alfalfa; oak; sugar beet; watermelon	Seed/ovule; root; woody; root; fruit	Mundt and Hinkle, 1976; Gagné et al., 1987; Brooks et al., 1994; Jacobs et al., 1985; Hopkins and Elmstrom, 1977
salicis	Aspen	Woody	Bacon and Mead, 1971
spp.	Alfalfa; aspen; cotton; cucumber	Root; woody; seed, root, stem, flower and boll; fruit	Gagné et al., 1987; Knutson, 1973; Misaghi and Donndelinger, 1990; Meneley and Stanghellini, 1974
Escherichia			
coli	Tomato	Fruit	Samish et al., 1963
spp.	corn; cotton	Root and stem; root and stem	McInroy and Kloepper, 1995b; McInroy and Kloepper, 1995b
Flavimonas			
oryzihabitans	Corn	Root	McInroy and Kloepper, 1995b
rhenanus	Clover	Root	Philipson and Blair, 1957
Flavobacterium			
capsulatum	?	Seed/ovule	Mundt and Hinkle, 1976
devorans	?	Seed/ovule	Mundt and Hinkle, 1976
lutescens	?	Seed/ovule	Mundt and Hinkle, 1976
rigense	?	Seed/ovule	Mundt and Hinkle, 1976
spp.	?; corn; cotton; lemon; potato; tomato	Seed/ovule; root and stem; root; root; tuber; fruit	Mundt and Hinkle, 1976; McInroy and Kloepper, 1995b; McInroy and Kloepper, 1995b; Gardner et al., 1982; De Boer and Copeman, 1974; Samish et al., 1963

Hydrogenophaga spp.	Corn; cotton	Root and stem; root and stem	McInroy and Kloepper, 1995b; McInroy and Kloepper, 1995b
Kingella kingae	Potato	Tuber	Sturz and Matheson, 1996
Klebsiella			
ozaenae	Grape	Stem	Bell et al., 1995
pneumoniae	Grape	Stem	Bell et al., 1995
terrigena	Corn; grape	Stem; stem	Fisher et al., 1992; Bell et al., 1995
spp.	Corn; cotton; lemon	Root and stem; root; root	McInroy and Kloepper, 1995b; McInroy and Kloepper, 1995b; Gardner et al., 1982
Kluyvera spp.	Corn; cotton	Root and stem; root	McInroy and Kloepper, 1995b; McInroy and Kloepper, 1995b
Methylobacterium spp.	Corn; cotton	Root and stem; root and stem	McInroy and Kloepper, 1995b; McInroy and Kloepper, 1995b
Moraxella bovis	Grape	Stem	Bell et al., 1995
Ochrobactrum anthropi	Corn; cotton	Stem; root	McInroy and Kloepper, 1995b; McInroy and Kloepper, 1995b
Pantoea			
agglomerans	Grape; potato	Stem; tuber	Bell et al., 1995; Sturz and Matheson, 1996
spp.	Corn; cotton	Root and stem; root and stem	McInroy and Kloepper, 1995b; McInroy and Kloepper, 1995b
Phyllobacterium spp.	Corn; cotton	Root and stem; root and stem	McInroy and Kloepper, 1995b; McInroy and Kloepper, 1995b
Proteus			
mirabilis	Cucumber	Fruit	Meneley and Stanghellini, 1974
vulgaris	?	Seed/ovule	Mundt and Hinkle, 1976

Table 1 Continued

Species[b]	Host[c]	Tissue	Ref.
Providencia sp.	Lemon	Root	Gardner et al., 1982
Pseudomonas			
acidovorans	?	Seed/ovule	Mundt and Hinkle, 1976
aeruginosa	Lemon; sugar beet	Root; root	Gardner et al., 1982; Jacobs et al., 1985
alcaligenes	?	Seed/ovule	Mundt and Hinkle, 1976
caryophilli	?	Seed/ovule	Mundt and Hinkle, 1976
chlororaphis	Corn; cotton	Root and stem; root and stem	McInroy and Kloepper, 1995b; McInroy and Kloepper, 1995b
corrugata	Corn; grape	Stem; stem	Fisher et al., 1992; Bell et al., 1995
cichorii	Grape	Stem	Bell et al., 1995
facilis	?	Seed/ovule	Mundt and Hinkle, 1976
fluorescens	?; bean; corn; lemon; sugar beet	Seed/ovule; seed; stem; root; root	Mundt and Hinkle, 1976; Pleban et al., 1995; Fisher et al., 1992; Gardner et al., 1982; Jacobs et al., 1985
marginata	?	Seed/ovule	Mundt and Hinkle, 1976
marginalis	Corn; grape	Stem; stem	Fisher et al. 1992; Bell et al., 1995
palleroni	?	Seed/ovule	Mundt and Hinkle, 1976
putida	?; corn; cotton; cotton; grape; lemon	Seed/ovule; root and stem; root; seed; root, stem, flower and boll; stem; root	Mundt and Hinkle, 1976; McInroy and Kloepper, 1995b; McInroy and Kloepper, 1995b; Misaghi and Donndelinger, 1990; Bell et al., 1995; Gardner et al., 1982
saccharophila	Corn; cotton	Root and stem; root and stem	McInroy and Kloepper, 1995b; McInroy and Kloepper, 1995b
stutzeri	?	Seed/ovule	Mundt and Hinkle, 1976

syringae	?; alfalfa; grape; peach; pear	Seed/ovule; root; stem; woody; root and stem	Mundt and Hinkle, 1976; Gagné et al., 1987; Bell et al., 1995; Dowler and Weaver, 1975; Whitesides and Spotts, 1991
tolaasii	Potato	Tuber	Sturz and Matheson, 1996
vesicularis	?	Seed/ovule	Mundt and Hinkle, 1976
spp.	Alfalfa; corn; corn; cotton; grape; peach; potato; potato; tomato; watermelon	Root; stem; root; root and stem; root and stem; stem; woody; tuber; tuber; fruit; fruit	Gagné et al., 1987; Fisher et al., 1992; Lalande et al., 1989; McInroy and Kloepper, 1995b; McInroy and Kloepper, 1995b; Bell et al., 1995; Dowler and Weaver, 1975; De Boer and Copeman, 1974; Sturz, 1995; Samish et al., 1963; Hopkins and Elstrom, 1977
Rahnella agquatilis	Grape	Stem	Bell et al., 1995
Ralstonia			
solanacearum	Corn; cotton; cotton	Root and stem; root and stem; seed, root, stem, flower and boll	McInroy and Kloepper, 1995b; McInroy and Kloepper, 1995b; Misaghi and Donndelinger, 1990
Rhizobium			
japonicum	Corn; cotton	Root; root	McInroy and Kloepper, 1995b; McInroy and Kloepper, 1995b
leguminosarum	Rice	Root	Yanni et al., 1997
Serratia			
liquefaciens	Lemon; potato	Root; tuber	Gardner et al., 1982; Sturz and Matheson, 1996
marcescens	?; lemon	Seed/ovule; root	Mundt and Hinkle, 1976; Gardner et al., 1982
plymuthica	Potato	Tuber	Sturz and Matheson, 1996
proteamaculans	Potato	Tuber	Sturz and Matheson, 1996
spp.	Corn; cotton	Root and stem; root and stem	McInroy and Kloepper, 1995b; McInroy and Kloepper, 1995b

Table 1 Continued

Species[b]	Host[c]	Tissue	Ref.
Shigella spp.	Lemon	Root	Gardner et al., 1982
Sphingomonas paucimobilis	Corn; cotton	Root and stem; root and stem	McInroy and Kloepper, 1995b; McInroy and Kloepper, 1995b
Stenotrophomonas maltophilia	Corn	Root	McInroy and Kloepper, 1995b
Variovorax paradoxus	Corn	Root and stem	McInroy and Kloepper, 1995b
Vibrio spp.	Corn; lemon	Stem; root	Fisher et al. 1992; Gardner et al., 1982
Xanthomonas campestris	Corn; cotton; grape; potato; pinto bean	Root and stem; root and stem; stem; tuber; stem	McInroy and Kloepper, 1995; McInroy and Kloepper, 1995b; Bell et al., 1995; Sturtz and Matheson, 1996; Thomas and Graham, 1952
spp.	?; cotton; potato; sugar beet; tomato; watermelon	Seed/ovule; seed, root, stem, flower and boll; tuber; root; fruit; fruit	Mundt and Hinkle, 1976; Misaghi and Donndelinger, 1990; De Boer and Copeman, 1974; Jacobs et al. 1985; Samish et al., 1963; Hopkins et al., 1977
Yersinia frederiksenii	Corn; cotton	Root and stem; stem	McInroy and Kloepper, 1995b; McInroy and Kloepper, 1995b
spp.	Lemon	Root	Gardner et al., 1982

[a] Excludes bacteria known to be pathogenic to the host and bacteria isolated from diazotrophic studies.
[b] Isolates grouped as spp. were not identified to species level, with the exception of McInroy and Kloepper (1995b) in which more than one species were grouped together.
[c] ?: Refers to unknown species from a group of various host plants.

species colonized but also in bacterial taxa. Variation in bacteria that have been reported as endophytes span a significant range of both gram-positive and gram-negative species (Table 1). In the original study by Mundt and Hinkle (1976), as many as 46 different bacterial species were identified from the 27 plant species surveyed. In a single, extensive study conducted by McInroy and Kloepper (1995b), close to 50 different bacterial species, with an additional 46 unidentifiable isolates, were recovered from the roots and stems of cotton and sweet corn. It is difficult to compare studies that identify bacteria in earlier studies with more recent studies due to the changing nature and incorporation of new methodologies for bacterial taxonomy (Vandamme et al., 1996). Nonetheless, certain trends are apparent with predominant bacterial types isolated as endophytes. Many belong to species previously identified as plant-associated bacteria and thus provide some insight into their evolved relationships with plants.

2. DEFINITIONS OF ENDOPHYTES AS PERTAINING TO BACTERIA

As emphasized by Chanway (1996), there are a variety of definitions given for the term endophyte, and consideration of each leads to different interpretations. The general definition for the term provided in this book by Stone et al. (see previous chapter, this book) states that endophytes are those organisms in which "infections are inconspicuous, the infected host tissues are at least transiently symptomless, and the microbial colonization can be demonstrated to be internal." This definition is somewhat confusing when applied to bacteria, however, since a clear definition for "symptomless" is required. For example, disease symptoms may refer specifically to macroscopic observations. However, less obvious microscopic symptoms may occur during early or latent stages of bacterial infections that can be initially categorized as "inconspicuous" or "asymptomatic." Likewise, "symptoms" may extend beyond obvious macroscopic or microscopic disease symptoms; host responses to endophyte infections may result in more subtle effects at the molecular or biochemical level. For example, stresses inflicted by the presence of bacteria may result in an increased energy load on a plant that does not result in typical disease symptoms, but might be detectable by loss of fitness as measured by reduced growth rates or reduced crop yield. In a second example, abnormal physiological symptoms inflicted by phytohormone-like compounds produced by bacteria may be considered disease in some cases but beneficial in other cases.

Questions pertaining to what constitutes true endophytic relationships center around the types of outcomes that may occur with endophytic infections, since overall outcomes of infections can be beneficial, detrimental, or neutral. Virtually every bacterial pathogen that infects its normal host will go through a "transiently symptomless stage,'' if the definition of symptoms refers specifically to macroscopic observations. Thus, based on the definition provided by Stone et al. (discussed in the previous chapter), all bacterial pathogens should be considered endophytes. Since latent infections are prevalent for bacterial pathogens (reviewed in Hayward, 1974), it is often difficult to discern in the literature reports of what represents nonsymptomatic, endophytic colonization by bacterial pathogens from bacterial infections simply delayed in symptom production. This issue is further complicated by the nature and understanding of general bacterial pathogen life cycles. Unlike certain fungal pathogens, such as *Colletotrichum* and *Acremonium* spp. that have distinct endophytic stages as a part of their general life cycle (Bailey et al., 1992; Bacon and Hill, 1996), precise endophytic stages have not been established in bacterial life cycles.

A more simplistic but resultantly more accurate definition for bacterial endophytes was proposed by Kado (1991), in which endophytes are ''bacteria living in plant tissues without doing substantive harm or gaining benefit other than residency.'' This definition restricts inclusion of bacteria that lead to disease symptoms or have detrimental effects on host plants, but still defines the term to describe bacterial endophytes that provide net beneficial effects for practical uses on plants. Since the general trend for the term endophyte has been toward agricultural benefit, the focus of this chapter is primarily on bacteria that fit this definition.

3. SPECULATION ON THE EVOLUTIONARY DEVELOPMENT OF ENDOPHYTIC BACTERIA

Regardless of the definition used, the variety of bacterial species identified in the literature as endophytes cover a wide range of bacteria, including several species that are typically associated with plants in other manners besides endophytic or pathogenic relationships. Such diversity leads to difficulties in speculating a single model for evolution of endophytic relationships. Therefore, we will consider parasitism as it relates to pathogens as a starting point. Heath (1986) described a generally accepted proposal for pathogen evolution, of which variations have been extended specifically for bacteria (e.g., Djordjevic et al., 1987). In this general model, proposed evolution of plant parasitism originates from saprotrophy. Increased complexity of an interaction between a parasite and host reflects a more evolved relationship, and thus a more evolved parasite. As a consequence, as parasites evolve with their hosts, so does the reliance of both partners on each other. Using this model, the least evolved parasites may be represented by saprotrophic bacteria that typically reside ectotrophically and survive off of nutrients exuded by the plant, such as fluorescent pseudomonads that comprise a significant population of rhizobacteria. More evolved parasites may be represented by those microbes that are pathogenic in nature but can survive effectively as saprophytes. The soft rot bacterium, *Erwinia carotovora*, provides an example of such a "facultative" pathogen. More evolved parasites may be represented by "semiobligate" and "obligate" pathogens, such as the specialized pathogen, *Pseudomonas syringae*, and the fastidious prokaryote, *Xylella fastidiosa*, respectively. Finally, the most evolved parasites are represented not by pathogens but by organisms occurring in mutualistic relationships such as the fastidious diazotroph, *Azotobacter*. It is important to note, as stated by Kado (1991), that plant-associated bacteria "have evolved to best survive on or within particular plant surfaces or tissues." This statement emphasizes that every bacterium selectively evolves toward enhancing its own fitness or survival, which does not necessarily mean evolving the interaction with the plant host. Thus, in many cases, evolution toward saprotrophy can be considered. With this assumption in mind, and keeping with the proposed evolution of parasitism, bacteria reported as endophytes have been identified from essentially every example of evolved interactions. Based on this evolutionary model and their interactive relationship with their host, bacterial endophytes can be separated into two generally different types, evolved and nonevolved.

As mentioned previously, depending on the definition used, bacterial pathogens can be classified as endophytes if latent periods of infection are considered. Critical population mass for successful infection and disease symptom production is obviously an important criterion for disease in many if not all pathosystems. One obvious reason for delayed symptom production may result from the necessity of pathogens to increase their populations sufficiently to express factors required to manifest disease symptoms. For example, infection by fastidious and cell wall-less prokaryotes, such as *Xylella fastidiosa*, *Phytoplasma* spp. and *Spiroplasma* spp., can often be observed in healthy hosts (e.g., Feldman et al., 1977). Infections by these bacteria typically require an incubation period within the plant host prior to symptom production. Such delay periods can last for weeks before any visual symptoms are observed. In a second, more molecularly defined example, studies have identified complex regulatory pathways in bacterial pathogens involved in expression of traits necessary for pathogen virulence. Quorum sensing is a form of positive gene regulation in bacteria that is prevalent in plant-associated bacteria, including many plant pathogens such as *Agrobacterium* and *Erwinia* (Pirhonen et al., 1993; Faqua et al., 1994). This form of gene regulation requires sufficient bacterial populations to be established before certain traits associated with virulence are expressed. This inherent regulatory pathway to control gene expression may have evolved from the necessity to develop a mechanism that avoids inducing host plant resistance responses. This clearly appears to be the case with the potato soft rot pathogen, *Erwinia carotovora*. In this pathosystem, the degradative enzyme pectate lyase is a major virulence factor involved in soft rot symptom production. Biochemical studies have indicated that low-level activity of pectate lyase results in the production of degraded host polymeric products of sizes capable of eliciting defense responses in the host plant (Barras et al., 1994). High-level activity of the same enzyme, however, results in production of degraded host products of smaller sizes that do not elicit plant defense responses but rather enhance virulence in the bacterium. It is likely that upon invasion of host tissue this bacterium resides asymptomatically within plant tissue with repressed virulence factors. These virulence factors are expressed only when populations of the pathogen reach high enough levels in the tissue to produce an abundance of the enzyme, under the relative assurance that degraded host polymeric fragments that elicit host defense responses are not generated.

Reduced virulence traits based on lack of appropriate host signals that induces their expression may also be a part of symptomless infections. For example, plant signals such as specific organic acids which appear to regulate the expression of the phytotoxin coronatine in *P. syringae* pv. *tomato* are less abundant in some "nonhost" plants such as cucumber as opposed to the "host" plant tomato (Li et al., 1998). Since production of the toxin appears to be a major factor in virulence of the pathogen (Bender et al., 1987), repression of toxin production in certain nonhost plants may provide a means for colonization to occur but prevents disease symptom manifestation.

Colonization of host tissues by pathogenic bacteria without symptom production commonly occurs. Such occurrences, however, can provide inoculum sources for disease to occur within the same season or in subsequent seasons when conditions support symptom manifestation. Several studies have been conducted with asymptomatic infection of fruit trees with *Pseudomonas syringae* pathovars (e.g., Cameron, 1970; Roos and Hattingh, 1987; Sundlin et al., 1988; Whitesides and Spotts, 1991). In these studies, *P. syringae* isolates were found to effectively colonize stem tissue as well as dormant leaf buds and leaf scars, and were thought to be the primary overwintering site that serves as the primary inoculum source for infections in subsequent years.

In the proposed evolution of plant-associated bacteria, the most evolved pathogens are those that have established obligate relationships with their hosts. Thus, the fastidious prokaryotes would appear to be among the most evolved pathogens, since their survival as plant pathogens require the plant host; their fastidious nature is exemplified by the inability to survive epiphytically or saprophytically on the plant. Since the most intimate associations for endophytes are those microbes that have lost their ability to cause disease symptoms but retain the ability to colonize host tissue in significant populations, examples of fastidious prokaryotic pathogens may provide evolutionary links between pathogens and evolved endophytes. *Clavibacter xyli* subsp. *cynodontis* (Cxc), the bacterium associated with Bermuda grass stunt disease (Davis et al., 1983; Liao and Chen, 1981), is a prime example of such a potential link. Cxc is a close relative of *Clavibacter xyli* subsp. *xyli* (Cxx), the causative agent of ratoon stunt disease of sugar cane (Davis et al., 1984). Although both Cxc and Cxx are significant vascular pathogens to their respective hosts, Cxc is also capable of colonizing the vascular tissue of maize without inflicting any apparent disease symptoms. Thus, Cxc provides an example of a plant pathogen that can endophytically colonize a nonhost plant without displaying disease symptoms.

The genetics that dictate differences between pathogens and endophytes have not been clearly defined. The best genetic study to date on this subject was performed with the fungal pathogen *Colletotrichum* by Freeman and Rodriguiz (1993). In this study, a single mutation resulted in loss of disease symptom manifestation, but not loss of colonization and growth ability within the host plant. Although no such mutant has been experimentally identified in bacterial pathogens, it is also possible that some bacteria may only require one or a few mutations prior to evolving into true endophytes.

To date, many genetic studies defining pathogenicity and virulence factors within plant pathogens are phenotypically linked to symptom production (Lindgren et al., 1986; Rahme et al., 1991; Alfano and Collmer, 1996, 1997). Pathogens mutated in specific genes located in the *hrp* gene cluster, a region within plant pathogen genomes important for basic pathogenic functions, still have the capability to reside and grow to some degree in host tissue without manifesting disease symptoms (Lindgren et al., 1986; Rahme et al., 1991). These bacteria may still have the basic machinery to catabolize plant-derived nutrients, and thus can survive in plant tissue, but no longer are capable of inducing visible disease symptoms.

Hrp gene clusters comprise a major genetic portion encoding pathogenicity of several major groups of bacterial plant pathogens (Lindgren et al., 1986; Rahme et al., 1991; Alfano and Collmer, 1996, 1997; Hueck, 1998). Genes that are found within these clusters are categorized into three general groups. Regulatory genes are involved in the expression of other genes within the *hrp* cluster, which primarily are induced by signals originating from the host plant. A second group of genes, referred to as *hsc* genes, includes homologs to structural genes for the type III protein secretion pathway originally described in mammalian pathogens (Alfano and Collmer, 1996, 1997; Bogdanove et al., 1996). The third group is composed of genes encoding outer membrane proteins (referred to as hop proteins) that are secreted through the type III secretion pathway. This group plays an important role in interactions between pathogens and hosts, since it includes avirulence genes (Alfano and Collmer, 1996, 1997, Leach and White, 1996), which are centrally responsible for elicitation of defense responses in plants. Avirulence genes and the more basic pathogenicity machinery of *hrc/hrp* genes encoding type III secretion products responsible for pathogenicity are likely to be lacking or significantly suppressed in endophytes. There is evidence that this may also be true of bacterial species. In at least one example, a nonpathogenic, "opportunistic" *Xanthomonas* strain was demonstrated to be lacking a major portion of the *hrp* gene cluster (Bonas et al., 1991). Molecular studies have demonstrated that pathogenic strains mutated in certain *hrp* genes are no longer capable of invoking necrogenic responses in plants that are responsible for disease symptom production. Depending on the location of the mutation, some strains are still capable of multiplying within plant tissue to some degree, albeit at lower levels than the wild-type pathogen (Lindgren et al., 1986; Rahme et al., 1991). These observations suggest that a more complex, evolved interaction is necessary to obtain endophytic relationships that are capable of achieving high populations, similar to pathogens, without invoking disease symptoms.

In addition to the active defense responses induced by plant pathogens, many nonpathogenic, endophytic bacteria have been reported to induce other forms of defense responses in plants, including induced systemic resistance (e.g., Kloepper et al., 1992; Chen et al., 1997; Benhamou et al., 1996a, b, 1998; Duijff et al., 1997). Although the molecular events involved in triggering systemic resistance have not yet been elucidated, such responses are likely designed for the plant to respond to the presence of a potential pathogen. As mentioned previously, bacteria inducing systemic responses may cause significant energy loads that may have a detrimental effect on the host plant. Bacteria capable of inducing systemic resistance, however, have not been restricted to pathogens or potential pathogens. For example, in a study by Benhamou et al. (1996b), pea plants treated with the rhizobacterium *Bacillus pumilus* were protected from infection by *Fusarium*. Ultrastructure studies indicated that plants pretreated with the bacterium responded rapidly to challenge by *Fusarium* infections and had numerous host responses classically associated with pathogen defense, including callose deposition with cell walls and the accumulation of phenolic compounds. These resistance structures were not apparent in plants that were pretreated with endophytic bacteria and not challenged by the fungal pathogen. In either case, bacteria are most likely utilizing plant-derived nutrients, and are thus inducing biochemical activity that may be depriving the plant of full acquisition of its energy expenditure and consequently result in some detrimental effects. Nonetheless, in this and other studies, whether significant energy losses of the host plant results detrimentally from the induced response, the net result of the interaction is most likely beneficial, and thus a consideration of overall gains to the plant may be more appropriate in considering endophytic relationships.

On the opposite end of the spectrum, several studies have demonstrated that bacterial species commonly associated with plants, but not considered pathogens and not originally isolated as endophytes, can also effectively colonize plant tissue without producing obvious detrimental effects on the plant (e.g., van Peer et al., 1989; Kloepper et al., 1992; Wiehe et al., 1994; Duijff et al., 1997; Pan et al., 1997). These microbes represent potential endophytes that are not as evolved in their interaction with their plant host compared to pathogens. Instead, they normally reside in association with the plant and under appropriate conditions can invade and colonize interior portions. In this case, a nonevolved endophyte is a bacterium capable of utilizing the organic compounds within plants as carbon and energy sources for growth and survival, withstanding the physiological conditions encountered within plant tissues, and avoiding inducing host defense responses that would prevent its growth and survival within the host tissue.

3. GROUPINGS OF BACTERIAL TYPES CONSIDERED AS ENDOPHYTES

A diversity of bacterial types have been isolated as endophytes, as indicated by Table 1. Two extensive studies documenting the types of bacteria in plant tissue were conducted by Mundt and Hinkle (1976) and by McInroy and Kloepper (1995b). The study by Mundt and Hinkle demonstrated a significant number of different bacterial types in several host species seeds and ovules, whereas the study by McInroy and Kleopper demonstrated that several different bacterial types could be isolated from just two plant species. Although other studies concerning bacterial endophytes are not nearly as extensive as these two examples, a significant number of bacteria representing several species have been isolated within interior portions of plants. Many of these bacteria can be categorized according to their potential interaction with the host plant.

3.1. Bacterial Plant Pathogens

A broad representation of bacterial strains that can be categorized into major pathogenic groups have been isolated as endophytes (Table 1). These major groups include genera comprising the gram-positive coryneform pathogens, including *Clavibacter* and *Curtobacterium* species. In addition, isolates of *Agrobacterium* sp., pathogenic *Erwinia* spp., and xanthomonads also have been isolated as endophytes. A number of strains classified as, or formerly classified as, *Pseudomonas* spp. have been isolated as endophytes. These include fluorescent as well as nonfluorescent species, such as *P. syringae*, *Burkholderia* spp., and *Ralstonia solanacearum*. It is unclear as to whether many of these isolates represent nonpathogenic strains of species normally regarded as causative agents for disease. It is possible that many of these isolates represent pathogenic strains residing in plants that are not their normal host for disease, similar to the example presented with *C. xyli* subsp. *cynodontis* residing in maize.

3.2. Other Detrimental Bacterial Endophytes

Bacterial endophytes that do not cause disease symptoms or other noticeable effects may have detrimental effects as a result of their infection of plants. For example, bacteria that naturally reside in plants can cause a variety of contamination problems, such as in tissue culture propagation, a common occurrence with *Bacillus* spp., or in molecular assays with the use of reporter genes such as β-glucuronidase in transformed plants (Tör et al., 1992).

3.3. Nonpathogenic Endophytes

Several bacteria, especially those taxonomically classified as species commonly associated with plants in nonpathogenic manners, have been isolated as endophytes from a wide variety of plant species and plant tissues. These include, for example, common rhizosphere-colonizing bacteria such as, among others, *Bacillus* spp., *Enterobacter* spp., fluorescent pseudomonads, *Serratia* spp., and *Stenotrophomonas* sp. In many of these cases, strains previously isolated as rhizosphere colonizers have been subsequently demonstrated to colonize interior portions of plant roots (Lalande et al., 1989; Kloepper et al., 1992; Benhamou et al., 1996a, b, 1998; Duijff et al., 1997; Pan et al., 1997). It is likely that in several of these examples the extent of endophytic colonization is restricted to subsurface root tissue, with little to no colonization of more interior sites such as the cortex or vascular tissues.

3.4. Diazotrophic Bacteria

Diazotrophs are capable of fixing atmospheric dinitrogen into combined forms utilizable by plants. Plant-associated diazotrophs have therefore been studied as a means of increasing biologically available nitrogen to the plant and thus decreasing the amount of nitrogen fertilizers to agricultural crops. Diazotrophs that endophytically interact with plants can broadly be regarded as endosymbionts, and further differentiated into groups that form nitrogen-fixing root and/or stem nodules and those that do not. Nodule-forming bacteria include members of the Rhizobiaceae, including species of *Rhizobium*, *Bradyrhizobium*, and *Sinorhizobium*, and species of *Frankia*, which are symbionts of leguminous and nonleguminous plants, respectively. These systems have been extensively reviewed (see van Rhijn and Vanderleyden, 1995; Pawlowski and Bisselin, 1996; Pueppke, 1996; Bladergroen and Spaink, 1998; Guan et al., 1998) and will not be discussed in detail here.

Diazotrophic endophytes that do not form nodules have been isolated primarily from grasses such as sugar cane (*Saccharum* spp.), rice (*Oryza sativa*), wheat (*Triticum aestivum*), sorghum (*Sorghum bicolor*), maize (*Zea mays*), and a range of tropical grasses. Bacterial endophytes isolated from these plants include species from the genera *Herbaspirillum*, *Azorhizobium*, *Azoarcus*, *Azospirillum*, and *Acetobacter* (reviewed in Baldani et al., 1997; James and Olivares, 1998; Reinhod-Hurek and Hurek, 1998). In several cases, the endophytes are obligately associated with the interior of the plant host and survive poorly in soil.

4. GENERAL ECOLOGICAL STUDIES

4.1. Endophytic Populations in Plants

Evaluations of average total populations of endophytic bacteria within host tissues have been attempted, however, significant variation has been reported within studies. Such variation undoubtedly arises from incidence of variable endophyte infection between samples, which can result from a number of factors including tissue type, plant source, plant age, and environmental influences. It is apparent, however, that variation may also arise from the methods used for enumeration. As a result, comparisons between studies are difficult. Enumeration of total populations has been reported on a per-gram-tissue basis, or a per-cm^3-tissue basis, and on a per-mL basis of sap or xylem fluids. Furthermore, attempts to evaluate total populations may produce varied results according to the growth medium used for isolation, variations in growth conditions of the host plant, and the state in which the plant tissue was used (e.g., Samish et al., 1989; Meleney and Stanghellini, 1974; Cole and Bugbee, 1976; Gardner et al., 1982; Sharrock et al., 1991; Bell et al., 1995; Hallmann et al., 1996). Nonetheless, endophytic populations appear to be under the same general influences dictating pathogenic bacterial populations.

In most studies, total colonization by nonpathogenic endophytes rarely have been documented to reside at population levels commonly observed for pathogenic bacteria in diseased tissue. However, on occasion, total populations have been recorded to reach fairly high levels. In a survey of cotton, Misaghi and Donndelinger (1990) detected average population sizes from 10^2 to greater than 10^4 g/tissue of root, stem, unopened flowers, and bolls. Similarly, McInroy and Kloepper (1995b) surveyed endophytic populations in field-grown sweet corn and cotton, and reported variation between 10^0 and 10^7 cfu/g tissue, depending on the type of tissue and the time at which samples were collected. Studies determining average populations in fruit tissue and storage organs of various crops were observed at lower than 10^3 cfu/g tissue of tomato and squash (Samish et al., 1963; Sharrock et al., 1991), and up to 10^4 cfu/cm^3 in potato tuber tissue (De Boer and Copeman, 1974). Some of the highest endophytic populations documented, however, appear to reside in the vascular tissues of host plants. Many studies have attempted to document total bacterial populations in xylem tissue of several species, especially in woody plants. De Boer and Copeman (1974) detected between 10^3 and as much as 10^7 cfu/cm^3 in potato stem tissue. Total colonization of woody tissue ranged from 10^2 cfu/mL sap in aspen to as much as 10^5 cfu/mL sap in heartwood of aspen (Knutson, 1973). The detected range of total bacterial populations in grape stems was similar to that detected in woody trees. Populations ranged from 10^2 to almost 10^5 cfu/mL sap of xylem (Bell et al., 1995). In contrast, when populations were determined on a per-gram-tissue basis, between 10^3 and 10^4 cfu/g xylem tissue was detected (Bell et al., 1995).

Endophytic colonization appears especially abundant in root tissue, which may be a reflection of a primary site where endophytes gain entry into plants. In a study performed with alfalfa roots, populations in root xylem were detected from 10^3 to greater than 10^4 cfu/g tissue (Gagné et al., 1987). In contrast, populations have been reported to reach greater than 10^6 cfu/g root tissue in sugar beet (Jacobs et al., 1985). Maximum population levels similar to that observed in sugar beet roots were also reported in citrus xylem root tissue, reflecting a value that was much greater than that detected in twig xylem, which was determined as 5 $\times$ 10^2 cfu/g tissue (Gardner et al., 1982).

Total endophytic populations within host plants appear to be under the influence of both the plant and the environment. For example, population fluctuations have been detected in genotypically different plants. Fluctuations have also been detected seasonally, and in plants grown at different sites (e.g., Samish et al., 1963; Dowler and Weaver, 1975; Gardner et al., 1982; McInroy and Kloepper, 1995a; Sturz and Christie, 1995). Dowler and Weaver (1975) observed an inability to isolate both pathogenic and nonpathogenic *P. syringae* strains from healthy peach during late spring and summer months. Gardner et al. (1982) observed fluctuations in total populations within roots of lemon throughout an entire year, and attributed the fluctuation to environmental factors, as well as inherent populations. In a study measuring total populations of culturable endophytes over two field seasons in both sweet corn and cotton, McInroy and Kloepper (1995a) detected significant population fluctuations within both crop plants. The location of bacterial populations in fruit also varies. In tomato, populations were found to be greatest in the middle of the fruit and less at the periphery, except at the stem scar. In contrast, populations in cucumber fruit were greatest at the periphery and less in the middle. Pillay and Nowak (1997) indicated that host genotype differences of tomato resulted in improved epiphytic colonization by endophytes. In this case, increased epiphytic colonization likely had an effect on the incidence of endophytic infection of host plants; however, experimental evidence to prove this was not provided. Pillay and Nowak (1997) also investigated the influence of temperature on endophytic populations in tomato. Significantly greater endophytic populations were detected in both roots and shoots at a lower temperature of 10°C compared to 30°C. The types of endophytic bacteria present in tissue may also influence colonization of others (Quadt-Hallmann et al., 1997b).

4.2. Endophytic Growth and Movement Within Plants

Endophytic growth has been monitored over extended periods after inoculation into host plants. Pleban et al. (1995) detected the presence of *B. cereus* inoculated into cotton over a 72-day period using a method incorporating radioactive labeling. Colonization studies were performed in cotton stems using six different bacterial strains labeled with rifampicin resistance (Chen et al., 1995). In this study, all six strains were detected 21 days after inoculation. Four of the strains initially established in cotton stems at populations above 10^6 cfu/g tissue. These four strains, and an additional strain, appeared to reach stabilizing populations between 10^3 and 10^5 cfu/g tissue within 7 days that were maintained for the remainder of the 21-day experiment. In this same study but in a separate experiment, a significant increase in populations of one bacterial strain increased from less than 10^1 cfu/g tissue to above 10^6 cfu/g tissue when inoculated at a concentration of 10^3 cfu/plant.

Myers and Strobel (1983) detected a *P. syringae* strain up to one year following inoculation into stems of elm seedlings. The presence of a *Bacillus subtilis* strain marked with resistance to the antibiotic rifampicin was detected over a 2-year period after inoculation into sugar maple (Hall and Davis, 1990). This study demonstrated the potential of long-term survival (over 2 years) from a point inoculation in woody tissue. Although rifampicin resistance is the predominant method for detecting bacteria in colonization studies, it should be cautioned that use of antibiotic resistance may not be the best method for detecting endophytic colonization since variability as a result of antibiotic "masking" has been reported with endophytic bacteria (McInroy et al., 1996).

Although it is apparent that endophytes are capable of movement upon entry within their hosts, few studies have been conducted to specifically evaluate movement of endophytic bacteria within plant tissue. Hall et al. (1990) detected movement of *B. subtilis* from the point of inoculation as much as 75% of the stem height of inoculated sugar maple seedlings. In this study, it was suggested that movement of *B. subtilis* occurred through the xylem tissue. Whitesides and Spotts (1991) detected movement of *P. syringae* inoculated into stems of pear, although distance was limited to about 3 cm 20 days post inoculation. In this study, movement was detected using strains originally isolated as epiphytes as well as strains isolated as endophytes. Inoculation of roots was also performed with these isolates, but endophytic populations were not detected above the crown of the plant. In the study performed by Myers and Strobel (1983), movement of *P. syringae* strains from the point of inoculation in elm seedlings after 1 year ranged from 0 to 64 cm. Chen et al. (1995) detected variability in movement between bacterial strains inoculated into cotton stems. Movement of a maximum distance of 3–5 cm from the point of inoculation was detected with two of five strains within 14 days, while no movement was detected for the remaining three bacteria after 21 days post inoculation.

4.3. Gain of Entry into Host Plants

There are some general aspects of pathogen infections that can be compared to infections by symptomless endophytes. Like bacterial pathogens, endophytes must possess the ability to infect and colonize host plant tissue. Bacterial pathogens do not invade plant tissue directly; infection typically occurs passively, with entry obtained through natural openings or wounds (Huang, 1986). Similarly, passive invasion is apparent with bacterial endophytes.

A primary site by which endophytes are proposed to gain entry into plants is within the root zone, where the preponderance of endophytic colonization has been detected compared to that of other host tissues (e.g., Gardner et al., 1982; Gagné et al., 1987; Whitesides and Spotts, 1991; McInroy and Kloepper, 1995b). Several studies utilizing microscopy have been conducted to verify endophytic colonization of host roots (Old and Nicolson, 1975; Jacobs et al., 1985; Frommel et al., 1991; Wiehe et al., 1995; Benhamou et al., 1996a, b, 1997; Pan et al., 1997), indicating entry at open sites on roots, such as lateral root emergence or other sites such as wounds. In a study by Gagné et al. (1987), increased endophytic colonization was observed in roots of plants that were artificially wounded compared to intact roots. In a study specifically designed to investigate entry mechanisms, Quadt-Hallmann et al. (1997a) demonstrated passive uptake of two different bacteria, an *Enterobacter asburiae* strain and a *P. fluorescens* strain, into cotton roots.

In many studies, bacteria that have been surface-inoculated onto plants have been observed to invade and colonize internal host tissue. In some cases, systemic colonization has resulted from a single inoculation point. An *E. cloacae* strain was recovered from the endosperm, stems, and leaves of corn following treatment of seed (Hinton and Bacon, 1995). Similarly, after treatment of germinated seed, an *Erwinia* sp. was recovered from stems, flowers, bolls, and roots of cotton (Misaghi and Donndelinger, 1990). Several other studies have demonstrated endophytic colonization on a more localized level. For example, in several studies, application of bacteria as drenches onto several different host plants resulted in colonization of intercellular spaces of root tissue (e.g., Jacobs et al., 1985; Gagné et al., 1987; Frommel et al. 1991; Kloepper et al., 1992; Benhamou et al., 1996a, b, 1998; Pan et al., 1997; Pillay and Nowak, 1997).

Although the root zone offers the most obvious site of entry for many endophytes, entry may also occur at sites on aerial portions of plants. Sharrock et al. (1991) suggested that in some cases endophytic populations within fruit may arise by entry through flowers. In this study, higher endophytic populations were observed within squash at positions near the flower end of the fruit compared to the stem end. The entry study conducted by Quadt-Hallmann (1997a) also investigated passive uptake of the rhizobacterium *Enterobacter asburiae* through the stem and cotyledon tissue after drenching the aerial portions of the plant.

Investigators have suggested that endophytes may gain entry into host plants by infection of seed (e.g., Hinton and Bacon, 1995; Misaghi and Donndelinger, 1990). Misaghi and Donndelinger (1990) were able to reduce the total numbers of endophytes by treating seeds with antibiotics or heat, suggesting seeds as a primary inoculum source for endophytic bacteria. However, seed as the actual entry site remains to be verified. In these cases, seeds may have provided the primary inoculum source, and gain of entry may result in invasion of roots after seed germination, as suggested by McInroy and Kloepper (1995a).

4.4. Host Specificities

Outside of bacterial pathogens, there is no evidence for host-specific interactions between endophytic bacteria and colonized hosts. There are a variety of reports in which bacteria originally isolated from one host are able to endophytically colonize one or several different plant species (Chen et al., 1995, Benhamou et al., 1996a, b, 1998; Benhamou et al., 1996b; Pan et al., 1997; Quadt-Hallmann et al., 1997a). For example, both an *Enterobacter asburiae* strain originally isolated from cotton and a *P. fluorescens* strain originally isolated from canola colonized cotton root tissue, reflecting the fact that endophytic colonization is not under strict control of host specificities that is prevalent for many pathogens (Quadt-Hallmann et al., 1997a). Lack of host specificity has also been demonstrated with other systems. Pan et al. (1997) demonstrated that inoculation of banana roots with a *B. cepacia* strain, originally isolated from asparagus rhizosphere, resulted in colonization of interior portions of roots, although colonization of the vascular tissue was not determined.

No distinct conclusions can be made for host range relationships with endophytes, although some effects may be apparent. There are some differences between different plants colonized by the same organism, but many endophytes isolated from one plant can endophytically colonize another (e.g., Frommel et al., 1991, 1993; Pillay and Nowak, 1997). And, as with the case of Cxc, bacteria that cause disease on one host (Burmuda grass) are capable of endophytic colonization of a nonhost plant (maize) in significant populations.

5. BENEFICIAL ACTIVITIES OF ENDOPHYTES IN AGRICULTURE

The preponderance of studies characterizing endophytes for potential use in agriculture has focused on surveying the types of endophytes in crop plants and studying population dynamics in various tissues throughout the plant. Within many of these studies, clear examples of their potential use for agricultural gains are demonstrated. Their use can be divided into two categories based on the proposed types of activities: growth promotion, based on activities of the bacterium that enhance growth potential of host plants, and disease control, based on suppression of pathogen infection.

5.1. Growth-Promoting Bacteria

As mentioned previously, atmospheric nitrogen fixers such as members of the Rhizobiaceae and other nonnodule diazotrophs reflect one aspect in which endophytes can benefit plants through growth promotion. In some cases, host-specific interactions can be breached, as demonstrated by Yanni et al. (1997), who showed that *Rhizobium leguminosarum* bv. *trifolii*, a natural clover root symbiont, can endophytically colonize rice roots. Other growth promotion factors may be linked to stimulation of better growth of the plant (Lalande et al., 1989; Frommel et al., 1991, 1993; Wiehe et al., 1994; Pillay and Nowak, 1997).

5.2. Biocontrol of Plant Diseases

The second group of bacteria that have been designated as potentially beneficial microbes includes endophytic bacteria with biocontrol potential. This group includes bacteria proposed to function by controlling plant diseases through traditional mechanisms, including antibiotic production and stimulation of host systemic defense responses, or through the utilization of genetically modified bacteria that colonize plants endophytically as delivery methods for expression of specific beneficial traits.

Biocontrol experiments indicate that bacteria characterized as biocontrol agents, either originally isolated as endophytes from the host plant or obtained from other sources such as other hosts or from the rhizosphere, are capable of suppressing diseases (e.g., Myers and Strobel, 1983; Hall et al., 1986; Brooks et al., 1994; Pleban et al., 1995; Chen et al., 1995; Sturz and Matheson, 1996). Many endophytic strains have demonstrated in vitro growth inhibition activity against fungal plant pathogens (Myers and Strobel, 1983; O'Brien et al., 1984; Brooks et al., 1994; Hinton and Bacon, 1995; Pleban et al., 1995; Pan et al., 1997). In several of these studies, a correlation was made with proposed biocontrol traits such as antibiotic activity, siderophore production, or production of lytic enzymes such as chitinases that degrade fungal cell wall components. However, as mentioned previously, the ability to induce systemic resistance is another proposed mechanism (Kloepper et al., 1992; Chen et al., 1995; Benhamou et al., 1996a; b, 1998; Duijff et al., 1997). that appears to have significant potential with the use of endophytes. Although there is much more to learn about induced systemic resistance, there is the premise that systemic colonization by the endophyte is not necessary to induce a sufficient systemic response in plants to achieve some level of disease control that can potentially have longlasting effects.

Endophytic colonization of roots by bacteria theoretically positions them at sites where initial infections by root pathogens occur. In addition to utilizing and harnessing natural biocontrol potentials of strains colonizing specific tissues, these bacteria also provide ideal strains to utilize for enhanced biocontrol activity using genetic approaches. These studies clearly point out the potential for biocontrol, but they also raise questions about issues pertaining to practicality of their use and thus provide impetus for future research.

The potential that endophytes have to offer for agricultural gains has been realized beyond their use as natural biocontrol agents. The inherent nature of certain endophytes to potentially colonize plants in a systemic manner provides a novel approach as a delivery system to plants for various beneficial traits. The use of genetically engineered strains of endophytic bacteria for enhanced pest control has been envisioned for some time (Dimock et al., 1988; Hackett et al., 1988; Misaghi and Donndelinger, 1990). However, several limitations hinder their rapid development. The selection of endophytes to construct genetically engineered organisms requires a variety of traits necessary for the construction of effective strains. First, the endophyte must reside at or deliver the pest-controlling factor to a site within the plant that is accessible to the pest. Second, the endophyte must not have significant crop-threatening disease capabilities. Third, the endophyte must be readily amenable to genetic manipulation. Several additional factors must be considered in the construction of recombinant microbes, including the ability to readily express foreign genes and the long-term maintenance of the gene(s) within the endophyte. This is most easily performed by stable integration into the chromosome of the endophyte, but it requires integration of a compatible form of the gene at a site that does not affect the fitness of the microbe. Once incorporated, the gene product must be expressed in a form that is in high enough concentration and functionally accessible to the pest. The most intensive study of a recombinant bacterial endophyte constructed for pest control purposes was performed with *Clavibacter xyli* subsp. *cynodontis* expressing the *cryIA* gene from *Bacillus thuringiensis* (Lampel et al., 1994; Tomasino et al., 1995). In this example, the resultant protein was an endotoxin that is active against the European corn borer. The recombinant bacterium was effective in reducing insect boring in corn in laboratory, greenhouse, and field studies; however, significant increases in overall yield were not detected compared to plants that were not treated with the endophyte.

Cell wall-less prokaryotes, such as *Spiroplasma* spp., provide an interesting system for potential use of genetic manipulation for control of problems such as those created with insect pests (Hackett et al., 1988, 1992). Spiroplasmas can systemically colonize the vascular tissue (phloem) of their plant host and therefore colonize areas of plants that are accessible to target pests. Spiroplasmas also invade and reside within insect hosts, and some strains are known to have deleterious effects on insect pests. However, much more work is necessary to effectively develop systems utilizing spiroplasmas. Problems associated with the development of such systems include the lack of well-defined genetic systems that are amenable to manipulation. Furthermore, life cycles of ideal candidate *Spiroplasma* strains for potential development are not well defined. As advances are made in our understanding of biological control mechanisms and endophytic interactions, genetic traits can be envisioned to be transferred between bacteria to enhance and deliver these capabilities.

6. FUTURE DIRECTIONS

Descriptions of diazotrophs, naturally occurring biocontrol bacteria, and genetically modified bacteria clearly point out the potential use of bacterial endophytes for plant beneficial purposes. However, further research is necessary to advance the science for practical uses. Demonstration of field efficacy is among the most important. However, a variety of additional factors that also provide hindrances for effective development of the use of beneficial endophytes can be addressed directly in the laboratory. Among others, a better understanding and improved strategies to enhance inoculation and delivery methods, long-term persistence, and stability of disease-controlling factors within host plants are necessary. The use of point inoculations (e.g., Myers and Strobel, 1983; Misaghi and Donndelinger, 1990; Brooks et al., 1994; Chen et al., 1995) seems impractical for large-scale crop production. In contrast, demonstration of systemic endophytic colonization by simple seed application (Hinton and Bacon, 1995) provides encouraging results. As demonstrated with studies in traditional biocontrol, the use of endophytes may be limited until we have a better understanding of the ecology of bacterial endophytes. However, as the field of biological control of plant pests advances, invariably additional beneficial traits with genetic potential for use in endophytes will become available. One example includes genetic manipulation of endophytes for delivery of known pathogen-antagonistic traits, such as cell wall–degradative enzymes (e.g., Shapira et al., 1989; Pleban et al., 1997), or antibiotics toxic to plant pathogens (e.g., Fenton et al., 1992). Such construction in selectively identified endophytes may prove useful in the future.

Among ecological studies, there is a distinct lack of knowledge of any limitations governing host specificities. It is unclear as to what factors, if any, within endophytes or host plants dictate specific host colonization potential, posing important questions. For example, are there specific host ranges for endophytic strains? Do endophytic strains possess basic or specific genetic differences that govern endophytic colonization compared to nonendophytic strains of the same species? Answers to these and other questions should enhance the potential utility of bacterial endophytes for future applications.

REFERENCES

Alfano, J. R. and Collmer, A. (1997). The type III (Hrp) secretion pathway of plant pathogenic bacteria: trafficking harpins, avr proteins, and death. *Journal of Bacteriology* **179**, 5655–5662.

Alfano, J. R. and Collmer, A. (1996). Bacterial pathogens in plants: life up against the wall. *Plant Cell* **8**, 1683–1698.

Bacon, C. W. and Hill, N. S. (1996). Symptomless grass endophytes: production of coevolutionary symbioses and their role in the ecological adaptations of infected grasses. *In* S. C. Redlin and L. M. Carris, (eds.), *Endophytic Fungi in Grasses and Woody Plants.* APS Press, St. Paul, MN, pp. 155–178.

Bacon, M. and Mead, C. E. (1971). Bacteria in the wood of living aspen, pine and alder. *Northwest Science* **45**, 270–275.

Bailey, J. A., O'Connell, R. J., Pring, R. J. and Nash, C. (1992). Infection strategies of *Colletotrichum* species. *In* J. A. Bailey and M J. Jeger (eds.), *Colletotrichum: Biology, Pathology and Control.* CAB Int., Wallingford, UK, pp. 88–120.

Baldani, J. I., Caruso, L., Baldani, V. L. D., Goi, S. R. and Döbereinger, J. (1997). Recent advances in BNF with non-legume plants. *Soil Biology and Biochemistry* **29**, 911–922.

Barras, F., F. van Gijsegem, F. and Chatterjee, A. K. (1994). Extracellular enzymes and pathogenesis of soft-rot erwinia. *Annual Review of Phytopathology* **32**, 201–234.

Bell, C. R., Dickie, G. A. Harvey, W. L. G. and Chan, J. W. Y. F. (1995). Endophytic bacteria in grapevine. Canadian Journal of Microbiology **41**, 46–53.

Bender, C. L., Stone, H. E. Sims, J. J. and Cooksey, D. A. 1987. Reduced pathogen fitness of *Pseudomonas syringae* pv. *tomato* Tn5 insertions defective in coronatine production. *Physiology Molecular Plant Pathology* **30**, 273–283.

Benhamou, N., Bélanger, R. R. and Paulitz, T. C. (1996a). Pre-inoculation of Ri T-DNA-transformed pea roots with *Pseudomonas fluorescens* inhibits colonization by *Pythium ultimum* Trow: an ultrastructural and cytochemical study. *Planta* **199**, 105–117.

Benhamou, N., Kloepper, J. W., Quadt-Hallmann, A. and Tuzun, S. (1996b). Induction of defense-related ultrastructural modifications in pea root tissues inoculated with endophytic bacteria. Plant Physiology **112**, 919–929.

Benhamou, N., Kloepper, J. W. and Tuzun, S. (1998). Induction of resistance against Fusarium wilt of tomato by combination of chitosan with an endophytic bacteria strain: ultrastructure and cytochemistry of the host response. *Planta* **204**, 153–168.

Bladergroen, M. R. and Spaink, H. P. (1998). Genes and signal molecules involved in *Rhizobia*-Leguminoseae symbiosis. *Symbiosis* **1**, 353–359.

Bogdanove, A. J., Beer, S. V., Bonas, U., Boucher, C. A., Collmer, A., Coplin, D. L., Cornelis, G. R., Huang, H.-C., Hutcheson, S. W., Panopoulos, N. J. and Van Gijsegem, (1996). Unified nomenclature for broadly conserved *hrp* genes of phytopathogenic bacteria. *Molecular Microbiology* **20**, 681–683.

Bonas, U., Schulte, R., Fenselau, S., Minsavage, G. V., Staskawicz, B. J. and Stall, R. E. (1991). Isolation of a gene cluster from *Xanthomonas campestris* pv. *vesicatoria* that determines pathogenicity and the hypersensitive response on pepper and tomato. Molecular Plant–Microbe Interaction **4**, 81–88.

Brooks, D. S., Gonzalez, C. F., Appel, D. N. and Filer, T. H. (1994). Evaluation of endophytic bacteria as potential biological control agents for oak wilt. *Biol. Con.* **4**, 373–381.

Cameron, H. R. (1970). Pseudomonas content of cherry trees. *Phytopathology* **60**, 1343–1346.

Chanway, C. P. (1996). Endophytes: they're not just fungi. *Canadian Journal of Botany* **74**, 321–322.

Chen, C., Bauske, E. M., Musson, G., Rodríguez-Kábana, R. and Kloepper, J. W. (1995). Biological control of *Fusarium* wilt on cotton by use of endophytic bacteria. *Biol. Con.* **6**, 83–91.

Cole, D. F. and Bugbee, W. M. (1976). Changes in resident bacteria, pH, sucrose, and invert sugar levels in sugarbeet roots during storage. *Applied and Environmental Microbiology* **31**, 754–757.

Davis, M. J., Gillaspie, A. G., Jr., Vidaver, A. K. and Harris, R. W. (1984). *Clavibacter*: a new genus containing some phytopathogenic coryneform bacteria, including *Clavibacter xyli* subsp. *xyli* sp. nov., subsp. nov. and *Clavibacter xyli* subsp. *cynodontis* subsp. nov., pathogens that cause ratoon stunting disease of sugarcane and Burmudagrass stunting disease. International Journal of Systematic Bacteriology **34**, 107–117.

Davis, M. J., Lawson, R. H., Gillaspie, A. G., Jr. and Harris, R. W. (1983). Properties and relationships of two xylem-limited bacteria and a mycoplasma like organism infecting Bermuda grass. *Phytopathology* **73**, 341–346.

De Boer, S. H., and Copeman, R. J. (1974). Endophytic bacterial flora in *Solanum tuberosum* and its significance in bacterial ring rot diagnosis. *Canadian Journal of Plant Science* **54**, 115–122.

Dimock, M. B, Beach, R. M. and Carlson, P. S. (1988). Endophytic bacteria for the delivery of crop protection agents. *In* D. W. Roberts and R. R. Granados (eds.), *Biotechnology, Biological Pesticides and Novel Plant-Pest Resistance for Insect Pest Management.* Boyce Thompson Institute for Plant Research at Cornell University, Ithaca. pp. 88–92.

Djordjevic, M. A., Gabriel, G. W. and Rolfe, B. G. (1987). Rhizobium—the refined parasite of legumes. *Annual Review of Phytopatholgy* **25**, 145–168.

Dowler, W. M. and Weaver, D. J. (1975). Isolation and characterization of fluorescent psuedomonads from apparently healthy peach trees. *Phytopathology* **65**, 233–236.

Duijff, B. J., Gianinazzi-Pearson, V. and Lemanceau, P. (1997). Involvement of the outer membrane lipopolysaccharides in the endophytic colonization of tomato roots by biocontrol *Pseudomonas fluorescens* strain WCS417r. *New Phytologist* **135**, 325–334.

Feldman, A. W., Hanks, R. W., Good, G. E. and Brown, G. E. (1977). Occurrence of a bacterium in YTD-affected as well as in some apparently healthy citrus trees. Plant Disease Report **61**, 546–550.

Fenton, A. M., Stephens, P. M., Crowley, J., O'Callaghan, M. and O'Gara, F. (1992). Exploitation of gene(s) involved in 2,4-diacetylphloroglucinol biosynthesis to confer a new biocontrol capability to a *Pseudomonas* strain. *Applied Environmental Microbiology* **58**, 3873–3878.

Fisher, P. J., Petrini, O. and Lappin Scott, H. M. (1992). The distribution of some fungal and bacterial endophytes in maize (*Zea mays* L.). *New Phytologist* **122**, 299–305.

Freeman, S. and Rodriguez, R. J. (1993). Genetic conversion of a fungal plant pathogen to a nonpathogenic, endophytic mutualist. *Science* **260**, 75–78.

Frommel, M. I., Nowak, J. and Lazarovits, G. (1993). Treatment of potato tubers with a growth promoting *Pseudomonas* sp.: plant growth responses and bacterium distribution in the rhizosphere. *Plant and Soil* **150,** 51–60.

Frommel, M. I., Nowak, J. and Lazarovits, G. (1991). Growth enhancement and developmental modifications of in vitro grown potato (*Solanum tuberosum* ssp. *tuberosum*) as affected by a nonfluorescent *Pseudomonas* sp. *Plant Physiology* **96,** 928–936.

Fuqua, W. C., Winans, S. C. and Greenberg, E. P. (1994). Quorum sensing in bacteria: the LuxR-LuxI family of cell density-responsive transcriptional regulators. *Journal of Bacteriology* **176,** 269–275.

Gagné, S., Richard, C., Rousseau, H. and Antoun, H. (1987). Xylem-residing bacteria in alfalfa roots. Canadian Journal of Microbiology **33,** 996–1000.

Gardner, J. M., Feldman, A. W. and Zablotowicz, R. M. (1982). Identity and behavior of xylem-residing bacteria in rough lemon of Florida citrus trees. *Applied Environmental Microbiology* **43,** 1335–1342.

Guan, C., Pawlowski, K. and Bisseling, T. (1998). Interaction between *Frankia* and actinorhizal plants. *Subcellular Biochemistry* **29,** 165–189.

Hackett, K. J., Henegar, R. B., Whitcomb, R. F., Lynn, D. E., Konai, M., Schroder, R. F., Gasparich, G. E., Vaughn, J. L. and Cantelo, W. W. (1992). Distribution and biological control significance of Colorado potato beetle spiroplasmas in North America. Biol. Con. **2,** 218–225.

Hackett, K. J., Lynn, D. E. and Whitcomb, R. F. (1988). Spiroplasmas and other mollicutes: possible application to plant pest control. *In* D. W. Roberts and R. R. Granados (eds.), *Biotechnology, Biological Pesticides and Novel Plant-Pest Resistance for Insect Pest Management.* Boyce Thompson Institute for Plant Research at Cornell University, Ithaca, NY, pp. 93–98.

Hall, T. J. and Davis, W. E. E. (1990). Survival of *Bacillus subtilis* in silver and sugar maple seedlings over a two-year period. *Plant Disease* **74,** 608–609.

Hall, T. J., Schreiber, L. R. and Leben, C. (1986). Effects of xylem-colonizing *Bacillus* spp. on Verticillium wilt in maples. *Plant Disease* **70,** 521–524.

Hallman, J., Kloepper, J. W. and Rodriguez-Kábana, R. (1997). Application of the Scholander pressure bomb to studies on endophytic bacteria of plants. *Canadian Journal of Microbiology* **43,** 411–416.

Hayward, A.C. (1974). Latent infections by bacteria. *Annual Review of Phytopathology* **12,** 87–97.

Heath, M. C. (1986). Evolution of parasitism in the fungi. In A. D. M. Rayner, C. M. Brasier and D. Moore (eds.), *Evolutionary Biology of the Fungi.* Cambridge University Press, New York, pp. 149–160.

Hinton, D. M. and Bacon, C. W. (1995). *Enterobacter cloacae* is an endophytic symbiont of corn. *Mycopathologia* **129,** 117–125.

Hollis, J. P. (1951). Bacteria in healthy potato tissue. *Phytopathology* **41,** 350–366.

Hopkins, D. L. and Elmstrom, G. W. (1977). Etiology of watermelon rind necrosis. *Phytopathology* **67,** 961–964.

Huang, J.-S. (1986). Ultrastructure of bacterial penetration in plants. *Annual Review of Phytopatholgy* **24,** 141–157.

Hueck, C. J. (1998). Type III protein secretion systems in bacterial pathogens of animals and plants. *Microbiolgical Molecular Review* **62,** 379–433.

Jacobs, M. J., Bugbee, W. M. and Gabrielson, D. A. (1985). Enumeration, location, and characterization of endophytic bacteria within sugar beet roots. *Canadian Journal of Botany* **63,** 1262–1265.

James, E. K. and Olivares, F. L. (1998). Infection and colonization of sugar cane and other graminaceous plants by endophytic diazotrophs. Critical Reviews in Plant Science **17,** 77–119.

Kado, C.I. (1991). Plant pathogenic bacteria. *In* A. Balows, H. G. Trüper, M. Dworkin, W. Harder and K.-H. Schleifer (eds.), *The Prokaryotes,* Vol. 1, 2nd Ed. Springer-Verlag, New York, pp. 659–674.

Kloepper, J. W., Wei, G. and Tuzun, S. (1992). Rhizosphere population dynamics and internal colonization of cucumber by plant growth-promoting rhizobacteria which induce systemic resistance to *Colletotrichum orbiculare*. *In* E. C. Tjamos, G. C. Papavizas and R. J. Cook (eds.), *Biological Control of Plant Diseases*. Plenum Press, New York, pp. 185–191.

Knutson, D. M. (1973). The bacteria in sapwood, wetwood, and heartwood of trembling aspen (*Populus tremuloides*). *Canadian Journal of Botany* **51,** 498–500).

Lalande, R., Bissonnette, N., Coutlée, D. and Antoun, H. (1989). Identification of rhizo-bacteria from maize and determination of their plant-growth promoting potential. *Plant and Soil* **115,** 7–11.

Lampel, J. S., Canter, G. L., Dimock, M. B., Kelly, J. L., Anderson, J. J., Uratani, B. B., Foulke, J. S. Jr., and Turner, J. T. (1994). Integrative cloning, expression, and stability of the *cryIA©* gene from *Bacillus thuringiensis* subsp. *kurstaki* in a recombinant strain of *Clavibacter xyli* subsp. *cynodontis*. *Applied and Environmental Microbiology* **60,** 501–508.

Leach, J. E. and White, F. F. (1996). Bacterial avirulence genes. Annual Review of Phyto-patholgy **34,** 153–179.

Li, X.-Z., Starratt, A. N. and Cuppels, D. A. (1998). Identification of tomato leaf factors that activate toxin gene expression in *Pseudomonas syringae* pv. *tomato* DC3000. *Phytopathology* **88,** 1094–1100.

Liao, C. H., and Chen, T. A. (1981). Isolation, culture, and pathogenicity to Sudan grass of a corynebacterium associated with ratoon stunting sugarcane and with Bermuda grass. *Phytopathology* **71,** 1303–1306.

Lindgren, P. B., Peet, R. C. and Panopoulos, N. J. (1986). Gene cluster of Pseudomonas syringae pv. "phaseolicola" controls pathogenicity of bean plants and hypersensi-tivity on nonhost plants. *Journal of Bacteriology* **168,** 512–522.

Liyanage, H., Palmer, D. A., Ullrich, M. and Bender, C. L. (1995). Characterization and transcriptional analysis of the gene cluster for coronafacic acid, the polyketide com-ponent of the phytotoxin coronatine. *Applied Environmental Microbiology* **61,** 3843–3848.

McInroy, J. A. and Kloepper, J. W. (1995a). Population dynamics of endophytic bacteria in field-grown sweet corn and cotton. *Canadian Journal of Microbiology* **41,** 895–901.

McInroy, J. A., and Kloepper, J. W. (1995b). Survey of indigenous bacterial endophytes from cotton and sweet corn. *Plant and Soil* **173,** 337–342.

McInroy, J. A., Musson, G., Wei, G. and Kloepper, J. W. (1996). Masking of antibiotic-resistance upon recovery of endophytic bacteria. *Plant and Soil* **186,** 213–218.

Meneley, J. C. and Stanghellini, M. E. (1974). Detection of enteric bacteria within locular tissue of healthy cucumbers. *Journal of Food Science* **39**, 1267–1268.

Meneley, J. C., and Stanghellini, M. E. (1975). Establishment of an inactive population of *Erwinia carotovora* in healthy cucumber fruit. *Phytopathology* **65**, 670–673.

Misaghi, I. J., and Donndelinger, C. R. (1990). Endophytic bacteria in symptom-free cotton plants. *Phytopathology* **80**, 808–811.

Mundt, J. O., and Hinkle, J. O. (1976). Bacteria within ovules and seeds. *Applied Environmental Microbiology* **32**, 694–698.

Myers, D. F. and Strobel, G. A. (1983). *Pseudomonas syringae* as a microbial antagonist of *Ceratocystis ulmi* in the apoplast of American elm. *Transactions of the British Mycological Society* **80**, 389–394.

O'Brien, J. G., Blanchette, R. A. and Sutherland, J. B. (1984). Assessment of *Streptomyces* spp. from elms for biological control of Dutch elm disease. *Plant Diseases* **68**, 104–106.

Old, K. M. and Nicolson, T. H. (1975). Electron microscopical studies of the microflora of roots of sand dune grasses. *New Phytologist* **74**, 51–58.

Pan, M. J., Rademan, S., Kuner, K. and Hastings, J. W. (1997). Ultrastructural studies on the colonization of banana tissue and *Fusarium oxysporum* f. sp. *cubense* race 4 by the endophytic bacterium *Burkholderia cepacia. Journal of Phytopathology* **145**, 479–486.

Pawlowski, K. and Bisseling, T. (1996). Rhizobial and actinorhizal symbioses: what are the shared features? *Plant Cell* **8**, 1899–1913.

Philipson, M. N. and Blair, I. D. (1957). Bacteria in clover root tissue. *Canadian Journal of Microbiology* **3**, 125–129.

Pillay, V. K. and Nowak, J. (1997). Inoculum density, temperature, and genotype effects on in vitro growth promotion and epiphytic and endophytic colonization of tomato (*Lycopersicon esculentum* L.) seedlings inoculated with a pseudomonad bacterium. Canadian Journal of Microbiolology 43:354-361.

Pirhonen, M., Flego, D., Heikinheimo, R. and Palva, E. T. (1993). A small diffusible signal molecule is responsible for the global control of virulence and exoenzyme production in the plant pathogen *Erwinia carotovora. EMBO Journal* **12**, 2467–2476.

Pleban, S., Chernin, L. and Chet, I. (1997). Chitinolytic activity of an endophytic strain of *Bacillus cereus. Letters in Applied Microbiolgy* **25**, 284–288.

Pleban, S., Ingel, F. and Chet, I. (1995). Control of *Rhizoctonia solani* and *Sclerotium rolfsii* in the greenhouse using endophytic *Bacillus* spp. *European Journal of Plant Pathology* **101**, 665–672.

Pueppke, S. G. (1996). The genetic and biochemical basis for nodulation of legumes by rhizobia. *Critical Reviews in Biotechnology* **16**, 1–51.

Quadt-Hallmann, A., Benhamou, N. and Kloepper, J. W. (1997a). Bacterial endophytes in cotton: mechanisms of entering the plant. *Canadian Journal of Microbiology* **43**, 577–582.

Quadt-Hallmann, A., Hallmann, J. and Kloepper, J. W. (1997b). Bacterial endophytes in cotton: location and interaction with other plant-associated bacteria. *Canadian Journal of Microbiology* **43**, 254–259.

Rahme, L. G., Mindrinos, M. N. and Panopoulos, N.J. (1991). Genetic and transcriptional organization of the hrp cluster of *Pseudomonas syringae* pv. *phaseolicola*. *Journal of Bacteriology* **173**, 575–586.

Reinhold-Hurek, B. and Hurek, T. (1998). Interactions of gramineous plants with *Azoarcus* spp. and other diazotrophs: identification, localization, and perspectives to study their function. *Critical Reviews in Plant Science* **17**, 29–54.

Roos, I. M. M. and Hattingh, M. J. (1987). Systemic invasion of plum leaves and shoots by *Pseudomonas syringae* pv. *syringae* introduced into petioles. Phytopathology **77**, 1253–1257.

Samish, Z., Etinger-Tulczynska, R. and Bick, M. (1963). The microflora within the tissue of fruits and vegetables. *Journal of Food Science* **28**, 259–266.

Shapira, R., Ordentlich, A., Chet, I. and Oppenheim, A. B. (1989). Control of plant diseases by chitinase expressed from cloned DNA in *Escherichia coli*. *Phytopathology* **79**, 1246–1249.

Sharrock, K. R., Parkes, S. L., Jack, H. K., Rees-George, J. and Hawthorne, B. T. (1991). Involvement of bacterial endophytes in storage rots of buttercup squash (*Curcurbita maxima* D. hybrid 'Delica'). *New Zealand Journal of Crop and Horticultural Science* **19**, 157–165.

Sturz, A. V. (1995). The role of endophytic bacteria during seed piece decay and potato turberization. *Plant and Soil* **175**, 257–263.

Sturz, A. V. and Christie, B. R. (1995). Endophytic bacterial systems governing red clover growth and development. *Annals of Applied Biology* **126**, 285–290.

Sturz, A. V., and Matheson, B. G. (1996). Populations of endophytic bacteria which influence host resistance to *Erwinia*-induced bacterial soft rot in potato tubers. *Plant and Soil* **184**, 265–271.

Sundin, G. W., Jones, A. L. and Olson, B. D. (1988). Overwintering and population dynamics of *Pseudomonas syringae* pv. *syringae* and *P.s.* pv. *morsprunorum* on sweet and sour cherry trees. *Canadian Journal of Plant Pathology* **10**, 281–288.

Tervet, I. W. and Hollis, J. P. (1948). Bacteria in the storage organs of healthy plants. *Phytopathology* **38**, 960–967.

Thomas, W. D., Jr. and Graham, R. W. (1952). Bacteria in apparently healthy pinto beans. *Phytopathology* **42**, 214.

Tomasino, S. F., Leister, R. T., Dimock, M. B., Beach, R. M. and Kelly, J. L. (1995). Field performance of *Clavibacter xyli* subsp. *cynodontis* expressing the insecticidal protein gene *crylA©* of *Bacillus thuringiensis* against European corn borer in field corn. *Biol. Con.* **5**, 442–448.

Tör, M., Mantell, S. H., and Ainsworth, C. (1992). Endophytic bacteria expressing b-glucuronidase cause false positives in transformation of *Dioscorea* species. *Plant Cell Report* **11**, 452–456.

Ullrich, M., Peñaloza-Vázquez, A., Bailey, A.-M. and Bender, C. L. (1995). A modified two-component regulatory system is involved in temperature-dependent biosynthesis of the *Pseudomonas syringae* phytotoxin coronatine. *Journal of Bacteriology* **177**, 6160–6169.

van Peer, R., Punte, H. L. M., de Wager, L. A. and Schippers, B. (1990). Characterization of root surface and endorhizosphere pseudomonads in relation to their colonization of roots. *Applied and Environmental Microbiology* **56**, 2462–2470.

van Peer, R. and Schippers, B. (1989). Plant growth responses to bacterization with selected *Pseudomonas* spp. strains and rhizosphere microbial development in hydroponic cultures. Canadian Journal of Microbiology **35,** 456–463.

van Rhijn, P. and Vanderleyden, J. (1995). The *Rhizobium*-plant symbiosis. *Microbiological Review* **59,** 124–142.

Vandamme, P., Pot, B., Gillis, M., De Vos, P., Kersters, K. and Swings, J. (1996). Polyphasic Taxonomy, a Consensus Approach to Bacterial Systematics. *Microbiology Review* **60,** 407–438.

Whitesides, S. K., and Spotts, R. A. (1991). Frequency, distribution, and characteristics of endophytic *Pseudomonas syringae* in pear trees. *Phytopathology* **81,** 453–457.

Wiehe, W., Hecht-Buchholz, C. and Höflich, G. (1994). Electron microscopic investigations on root colonization of *Lupinus albus* and *Pisum sativum* with two associative plant growth promoting rhizobacteria, *Pseudomonas fluorescens* and *Rhizobium leguminosarum* bv. *trifolii. Symbiosis* **17,** 15–31.

Yanni, Y. G., Rizk, R. Y., Corich, V., Squartini, A., Ninke, K., Philip-Hollingsworth, S., Orgambide, G., De Bruijn, F., Stoltzfus, J., Buckley, D., Schmidt, T. M., Mateos, P. F., Ladha, J. K. and Dazzo, F. B. (1997). Natural endophytic association between *Rhizobium leguminosarum* bv. *trifolii* and rice roots and assessment of its potential to promote rice growth. *Plant and Soil* **194,** 99–114.

III
ENDOPHYTE PHYSIOLOGY AND THE BIOCHEMICAL ARSENAL

10

Physiological Adaptations in the Evolution of Endophytism in the Clavicipitaceae

Charles W. Bacon
Russell Research Center, Agricultural Research Service,
U.S. Department of Agriculture, Athens, Georgia

James F. White, Jr.
Cook College, Rutgers University, New Brunswick, New Jersey

1. INTRODUCTION

We are concerned with physiological adaptations in a relatively small group of closely related fungi in the family Clavicipitaceae. These organisms are broadly distinguished as being symbiotic associates of either insect and fungi (*Cordyceps* spp.), or grasses, rushes, and sedges (*Balansia*, *Claviceps*, *Myriogenospora*, *Epichloe*, *Echinodothis*, *Atkinsonella*, and *Balansiopsis*). The species of the genus *Cordyceps* will not be considered in this discussion, which will focus instead on symbiotic fungal associations with grasses. These fungi are either entirely systemic and endophytic, or epibiotic symbionts of their hosts. Also included in this discussion are the species of *Neotyphodium*, the anamorphic states of *Epichloe* (Glenn et al., 1996). Many are intercellular in their distribution within the host, and this distribution is usually restricted to the ovary-floret, or foliage and above-ground portions of grasses and their allies. Thus, we are dealing with intercellular foliage symbionts which should be contrasted with the examples of intracellular root and woody stem symbionts presented elsewhere in these proceedings. Intercellular fungi apparently have an immense biological consequence on grasses since they are found associated with most of the tribes of the Gramineae. Endophytic associations are established at the seedling stage and are re-markably long-lived; for instance, clumps of endophyte-infected tall fescue per-sists for decades.

In most instances these grasses are referred to as defensive mutualisms (Siegel et al., 1985; Clay, 1988; Schardl and Tsai, 1992), although in other in-stances some species or members of a population apparently contribute nothing to the fitness of the association and may be antagonistic. The fact that the most defensive and successful fungal endophytes are symptomless during the entire life of their host emphasizes the importance of this habit, and suggests that this is the focal point from which relevant discussions should be directed for salient features essential in the evolution of these fungi. Morphological changes are char-acteristically linked with specific symbioses, especially in the regions of nutrient exchanges and the points where there is access to metabolic products responsible for any regulatory and beneficial aspects within the symbiosis. Thus, physiologi-cal adaptations in endophyte-grass species should consider the anatomical struc-ture of this interphase, any specificities for nutrient exchange, and any resulting biochemical and morphological interactions with the hosts. Similarly, we should study examples of specific nutrients exchanged along the interphase and the re-sulting novel metabolites produced within the association. The three basic mor-phological types found within endophyte-grass associations will be described rel-ative to phylogeny within the Clavicipitaceae. The few known physiological interactions, primarily nutritional, and adaptations will be discussed along with the production of novel metabolites derived from fungi, which most feel were the driving force for the coevolution of fungal endophytes within grasses.

2. EVOLUTIONARY RELATIONSHIP WITHIN THE CLAVICIPITACEAE

A cladogram of fungi within the Clavicipitaceae is presented in Glenn (1985) and Glenn et al. (1996). The organisms within this family may be grouped into three basic types according to the nature of the interaction with grasses. Thus, we have the epibiotic genera including *Myriogenospora*, *Echinodothis*, *Atkinso-nella*, and some species of *Balansia* (White and Glenn, 1994; Leuchtmann and Clay, 1988; White, 1994); the endophytic genera including species of *Balansia*, and *Epichloe*; and the closely related entirely endophytic *Neotyphodium* species (Glenn et al., 1996; White, 1994). The later group is distinguished from the others in producing a completely symptomless association with their hosts.

While we distinguish between these genera in terms of their symptoms, in all instances the hyphal cells that are in intimate contact with the host cells are all intercellular. These fungi are associated with grasses and sedges but this association is not host-specific, although these fungi are found with distinct groups of grasses. For example, species of *Balansia*, *Echinodothis*, and *Myriogenospora* are associated with warm-season perennial Poaceae species with C_4 acid decarboxylase activity, although there apparently is no distinction in the acid of the C_4 carbon fixation pathways, e.g., NAD-malate, or NADH-malate, and their associated leaf anatomies. The epibiont *Myriogenospora* is very narrow in its distribution as it is found only on two tribes, Andropogoneae and Paniceae, that are closely related (Supertribe Panicoideae), and both are considered to be the most advanced and most specialized of the grasses (Hattersley, 1986). The occurrence of *M. atramentosa* on this relatively small group of grasses with very little phylogenetic variation (Sobral et al., 1994), suggests that the epibiotic habit may be derived from the endophytic habit which is common to most Balansiae (Glenn et al.,1997).

The species of *Epichloe* and *Neotyphodium* are associated with several tribes and species of cool-season Poaceae all with the C_3 carbon fixation pathway. The essential difference between these two species is that the former is pathogenic with symptoms including a bright yellow stromata that envelope inflorescences and one or more culm leaves, while the later is a symptomless asexual endophyte. They both are related according to DNA-based phylogenetic techniques (Glenn et al., 1996).

In summary, the Clavicipitaceae consists of fungi that produce three generalized morphological structures across which physiological interactions occur:

1. Some fungi are ovarian replacement diseases. The only interaction with the grass is that at the base of the grass ovary, and this infection is intracellular and intercellular. Examples of these are represented by species of *Claviceps*. These species are in the tribe Clavicipitaceae, are not endophytically associated with its host foliage cells, and are not a part of this discussion, although references will be made to them for the sake of comparison.

2. Other grass symbionts are epibiotic species, and all fungi in this group belong to the tribe Balansiae, which are characterized as having simple stromata that are located on stems, florets, or leaves of hosts. The hyphae are not intercellular but are epicuticular, and very seldom is there deep penetration into the hosts, below the epidermal cells. The perennial nature of infection by this group is due to the presence of hyphae within close apposition to the apical meristem of each host.

3. The final group includes strictly endophytic species, which include those fungi that belong to the tribe Balansiae and produce stromata that are located on leaves, stems, and florets of hosts; the hyphae are intercellular and only rarely epicuticular, except at points of egress.

3. FUNGUS-GRASS PHYSIOLOGICAL INTERACTIONS

3.1. *In Vitro* Nutritional Requirements

In vitro studies indicate that all grass endophytes are capable of growth on a wide range of simple carbon sources (Davis et al., 1986b; Kulkarni and Nielsen, 1986; Bacon, 1990). It is unknown as to whether complex carbohydrates can be utilized directly, although media utilizing these have been reported (Bacon and White, 1994). Kulkarni and Nielsen (1986) reported that for one strain of *N. coenophialum* the pentoses arabinose, ribose, and xylose and the hexoses galactose, sorbose, and rhamnose were not utilized for growth. However, sucrose, trehalose, mannitol, and sorbitol were equally good carbon sources for growth of this fungus. Similar results were obtained for six species of *Balansia* (Bacon, 1985, 1990). The nitrogen requirements of endophytic fungi have been studied the least. Isolates of *Neotyphodium* spp. utilize ammonia more efficiently than nitrate (Kulkarni and Nielsen, 1986), and several amino acids are used by *Balansia* spp. and *N. coenophialum* that can influence growth rates, pigment production, indole and ergot alkaloid biosynthesis (Kulkarni and Nielsen, 1986; Bacon, 1985; Lyons et al., 1986; Porter et al., 1985).

Both complex and chemically defined media are used to demonstrate the ability of specific endophytes to produce secondary metabolites *in vitro*. Most of these media are based on media initially defined for the growth and ergot alkaloid production by *Claviceps* spp. Most laboratory media contain yeast extract, or some other complex nutrient source that probably supplies, in addition to nitrogen factors, vitamins. Fungal endophytes require vitamins, which were first established as thiamine, nicotinic acid, and pyridoxine for the *Balansia* spp (Bacon, 1985). The vitamin requirement for *Neotyphodium* species includes biotin, pyridoxine, and thiamine (Kulkarni & Nielsen, 1986). Generally, these requirements are not unique but rather are similar to those of other free-living ascomycetous fungi. However, grass endophytes are never found in the free-living state, indicating a total dependence on grasses for survival.

3.2. *In Planta* Interactions

Since the endophytic hyphae are always intercellular, nutrients are derived from the apoplasm which contains a variety of substances derived from the host, although the concentration of such substances may be low. The nutrients include amino acids, simple sugars, and a variety of vitamins and vitamin precursors. There are no data on the apoplastic contents of grasses reported as hosts for endophytes; neither is there information on the concentration of nutrients found in the apoplasm of infected plants compared to uninfected plants. This information would help establish if there is a mechanism resulting in the accumulation of nutrients beyond that normally found at this location and indicate that there is a directed flow of nutrients into the apoplasm, induced by the fungus. The mechanism might consist of specific chemical signals or substances that can alter membrane permeability, resulting in, for example, "leaky cell membranes." In the case of the symptom-inducing endophytes, the amount of stromata produced and the resulting thousands of conidia and ascospores produced therein would suggest that nutrients in the apoplasm are not lacking, and that the amount of nutrients within the apoplasm is ample for growth and reproduction of the fungus, with ample amounts left over for plant growth in general.

However, there appears to be a correlation between stromata production and the failure of seed to set. Most host infections, especially those induced by species of *Balansia* and *Epichloe*, do not produce seed. This suggests that competition for nutrients is in the favor of the fungi and that perhaps there is a directed movement from the grass to the fungus. An alternative explanation for this correlation, and one totally distinct from competition, is that there might be specific fungus-produced substances that inhibit flowering, resulting in an accumulation of nutrients within the host that now may be used for stromata production. Furthermore, in certain fungal species there is a physical binding of inflorescences, preventing the emergence of flowers. Species of *Epichloe*, *Atkinsonella*, and *Myriogenospora* interact with host in this fashion.

There are only a few studies dealing with translocation and specific physiological information dealing with nutrient exchange and accumulation *in vivo*, and these are limited to species of *Myriogenospora* and *Neotyphodium*, usually in infected parts (Smith, 1982; Smith et al., 1985; Lyons et al., 1990a; Lyons, 1985). Other studies deals with basic processes such as photosynthesis, respiration, translocation in symbiotic plants, and soil–symbiotic plant interaction (Belesky et al.,1987; West, Oosterhuis and Wullschleger, 1990; Richardson, Bacon and Hoveland, 1990; Richardson, Hoveland and Bacon, 1993).

In *Neotyphodium*-infected tall fescue, the hyphae are not found in the leaf blade, and their presence in the leaf sheath provided a convenient means of determining the effects of infection in both infected and uninfected parts of the hosts. In this case, blade and sheath were examined separately for assimilation of $^{14}CO_2$ into amino acids in blades and the transport of the labeled amino acids to the sheath were compared in infected and uninfected plants. The results indicated that the assimilation of ^{14}C into amino acids in blades was increased by infection an average of 85% in three paired experiments with infected and uninfected plants (Table 1). Furthermore, the incorporation of ^{14}C into amino acids in infected plants was significantly greater than in uninfected plants. However, there were no differences in the radioactivity in sugars, organic acids, or the residue. In sheaths, no differences were found between infected and uninfected plants in radioactivity of amino acids, sugars, organic acids, or residue fractions. The major amino acid labeled was alanine (40%) in both infected plant and uninfected plant. Other amino acids labeled included serine, glycine, glutamic acid,

Table 1 Incorporation of ^{14}C into Amino Acids in Blades of
Uninfected and Endophyte-Infected KY31 Tall Fescue

Sample	d min^{-1} × 10^3 g^{-1}	% Total radioactivity
Uninfected	464[a]	0.05[b]
Infected	657	0.09

[a] Values are from one paired experiment with two replications; pooled differences of three experiments are significantly different according to t test, $p = 0.01$.
[b] Values are means of three paired experiments; means are significantly different according to t test, $p = 0.01$ (Lyons, 1985).

aspartic acid, valine, and γ-amino butyric acid. Histidine and asparagine were the only two amino acids found to show a qualitative difference in blades of infected and uninfected plants (Lyons, 1985).

The activities of specific enzymes also varied in whole plant experiments in which plants were fertilized with either high or low ammonium or nitrate nitrogen fertilizers (Lyons et al., 1990a; Lyons, 1985). In the leaf blade, the activity of one enzyme, glutamine synthetase, an enzyme that is central in plant nitrogen assimilation, was significantly higher in infected plants compared to uninfected plants with each of the four nitrogen treatments. Glutamine synthetase activities also were increased by the high rates of nitrogen, but ammonium or nitrate nitrogen had no effect on this enzyme. Total protein was unaffected. However, the NH_4^+ concentration was nearly doubled by infection in the sheath, providing such plants were fertilized with a high rate of ammonium fertilizations (Table 2). There were no effects on the amounts of ammonium in the blade,

Table 2 Concentrations of Ammonium in Blades and Sheaths of Uninfected and Endophyte-Infected KY31 Tall Fescue Grown at High Rates of NO^{3-} and NH_4^+

	NH_4^+ (mg N g^{-1} dry weight)			
	NO^{3-} treatment		NH$_4^+$ treatment	
Samples	Blade	Sheath	Blade	Sheath
Uninfected	67a[a]	55a	76a	106a
Infected	71a	54a	126a	202a

[a] Values are means of three replicates and means within each nitrogen treatment followed by the same letter are not significantly different according to Duncan's new multiple range test, $p = 0.05$ (Lyons, 1985).

regardless of infection status, and the use of nitrate fertilization had no effect on the concentrations of NH_4^+ in plants.

These studies suggest that the interaction of the endophyte with the grass significantly alters at least nitrogen metabolism and is complex. Quantitation of the endophyte in sheaths by serological methods indicated that it is present to the extent of about 0.5% of plant dry weight in plants (Siegel et al.,1984). It is apparent that this small amount of hyphae exerts an enormous effect on the host metabolism. For example, the total amino acid concentration in the endophyte grown *in vitro* in high nitrogen nutrient solution is about 2.5 mmol/g^{-1} dry weight, as compared to the total amino acid concentrations in the grass, which is in excess of 100 mmol/g dry weight. The contribution of the fungus to the free amino acid pools in the infected grass is probably insignificant, and similarly the amount of ammonium is also not extracted from the fungus since the high concentration observed is toxic to the fungus.

In *M. atramentosa* nutrient movement was studied following the incorporation of $^{14}CO_2$ and translocation of ^{14}C-sucrose in intact plants or in excised leaf pairs of bahiagrass (Smith, 1982; Smith et al., 1985). Analysis of such experiments revealed that stromata of the fungus receiving labeled CO_2 contained large amounts of mannitol, arabitol, and some trehalose; but only trace amounts of sucrose and glucose. However, leaves from these experiments contained large amounts of galactose, glucose, sucrose, and fructose. These experiments did not establish the major translocate, which was shown to be an unidentified carbohydrate and was present in high amount in the stromata, but absent in leaves (Smith, 1982). This study established that significantly greater amounts of ^{14}C-labeled compounds occurred in infected leaves as opposed to uninfected leaves (Smith et al., 1985). The increased incorporation may have been due to either enhanced photosynthesis in infected leaves or changes in the translocation pattern. Increased translocation has been reported in plants infected with *Epichloe* and *Balansia* endophytes (White and Camp, 1995). Photosynthesis is not enhanced in *Neotyphodium*-infected tall fescue (Belesky et al., 1987; Richardson, Hoveland and Bacon, 1993), and nonnecrotic leaves from diseased plants (Daly, 1976). However, when photosynthesis is measured by incorporation of labeled CO_2 on individual leaves, there is a decided difference between infected and noninfected leaves (Lyons, 1985).

These results indicate that there is translocation and a conversion of sugars produced during photosynthesis into fungal carbohydrates. This process is central to the development of fungal nutrition and final development of stromata in several endophytes. This process is enzymatic and generally results in the conversion of host sucrose to fungal sugar alcohols and the diglucoside trehalose, generally by the pentose phosphate pathway. A number of roles are suggested for the production of these compounds by biotrophic fungi: (1). The conversion of host sugars into fungal sugars and derivatives maintains a concentration gradient favorable for the passive diffusion of host nutrients into the fungus; (2) sugar alcohols are more reduced than their parent monosaccharides, providing for more efficient energy storage; (3) the presence of mannitol and other solutes in the cytoplasm of fungal cells serves to lower the water potential within the fungus, favoring diffusion of water from the host into the fungus; (4) sugar alcohols can serve as intermediates for aldose–ketose interconversions; (5) sugar alcohols can serve as intermediates for the transhydrogenation of coenzymes NAD and NADP. The conversion of xylose to xylitol requires the oxidation of NADPH to NADP. The significance of transhydrogenation may reside in the overall regulation of metabolism, as NADH is generally the result of degradative processes and NADPH is associated with biosynthetic activities (Stacy, 1973).

Studies dealing with the biotrophic habit indicated that photosynthates are translocated across stromatic bridges that join pairs of leaves (Smith et al., 1985). The negligible incorporation to the unlabeled control tissue indicated that the stromata have an active role in translocation between leaves. Thus, in the *Myriogenospora*-infected plants there is a translocation of carbohydrates from host to fungus and from fungus to host, although it is not clear as to the nature of the product(s) translocated.

The basis for these effects presents the biggest dilemma. Biotrophic pathogens have been shown to exert a sink effect on their hosts in many cases (Hancock and Huisman, 1981; Walters, 1985), and in some cases uninfected leaves in infected plants exhibit increased photosynthesis presumably to compensate for loss of photosynthates to the pathogen in the infected leaves (Daly, 1976; Walters, 1985). Undoubtedly, endophytic fungi depend on the apoplastic solutes of its host for growth and the final effect is influenced by the composition of these solutes. Whether endophytic fungi exert a significant enough sink effect on the host to influence nitrogen uptake is not established. In the few experiments dealing with *in vivo* effects from endophytes (Lyons et al., 1990b; Lyons, 1985; Smith et al., 1985) in which transport of labeled photosynthates was measured, there were no differences between infected and uninfected plants during the short duration of an experimental time period. Longer periods might have given different results. A better understanding of others aspects of the relationship, such as effects of infection on respiration and growth of the host, might provide a better framework for interpreting results; and additional studies on nitrogen uptake and on the synthesis of large quantities of specific fungus compounds such as *N*-acetyl lolines should prove informative.

3.3. Nutrient Exchange and Translocation Across the Epibiotic and Endophytic Hyphae

There are two requirements that must be met if heterotrophic parasites are to be supplied with carbohydrates by autotrophs (Smith et al., 1969): (1) surplus carbohydrate must be present in the autotrophic donor at the site of contact with the recipient and (2) mechanisms for the selective and efficient transfer of carbohydrate to the heterotroph from the autotroph must be present. It is possible to present a hypothetical explanation for the carbohydrate uptake by both epibiotic and endophytic endophytes, which is consistent with that indicated by Smith et al. (1969). This explanation is in some respects similar to the mechanism outlined by Bushnell and Gay (1978) for accumulation of solutes in the haustoria of powdery mildews.

For example, consider nutrient acquisition by the epibiotic species *Myriogenospora atementosa* symbiotic with bahaiagrass. The carbohydrate to be taken up by *Myriogenospora* is produced by carbon fixation in the chloroplasts of mesophyll cells in the grass plant. Bahiagrass is a tropical species: both the Calvin cycle (C_3) and Hatch-Slack (C_4) carbon fixation pathways are operative (Chen et al., 1970; Mullis and Fallona, 1987). The four-carbon acids (malate and aspartate) produced during the fixation of CO_2 in the C_4 pathway are transported through plasmodesmata from mesophyll to bundle sheath cells where the acids are decarboxylated and the released carbon refixed into sugars by the C_3 pathway. This transport is said to occur in the symplasm, i.e., transport occurs within the interconnected cytoplasm of host cells. Transport within the symplasm may be rapid and often occurs counter to concentration gradients through the expenditure of metabolic energy. Symplasmic transport is under direct metabolic control, which may involve cytoplasmic streaming, hormonal responses, or other active mechanisms (Luttge and Higginbotham, 1979). The products of C_3 carbon fixation are important to epibiotic and endophytic associations. Sugars formed in the mesophyll cells are transported outside of the mesophyll plasmalemma and into the host apoplasm, which composes the plant's cell walls and is continuous throughout the plant except for the endodermal Casparian strip and other suberized structures. Transport in the apoplasm is outside of direct metabolic control and follows concentration gradients. Indirect metabolic control over transport in the apoplasm is exercised by changes in permeability in the plasmalemma that promote or restrict the flow of solutes between the symplasm and apoplasm. Movement of solutes in the apoplasm may be thought of as following a source and sink relationship. Sugars are produced in the mesophyll cells (the source) and flow in the apoplasm to their areas of utilization and transport (the sink). The sink for short-distance transport in healthy leaves is the conducting elements of the phloem. Loading of the phloem sieve tubes by the bundle sheath cells results in the entry of carbohydrate, generally in the form of sucrose, into the symplasm for long-distance translocation (Geiger, 1976; Giaquinta, 1976). However, the manner of phloem loading may vary and this too will affect the concentration of solutes in the apoplasm.

Because short-distance transport of sucrose between mesophyll cells and bundle sheath cells occurs within the apoplasm, potentially large amounts of soluble carbohydrates may be present in the apoplasm, and these carbohydrates are in a "dynamic equilibrium" with the carbohydrates in the symplasm (Hancock,

1977). Endophytes and epibionts take advantage of the carbohydrates present in the apoplasm. Additionally, Hancock and Huisman (1981) state that the most likely means of nutrient uptake by intercellular, nonhaustoriate biotrophs and hemibiotrophs are through the interception of solutes present in the hydrated cell walls of the host. Bushnell and Gay (1978) state that the uptake of solutes by powdery mildews is from the apoplasm as the plasmalemmas of the host and pathogen are separated by the extrahaustorial matrix and pathogen cell wall. Several species of the parasitic angiosperm *Cuscuta*, including *C. europea* L., withdraw nutrients from the apoplasm of *Vicia faba* L. (Wolswinkel, 1997). Infection by *Cuscuta* increases levels of soluble carbohydrate in the apoplasm; phloem unloading is stimulated in the area of the host adjacent to the pathogen, replenishing the carbohydrate withdrawn from the apoplasm. Wolswinkel (1997) suggested that phloem unloading is due to increased permeability of the host phloem elements and bundle sheath cells caused by the release of a growth regulator by *Cuscuta* into the host bean stem. Thus, there are examples of nutrient acquisition from the apoplasm, indicating that the mechanism has been successfully exploited by a large number of fungi.

In infections caused by the epibiotic species, *Myriogenospora*, the changes in epidermal cell size and shape suggests that either a growth regulatory substance is either produced by *Myriogenospora* and secreted into the host or the growth substance in produced by the host in response to the presence of the pathogen. Indoleacetic acid and other auxin-type compounds are produced by grass endophytes (Porter et al., 1977, 1985; De Battista et al., 1990) and their role might figure in the flow of nutrients into the apoplasm.

Long-distance transport of carbohydrates in healthy plants is under the control of plant growth regulators such as auxins and cytokinins. Amounts of growth regulators are expected to be enhanced in infected compared to uninfected host tissue. However, there are problems in relating these changes in amounts of growth regulators to the increased availability of carbohydrates for endophyte use. The artificial application of growth regulators generally causes the translocation of substances away from the point of application (Wardlaw, 1974; Lepp and Peel, 1971). Although plant growth regulators may have a role in the alteration of normal translocation patterns in grasses infected with *Myriogenospora*, *Neotyphodium*, and in other endophyte–host combinations, there are no conclusive data that long-distance transport is stimulated by pathogen export or enhanced host production of regulatory substances.

The first requirement (Lewis, 1967) for the supply of carbohydrate from the grass host to the fungal pathogen is met in epibiotic and endophytic infections caused by *Myriogenospora*: a surplus of mobile carbohydrate in the form of sucrose is present in the hydrated walls of the host epidermis. However, in *Myriogenospora* and other epibiotic species, the fungus is separated from the host apoplasm by the host cuticle. The second requirement stated by Lewis (1967) is that there must be a mechanism for the efficient transfer of carbohydrate from the plant host into the pathogen. This must include a consideration of the barrier to uptake posed by the cuticle since this species is primarily epibiotic. The mechanism for the transfer of carbohydrate from infected grass to *Myriogenospora* is similar in certain respects to the uptake of carbohydrate by the sphacelial or conidial stroma of *Claviceps purpurea* (Mower and Handcock, 1975). The base or foot of the stroma in *Claviceps* is subtended by a core of hyphae that extends downward into the vascular bundle of the host rachilla (Luttrell, 1979). The presence of mannitol, sucrose, and other carbohydrates in the stroma decreases the water potential in the area of the fungus relative to the water potential in the adjacent host tissue. Consequently, plant sap flows into the area of the host pathogen interface (Mower and Handcock, 1975), i.e., diffusional transport takes place. The sucrose present in the plant sap is converted to monosaccharides and other fungal carbohydrates (Smith, 1982; Smith et al., 1985). This conversion causes a "sucrose deficit," which the host has a tendency to fill by increased translocation of sucrose into the conidial stroma.

Obvious differences exist between infections produced by *Claviceps* and epibiotic species such as *Myriogenospora*, but there are similarities between *Claviceps* and endophytic endophytes such as *Neotyphodium* spp. Hyphae arising from the foot of the conidial stroma of *Claviceps* are in direct contact with vascular elements of the host rachilla. The growth of *Myriogenospora* and other epibiotic species is superficial and lacks penetrative or specialized assimilative structures whereas the hyphae of *Balansia* spp. and *Neotyphodium* spp are in direct contact with host cell walls. Sucrose present in the hydrated cell walls of the host epidermis is separated from the stroma of epibiotic fungi by the hydrophobic cuticle. However, the host cuticle may not provide such a barrier for the flow of carbohydrates and water as it may seem at first glance (White and Camp, 1995). The cuticle of the host is hydrophobic because of the presence of lipids and waxes, and therefore is resistant to the bulk flow of water (Kolattukudy et al., 1981). The cuticle of *Citrus aureum* L. has pores of sufficient diameter to permit the flow of water and glucose molecules, but not of sucrose molecules (Schoonherr, 1976). It is not known if the cuticles of grasses have pores or, if so, whether they are of sufficient diameter for the translocation of either sucrose or the unidentified low molecular weight carbohydrate present in uninfected leaves and stromata. The leakiness of cuticles is well documented in studies of communities of leaf epiphytes and of leaf leachates (Tukey, 1970). The presence of the stromata may influence the formation of cuticular pores. Cuticles of *Vitus vinifera* L. are isolated free of pores (Baker et al., 1982). However, cutin esterase derived from culture filtrates of *Fusarium solani* (Mart.) Sacc. f. sp. *pisi* (F. R. Jones) Snyd. & Hans. added to cuticles of *V. vinifera* produced pores in the cuticle of sufficient diameter to permit the passage of high molecular weight enzymes as well as monosaccharides.

It is conceivable that the presence of the tightly appressed epibiotic stroma of *Myriogenospora*, and stromata of *Balansia* species (White and Owens, 1992;

White et al., 1995, 1996), which have lowered water potentials due to the presence of mannitol and other solutes, causes the flow of water from the hydrated cell walls of the host through the cuticle or cell wall and into the fungus. Thus, sugar alcohol metabolism and the level of activity of the pentose phosphate pathway may quantitatively determine the level of sugar alcohol synthesis. Regulation of translocation may reside at this point in metabolism. Plant sap flowing to the area of infection due to the diffusion of water into the stromata serves to maintain levels of sucrose in the host tissue appressed to the fungus. In the absence of a demonstration of cuticular pores of sufficient diameter, the mechanism of carbohydrate uptake by *Myriogenospora* from the epidermal cells of the host and the mechanism that permits reciprocal translocation are yet to be determined. The ^{14}C-labeled compounds (Smith, 1982; Smith et al., 1985) indicate that translocation from host to fungus and from fungus to host may remain as sucrose. Sucrose, present in trace amounts in the fungus (Smith, 1982), could be translocated without requiring enzymes to convert fungal carbohydrates back to sucrose for translocation in the host.

Host sucrose requires hydrolysis prior to further utilization by the fungus. Sucrose hydrolytic enzymes have been isolated from the cell walls and cytoplasm of a number of plant pathogenic fungi and endophytic fungi (Bilgrami and Verma, 1978; Lam et al., 1995), and in response to the levels of soluble sugars. Further, increases in invertase activity in infected tissue appears to occur frequently (Clancy and Coffey, 1980; Lam et al., 1995). In infections caused by *Claviceps*, hydrolysis of sucrose has been reported as occurring at the cell walls of the fungus (Dickersont et al., 1976) or in the cytoplasm of the cells that make up the conidial stroma (Mower and Handcock, 1975). It is unclear as to where hydrolysis of sucrose takes place in endophytic hyphae or stroma. The small amounts of sucrose present in *Myriogenospora*-infected leaves suggest that sucrose is a major carbohydrate translocated to this fungus (Smith, 1982). Glucose derived from the hydrolysis of sucrose is respired by fungi or used in the production of trehalose. Fructose derived from sucrose hydrolysis is converted to mannitol through a series of enzymatic steps of the pentose phosphate pathway (Lewis, 1967). Most of the sucrose that enters the conidial stroma of *Claviceps* remains as sucrose, and invertase activity is subject to catabolite repression when glucose concentrations are greater than 10% (Mower and Handcock, 1975), although in *Epichloe* and related species invertase activity was induced by 2% sucrose. However, the presence of other monosaccharides might contribute to the effect of invertase activity since in several *Epichloe/Neotyphodium* isolates there were reports of variation in this response to sucrose. Concentrations of free glucose in the conidial stroma of *Claviceps* are sufficient to repress invertase activity (Mower and Handcock, 1975). However, because of the availability of the host vasculature to hyphae of *Claviceps*, immediate hydrolysis of sucrose is less important in endophytic hyphae than in epibiotic fungi in which sucrose hydrolysis is the driving force for maintaining a concentration gradient favorable for the flow of carbohydrate and water to the pathogen. Because essentially no free glucose or other monosaccharides are present in the stroma of *Myriogenospora* (Smith, 1982), catabolite repression of its invertases might not occur. Since considerable invertase activities and soluble sugars have been reported in the endophytic species *Neotyphodium* and *Epichloe* (Lam et al., 1995), and that the invertases of these species are induced by sucrose, catabolite repression is possible and might represent a difference in the movement of sugar between the two basic groups and *Claviceps*. In either case, mannitol should function as a store of reducing power, osmoregulatory agent, and control for water influx necessary for stroma growth and hyphal expansion within the stroma.

3.4. Nutrient Exchange and Translocation: The Stromata

The reproductive structures, i.e., the stromata, such as that produced by *E. amaril-lans* and other species with similar endophytic habits, acquire nutrients by surface layers of mycelium and the process is greatly facilitated by the bulk flow of water to the superficial mycelial layer, an area where water evaporates (White and Camp, 1995; White et al., 1997). It seems likely that sugars that leak into the apoplast are carried with the evaporating water into the mycelial cortex layer. The resulting proposed mechanism for facilitating nutrient acquisition by stromata of endophytes is referred to as facilitated evaporative-bulk flow movement of nutrients (White et al., 1997) and is yet another example of bulk flow movement of water. In addition to this mechanism, diffusional transport and active transport of nutrients by endophytic hyphae may also be possible.

4. PHYSIOLOGY AND COEVOLUTION

4.1. Novel Metabolites

The variety of defensive compounds (Table 3) found within symbiotic individuals of a population of endophyte-infected forage grasses are numerous and will vary because of the process of natural selection under herbivory. Long-term herbivory, and specific types of herbivory (insect or mammalian), should lead to subpopulations of toxic grasses, each chemically defined and based on its specific mixture of deterring and toxic compounds. Since several biotrophic and endophytic species are maternally transmitted, seed-sown pastures will reflect this genetic diversity. The plant and fungus have distinct physiologies and it is not expected that they interacted directly. However, the results of specific biochemical pathways are important to the process of selection, resulting in a system of defensive mutualism.

    The degree of chemical expression by the fungal mutualist depends not only on the biochemical competence or genotype of the fungus (Bacon et al.,

Table 3 Novel Secondary Metabolites Produced by Endophytes, Grouped According to Their Occurrence in *Balansia*-Infected (Balansioid) and *Neotyphodium*-Infected Grasses

Neotyphodium species symbiota	*Balansia* species symbiota
Ergovaline	Elymoclavine
Ergonovine	Ergonovine
Ergosine	Ergonine
Chanoclavine I	Chanoclavine I
Peramine	Isochanoclavine I
Indole acetic acid	Agroclavine
Ergosterol	Dihydroelymoclavine
Ergosinine	6,7-Secoagroclavine
Cyclopentanoid	Penniclavine
Sesquiterpenoids[a]	Erytho 1-(3-indoly)propane-1,2,3-triol
Ergonovine	Threo-1-(3-indoly)propane-1,2,3-triol
Caffeic acid[a]	3-Indole acetic acid
p-Coumaric acid	Methyl-3-indolecarboxylate
p-Hydroxybenzoic acid	3-Indoleacetamide
Loline	3-Indole ethanol
N-Acetylloline	Ergobalansine
N-Formylloline	Ergobalansinine
N-Acetylnorloline	
Paxilline	
Lolitrems A, B, C, and D	

[a] These compounds were isolated from *Epichloe typhina* and are grouped with *Neotyphodium* spp. Other metabolites such as hormane, norharmane, and halostachine were isolated from tall fescue (Yates, 1983), but the infection status of the grasses was not reported; therefore, they have been excluded from the list.
Source: Bacon, 1994.

1975; Bacon, 1988), but also the genotype of each plant (Hill et al., 1990; Hill, Parrott and Pope, 1991b), and the environmental factors of soil N (Gaynor and Hunt, 1983; Arechavaleta et al., 1989; Lyons et al., 1990b) and moisture (Arechavaleta et al., 1991). This raises the issue as to whether the controlling mechanism is associated with the plant, the endophyte, or an interaction between the two. By inserting endophytes into a common tall fescue genotype and by conducting genetic studies between high- and low-ergovaline producing symbiotic plant genotypes (Agee and Hill, 1992, 1994), it has been documented that at least the plant can regulate the expression of ergovaline production by the endophyte. In this regard, regulation well reflect the individual variation of toxic precursors and primary metabolites released in the apoplasm from the plant.

The known compounds associated with either insect deterrence, mammalian toxicity, or fungal resistance are secondary metabolites, most of which originate from primary metabolism. Thus, these metabolites are in part under the regulation of catabolic pathways. Many genes in microorganisms encoding for catabolic activities are maximally expressed only under conditions of nutritional stress (low nitrogen, carbon, sulfur, or phosphorus), although such enzymes are also influenced by pH changes, oxygen status, and other abiotic factors. In general, nutrients in the apoplasm are reported for most plants as being sparse (Morgan, 1984). Therefore, the nutritional stress within the apoplasm may account for the final quantitative expression of specific toxic and fitness-enhancing metabolites associated with symbiotic grasses. The level of expression may be confounded by the interaction of environmental or host nutritional factors (Arechavaleta et al.,1992; Richardson et al.,1993a; Agee and Hill, 1994; Hill et al., 1996).

The ergot alkaloids (Table 3) found in symbiotic grasses consist of both the ergopeptine and clavine types (Porter et al., 1979b, 1981; Yates et al., 1985; Lyons et al., 1986). These two groups differ from each other in the presence (ergopeptide bond) or absence (clavine) of a peptide bond and attached amino acids. There is considerable controversy over the biological activity of each ergot alkaloid, but it is generally agreed that the ergopeptine alkaloids are more active than the clavine alkaloids. The tremorgenic neurotoxins consist of four biologically active compounds, all containing a complex indole isoprenoid ring system (Gallagher et al.,1984), commonly called the lolitrems. These have been isolated only from the perennial ryegrass symbiotum and are considered responsible for ryegrass staggers of sheep (Gallagher et al.,1982, 1984). This group of toxins is apparently absent in symbiotic tall fescue and has not been examined for in the *Balansia*-infected grasses. The major lolitrem is lolitrem B, which ranges from 3 to 25 µg/g dry weight in perennial ryegrass herbage, and a smaller concentration also occurs in ryegrass seed (Gallagher et al., 1987). The lolitrems, unlike the ergot alkaloids, have not been isolated from cultures of the fungus *A. lolii*, but its indole isoprenoid precursor, paxilline, has, indicating that it is synthesized by the fungus in culture (Christopher and Mantle, 1987; Weedon and Mantle, 1987, Mantle and Weedon, 1994). In addition to finding paxilline in fungus cultures, it has been detected in ryegrass seed (Weedon and Mantle, 1987). Thus, the fungus produces paxilline, which is then converted to the lolitrems in culture and *in planta*.

The loline alkaloids are pyrrolizidine bases that are found in the tall fescue symbiotum in concentrations as high as 0.8% of the dry weight of tall fescue plants (Bush et al., 1982). These alkaloids have only recently been isolated from cultures of the fungus (Bush et al., 1997). A limited number of studies suggests that the loline alkaloids are only mildly toxic (Davis et al., 1986b; Strahan et al., 1988; Eichenseer et al., 1991), especially when toxicity is compared with that of the usual pyrrolizidine alkaloids (McLean, 1970). This class of alkaloids is distinct in that it is found in very high concentrations, high enough to serve as the osmoticum responsible for drought tolerance within the *Neotyphodium* symbiotic grasses.

Other compounds reported in symbiotic grasses that may influence the physiology of the symbiota included among mammalian herbivores include 3-indoleacetic acid, indole ethanol, and related simple indoles. These compounds have plant growth–stimulating properties (Porter et al., 1988). Peramine is the insect deterrent (Rowan et al., 1986). The biological activity of tetraenone steroid (Porter et al., 1977, 1985), the ergosterols (Davis et al., 1986a), phenolic acid derivatives (Koshino et al., 1988), and sesquiterpenes (Yshihara et al., 1985) has not been determined. However, the effects of some of these on insects, laboratory animals, and fungi suggest that toxicity may be their primary action (Yshihara et al., 1985; Davis et al., 1986a; Dowd et al., 1988). The functions of these and other compounds have not been established within the overall physiology of the association.

4.2. Coevolution

Improved fitness resulting from endophytes symbiotic with grasses includes increased growth rate and tiller density (Bradshaw, 1988; Kackley et al., 1990; Hill, Belesky & Stringer, 1991), changes in morphology (Diehl, 1950; Hill et al., 1990), resistance, and toxicity to grazing animals (Byford, 1979; Wallner et al., 1983; Read and Camp, 1986), insect deterrence (Prestidge et al., 1982; Clay, 1988), nematode resistance (Kimmons et al., 1990; West et al., 1990a), disease resistance (White and Cole, 1985; Yshihara et al., 1985), and drought tolerance (Read and Camp, 1986; Arechavaleta et al., 1989; Elmi and West, 1989; West et al., 1993). The chemical basis for each component of fitness is not completely understood, but data defining cause-and-effect relationships are eminent. There may be a chemical similarity between compounds that are responsible for each aspect of improved fitness although different compounds which may be found responsible in each symbiotic species. Specific information on mutualisms is far from complete, as there are only a few experiments that document positive benefits from endophytes within associations, particularly the balansioid symbiota. Nevertheless improved fitness is considered the essential drive for coevolution. Plant–fungus associations appear to have entered close relationships very early in their evolution. The Gramineae probably originated in the upper Cretaceous Period of the Mesozoic Era, much later than dicotyledonous plants. The fossil records of the first authentic species of grasses, fruits of *Stipa* (Cockerell, 1956) and *Phalaris* (Beetle, 1958), were obtained from the Late Tertiary deposits, approximately 40–50 million years ago. The oldest fossil *Festuca* species was reported in the Miocene Epoch of the Tertiary Period (Thomasson, 1986). Thus, geologically speaking, we are dealing with a relatively young symbiota in the case of the *Neotyphodium*–fescue associations, which must have been initiated at least during the Early Pliocene of the Cenozoic Era, approximately 25 million years ago.

Natural populations of symbiotic grasses persist longer and compete better than nonsymbiotic populations (Clay, 1986; Kelley and Clay, 1987; Prestidge et al., 1982; West et al., 1993). However, as discussed earlier, the components of improved fitness of symbiotic populations are multifaceted and difficult to access. Most research on symbiotic grasses has emphasized variation of fitness at the population level, with relatively few studies partitioning that variation into its genetic and environmental components. It is relatively difficult or impossible to interpret which symbiotic component, physiological or otherwise, contributed to the phenotypic expression observed. It is equally difficult to distinguish any evolutionary significance to the fitness variations observed in natural populations. The alternative approach of using symbiotic and nonsymbiotic clones (Hill et al., 1990; Hill, Parrott & Pope, 1991; Agee and Hill, 1992) has greatly facilitated our approach to and understanding of several aspects of the variation of mutualistic responses within a population. This approach should allow us to extrapolate to specific physiological effects responsible for such variation of trophic interactions.

All of the positive benefits derived from the symbiota are expressed or measured when the association is subjected to stresses. Characteristics that are associated with drought tolerance are not expressed unless the plant is exposed to prolonged drought conditions. For example, roots of symbiotic plants will grow faster and deeper into a soil profile under drought stress than its cloned nonsymbiotic ramet. However, there is no difference when both are grown at field capacity (Richardson, Hill and Hoveland, 1990). Furthermore, total reserve carbohydrate content and structures are similar among plants, regardless of endophyte content, when soil water content is at field capacity (-0.1 bar), but sugar monomers increase and polymers decrease in symbiotic tall fescue once drought stress conditions are imposed (Richardson et al., 1992). Moreover, there is variation among symbiota as, for example, plant morphological changes are not constant among tall fescue genotypes (Hill et al.,1990) and *Balansia*-infected grasses (Diehl, 1950). In endophyte-infected tall fescue, and possibly other *Neotyphodium* species, variation is expressed specifically by tillering capacity, specific leaf weights, and crown weight which may increase, decrease, or remain the same depending on the genotype of the infected plant. The genetic plasticity necessary for evolution to take place is present in symbiotic populations. The expression of change is due either to the fungal component or the grass or both.

Fungi are rated as highly adaptable because they have biochemical abilities to utilize a wide variety of substrates and produce precursors from intermediary metabolism for use in primary and secondary metabolism. On the other hand, grasses are one of the few groups of plants that lack the ability to produce excessive secondary metabolites (Zahner et al., 1983). In theory, cohabitation and the establishment of a completely compatible association, although gradual, were apparently based partially on the need for secondary metabolites that were lacking in the grass but were contributed by the fungus. The evolutionary events that resulted in the cohabitation of grasses with this group of fungi will probably remain unknown. It is clear, however, that several of the clavicipitaceous fungal mutualists meet some of the criteria that Law (1985) proposed for mutualism: (1) the inhabitants are genetically similar while their hosts are genetically diverse; (2) the inhabitants rarely or never undergo sexual reproduction; and (3) the inhabitants lack strong specificity to a particular host species. The *Neotyphodium* species show highly successful defensive mutualism and express all three criteria.

To understand the need for a mutualistic relationship to develop between grasses and fungal endophytes, a complete understanding of the systematics and evolution of both the endophyte and species, as well as their biochemical relatedness and requirements, must be known. Briefly, the major impetus considered responsible for extending the adaptability of symbiotic grasses is both the physiological competence of the fungus and the resulting variety of secondary metabolites within the association, most of which are produced by the fungus. For more detailed discussions of this evolutionary aspect, as well as a model for fitness and other salient but theoretical evolutionary events within these mutualisms, the reader is referred to earlier reviews (White, 1988; Bacon and Hill, 1996).

REFERENCES

Agee, C. S. and Hill, N. S. (1992). Variability in progeny from high- and low-ergovaline producing tall fescue parents. *In Agronomy Abstracts.* ASA, Madison, WI., p. 82.

Agee, C. S. and Hill, N. S. (1994). Ergovaline variability in *Acremonium*-infected tall fescue due to environment and plant genotype. *Crop Science* 34, 221–226.

Arechavaleta, M., Bacon, C. W., Hoveland, C. S. and Radcliffe, D. E. (1989). Effect of the tall fescue endophyte on plant response to environmental stress. *Agronomy Journal* 81, 83–90.

Arechavaleta, M., Bacon, C. W., Plattner, R. D., Hoveland, C. S. and Radcliffe, D. E. (1992). Accumulation of ergopeptide alkaloids in symbiotic tall fescue grown under deficits of soil water and nitrogen fertilizer. *Applied and Environmental Microbiology* 58, 857–861.

Bacon, C. W. (1985). A chemically defined medium for the growth and synthesis of ergot alkaloids by the species of Balansia. *Mycologia* 77, 418–423.

Bacon, C. W. (1988). Procedure for isolating the endophyte from tall fescue and screening isolates for ergot alkaloids. *Applied Environmental Microbiology.* 54, 2615–2618.

Bacon, C. W. (1990). Isolation, Culture, and Maintenance of Endophytic Fungi of Grasses. *In* D. P. Labeda (ed.), *Isolation of Biotechnological Organisms from Nature.* McGraw-Hill, New York, pp. 259–282.

Bacon, C. W. (1994). Fungal endophytes, other fungi, and their metabolites as extrinsic factors of grass quality. *In* G. C. Fahey, ed. (ed.), *Forage Quality, Evaluation and Utilization.* American Society of Agronomy, Madison, WI, pp. 318–366.

Bacon, C. W. and Hill, N. S. (1996). Symptomless grass endophytes: products of coevolutionary symbioses and their role in the ecological adaptation of infected grasses. *In* S. C. Redlin (eds.), *Systematics, Ecology and Evolution of Endophytic Fungi in Grasses and Woody Plants*, American Phytopathological Society, St. Paul, MN, pp. 155–178.

Bacon, C. W., Porter, J. K. and Robbins, J. D. (1975). Toxicity and occurrence of Balansia on grasses from toxic fescue pastures. *Applied Microbiology* 29, 553–556.

Bacon, C. W. and White, J. F., Jr. (1994). Stains, media, and procedures for analyzing endophytes. *In* C. W. Bacon and J. F. White, Jr. (eds.), *Biotechnology of Endophytic Fungi.* CRC Press, Boca Raton, FL, pp. 47–56.

Baker, C. J., McCormick, S. L. and Bateman, D. F. (1982). Effects of purified cutin esterase upon the permeability and mechanical strength of cutin membranes. *Phytopathology* 72, 420–423.

Beetle, A. A. (1958). *Piptochaetium* and *Philaris* in the fossil record. *Bulletin Torrey Botanical Club* 85, 179–181.

Belesky, D. P., Devine, O. J., Pallas, J. E., Jr. and Stringer, W. C. (1987). Photosynthetic activity of tall fescue as influenced by a fungal endophyte. *Photosynthetica* 21, 82–87.

Bilgrami K. S. and Verma R. N. (1978). *Physiology of Fungi.* Vikas Press: New Delhi.

Bradshaw, A. D. (1988). Population differentiation in *Agrostis tenuis* Sibth. II. The incidence and significance of infection by *Epichloe typhina. New Phytologist* 58, 310–315.

Bush, L. P., Cornelius, P. C., Buckner, R. C., Varney, D. R., Chapman, R. A., Burrus, P. B., Kennedy, C. W., Jones, T. A. and Saunders, M.J. (1982). Association of N-acetyl loline and N-formyl loline with *Epichloe typhina* in tall fescue. *Crop Science* 22, 941–943.

Bush, L. P., Wilkerson, H. H., and Schardl, C. L. (1997). Bioprotective alkaloids of grass-fungal endophyte symbioses. *Plant Physiology* 114, 1–7.

Bushnell, E. R. and Gay, J. (1978). Accumulation of solutes in relation to the structure and function of haustoria in powdery mildews. *In* D. M. Spencer (eds.), *The Powdery Mildews.* Academic Press, New York, pp. 183–235.

Byford, M. J. (1979). Ryegrass staggers in sheep and cattle. *New Zealand Journal Agricultural Research* May 1979, 65.

Chen, T. M., Brown, R. H. and Black, C. C. (1970). CO_2 compensation, concentration, rate of photosynthesis, and carbonic anhydrase activity of plants. *Weed Science* 18, 399–403.

Christopher, W. M. and Mantle, P. G. (1987). Paxilline biosynthesis by *Acremonium loliae*, a step toward defining the origin of lolitrem neurotoxins. *Phytopathology* 26, 969–971.

Clancy, F. G. and Coffey, M. D. (1980). Patterns of translocation, changes in invertase activity, and polyol formation in flax infected with the rust fungus, *Melamspora lini. Physiolical Plant Pathology* 17, 41–52.

Clay, K. (1986). Grass endophytes. *In* N. J. Fokkema and J. Van Den Heuvel (eds.), *Micro-

biology of the Phyllosphere, Cambridge University Press, Cambridge UK, pp. 188–204.

Clay, K. (1988). Fungal endophytes of grasses: a defensive mutualism between plants and fungi. *Ecology* 69, 10–16.

Cockerell, T. D. A. (1956). The fossil flora of Florissant, Colorado. *American Museum of Natural History Bulletin* 24, 71–110.

Daly, J. M. (1976). The carbon balance of diseased plants: changes in respiration, photosynthesis and translocation. *In* R. Heitefuss and P. H. Williams (eds.), *Physiological Plant Pathology*, Springer-Verlag, Berlin, pp. 450–479.

Davis, N. D., Clark, E. M., Schrey, K. A. and Diener, U. L. (1986a). Steriod metabolites of *Acremonium coenophialum*, an endophyte of tall fescue. *Journal of Agricultural and Food Chemistry* 34, 105–108.

Davis, N. D., Clark, E. M., Schrey, K. A. and Diener, U. L. (1986b). In vitro growth of *Acremonium coenophialum*, an endophyte of toxic tall fescue grass. *Applied and Environmental Microbiology* 52, 888–891.

De Battista, J. P., Bacon, C. W., Severson, R. F., Plattner, R. D. and Bouton, J. H. (1990). Indole acetic acid production by the fungal endophyte of tall fescue. *Agronomy Journal* 82, 878–880.

Dickerson, A. G., Mantle, P. G. and Nisbet, L. J. (1976). Carbon assimilation by *Claviceps purpurea* growing as a parasite. *Journal of General Microbiology* 97, 267–276.

Diehl, W. W. (1950). *Balansia* and Balansiae in America. In *USDA Agric. Monograph* 4 (ed. U.S.Government Printing Office), U.S. Govt. Print. Office, Washington, DC, pp. 1–82.

Dowd, P. F., Cole, R. J. and Vesonder, R. F. (1988). Toxicity of selected tremogenic mycotoxins and related compounds to *Spodoptera frugiperda* and *Heliothis zea*. *Journal of Antibiotics* 61, 1868–1872.

Eichenseer, H., Dahlman, D. L. and Bush, L. P. (1991). Influence of endophyte infection, plant age and harvest interval on *Rhopalosiphum padi* survival and its relation to quantity of N-formyl and N-acetyl loline in tall fescue. *Entomologia Experimentatis et Applicata* 60, 29–38.

Gallagher, R. T., Smith, G. S., Di Menna, M. E. and Young, P. W. (1982). Some observations on neurotoxin production in perennial ryegrass. *New Zealand Veterinary Journal* 30, 203–204.

Gallagher, R. T., Hawkes, A. D., Steyn, P. S. and Vleggaar, R. (1984). Tremorgenic neurotoxins from perennial ryegrass causing ryegrass staggers disorder of livestock: structure and elucidation of lolitrem B. *Chemical Communications* 614–616.

Gallagher, R. T., Smith, G. S. and Sprosen, J. M. (1987). Distribution and accumulation of lolitrem B neurotoxin in perennial ryegrass plants. *4th Animal Science Congress of the Asian-Australasian Association of Animal Production*, Hamilton, New Zealand, p. 404.

Gaynor, D. L. and Hunt, W. F. (1983). The relationship between nitrogen supply, endophytic fungus, and Argentine stem weevil resistance in ryegrass. *In Proceeding of N.Z. Grassland Association*, Ag Research, Grasslands Research Center, Palmerston North, NZ, pp. 257–263.

Geiger, D. R. (1976). Phloem loading in source leaves. *In* I. F. Wardlaw and J. B. Passioura (eds.), *Transport and Transfer Processes in Plants*. Academic Press, New York, pp. 167–183.

Giaquinta, R. (1976). Evidence for phloem loading from the apoplast: chemical modification of membrane sulfhydryl groups. *Plant Physiology* 57, 872–875.

Glenn, A. E. (1995). *Molecular Phylogeny of Acremonium and Its Taxonomic Implications*. University of Georgia, M.S. Thesis, Athens, Ga.

Glenn, A. E., Bacon, C. W., Price, R. and Hanlin, R. T. (1996). Molecular phylogeny of *Acremonium* and its taxonomic implications. *Mycologia* 88, 369–383.

Glenn, A. E., Rykard, D. M., Bacon, C. W. and Hanlin, R. T. (1997). Molecular characterization of *Myriogenospora atramentosa* and its occurrence on some new hosts. *Mycological Research* 102, 483–490.

Hancock, J. G. (1977). Soluble metabolites in intercellular regions of squash hypocotyl tissue: implications for exudation. *Plant and Soil* 47, 103–112.

Hancock, J. G. and Huisman, O. C. (1981). Nutrient movement in host–pathogen systems. *Annual Review of Phytopathology* 19, 309–331.

Hattersley, P. W. (1986). Variations in photosynthetic pathway. *In* T. R. Soderstrom, K. W. Hilu and M. E. Barkworth Campbell (eds.), *Grass Systematics and Evolution*. Smithsonian Institution Press, Washington, DC, pp. 48–69.

Hill, N. S., Stringer, W. C., Rottinghaus, G. E., Belesky, D. P., Parrott, W. A. and Pope, D. D. (1990). Growth, morphological, and chemical component responses of tall fescue to *Acremonium coenophialum*. *Crop Science* 30, 156–161.

Hill, N. S., Belesky, D. P. and Stringer, W. C. (1991). Competitiveness of tall fescue as influenced by *Acremonium coenophialum*. *Crop Science* 31, 185–190.

Hill, N. S., Pachon, J. G. and Bacon, C. W. (1996). *Acremonium coenophialum*–mediated short- and long-term drought acclimation in tall fescue. *Crop Science* 36, 665–672.

Hill, N. S., Parrott, W. A. and Pope, D. D. (1991). Ergopeptine alkaloid production by endophytes in a common tall fescue genotype. *Crop Science* 31, 1545–1547.

Kackley, K. E., Grybauskas, A. P., Dernoeden, P. H. and Hill, R. L. (1990). Role of drought stress in the development of summer patch in field-inoculated Kentucky bluegrass. *Phytopathology* 80, 655–658.

Kelley, S. E. and Clay, K. (1987). Interspecific competitive interactions and the maintenance of genotypic variation within two perennial grasses. *Evolution* 41, 92–103.

Kimmons, C. A., Gwinn, K. D. and Bernard, E. C. (1990). Nematode reproduction on endophyte-infected and endophyte-free tall fescue. *Plant Disease* 74, 757–761.

Kolattukudy, P. E., Espelie, K. E. and Soliday, C. L. (1981). Hydrophobic layers attached to cell walls. *In* W. Tanner (ed.), *Plant Carbohydrates. II. Extracellular Carbohydrates*. Berlin: Springer-Verlag, Berlin, pp. 225–250

Koshino, H., Terada, S., Yoshihara, T., Sakamura, S., Shimanuki, T., Sato, T. and Tajimi, A. (1988). Three phenolic acid derivatives from stromata of *Epichloe typhina* on *Phleum pratense*. *Phytochemistry* 27, 1333–1338.

Kulkarni, R. K. and Nielsen, B. (1986). Nutritional requirements for growth of fungus endophyte of tall fescue. *Mycologia* 78, 781–786.

Lam, C. K., Belanger, F. C., White, J. F., Jr. and Daie, J. (1995). Invertase activity in *Epichloë/Acremonium* fungal endophytes and its possible role in choke disease. *Mycological Research* 99, 867–873.

Law, R. (1985). Evolution in a mutualistic environment. *In* D. H. Bougher (ed.), *The*

Biology of Mutualism: Ecology and Evolution Croom Helm, London, pp. 145–170.

Lepp, N. W. and Peel, A. J. (1971). Influence of IAA upon the longitudinal and tangential movement of labeled sugars in the phloem of willow. *Planta* 97, 50–61.

Leuchtmann, A. and Clay, K. (1988). *Atkinsonella hypoxylon* and *Balansia cyperi*, epiphytic members of the Balansiae. *Mycologia* 80, 192–199.

Lewis, D. H. (1967). Sugar alcohols (polyols) in fungi and green plants. *New Phytologist* 66, 143–184.

Luttge U. and Higginbotham N. (1979). *Transport in Plants*. Springer-Verlag: Berlin.

Luttrell, E. S. (1979). Host-parasite relationships and development of the ergot sclerotium in *Claviceps purpurea*. *Canadian Journal of Botany* 58, 942–958.

Lyons P. C. (1985). *Infection and in vitro ergot alkaloid synthesis by the tall fescue endophyte and effects of the fungus on host nitrogen metabolism.* University of Georgia, Athens.

Lyons, P. C., Evans, J. J. and Bacon, C. W. (1990a). Effects of the fungal endophyte *Acremonium coenophialum* on nitrogen accumulation and metabolism in tall fescue. *Plant Physiology* 92, 726–732.

Lyons, P. C., Evans, J. J. and Bacon, C. W. (1990b). Effects of the fungal endophyte *Acremonium coenophialum* on nitrogen accumulation and metabolism in tall fescue. *Plant Physiology* 92, 726–732.

Lyons, P. C., Plattner, R. D. and Bacon, C. W. (1986). Occurrence of peptide and clavine ergot alkaloids in tall fescue. *Science* 232, 487–489.

Mantle, P. G. and Weedon, C. M. (1994). Biosynthesis and transformation of tremorgenic indole-diterpenoids by *Penicillium paxilli* and *Acremonium lolii*. *Phytochemistry* 36, 1209–1217.

McLean, E. K. (1970). The toxic actions of pyrrolizidine (*Senecio*) alkaloids. *Pharmacological Reviews* 22, 429–483.

Morgan, J. M. (1984). Osmoregulation and water stress in higher plants. *Annual Review of Plant Physiology* 35, 299–319.

Mower, R. L. and Handcock, J. G. (1975). Mechanisms of honeydew formation by *Claviceps* species. *Canadian Journal of Botany* 53, 2826–2834.

Mullis, K. B. and Fallona, F. A. (1987). Specific synthesis of DNA in vitro via a polymerase-catalyzed chain reaction. *Methods in Enzymology* 155, 335 -350.

Porter, J. K., Bacon, C. W., Robbins, J. D., Himmelsbach, D. S. and Higman, H. C. (1977). Indole alkaloids from *Balansia epichloe* (Weese). *Journal of Agriculture and Food Chemistry* 25, 88–93.

Porter, J. K., Robbins, J. D., Bacon, C. W. and Himmelsbach, D. S. (1978). Determination of epimeric 1-(3-indolyl)propane-1,2,3,triol isolated from *Balansia epichloe*. *Lloydia* 41, 43–49.

Porter, J. K., Bacon, C. W., Cutler, H. G., Arrendale, R. F. and Robbins, J. D. (1985). In vitro auxin production by *Balansia epichloe*. *Phytochemistry* 24, 1429–1431.

Prestidge, R. A., Pottinger, R. P. and Barker, G. M. (1982). An association of *Lolium* endophyte with ryegrass resistance to Argentine stem weevil. In: *Proceedings of the New Zealand Weed Pest Control Conference*, pp. 119–122.

Read, J. C. and Camp, B. J. (1986). The effect of the fungal endophyte *Acremonium coenophialum* in tall fescue on animal performance, toxicity, and stand maintenance. *Agronomy Journal* 78, 848–850.

Richardson, M. D., Bacon, C. W., Hill, N. S. and Hinton, D. M. (1993a). Growth and water relations of *Acremonium coenophialum*. *In* D. E. Hume, G. C. M. Latch, and H. S. Easton (eds.), *Proceeding of the Second International Symposium on Acremonium/Grass Interactions*, AgResearch, Grasslands Research Center, Palmerston North, New Zealand, pp. 181–184.

Richardson, M. D., Bacon, C. W. and Hoveland, C. S. (1990). The effect of endophyte removal on gas exchange in tall fescue. *In* S. Quisenberry and R. Joost (eds.), *Proceeding of International Symposium on Acremonium/Grass Interactions*. Baton Rouge, LA: Louisiana Agricultural Experiment Station, Baton Rouge, LA.

Richardson, M. D., Chapman, G. W., Jr., Hoveland, C. S. and Bacon, C. W. (1992). Sugar alcohols in endophyte-infected tall fescue under drought. *Crop Science* 32, 1060–1061.

Richardson, M. D., Hill, N. S. and Hoveland, C. S. (1990). Rooting patterns of endophyte infected tall fescue grown under drought stress. In: *Agronomy Abstracts*. American Society of Agronomy, Madison, WI, p. 129.

Richardson, M. D., Hoveland, C. S. and Bacon, C. W. (1993b). Photosynthesis and stomatal conductance of symbiotic and nonsymbiotic tall fescue. *Crop Science* 33, 145–149.

Rowan, D. D., Hunt, M. B. and Gaynor, D. L. (1986). Peramine, a novel insect feeding deterrent from ryegrass infected with the endophyte *Acremonium loliae*. Chemical Communications 935–936.

Schardl, C. L. and Tsai, H.-F.Tsai (1992). Molecular biology and evolution of the grass endophytes. *Natural Toxins* 1, 171–184.

Schoonherr, J. (1976). Water permeability of isolated cuticular membranes: the effect of pH and cations on diffusion, permeability, and size of polar pores in the cutin matrix. *Planta* 128, 113–126.

Siegel, M. R., Johnson, M. C., Varney, D. R., Nesmith, W. C., Buckner, R. C., Bush, L. P., Burrus, P. B., II, Jones, T. A. and Boling, J. A. (1984). A fungal endophyte in tall fescue: incidence and dissemination. *Phytopathology* 74, 932–937.

Siegel, M. R., Latch, G. C. M., Bush, L. P., Fammin, N. F., Rowen, D. D., Tapper, B. A., Bacon, C. W. and Johnson, M. C. (1991). Alkaloids and insecticidal activity of grasses infected with fungal endophytes. *Journal of Chemical Ecology* 16, 3301–3315.

Siegel, M. R., Latch, G. C. M. and Johnson, M. C. (1985). *Acremonium* fungal endophytes of tall fescue and perennial ryegrass: significance and control. *Plant Disease* 69, 179–183.

Smith, D., Muscatine, L. and Lewis, D. H. (1969). Carbohydrate movement from autotrophs to heterotrophs in parasitic and mutualistic symbioses. *Biological Review* 44, 17–90.

Smith, K. T. (1982). Uptake, translocation, and conversion of carbohydrates between host and fungus in bahiagrass infected with *Myriogenospora atramentosa*. University of Georgia, Ph.D. dissertation, Athens.

Smith, K. T., Bacon, C. W. and Luttrell, E. S. (1985). Reciprocal translocation of carbohydrates between host and fungus in Bahia grass infected with *Myriogenospora atramentosa*. *Phytopathology* 75, 407–411.

Sobral, W. S., Braga, D. P. V., LaHood, E. S. and Keim, P. (1994). Phylogenetic analysis of chloroplast restriction enzyme site mutations in the *Saccharinae* Griseb. subtribe of the *Andropogoneae* Dumort. tribe. *Theoretical and Applied Genetics.* 87, 843–853.

Stacy, B. E. (1973). Plant polyolys. *In* J. B. Pridham (ed.), *Plant Carbohydrate Chemistry.* Academic Press, New York, pp. 47–59.

Thomasson, J. R. (1986). Fossil grasses: 1820–1986 and beyond. *In* T. R. Soderstrom, K. W. Hilu, C. S. Campbell and M. E. Barkworth (eds.), *Grass Systematics and Evolution.* Smithsonian Institution Press, Washington, DC, pp. 159–167.

Tukey, H. B. (1970). The leaching of substances from plants. *Annual Review of Plant Physiology* 21, 305–324.

Wallner, B. M., Booth, N. H., Robbins, J. D., Bacon, C. W., Porter, J. K., Kiser, T. E., Wilson, R. W. and Johnson, B. (1983). Effect of an endophytic fungus isolated from toxic pasture grass on serum prolactin concentrations in the lactating cow. *American Journal of Veterinary Research* 44, 1317–1322.

Walters, D. R. (1985). Shoot:root interrelationships: The effects of obligately biotrophic fungal pathogens. *Biological Reviews* 60, 47–79.

Wardlaw, I. F. (1974). Phloem transport: physical, chemical, or impossible? *Annual Review of Plant Physiology* 25, 515–539.

Weedon, C. M. and Mantle, P. G. (1987). Paxilline biosynthesis by *Acremonium loliae*; a step towards defining the origin of lolitrem neurotoxins. *Phytochemistry* 26, 969–971.

West, C. P., Izekor, E., Oosterhuis, D. M. and Robbins, R. T. (1990a). The effect of *Acremonium coenophialum* on the growth and nematode infestation of tall fescue. *Plant and Soil* 112, 3–6.

West, C. P., Izekor, E., Turner, K. E. and Elmi, A. A. (1993). Endophyte effects on growth and persistence of tall fescue along a water-supply gradient. *Agronomy Journal* 85, 264–270.

West, C. P., Oosterhuis, D. M. and Wullschleger, S. D. (1990b). Osmotic adjustment in tissues of tall fescue in response to water deficit. *Environmental and Experimental Botany* 30, 1–8.

White, J. F., Jr. (1988). Endophyte-host associations in forage grasses. XI. A proposal concerning origin and evolution. *Mycologia* 80, 442–446.

White, J. F., Jr. (1994). Taxonomic relationships among the members of the Balansieae (Clavicipitales). *In* C. W. Bacon and J. F. White, Jr. (eds.), *Biotechnology of Endophytic Fungi of Grasses.* CRC Press, Boca Raton, FL, pp. 3–20.

White, J. F., Jr., Sharp, L. T., Martin, T. I. and Glenn, A. E. (1995). Endophyte-host associations in grasses. XXI. Studies on the structure and development of *Balansia obtecta. Mycologia* 87, 172–181.

White, J. F., Jr., Bacon, C. W. and Hinton, D. M. (1997). Modification of host cells and tissues by the biotrophic endophyte *Epichloe amarillans* (Clavicipitaceae; Ascomycotina) *Canadian Journal of Botany* 75, 1061–1069.

White, J. F., Jr. and Camp, C. R. (1995). A study of water relations of *Epichloe amarillans* White, an endophyte of the grass *Agrostis hiemalis* (Walt.) B.S.P. *Symbiosis* 18, 15-25.

White, J. F., Jr. and Cole, G. T. (1985). Endophyte-host associations in forage grasses.

III. In vitro inhibition of fungi by *Acremonium coenophialum*. *Mycologia* 77, 487–489.

White, J. F., Jr., Drake, T. E. and Martin, T. I. (1996). Endophyte-host associations in grasses 23. A study of two species of *Balansia* that form stromata on nodes of grasses. *Mycologia* 88, 89–97.

White, J. F., Jr. and Glenn, A. E. (1994). A study of two fungal epibionts of grasses: Structural features, host relationships, and classification in the genus *Myriogenospora* (Clavicipitales). *American Journal of Botany* 81, 216–223.

White, J. F., Jr. and Owens, J. R. (1992). Stromal development and mating system of *Balansia epichloe*, a leaf-colonizing endophyte of warm-season grasses. *Applied and Environmental Microbiology* 58, 513–519.

Wolswinkel, P. (1997). Enhanced rate of 14C-solute release to the free space by the phloem of *Vicia faba* stems parasitized by *Cuscuta*. *Acta Botanical Nerrl* 23, 177–188.

Yates, S. G. (1983). Tall fescue toxins. *In* M. Recheigel (ed.), *Handbook of Naturally Occurring Food Toxicants*. CRC Press, Boca Raton, FL, pp. 249–273.

Yates, S. G., Plattner, R. D., and Garner, G. B. (1985). Detection of ergopeptine alkaloids in endophyte infected, toxic Ky-31 tall fescue by mass spectrometry/mass spectrometry. *Journal of Agriculture and Food Chemistry* 33, 719–722.

Yshihara, T., Togiya, S., Koshino, H., Sakamura, S., Shimanuki, T., Sato, T. and Tajimi, A. (1985). Three fungitoxic sesquiterpenes from stromata of *Epichloe typhina*. *Tetrahedron Letters* 26, 5551–5554.

Zahner, H., Anke, H. and Anke, T. (1983). Evolution and secondary pathways. *In* J. W. Bennett and A. Ciegler (eds.), *Secondary Metabolism and Differentiation in Fungi*. Marcel Dekker, New York, pp. 153–170.

11

Polyketide and Peptide Products of Endophytic Fungi: Variations on Two Biosynthetic Themes of Secondary Metabolism

Jan S. Tkacz
Merck Research Laboratories, Rahway, New Jersey

1. INTRODUCTION

Endophytic fungi, like fungi adapted to other ecological niches, are capable of producing a variety of relatively low molecular weight (less than 1500) metabolites that have no apparent role in their primary metabolism or growth *in vitro* and that have come to be known as *secondary* metabolites. Although these compounds display an extraordinary degree of structural diversity, they can be categorized according to their biosynthetic origins (Turner and Aldridge, 1983; Campbell, 1984). By far the largest groups are those made from acetate equivalents (the polyketides), those formed from mevalonate-derived C_5 units (the isoprenoids, also called terpenoids), and amino acid-derived compounds. Examples are shown in Table 1. The boundaries between these classes are not entirely distinct, however. Some compounds represent fusions of polyketides and peptides or amino acids. Others are composed of both amino acids and isoprenyl units. Less populated groups of fungal compounds derive more directly from glucose, shikimic acid, fatty acids, or tricarboxylic acid cycle intermediates.

Over the past few years, the cellular processes involved in the formation of members of two of the major classes, i.e., polyketides and peptides, have been substantially elucidated with respect to both biochemical mechanism and genetic organization. The picture that has emerged is one in which each fungal product is assembled from carboxylic acid, amino acid, or hydroxy acid units in a series

Table 1   Examples of Fungal Secondary Metabolites from Several Biosynthetic Classes

Polyketide

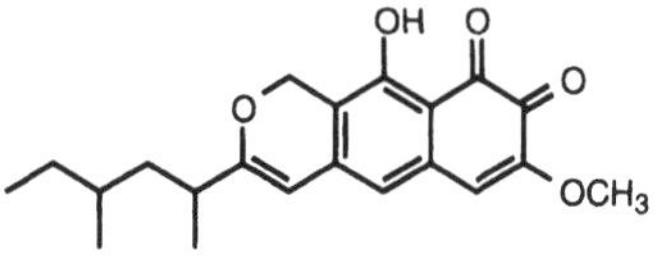

Obionin A
Leptosphaeria obiones
(Poch & Gloer, 1989)

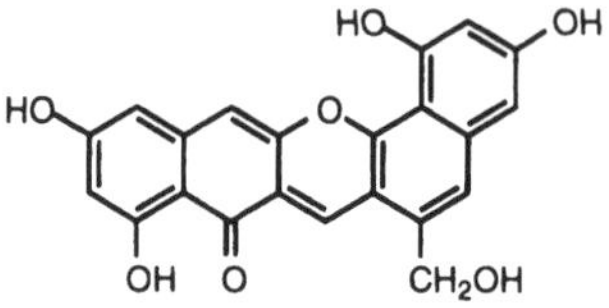

Hypoxyxylerone
Hypoxylon fragiforme
(Edwards *et al.*, 1991)

Isoprenoid

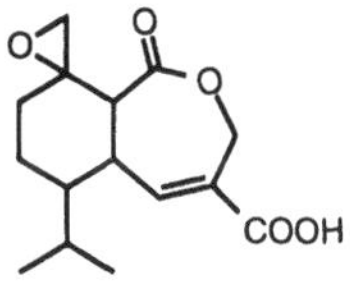

Heptelidic acid (Avocettin)
Chaetomium globosum
Gliocladium virens
Trichoderma viride
(Itoh *et al.*, 1980)

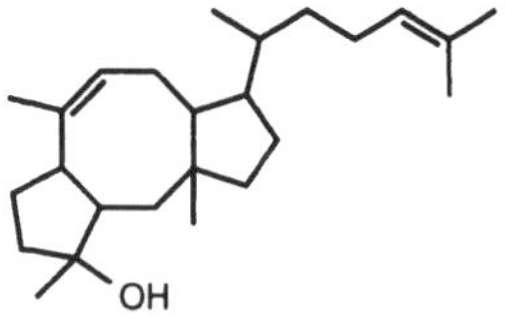

Cochliobolus heterostrophus
(Nozoe *et al.*, 1968)

Mixed Peptide/Polyketide

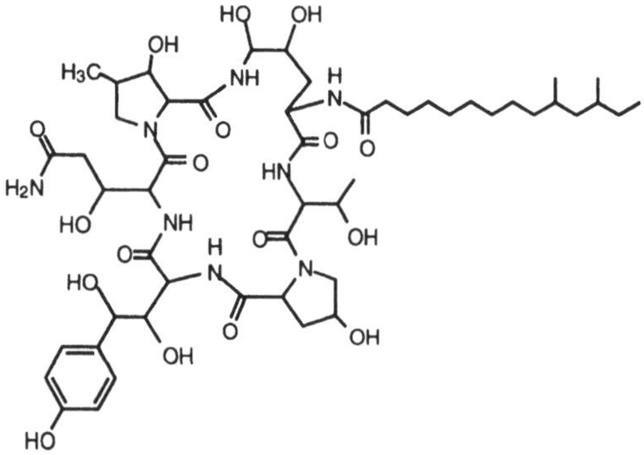

Pneumocandin A (L-671,329)
Zalerion arboricola
Pezicula sp. (anamorph: *Cryptosporiopsis* sp.)
(Adefarati *et al.*, 1991; Noble *et al.*, 1991)

of chemical steps mediated by a large, dedicated, multifunctional enzyme. Covalently bound to the enzyme, whether it is a polyketide synthase or a multicarrier peptide synthetase, is 4'-phosphopantetheine, which plays an essential part in the catalytic process by providing sulfhydryl sites for monomer unit thioesterification. Within the genes that encode these enzymes, modules that specify catalytic domains are arranged in an order corresponding to the overall sequence of catalytic steps performed by the protein. Ancillary genes provide proteins required to form unique monomer units, to transform the initial polyketide or peptide into its final form, or to protect the fungus from the bioactivity of its own product. These aspects of secondary metabolism will be illustrated in this chapter with examples provided by fungi that are associated with plants, either as mutualists or pathogens. Parallel findings from work with fungi from other natural habitats will also be mentioned.

In the organisms studied thus far, the genes required for the synthesis of polyketide or peptide products occur in clusters that occupy chromosomal loci of significant size (Keller and Hohn, 1997). The fact that these genomic regions are preserved in functional form generation after generation in wild populations serves as perhaps the clearest indication of the importance of these metabolites to the survival of the producing organisms in nature—if not to their propagation in the laboratory. Among microbiologists, there is growing acceptance of the view that the metabolites termed *secondary* are, in fact, *special*, with each providing adaptive value to the producing organism in its natural environment (Campbell, 1984; Williams et al., 1989; Vining, 1990, 1992). [Plant physiologists have long recognized this role for plant secondary metabolites (Harborne, 1993), but many alternative ideas have clouded the debate among microbiologists (Bennett and Bentley, 1989; Williams et al., 1989; Vining, 1990).] In the view of Bennett and Bentley (1989), the diversity seen among these metabolites is fundamentally a molecular manifestation of biological variation and natural selection.

2. POLYKETIDES

2.1. Mechanism of Biosynthesis—Theme 1

The term *polyketide* refers to a molecule in which multiple carbons bearing oxo or hydroxyl groups are separated from one another by single carbon units and which formally can be viewed as having been derived through the polymerization of acetate. With this in mind, it is not surprising that the biosynthetic mechanisms for polyketide assembly and long-chain fatty acid formation are very similar. Figure 1A illustrates the essential features of fatty acid synthesis (reviewed by Hopwood and Sherman, 1990). The process is initiated when acetate (the primer unit) is first moved from the 4'-phosphopantetheine arm of coenzyme A to a protein-bound 4'-phosphopantetheine residue and is subsequently transferred to

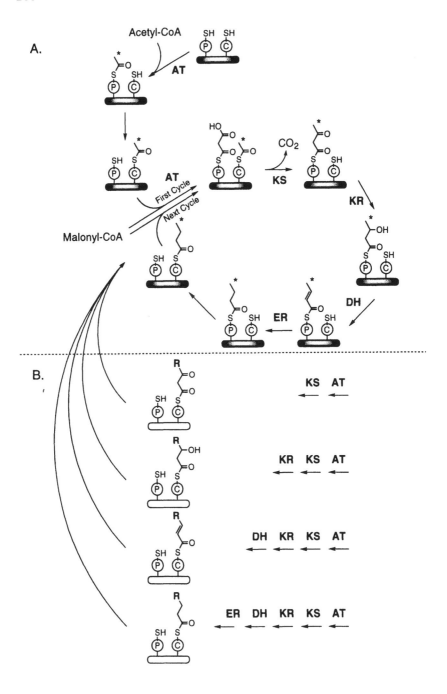

a cysteine residue in the active site of β-ketoacyl synthase (KS). The 4'-phospho-pantetheine residue is attached either to a discrete 77 amino-acid acyl carrier protein (ACP), as in the multisubunit bacterial or plant fatty acid synthases (type II enzymes), or to an acyl carrier domain (AC) of a multifunctional protein as in the fatty acid synthases of animals (type I enzymes). Malonate (an extender unit) is then transferred by acyltransferase (AT) from malonyl-CoA to the pro-tein-bound 4'-phosphopantetheine. Decarboxylation of the malonate provides a carbon nucleophile for the condensation of primer unit with the portion of the extender unit that remains at the ACP/AC site. The nascent β-ketoacyl chain is converted to a fully saturated thioester in three sequential processing steps: an NADPH-dependent reduction catalyzed by β-ketoacyl reductase (KR), dehydra-tion of the resulting β-hydroxythioester mediated by a dehydratase (DH), and further NADPH-dependent reduction by enoyl reductase (ER). The chain is then transferred to the active site cysteine of the ketoacyl synthase, which permits the process to be repeated with the next extender unit.

Polyketide synthesis departs from this paradigm in several distinctive ways (Fig. 1B; Hopwood and Sherman, 1990; Jordan and Spencer, 1993; Katz and Donadio, 1993; Hutchinson and Fujii, 1995). A polyketide synthase (PKS) is dedicated to the formation of a unique product and may employ a carboxylic acid primer unit other than acetate. The extent to which the nascent chain is processed during each elongation cycle determines whether a chain with a β-keto, β-hydroxyl, enoyl, or alkyl group will enter the next cycle. This is a key distinction because it has the potential to introduce chiral centers along with a great deal of structural variation as elongation proceeds. Release of the fully

---

Figure 1   Essentials of fatty acid and polyketide synthesis. (A) The shaded symbol repre-sents a generalized fatty acid synthase with two sulfhydryl sites, one on 4'-phosphopanteth-eine that is protein-bound (P) and the other provided by a cysteine residue (shown by C enclosed in a circle) in the active site of β-ketoacyl synthase (KS). Condensation of the acetate primer unit and an acetate equivalent (extender unit) provided by the decarboxyl-ation of mevalonic acid results in a nascent chain with carbon 2 of the primer unit (asterisk) at its distal terminus. Long-chain fatty acids are formed by repeated cycles of condensa-tion, reduction by β-ketoacyl reductase (KR), dehydration of the resulting β-hydroxyth-ioester by a dehydratase (DH), and further reduction by enoyl reductase (ER). (B) Polyke-tides are formed by an analogous process that has several distinct features. A polyketide synthase (unshaded symbol) may use carboxylic acids other than acetate as primer units (R—COOH). During each elongation cycle, the fate of the β-keto function depends on which of four series of processing reactions is followed; it may remain as a keto moiety or be converted to a hydroxyl, enoyl, or alkyl group for entry into the next elongation cycle. In the final step, the chain may be cyclized, lactonized, or aminated.

grown chain can be accompanied by cyclization, lactonization, or amide bond formation and may involve a thioesterase. Thus, the origin of structural diversity among polyketides can be seen in the staggering number of permutations possible for primer units, processing sequences, chiral centers, and cyclization reactions.

Polyketide synthases in which catalytic centers are combined as domains on a single or very few large polypeptide molecules are known as type I enzymes by analogy with type I fatty acid synthases. Similarly, multienzyme PKS systems composed of discrete largely mono- or bifunctional proteins that act in concert are termed type II enzymes. Both types are found in actinomycetes (Katz and Donadio, 1993). Generally, the products of bacterial type I PKSs are complex polyketides commonly known as macrolides (e.g., erythromycin, avermectin); their synthesis requires discrimination among potential malonyl, methylmalonyl, and ethylmalonyl extender units as well as considerable variation in the extent of processing that follows each condensation. Bacterial type II systems produce aromatic polyketides such as actinorhodin, tetracenomycin, and doxorubicin for which minimal processing is needed during condensation cycles and most β-keto functions are left unreduced. How a PKS is programmed to catalyze the series of chemical steps needed to form a unique product has been the central question in polyketide research for the past decade, and the answer now available is still incomplete. Genetic analysis of several bacterial type I systems has shown that there is a distinct catalytic domain for each step in the assembly process and that the active sites needed for one cycle of condensation and processing are encoded in DNA modules in the same order as required biochemically. More than one module occurs within an open reading frame that specifies a multifunctional polypeptide. In type II enzymes where there are fewer active sites and each type of site is specified only once among the genes that govern the synthesis of a polyketide product, some active sites are used repeatedly or iteratively while others work only during specific condensation cycles.

The PKSs of filamentous fungi that have been studied to date are multifunctional proteins and from this structural perspective are considered type I enzymes. Enzymatically, however, a fungal PKS resembles a bacterial type II system since it carries a site for a specific catalytic function only once within its structure and must use sites repeatedly during product formation. The fungal PKS that has been examined in greatest detail is the one that makes 6-methylsalicylic acid for patulin biosynthesis in *Penicillium patulum* (Jordan and Spencer, 1993). It is a 750-kDa tetramer consisting of identical subunits, approximately 180 kDa in size (Spencer and Jordan, 1992). Theoretically, to make 6-methylsalicylic acid from one acetyl and three malonyl residues, the enzyme performs 11 steps. Analysis of the gene for the PKS monomer indicates the presence of KS, AT, KR, DH, and AC domains in the translation product and a predicted size of 190,731 Da (Beck et al., 1990; Bevitt et al., 1992). The AT and KS domains are apparently used repeatedly in three rounds of elongation. The KR and DH domains, on the other hand, function in only one of the rounds, but the factors restricting their action exclusively to the C_6 stage are not presently understood. It is clear, however, that the enzyme does not proceed with the third condensation to form the C_8 chain if reduction is prevented by withholding of NADPH; under these conditions the enzyme releases a triacetic acid lactone as its product (Spencer and Jordan, 1992).

In the following sections, the polyketides that are discussed are fairly uncomplicated structures. However, endophytes are capable of producing more complex polyketides, examples of which are shown in Table 1 and elsewhere (Turner and Aldridge, 1983; Huang and Kaneko, 1996; Walton, 1996). A description of the organization of genes involved in the synthesis of polyketide mycotoxins, aflatoxin and sterigmatocystin, in saprobic aspergilli lies outside the purview of this chapter. Recent articles by Yu and Leonard (1995) and Brown et al. (1996) can be consulted for further details.

2.2. T-Toxin

A family of linear (C_{37}–C_{45}) polyketide molecules collectively known as T-toxin (Fig. 2) is produced by *Cochliobolus heterostrophus* (anamorph: *Bipolaris maydis*) belonging to the biotype designated as race T (Yoder et al., 1993). Organisms of this race are responsible for Southern corn leaf blight, a disease affecting maize plants with type T (Texas) cytoplasmic male sterility (*cms*-T) which have been widely cultivated since the 1960s. T-Toxin is not made by *C. heterostrophus* strains of race O and is responsible for the extreme virulence of race T strains. In genetic crosses of T and O strains, toxin production segregates as a single locus, *Tox1* (Yang et al., 1994). The molecular basis of the toxicity of these polyketides in susceptible plants is well understood. T-toxin interacts strongly with a 13-kDa integral membrane protein (product of the T-*urf13* gene) that occurs in the mitochondria of *cms*-T corn but not in mitochondria of plants from the other two *cms* groups (Levings and Siedow, 1992; Levings et al., 1995; Rhoads et al., 1995). Toxin binding causes URF13 proteins to form pores in the

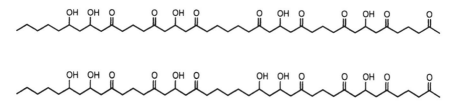

Figure 2   T-toxin is a family of linear polyketides with odd-numbered chain lengths ranging from C_{37} to C_{45}. The two most predominant species are the C_{41} compounds shown.

membrane, resulting in ion leakage and loss of mitochondrial function (Siedow et al., 1995).

The polyketide synthase that produces T-toxin has not been studied biochemically, but its structure has been deduced by analysis of the gene identified in the following way. Insertionally tagged mutants of *C. heterostrophus* race T were generated by restriction enzyme–mediated integration (Lu et al., 1994). To screen among them for toxin-deficient strains, a bioassay was used that employed *Escherichia coli* made sensitive to T-toxin by transformation with the T-*urf13* gene under the control of the bacterial *trpE* promoter (Dewey et al., 1988). The defect in mutants with a *Tox⁻* phenotype mapped at the *Tox1* locus. Sequencing of the DNA flanking the insert in one of the mutants showed that integration had occurred within a gene encoding a polyketide synthase, presumably the T-toxin synthase (Yang et al., 1996). The gene (*PKS1*) has four small introns and 7.6 kb of coding sequence, corresponding to a protein of 2530 amino acids. Since consensus sequences for each of six domains, i.e., AT, KS, KR, DH, ER, and AC, occur in the predicted amino acid sequence, T-toxin synthase fits the definition of a type I (multifunctional) PKS.

PKS1 is present as a single copy within the *Tox1* locus of *C. heterostrophus*. To permit further analysis of this locus, an enrichment strategy has been used to isolate chemically induced *Tox⁻* mutants (Yang et al., 1994). A *Tox1⁺* strain of *C. heterostrophus* was transformed with the T-*urf13* gene under the control of an inducible *Aspergillus nidulans* promoter to produce a strain that was conditionally sensitive to its own toxin. Following mutation of this strain and induction of T-*urf13*, *Tox⁻* mutants were recovered at high frequency among the survivors. These mutations will help define other genes whose involvement in T-toxin formation may be anticipated from the fact that the toxin is not a single compound but a family of linear polyketides with each member having an odd rather than even number of carbon atoms.

2.3. Dihydroxynaphthalene Melanin

Melanin pigments of fungi can arise by four distinct biochemical routes (reviewed by Bell and Wheeler, 1986). Ascomycetes commonly produce dihydroxynaphthalene (DHN) melanin, and the first metabolic precursor specifically committed to its formation is 1,3,6,8-tetrahydroxynaphthalene (THN; Fig. 3), which is the product of a pentaketide synthase. An albino phenotype results from loss-of-function mutations in the gene for THN synthase. The Alm⁻ mutation of *Alternaria alternata* leads to albinism, and the *ALM* gene resides in a 30-kb genomic region, clustered together with the *BRM1* and *BRM2* genes which are thought to encode two other melanin pathway enzymes, scytalone dehydratase and 1,3,8-trihydroxynaphthalene reductase, respectively (Kimura and Tsuge, 1993; Shiotani and Tsuge, 1995). The DNA sequence of *ALM* has not yet been reported, but with

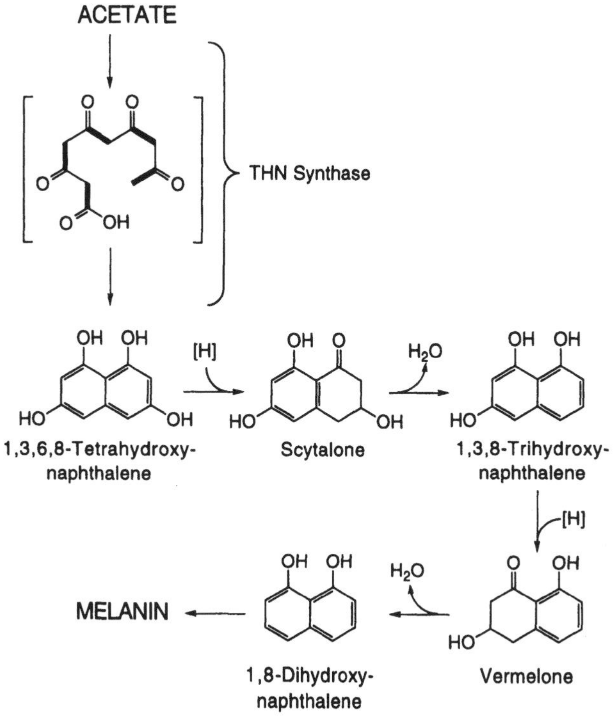

Figure 3 The biosynthesis of DHN melanin proceeds from acetate via a polyketide synthase that produces 1,3,6,8-tetrahydroxynaphthalene (THN). A hypothetical polyketide intermediate is shown with each of the five acetate residues indicated by the bonds in boldface. Two successive pairs of reduction and dehydration reactions convert the latter to 1,8-dihydroxynaphthalene (DHN), which is polymerized to melanin.

a transcript size of 7.2 kb, it is a reasonable candidate for the THN synthase. Cloning and sequencing of the *PKS1* gene of *Colletotrichum lagenarium*, which restores melanin synthesis when introduced into a *C. lagenarium* albino mutant, established it as the THN polyketide synthase (Takano et al., 1995). The gene was recovered on an 8-kb fragment of genomic DNA, has two small introns, and encodes a protein containing 2187 amino acids. The predicted amino acid sequence includes AT and KS domains along with two AC domains. The *C. lagenarium* genes for scytalone dehydratase (*SCD1*; Kubo et al., 1996) and 1,3,8-trihydroxynaphthalene reductase (*THR1*; Perpetua et al., 1996) have also been cloned but are not closely linked to *PKS1* or one another. Similarly, the melanin genes in *Magnaporthe grisea* are not clustered (Chumley and Valent, 1990), but *Cochliobolus* species are more like *A. alternata*, having genes for THN synthase and

1,3,8-trihydroxynaphthalene reductase that are linked to each other but not to the scytalone dehydratase gene (Kubo et al., 1989; Tanaka et al., 1991).

The *PKS1* gene product is highly homologous throughout its sequence to the protein encoded by the *wA* gene of *Aspergillus nidulans*, which is required for the green pigmentation of asexual spores (Mayorga and Timberlake, 1992). From this it might be inferred that a DHN-like pathway is involved in the formation of the conidial wall pigment. This is in accord with the reported structure of the putative green-pigment precursor (Brown et al., 1993) and the observation that tricyclazole and related reductase inhibitors, which were developed as crop protectants targeted upon DHN melanin synthesis, also affect green pigment formation in *Aspergillus* and *Penicillium* species (Wheeler and Klich, 1995).

3. PEPTIDES

3.1. Formation by the NonRibosomal Multicarrier Mechanism—Theme 2

The repertoire of amino acids used in the formation of peptidic secondary metabolites is at least 10 times larger than the selection of amino acids employed in protein synthesis. In addition to the usual L-amino acids, N-methylated amino acids, hydroxylated amino acids, D-amino acids, and amino acids that are unusual in other ways have been found as constituents of secondary metabolites. Hydroxy acids can also be combined with amino acids to form depsipeptides or lactones. Examples of this structural diversity will be presented in the sections that follow.

Ribonucleic acids (ribosomes, t-RNA, m-RNA) are not involved in the synthesis of peptidic secondary metabolites. In place of the generalized biosynthetic machinery that nucleic acids provide, large multifunctional enzymes, each dedicated to the synthesis of a unique product, are responsible for arranging the monomer units in the appropriate sequence and then linking them to form the peptide (Kleinkauf and von Döhren, 1990, 1996). For the cases which have been examined in fungi, each enzyme consists of a single protein that catalyzes all of the essential reactions, but in bacteria more than one multifunctional protein may be needed for the assembly of a product. The current hypothesis for the multistep assembly process is based on genetic as well as biochemical evidence (Stein et al., 1994) and is shown schematically in Fig. 4. The carboxyl groups of the amino acids to be polymerized are activated in a process that requires ATP and involves the formation and stabilization of aminoacyladenylates at sites on the enzyme. The aminoacyladenylate–enzyme complexes are noncovalent and can be reversed with excess pyrophosphate to regenerate ATP. This forms the basis of the pyrophosphate exchange reaction that can be used with nonprotein amino acid substrates for the detection of multifunctional enzymes in crude extracts and with any of the amino acid substrates once the enzyme preparation has been

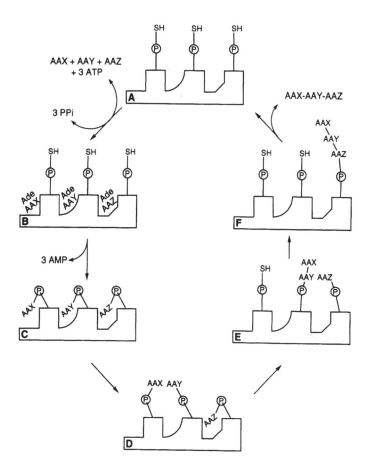

Figure 4 Steps in peptide assembly on a multicarrier enzyme. (A) A tripeptide synthetase using three different amino acid substrates (AAX, AAY, and AAZ) is depicted. It has activation sites specific for each amino acid along with three 4′-phosphopantetheine residues (P) linked to AC domains for the thioesterification of the amino acids. (B) At each activation site, the specified amino acid reacts with ATP releasing inorganic pyrophosphate (PPi) and forming an aminoacyladenylate (Ade-AA) complexed noncovalently with the peptide synthetase. (C) The activated amino acids become enzyme-bound when they are thioesterified to 4′-phosphopantetheine residues. (D–F) The amino acids are moved from their activation sites and are condensed. Termination of the assembly process may involve the release of a linear peptide as shown or the cyclization of the nascent peptide with the formation of an internal peptide or ester bond.

purified. Most activation sites exhibit high substrate selectivity, but certain activation sites are less discriminating. This can lead to the synthesis of a family of products and sometimes allows novel peptides to be formed from amino acid analogs that are provided exogenously. The discussions of enniatin and cyclosporin that follow will serve to illustrate this point. Each activated amino acid becomes enzyme-bound when it is thioesterified to a 4'-phosphopantetheine residue that is linked to an AC domain in the protein. Because there is a phosphopantetheine for each amino acid rather than one repeatedly used for all the incoming monomer units as in the fungal type I polyketide synthases, this mechanism has been called the multiple carrier model.

Modification of an amino acid by epimerization or N-methylation is accomplished while it is in its thioesterified form, and catalytic domains for these modifications are present within the multifunctional enzyme. However as we shall see in the case of isopenicillin N, epimerization may occur at a later stage of assembly. Epimerization reactions are independent of pyridoxal phosphate, but their exact mechanism is presently unclear. Methylation requires S-adenosylmethionine as the methyl donor. No domains for substrate hydroxylation have been found in multicarrier enzymes. The hydroxylated amino acids that are components of some peptides occur either because the specificity of the activation site demands a preformed hydroxylated amino acid or because the peptide becomes the substrate for hydroxylation once it is released from the multifunctional enzyme (Adefarati et al., 1991).

On the multicarrier enzyme, the 4'-phosphopantetheine residues are considered to act as tethers permitting movement of an amino acid away from its activation site toward an adjacent amino acid thioester. Proximity of the thioesters allows the formation of a peptide bond through the nucleophilic attack by the amino nitrogen of one amino acid on the thioester carboxyl of its neighbor. Although the peptide grows in a unidirectional manner, the factors that govern this aspect of the assembly process are ill defined at present. The full-length nascent peptide may be liberated as a linear product or may be cyclized through the formation of an internal peptide or ester bond as part of the release process.

Much of the support for this view of peptide formation has come from DNA sequence analysis and site-directed mutagenesis which has shown that the genes for multicarrier peptide synthetases, like their counterparts for type I polyketide synthases, are organized in modular sets of recurring motifs. Further details may be found in the recent comprehensive review by Kleinkauf and von Döhren (1996).

To illustrate how structural diversity arises from the basic multicarrier mechanism, we will now consider several examples: a linear tripeptide (ACV of isopenicillin N synthesis), two cyclic peptides (HC toxin and cyclosporin), and a cyclic hexadepsipeptide (enniatin). N-Methylation of amino acids is a common feature of enniatin and cyclosporin synthesis. D-Amino acids are constituents of

ACV, HC toxin, and cyclosporin, but in each case epimerization is achieved at a distinctively different point in the overall assembly processes.

3.2. Isopenicillin N

Acremonium chrysogenum is one of several eukaryotic and prokaryotic organisms with the capacity to make bicyclic β-lactam antibiotics (Aharonowitz et al., 1992). While the ultimate products (e.g., penicillin G, cephalosporin C, cephamycin C) differ among the producing organisms, isopenicillin N is uniformly used as the precursor (Fig. 5). Two enzymatic steps are needed to make isopenicillin N (IPN); in the first, L-α-aminoadipic acid, L-cysteine, and L-valine are polymerized to form a linear tripeptide δ-(L-α-aminoadipyl)-L-cysteinyl-D-valine known as ACV, and in the second, ACV is cyclized. A multicarrier enzyme known as ACV synthetase, catalyzes the polymerization, and ACV cyclase (synonym: IPN synthase) converts ACV to isopenicillin N.

The ACV synthetase of *A. chrysogenum* activates and binds L-α-aminoadipic acid, L-cysteine, and L-valine (Banko et al., 1987; Baldwin et al., 1993, 1994). The first peptide bond it forms is between the δ-carboxyl of L-α-aminoadipic acid and the α-amino of L-cysteine. The enzyme does not use D-valine as a substrate, though the tripeptide product is LLD-ACV. Epimerization of L-valine appears to take place as the tripeptide is released from the enzyme. The size of the enzyme was estimated to be 300 kDa (Baldwin et al., 1990). Cloning of the *A. chrysogenum* gene for ACV synthetase (*pcbAB*) was reported by two groups (Hoskins et al., 1990; Gutiérrez et al., 1991). The gene is free of introns, and its 11.1-kb open reading frame predicts a protein of 415 kDa with 3712 amino acids. Three regions in the deduced protein have homology with each other and with amino acid–activating regions of other multicarrier enzymes. The gene lies immediately upstream of the *pcbC* gene which encodes ACV cyclase (Samson et al., 1985; Kriauciunas et al., 1991). The two genes have oppositely oriented reading frames separated by a 1.2-kb region with potential promoter sequences. The promoter for *pcbC* appears to be 5 times stronger than that for *pcbAB* based on the expression of reporter genes (i.e., *gusA* and *lacZ*; Menne, et al., 1994). The genes have been localized on chromosome *VI* by molecular karyotyping (*A. chrysogenum* has eight chromosomes; Skatrud and Queener, 1989). In *Penicillium chrysogenum* and *Aspergillus nidulans*, fungi that produce the "acyl transfer" penicillins, homologous genes occur in 20 kb clusters along with *penDE*, the gene for the acyltransferase (Fig. 5.; Smith et al., 1990; Aharonowitz *et al.*, 1992; Martín and Gutiérrez, 1995). *A. chrysogenum*, on the other hand, produces cephalosporin C, a ring-expanded β-lactam. The *cefEF* and *cefG* genes that encode enzymes for the last two steps in the cephalosporin pathway are also clustered (Mathison et al., 1993), but this cluster resides on chromosome *II* rather than chromosome *VI*. The conservation of sequence and chromosomal organiza-

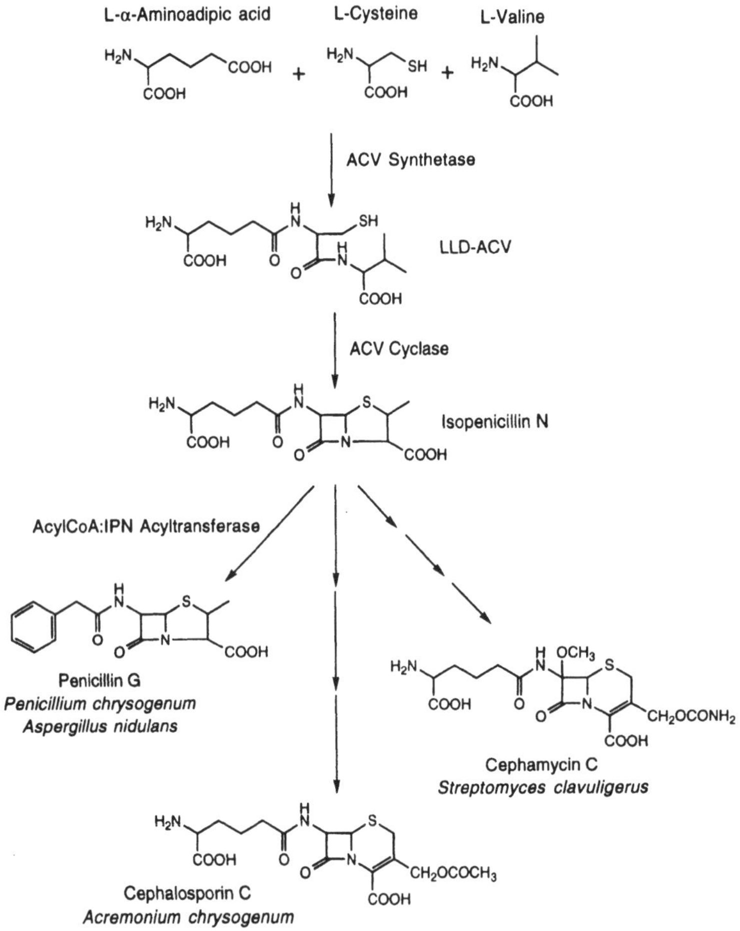

Figure 5 The biosynthesis of bicyclic β-lactam antibiotics in fungi and streptomycetes. Following the formation of the ACV tripeptide and its cyclization to isopenicillin N, different pathways are followed. For the formation of the "acyl transfer" penicillins, the L-α-aminoadipoyl side chain of isopenicillin N is exchanged by an acyltransferase. A variety of acyl-CoA substrates can be substrates, e.g., phenylacetyl-CoA yields penicillin G. The biosynthesis of cephalosporin C and cephamycin C involves an oxidative expansion of the thiazoline ring after the L-α-aminoadipyl moiety of isopenicillin N is epimerized to the D isomer to make penicillin N.

tion among the β-lactam pathway genes in the various producing microorganisms has been taken as evidence of horizontal gene transfer from prokaryotes to eukaryotes (Aharonowitz *et al.*, 1992; Martin and Gutiérrez, 1995; Buades and Moya, 1996).

3.3. Enniatin

Species of *Fusarium* that cause wilt diseases of vascular plants make a variety of phytotoxins. Among them are the enniatins, which are cyclic hexadepsipeptide ionophores composed of alternating residues of 2-D-hydroxyisovaleric acid and an *N*-methyl, branched chain, L-amino acid (Fig. 6). The major product of *F. sambucinum* is enniatin A in which the amino acid is isoleucine; *F. oxysporum* makes the valine compound, enniatin B, as its primary depsipeptide (Audhya and Russell, 1973; Zocher et al., 1976). The enniatin synthetase was the first fungal multicarrier enzyme that was characterized biochemically. As purified from *F. oxysporum*, it was capable of catalyzing the synthesis of enniatin from 2-D-hydroxyisovaleric acid, ATP, *S*-adenosyl-L-methionine, and one of the three branched chain L-amino acids. Supplementing the reaction mixture with more than one amino acid resulted in the formation of novel mixed-enniatin species (Zocher et al., 1982). With respect to the three amino acid substrates, the enzyme exhibited the lowest K_m for L-valine (80 µM) and the highest for L-leucine (260 µM), making enniatin B and enniatin C the most and least favored products, respectively. Like its amino acid counterpart, the hydroxy acid was activated as an adenylate that was subject to the pyrophosphate exchange reaction. Following activation, both substrates were thioesterified, and the amino acid was N-methylated. In the absence of the methyl donor, an unmethylated enniatin was formed at a rate about 10% of that for enniatin in complete reaction mixtures. Photoaffinity labeling of the enzyme with *S*-adenosyl-L-[^{14}C-methyl]methionine suggested that

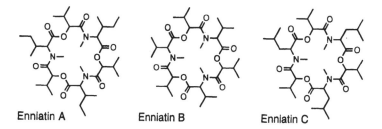

Enniatin A   Enniatin B   Enniatin C

Figure 6 Enniatins, cyclic hexadepsipeptides composed of alternating residues of 2-D-hydroxyisovaleric acid and *N*-methyl-L-isoleucine (enniatin A), *N*-methyl-L-valine (enniatin B), or *N*-methyl-L-leucine (enniatin C).

there is only one methyl binding site per enzyme molecule, though three amino acids in the cyclic product are N-methylated (Billich and Zocher, 1987). A molecular weight of 250 kDa was estimated for enniatin synthetase, both in its native state and in its reduced and denatured form (Zocher *et al.*, 1982). The data support the conclusion that the synthetase is a single polypeptide rather than a multimeric enzyme and that enniatin is produced by two successive condensations of dipeptidol units followed by cyclization of the hexapeptidol.

The gene encoding enniatin synthetase (*esyn1*) has been cloned from *F. scirpi* and found to be devoid of introns (Haese et al., 1993). In agreement with the mechanism envisioned from the biochemical evidence, the open reading frame predicts a 347-kDa protein product with two domains for amino acid/hydroxy acid activation. Expression of portions of the gene in *E. coli* has allowed the biochemical identification of the N-terminal domain as the 2-D-hydroxyisovaleric acid activation site and second region as the amino acid activation site (Haese et al., 1994). Adjacent to the second activation domain was a region of 434 amino acids that contained the photoaffinity labeling site and that showed homology to *S*-adenosyl-L-methionine-dependent methyltransferases from other organisms. As will be seen in the next section, similar methylation domains occur repeatedly in cyclosporin synthetase. Although the expressed recombinant *esyn1* domains could catalyze the formation of adenylated substrates, they were incapable of binding the amino acids through thioester linkages, presumably because the serine sites for 4'-phosphopantetheine attachment remained unmodified when the proteins were over expressed in *E. coli*. Similarly, attempts to express functional polyketide synthases in *E. coli* have yielded inactive apoproteins.

In an *F. scirpi* strain with enhanced enniatin production, the synthetase is expressed during the growth and stationary phases of vegetative cultures, as well as in spores (Billich and Zocher, 1988). Enniatin production, on the other hand, ceases at the end of the growth phase unless the culture is supplemented at that point with exogenous L-valine. Valine is not only a substrate for the synthetase but is also the precursor of 2-D-hydroxyisovalerate; it is deaminated to 2-ketoisovalerate, which is then reduced to 2-D-hydroxyisovaleric acid by a specific dehydrogenase (Lee et al., 1992). Thus, the formation of enniatin in this strain is regulated primarily by substrate availability.

4. POLYKETIDE/FATTY ACID–PEPTIDE HYBRIDS

4.1. Cyclosporin

Cyclosporin A is a cyclic peptide containing 11 amino acid residues made by the fungus *Tolypocladium inflatum* and other fungal species (Dreyfuss and Gams, 1994; Traber and Dreyfuss, 1996). With potent immunosuppressant activity, it is in worldwide clinical use under the trade name Sandimmune® to prevent graft rejection and to treat autoimmune disease. Cyclosporin A is composed of eight L-amino acids and three unusual amino acids: 2-amino-3-hydroxy-4-methyl-6-octenoic acid at position 1 (Fig. 7), 2-aminobutyric acid in position 2, and D-alanine in position 8. The multicarrier synthetase that produces cyclosporin is a single polypeptide that is remarkable for its size and enzymatic complexity, performing 40 discrete catalytic steps (Dittmann et al., 1990; Lawen and Zocher, 1990; Lawen et al., 1992). The synthetase not only activates and binds the amino acids but also N-methylates seven of them. After forming the peptide bonds starting from the D-alanine and proceeding in the direction of N-methyl-L-leucine (dotted line and arrow in Fig. 7), it joins the final L-alanyl residue with the D-alanyl residue to cyclize the peptide. The biochemical characterization of the enzyme led to a molecular weight estimate ranging from 800 kDa (Lawen and Zocher, 1990) to 1400 ± 160 kDa (Lawen et al., 1992; Schmidt et al., 1992). However, by probing a genomic library of *T. inflatum* with synthetic oligonucleotides based on amino acid sequences of synthetase fragments, Weber et al. (1994)

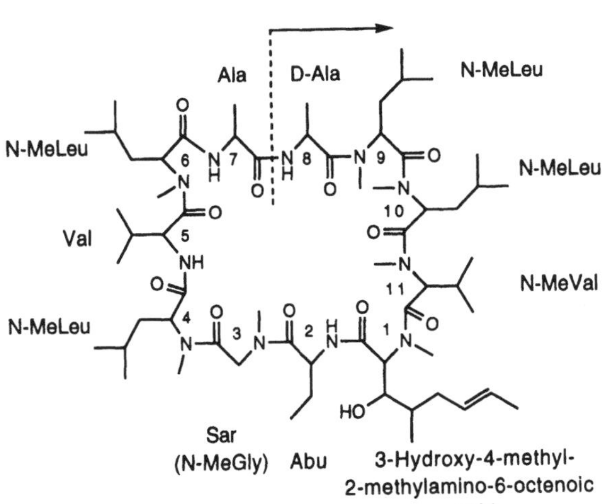

Figure 7 Cyclosporin A. The numbering of the positions is arbitrary and does not correspond to the order of polymerization. Biosynthesis is initiated from D-alanine (dotted line) and proceeds in the direction of N-methyl leucine (as indicated by the arrow). Abu denotes 2-aminobutyric acid, and Sar stands for sarcosine. In the final product, 2-amino-3-hydroxy-4-methyl-6-octenoic acid is present in its N-methyl form, 3-hydroxy-4-methyl-2-methyl-amino-6-octenoic acid. There are 25 naturally occurring cyclosporins that have amino acid substitutions in positions 1, 2, 4, 5, 7, or 11 or unmethylated peptide bonds in positions 1, 4, 6, 9, 10, or 11. Other variants have been produced by precursor-directed biosynthesis in vivo (positions 2 and 8; Traber et al., 1989) and in vitro (positions 2, 5, 8, and 11; Lawen et al., 1989).

identified a huge (45.8-kb) but intron-free gene (*simA*) encoding an extremely large protein with a calculated mass of 1689 kDa. The predicted amino acid sequence of the product had eleven domains with the characteristics of amino acid activation domains—seven with a postulated consensus sequence for methylation (domains 2, 3, 4, 5, 7, 8, 10, each about 1490 amino acids long) and four without (domains 1, 6, 9, 11, each with about 1070 amino acids). The order of these activation and activation/methylation domains corresponds with the biosynthetic order established biochemically.

Unlike the HT synthetase discussed in the next section, cyclosporin synthetase is unable to isomerize L-alanine, and initiation of a round of cyclosporin synthesis in *T. inflatum* depends on the formation of D-alanine by a novel racemase (Hoffmann et al., 1994). Although alanine racemases are well known in bacteria where they play crucial role in cell wall formation, the racemase that occurs in *T. inflatum* is the first example of such an enzyme in a eukaryote. Hoffmann et al. regard D-alanine production as the rate-limiting step for cyclosporin synthesis. Cyclosporin synthetase also requires the unusual amino acid, 2-amino-3-hydroxy-4-methyl-6-octenoic acid. The biosynthesis of this amino acid has not been completely elucidated, but a polyketide synthase has been found in the cyclosporin-producing organism that makes 3-hydroxy-4-methyl-6-octenoic acid, the likely precursor of the required amino acid (Offenzeller et al., 1993). On this basis, cyclosporin may be considered a polyketide/peptide hybrid. It is worthwhile mentioning that the pendent methyl of 3-hydroxy-4-methyl-6-octenoic acid arises via an *S*-adenosylmethionine-dependent methyl transfer on the PKS rather than through the incorporation of a methylmalonate extender unit, the route that would have been taken by a bacterial type I PKS. This mode of methylation is a general feature of methylated polyketides from fungi and is an outcome of the iterative nature of the fungal type I enzymes; selection of an extender unit other than malonic acid in a specific condensation cycle is precluded, making a specific methyl transfer mechanism necessary.

4.2. HC Toxin

Pathogenicity of *Cochliobolus carbonum* race 1 (anamorph: *Bipolaris zeicola* or *Helminthosporium carbonum*) on *hm/hm* corn requires the production of the host-selective phytotoxin HC toxin. The toxin is a cyclic tetrapeptide composed of residues of L-alanine, D-alanine, D-proline, and the unusual amino acid 2-amino-9,10-epoxy-8-oxodecanoic acid (AEO; Pope et al., 1983) and is a member of a group of structurally related cyclic tetrapeptides (Fig. 8). The group includes Cyl-2, a metabolite of *Cylindrocladium scoparium* first isolated as a root growth inhibitor, chlamydocin from *Diheterospora chlamydsporia*, a potent cytostatic agent, WF-3161 from *Petriella guttulata* and trapoxins A and B from *Helicoma ambiens*, compounds with antitumor activity, and apicidin from *Fusarium pal-*

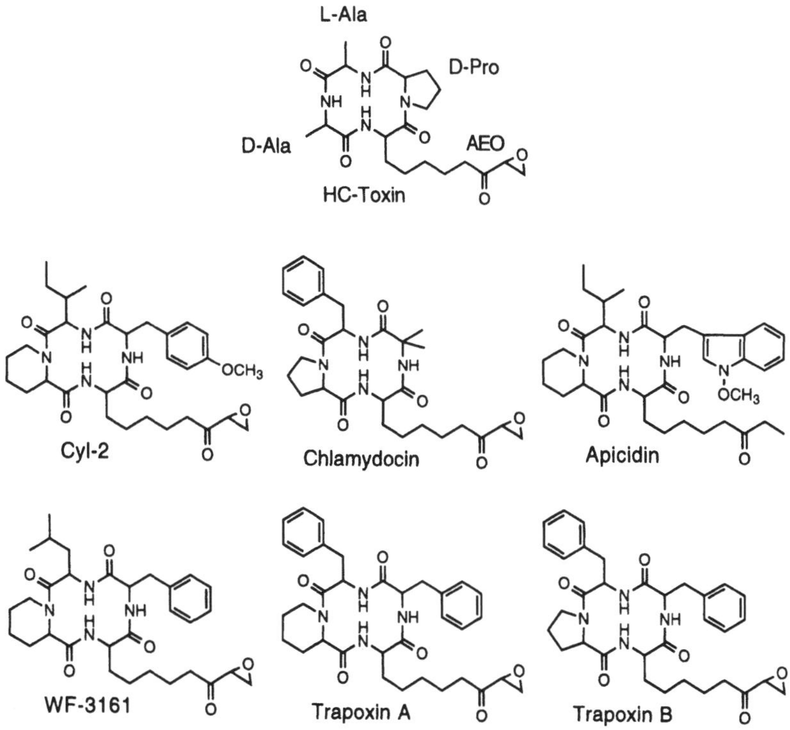

Figure 8 HC toxin and related cyclic tetrapeptides.

lidoroseum, an antiparasitic agent (Itazaki et al., 1990 and references therein; Singh et al., 1996). The diverse biological activities of these peptides have been attributed to their common ability to inhibit histone deacetylase; this mechanism of action has been demonstrated in mammalian, avian, plant, fungal, and protozoal cells (Kijima et al., 1993; Brosch et al., 1995; Darkin-Rattray et al., 1996).

The multicarrier enzyme that produces HC toxin is a 575-kDa protein that activates L-proline, L-alanine, and D-alanine and is capable of isomerizing L-proline and also L-alanine. Initial biochemical work suggested the involvement of two synthetases, one for the activation and epimerization of L-proline and the other for L- or D-alanine activation and L-alanine isomerization (Walton, 1987; Walton and Holden, 1988). However, cloning and sequencing of DNA from the *TOX2* locus, which governs HC toxin production, led to the realization that the two proteins represented segments within a 15.7-kb open reading frame, now called *HST1* (for *HC toxin synthetase*; Panaccione et al., 1992; Scott-Craig et al., 1992). The *HTS1* gene has no introns and its deduced product has the hallmarks of a multicarrier enzyme, including four amino acid activating domains. It is unclear as to why the enzyme should bind and activate D-alanine and apparently epimerize L-alanine as well. In the DNA sequence of *HTS1*, only one epimerizing motif can be recognized, and this is presumed to be involved in the isomerization of L-proline (Kleinkauf and von Döhren, 1996). The existence of four amino acid activating domains suggests that the intact HT synthetase is capable of activating AEO or an AEO precursor. Initial exploration of the origin of this amino acid, based on [^{14}C]acetate incorporation and cerulenin inhibition, suggested the involvement of a fatty acid or polyketide synthase (Wessel et al., 1988). In this connection, it is noteworthy that the *TOXC* gene, which is always present in HC toxin–producing strains of *C. carbonum*, encodes a protein of 2080 amino acids with homology to the β-subunit of fungal fatty acid synthases and domains for AT, ER, DH, and malonyl-palmityl transferase (Ahn and Walton, 1997). Eliminating the function of *TOXC* prevents HC-toxin synthesis but does not alter the intracellular levels of the HT synthetase. Examination of genes in the vicinity of *TOXC* may identify the putative α-subunit of the enzyme and lead an understanding of how the product of the fatty acid synthase might participate in AEO formation. An intriguing parallel is seen in the 60-kb cluster of genes for sterigmatocystin production in *Aspergillus nidulans* where homologs of fatty acid synthase α and β-subunits are found in addition to a polyketide synthase (Yu and Leonard, 1995; Brown et al., 1996). In that case, the fatty acid synthase may be responsible for producing a hexanoyl primer unit used by the polyketide synthase for the assembly of the 10-deoxynorsolorinic acid, the precursor of sterigmatocystin. Another gene involved in HC toxin formation is *TOXA*, which lies immediately adjacent to *HST1* in an orientation that permits divergent transcription from a common promoter region. The protein encoded by *TOXA* has 10 or more membrane-spanning regions and sequence similarity with efflux pumps that confer antibiotic resistance (Pitkin et al., 1996). The function of the *TOXA* product is likely to be protection of the producing cell from the inhibition by its own product since disruption of *TOXA* appears to be lethal.

The *HTS1* transcript can be detected at various growth phases of *C. carbonum* and in conidia (Jones and Dunkle, 1995), yet HC toxin synthesis commences only as spores germinate and produce appressoria (Weiergang et al., 1996). The toxin is not made in vegetative cells growing as agitated submerged cultures. Conceivably, the synthesis of HC toxin could be another example of peptide formation regulated not by the availability of the synthetase but by one of the required amino acids (e.g., AEO). Mention has already been made of analogous scenarios for the control of enniatin and cyclosporin production by L-valine and D-alanine, respectively.

The organization of the genes for HC toxin synthesis is quite complex (Walton, 1996). The *TOX2* locus spans a 540-kb segment of a 3.5-Mb chromosome, the largest of the approximately 13 chromosomes of *C. carbonum* (Ahn and Walton, 1996). The 22-kb region containing the neighboring *HST1* and *TOXA* genes is duplicated within this locus (Panaccione et al., 1992; Pitkin et al., 1996). The gene pairs are separated by approximately 0.3 Mb and are each flanked by copies of *TOXC* (three in all). Despite this complexity, *TOX2* is inherited in a simple Mendelian fashion.

If the fatty acid synthase encoded in part by *TOXC* is involved in generating AEO, then HC toxin, like cyclosporin, can be viewed as an hybrid product. It is interesting that the polyketide/fatty acid moieties play very important roles in biological activities of both hybrid peptides. The 8-oxo functionality on AEO is a prerequisite for the biological activities of the histone deacetylase inhibitors (Shute et al., 1987), and the resistance of *Hm/Hm or Hm/hm* strains of corn to *C. carbonum* race 1 is attributable to an active defense mechanism involving the enzymatic capacity to reduce this carbonyl moiety (Meeley et al., 1992). In the case of cyclosporin, the 2-amino-3-hydroxy-4-methyl-6-octenoic acid is crucial for cyclophilin binding (Pflügl et al., 1993; Thériault et al., 1993).

5. ISOPRENOIDS

For the last major biosynthetic category of fungal secondary metabolites, i.e., the isoprenoids, trichothecene biosynthesis in *Fusarium* species provides the example best characterized at the DNA level. The trichothecenes are a family of sesquiterpene mycotoxins capable of inhibiting protein synthesis in animal and human cells (Fig. 9; Desjardins et al., 1993). In *F. sporotrichioides*, there are nine trichothecene pathway genes clustered in a 25-kb region of genomic DNA (Hohn et al., 1993; Keller and Hohn, 1997). Of the eight genes with known functions, six encode biosynthetic enzymes (Hohn and Beremand, 1989; Hohn et al., 1995; McCormick et al., 1996), one is for a regulatory protein (Proctor et al., 1995), and the last produces an efflux protein. However, there is no gene in the cluster for a protein with a function analogous to that of a polyketide synthase or a multicarrier enzyme, namely, the formation of the carbon skeleton by polymerization of monomer units. The trichothecene pathway uses the C_{15} precursor, farnesyl

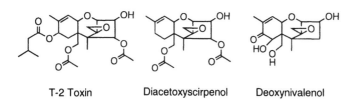

| T-2 Toxin | Diacetoxyscirpenol | Deoxynivalenol |

Figure 9 Trichothecene mycotoxins.

diphosphate, which has a major role in primary metabolism. Cyclization of the farnesyl moiety can be considered the first dedicated step in trichothecene formation and is performed by the product of the *Tri5* gene that lies within the cluster (Hohn and Beremand, 1989; Hohn and Plattner, 1989; Cane et al., 1995). If only a single copy of the farnesyl diphosphate synthase gene is present in *F. sporotrichioides*, then its product is needed for the carbon backbones of both primary and secondary metabolites.

Three mevalonate pathway genes (3-hydroxy-3-methylglutaryl-CoA reductase, farnesyl diphosphate synthase, and geranylgeranyl diphosphate synthase) have been cloned from *Gibberella fujikuroi*, a fungus that makes the diterpenoid gibberellins. Each of the genes is present as a single copy, and their expression is apparently not regulated by carbon or nitrogen metabolism (Homann et al., 1996; Mende et al., 1997; Woitek et al., 1997). These findings prompt an exploration of the possibility that the isoprenoid pathway is not highly regulated in fungi that require it for the biosynthesis of secondary as well as primary metabolites.

Recently, evidence has been presented that the C_5 isopentenyl diphosphate units utilized for taxol synthesis in the plant *Taxus chinensis* are generated by a novel nonmevalonate pathway in which glyceraldehyde 3-phosphate is condensed with a C_2 derivative of pyruvate (Eisenreich et al., 1996; Schwender et al., 1996). The existence of this pathway explains the insensitivity of chloroplast isoprenoid synthesis (carotenoids, plastoquinone, and the prenyl side chains of chlorophyll) to lovastatin, a potent inhibitor of 3-hydroxy-3-methylglutaryl-CoA reductase and sterol synthesis in vascular plants (Lichtenthaler et al., 1997). Whether any fungi have acquired this pathway is unknown, but reevaluations of previous isotopic incorporation experiments in light of the present findings could provide the answer.

6. CONCLUSION

During the evolution of secondary metabolism in fungi, a biosynthetic strategy based on the polymerization of monomer units that are activated as thioesters has been implemented in two configurations, both making use of the cofactor 4'-phosphopantetheine. The type I polyketide synthases represent one arrangement. In these enzymes, the cofactor is charged repeatedly with incoming carboxylic acid monomer units over the course of a single assembly cycle. Multicarrier peptide synthetases, on the other hand, have several covalently bound cofactor residues, each of which receives only one amino acid or hydroxy acid monomer unit per catalytic cycle. In either configuration, the enzyme constitutes a molecular scale assembly line for the synthesis of a specific product.

Although the ability to make bioactive secondary metabolites is not a distinction separating endophytes from other groups of fungi, it is clear that production of these metabolites is a prominent factor in the adaptation of a fungus as an endophytic mutualist, a plant pathogen, or a saprobe that colonizes the plant during senescence or after its death (Claydon et al., 1985; Siegel et al., 1987; Carroll, 1988; Petrini et al., 1992; Walton, 1996). Secondary metabolites that are plant growth regulators, that can protect against microbial pathogens or harmful insect pests, or that deter feeding by insects and herbivores are benefits which can accrue to a host plant from an endophytic mutualist. The balance may be shifted in favor of the fungus when one of its secondary metabolites acts as a virulence factor for pathogenesis. Through their polyketide synthases and multi-carrier peptide synthetases, fungi display a versatility in the creation of bioactive secondary metabolites that has served them well throughout the evolution of their relationships with plants.

Two of the metabolites discussed in this chapter (β-lactam antibiotics and cyclosporin) have bioactivities that make them useful as therapeutic agents in human medicine. Pneumocandin A shown in Table 1 belongs to a class of antifungal antibiotics that also has therapeutic potential, and semisynthetic derivatives of compounds from this class are currently in clinical trials (Georgopapadakou and Tkacz, 1995). Consideration of taxonomic diversity among the endophytes, the likelihood that each vascular plant species may harbor two to five host-specific endophytes, and the high variability in the nature of host–fungus relationships has led to the conclusion that endophytes represent a potentially rich source of novel secondary metabolites with possible applications in medicine or agriculture (Dreyfuss and Chapela, 1994; Bills, 1995).

ACKNOWLEDGMENTS: This chapter was finished in July 1997. I wish to thank my colleagues Gerald Bills, Kevin Byrne, Steven Gould, Myra Kurtz, Jennifer Nielsen, and Mary Anne Talle for helpful discussions and critically reviewing the manuscript.

NOTE IN PROOF: In the interval since the completion of this chapter, several related articles have appeared. The sequence and disruption of a PKS gene responsible for the formation of conidial pigment in *Aspergillus fumigatus* has been reported (Langfelder et al., 1998, *Med. Microbiol. Immunol.* 187: 79–89; Tsai et al., 1998, *J. Bacteriol.* 180: 3031–3038). A revision of the *A. nidulans wA* gene sequence and of the structure of its polyketide product was made (Watanabe et al., 1999, *Tetrahedron Lett.* 40: 91–94). A gene cluster in *A. terreus* governing the biosynthesis of the cholesterol-lowering drug, lovastatin, has been described and subjected to mutational analysis (Hendrickson et al., 1999, *Chem. Biol.* 6: 429–439; Kennedy et al., 1999, *Science* 284: 1368–1372). The cluster contains two polyketide synthases, one of which produces the major backbone of lovastatin. One additional gene of the cluster encodes a protein that interacts with this nonaketide synthase to guide the correct processing of the nascent polyketide chain. This is the first report of an accessory protein that functions in this way. The polyketide synthase gene mediating the production of the fumonisin backbone in *Gibberella fujikuroi* has been characterized (Proctor et al., 1999, *Fungal Genet. Biol.* 27: 100–112). Expression of the 6-methylsalicylic acid PKS from *Penicillium patulum* in *active* form was achieved in *Esch-*

erichia coli and *Saccharomyces cerevisiae* by introducing into each host a heterologous phosphopantetheinyl transferase capable of activating the fungal PKS apoprotein (Kealey et al., 1998, *Proc. Nat. Acad. Sci. USA* 95: 505–509). Phylogenetic analysis of the ketosynthase domains of fungal polyketide synthases suggests two subclasses, and primer pairs for the recovery of each class by PCR have been developed (Bingle et al., 1999, *Fungal Genet. Biol.* 26: 209–223). Further evidence that the 2-amino-9,10-epoxy-8-oxodecanoic acid required for HC toxin production is derived from a fatty acid has been provided by ^{13}C-acetate labeling studies (Cheng et al., 1999, *J. Nat. Prod.* 62: 143–145). The *srfA-A* gene that encodes a multicarrier peptide synthetase for surfactin production by *Bacillus subtilis* has been reprogrammed by substituting native modules in the protein with ones taken from the *acvA* gene (ACV synthetase) of *P. chrysogenum* (Schneider et al., 1998, *Mol. Gen. Genet.* 257: 308-318). Structure-function analysis of adenylation domains of peptide synthetases has permitted the proposal of "rules" governing the amino acid specificity of these domains (Stachelhaus et al., 1999, *Chem. Biol.* 6: 493–505). New reviews have appeared for fungal phytotoxins (Hohn, 1997, *In* Carroll and Tudzynski (eds.), *The Mycota V Part A*. Springer Verlag, Berlin Heidelberg, pp. 129–144), for the genetics of aflatoxin biosynthesis (Brown et al., 1999, *Fungal Genet. Biol.* 26: 81–98), for multicarrier peptide synthetases (Marahiel, 1997, *Chem. Biol.* 4: 561–567; Marahiel et al., 1997, *Chem. Rev.* 97: 2651–2673; von Döhren et al., 1997, *Chem. Rev.* 97: 2675–2705; Konz and Marahiel, 1999, *Chem. Biol.* 6: R39-R48) and for regulation of β-lactam biosynthesis in filamentous fungi (Brakhage, 1998, *Microbiol. Molec. Biol. Rev.* 62: 547–585). A cluster of genes responsible for the synthesis of trichothecenes in *Myrothecium roridum* has been found (Trapp et al., 1998, *Molec. Gen. Genet.* 257: 421–432); the organization and orientation of genes in this cluster differ from those encoding a very similar biosynthetic pathway in *Fusarium sporotrichioides*. The products of the *TRI6* gene of *Fusarium* and the homologous *MRTRI6* gene of *Myrothecium* act as positive regulators of the trichothecene biosynthetic genes (Hohn et al., 1999, *Fungal Genet. Biol.* 26: 224–235). The gene for the protein that mediates the cyclization of geranylgeranyl diphosphate to form *ent*-kaurene, the first cyclic diterpene intermediate of gibberellin biosynthesis, has been obtained from *Phaeosphaeria* sp. and from *G. fujikuroi* (Kawaide et al., 1997, *J. Biol. Chem.* 272: 21706–21712; Tudzynski et al., 1998, *Current Genet.* 34: 234–240). In *G. fujikuroi*, this gene is part of a cluster that includes a new gene for geranylgeranyl diphosphate synthase (the second in this organism) and three cytochrome P450 genes (Tudzynski and Hölter, 1998, *Fungal Genet. Biol.* 25: 157-170). The dimethylallyltryptophan synthase that performs the first step in ergot alkaloid synthesis in *Claviceps purpurea* is encoded by a gene that appears to be part of a cluster which includes a region encoding the modular tripeptide synthetase, lysergyl peptide synthetase (Tudzynski et al., 1999, *Molec. Gen. Genet.* 261: 133–141). To date, the nonmevalonate pathway of isoprenoid biosynthesis has been found only in plants and bacteria (Eisenreich et al., 1998, *Chem. Biol.* 5: R221-R223).

REFERENCES

Adefarati, A. A., Giacobbe, R. A., Hensens, O. D. and Tkacz, J. S. (1991). Biosynthesis of L-671,329, an echinocandin-type antibiotic produced by *Zalerion arboricola*: origins of some of the unusual amino acids and the dimethylmyristic acid side chain. *Journal of the American Chemical Society* 113, 3542–3554.

Aharonowitz, Y., Cohen, G. and Martín, J. F. (1992). Penicillin and cephalosporin biosynthetic genes: structure, organization, regulation, and evolution. *Annual Review of Microbiology* 46, 461–495.

Ahn, J.-H. and Walton, J. D. (1996). Chromosomal organization of *TOX2*, a complex locus controlling host-selective toxin biosynthesis in *Cochliobolus carbonum*. *Plant Cell* 8, 887–897.

Ahn, J.-H. and Walton, J. D. (1997). A fatty acid synthase gene in *Cochliobolus carbonum* required for production of HC-toxin, cyclo(D-prolyl-L-alanyl-D-alanyl-L-2-amino-9,10-epoxi-8-oxodecanoyl). *Molecular Plant–Microbe Interactions* 10, 207–214.

Audhya, T. K. and Russell, D. W. (1973). Production of enniatin A. *Canadian Journal of Microbiology* 19, 1051–1054.

Baldwin, J. E., Bird, J. W., Field, R. A., O'Callaghan, N. M. and Schofield, C. J. (1990). Isolation and partial characterization of ACV synthetase from *Cephalosporium acremonium* and *Streptomyces clavuligerus*. *Journal of Antibiotics* 43, 1055–1057.

Baldwin, J. E., Byford, M. F., Field, R. A., Shiau, C.-Y., Sobey, W. J. and Schofield, C. J. (1993). Exchange of the valine 2-H in the biosynthesis of L-δ-(α-aminoadipoyl)-L-cysteinyl-D-valine. *Tetrahedron* 49, 3221–3226.

Baldwin, J. E., Shiau, C. Y., Byford, M. F. and Schofield, C. J. (1994). Substrate specificity of L-δ-(α-aminoadipoyl)-L-cysteinyl-D-valine synthetase from *Cephalosporium acremonium*: demonstration of the structure of several unnatural tripeptide products. *Biochemical Journal* 301, 367–372.

Banko, G., Demain, A. L. and Wolfe, S. (1987). δ-(L-α-Aminoadipyl)-L-cysteinyl-D-valine synthetase (ACV synthetase): a multifunctional enzyme with broad substrate specificity for synthesis of penicillin and cephalosporin precursors. *Journal of the American Chemical Society* 109, 2858–2860.

Beck, J., Ripka, S., Siegner, A., Schiltz, E. and Schweizer, E. (1990). The multifunctional 6-methylsalicylic acid synthase gene of *Penicillium patulum*; its gene structure relative to that of other polyketide synthases. *European Journal of Biochemistry* 192, 487–498.

Bell, A. A. and Wheeler, M. H. (1986). Biosynthesis and functions of fungal melanins. *Annual Review of Phytopathology* 24, 411–451.

Bennett, J. W. and Bentley, R. (1989). What's in a name? Microbial secondary metabolism. *Advances in Applied Microbiology* 34, 1–28.

Bevitt, D. J., Cortes, J., Haydock, S. F. and Leadley, P. F. (1992). 6-Deoxyerythronolide-B synthase 2 from *Saccharopolyspora erythraea*; cloning of the structural gene, sequence analysis and inferred domain structure of the multifunctional enzyme. *European Journal of Biochemistry* 204, 39–49.

Billich, A. and Zocher, R. (1987). N-Methyltransferase function of the multifunctional enzyme enniatin synthetase. *Biochemistry* 26, 8417–8423.

Billich, A. and Zocher, R. (1988). Constitutive expression of enniatin synthetase during fermentative growth of *Fusarium scirpi*. *Applied and Environmental Microbiology* 54, 2504–2509.

Bills, G. F. (1995). Analyses of microfungal diversity from a user's perspective. *Canadian Journal of Botany* 73, S33–S41.

Brosch, G., Ransom, R., Lechner, T., Walton, J. D. and Loidl, P. (1995). Inhibition of maize histone deacetylases by HC toxin, the host-selective toxin of *Cochliobolus carbonum*. *Plant Cell* 7, 1941–1950.

Brown, D. W., Hauser, F. M., Tommasi, R., Corlett, S. and Salvo, J. J. (1993). Structural elucidation of a putative conidial pigment intermediate in *Aspergillus parasiticus*. *Tetrahedron Letters* 34, 419–422.

Brown, D. W., Yu, J.-H., Kelkar, H. S., Fernandes, M., Nesbitt, T. C., Keller, N. P., Adams, T. H. and Leonard, T. J. (1996). Twenty-five coregulated transcripts define a sterigmatocystin gene cluster in *Aspergillus nidulans*. *Proceedings of the National Academy of Sciences of the USA* 93, 1418–1422.

Buades, C. and Moya, A. (1996). Phylogenetic analysis of the isopenicillin-N-synthetase horizontal gene transfer. *Journal of Molecular Evolution* 42, 537–542.

Campbell, I. M. (1984). Secondary metabolism and microbial physiology. *Advances in Microbial Physiology* 25, 1–60.

Cane, D. E., Shim, J. H., Xue, Q., Fitzsimons, B. C. and Hohn, T. M. (1995). Trichodiene synthase: identification of active site residues by site-directed mutagenesis. *Biochemistry* 34, 2480–2488.

Carroll, G. (1988). Fungal endophytes in stems and leaves: from latent pathogen to mutualistic symbiont. *Ecology* 69, 2–9.

Chumley, F. G. and Valent, B. (1990). Genetic analysis of melanin-deficient, nonpathogenic mutants of *Magnaporthe grisea*. *Molecular Plant–Microbe Interactions* 3, 135–143.

Claydon, N., Grove, J. F. and Pople, M. (1985). Elm bark beetle boring and feeding deterrents form *Phomopsis oblonga*. *Phytochemistry* 24, 937–943.

Darkin-Rattray, S. J., Gurnett, A. M., Myers, R. W., Dulski, P. M., Crumley, T. M., Allocco, J. J., Cannova, C., Meinke, P. T., Colletti, S. L., Bednarek, M. A., Singh, S. B., Goetz, M. A., Dombrowski, A. W., Polishook, J. D. and Schmatz, D. M. (1996). Apicidin: a novel antiprotozoal agent that inhibits parasite histone deacetylase. *Proceedings of the National Academy of Sciences of the USA* 93, 13143–13147.

Desjardins, A. E., Hohn, T. M. and McCormick, S. P. (1993). Trichothecene biosynthesis in *Fusarium* species: chemistry, genetics, and significance. *Microbiological Review* 57, 595–604.

Dewey, R. E., Siedow, J. N., Timothy, D. H. and Levings III, C. S. (1988). A 13-kilodalton maize mitochondrial protein in *E. coli* confers sensitivity to *Bipolaris maydis* toxin. *Science* 239, 293–295.

Dittmann, J., Lawen, A., Zocher, R. and Kleinkauf, H. (1990). Isolation and partial characterization of cyclosporin synthetase from a cyclosporin non-producing mutant of *Beauveria nivea*. *Biological Chemistry Hoppe-Seyler* 371, 829–834.

Dreyfuss, M. M. and Chapela, I. H. (1994). Potential of fungi in the discovery of novel, low-molecular weight pharmaceuticals. *In* V. P. Gullo (ed.), *The Discovery of Natural Products with Therapeutic Potential*. Butterworth-Heinemann, Boston, pp. 49–80.

Dreyfuss, M. M. and Gams, W. (1994). Proposal to reject *Pachybasium niveum* Rostr. in order to retain the name *Tolypocladium inflatum* W. Gams for the fungus that produces cyclosporin. *Taxon* 43, 660–661.

Edwards, R. L., Fawcett, V., Maitland, D. J., Nettleton, R., Shields, L. and Whalley, A. J. S. (1991). Hypoxyxylerone: a novel green pigment from the fungus *Hypoxylon fragiforme*. *Chemical Communications* 1009–1010.

Eisenreich, W., Menhard, B., Hylands, P. J., Zenk, M. H. and Bacher, S. (1996). Studies on the biosynthesis of taxol: the taxane carbon skeleton is not of mevalonoid origin. *Proceedings of the National Academy of Sciences of the USA,* 93, 6431–6436.

Georgopapadakou, N. H. and Tkacz, J. S. (1995). The fungal cell wall as a drug target. *Trends in Microbiology* 3, 98–104.

Gutiérrez, S., Díez, B., Montenegro, E. and Martín, J. F. (1991). Characterization of the *Cephalosporium acremonium pcbAB* gene encoding α-aminoadipyl-cysteinyl-valine synthetase, a large multidomain peptide synthetase: linkage to the *pcbC* gene as a cluster of early cephalosporin biosynthetic genes and evidence of multiple functional domains. *Journal of Bacteriology* 173, 2354–2365.

Haese, A., Pieper, R., von Ostrowski, T. and Zocher, R. (1994). Bacterial expression of catalytically active fragments of the multifunctional enzyme enniatin synthetase. *Journal of Molecular Biology* 243, 116–122.

Haese, A., Schubert, M., Herrmann, M. and Zocher, R. (1993). Molecular characterization of the enniatin synthetase gene encoding a multifunctional enzyme catalysing N-methyldepsipeptide formation in *Fusarium scirpi. Molecular Microbiology* 7, 905–914.

Harborne, J. B. (1993). Advances in chemical ecology. *Natural Products Reports* 10, 327–348.

Hoffmann, K., Schneider-Scherzer, E., Kleinkauf, H. and Zocher, R. (1994). Purification and characterization of eucaryotic alanine racemase acting as key enzyme in cyclosporin biosynthesis. *Journal of Biological Chemistry* 269, 12710–12714.

Hohn, T. M. and Beremand, P. D. (1989). Isolation and nucleotide sequence of a sesquiterpene cyclase gene from the trichothecene-producing fungus *Fusarium sporotrichioides. Gene* 79, 131–138.

Hohn, T. M., Desjardins, A. E. and McCormick, S. P. (1995). The *Tri4* gene of *Fusarium sporotrichioides* encodes a cytochrome P450 monooxygenase involved in trichothecene biosynthesis. *Molecular and General Genetics* 248, 95–102.

Hohn, T. M., McCormick, S. P. and Desjardins, A. E. (1993). Evidence for a gene cluster involving trichothecene-pathway biosynthetic genes in *Fusarium sporotrichioides. Current Genetics* 24, 291–295.

Hohn, T. M. and Plattner, R. D. (1989). Expression of the trichodiene synthase gene of *Fusarium sporotrichioides* in *Escherichia coli* results in sesquiterpene production. *Archives of Biochemistry and Biophysics* 275, 92–97.

Homann, V., Mende, K., Arntz, C., Ilardi, V., Macino, G., Morelli, G., Böse, G. and Tudzynski, B. (1996). The isoprenoid pathway: cloning and characterization of fungal FPPS genes. *Current Genetics* 30, 232–239.

Hopwood, D. A. and Sherman, D. H. (1990). Molecular genetics of polyketides and its comparison to fatty acid biosynthesis. *Annual Review of Genetics* 24, 37–66.

Hoskins, J. A., O'Callaghan, N., Queener, S. W., Cantwell, C. A., Wood, J. S., Chen, V. J. and Skatrud, P. L. (1990). Gene disruption of the *pcb*AB gene encoding ACV synthetase in *Cephalosporium acremonium. Current Genetics* 18, 523–530.

Huang, L. H. and Kaneko, T. (1996). Pyrenomycetes and Loculoascomycetes as sources of secondary metabolites. *Journal of Industral Microbiology and Biotechnology* 17, 402–416.

Hutchinson, C. R. and Fujii, I. (1995). Polyketide synthase gene manipulation: a structure-function approach in engineering novel antibiotics. *Annual Review of Microbiology* 49, 201–238.

Itazaki, H., Nagashima, K., Sugita, K., Yoshida, H., Kawamura, Y., Yasuda, Y., Matsumoto, K., Ishii, K., Uotani, N., Nakai, H., Terui, A., Yoshimatsu, S., Ikenishi, Y. and Nakagawa, Y. (1990). Isolation and structural elucidation of new cyclotetrapeptides, trapoxins A and B, having detransformation activities as antitumor agents. *Journal of Antibiotics* 43, 1524–1532.

Itoh, Y., Kodama, K., Furuya, K., Takahashi, S., Haneishi, T., Takiguchi, Y. and Arai, M. (1980). A new sesquiterpene antibiotic, heptelidic acid: producing organisms, fermentation, isolation and characterization. *Journal of Antibiotics* 33, 468–473.

Jones, M. J. and Dunkle, L. D. (1995). Virulence gene expression during conidial germination in *Cochliobolus carbonum*. *Molecular Plant–Microbe Interactions* 8, 476–479.

Jordan, P. M. and Spencer, J. B. (1993). The biosynthesis of tetraketides: enzymology, mechanism and molecular programming. *Biochemical Society Transactions* 21, 222–228.

Katz, L. and Donadio, S. (1993). Polyketide synthesis: prospects for hybrid antibiotics. *Annual Review of Microbiology* 47, 875–912.

Keller, N. P. and Hohn, T. M. (1997). Metabolic pathway gene clusters in filamentous fungi. *Fungal Genetics and Biology* 21, 17–29.

Kijima, M., Yoshida, M., Sugita, K., Horinouchi, S. and Beppu, T. (1993). Trapoxin, an antitumor cyclic tetrapeptide, is an irreversible inhibitor of mammalian histone deacetylase. *Journal of Biological Chemistry* 268, 22429–22435.

Kimura, N. and Tsuge, T. (1993). Gene cluster involved in melanin biosynthesis of the filamentous fungus *Alternaria alternata*. *Journal of Bacteriology* 175, 4427–4435.

Kleinkauf, H. and von Döhren, H. (1990). Nonribosomal biosynthesis of peptide antibiotics. *European Journal of Biochemistry* 192, 1–15.

Kleinkauf, H. and von Döhren, H. (1996). A nonribosomal system of peptide biosynthesis. *European Journal of Biochemistry* 236, 335–351.

Kriauciunas, A., Frolik, C. A., Hassell, T. C., Skatrud, P. L., Johnson, M. G., Holbrook, N. L. and Chen, V. J. (1991). The functional role of cysteines in isopenicillin N synthase; correlation of cysteine reactivities toward sulfhydryl reagents with kinetic properties of cysteine mutants. *Journal of Biological Chemistry* 266, 11779–11788.

Kubo, Y., Takano, Y., Endo, N., Yasuda, N., Tajima, S. and Furusawa, I. (1996). Cloning and structural analysis of the melanin biosynthesis gene *SCD1* encoding scytalone dehydratase in *Colletotrichum lagenarium*. *Applied and Environmental Microbiology* 62, 4340–4344.

Kubo, Y., Tsuda, M., Furusawa, I. and Shishiyama, J. (1989). Genetic analysis of genes involved in melanin biosynthesis of *Cochliobolus miyabeanus*. *Experimental Mycology* 13, 77–84.

Lawen, A., Dittmann, J., Schmidt, B., Riesner, D. and Kleinkauf, H. (1992). Enzymatic biosynthesis of cyclosporin A and analogues. *Biochimie* 74, 511–516.

Lawen, A., Traber, R., Geyl, D., Zocher, R. and Kleinkauf, H. (1989). Cell-free biosynthesis of new cyclosporins. *Journal of Antibiotics* 42, 1283–1289.

Lawen, A. and Zocher, R. (1990). Cyclosporin synthetase, the most complex peptide synthesizing multienzyme polypeptide so far described. *Journal of Biological Chemistry* 265, 11355–11360.

Lee, C., Görisch, H., Kleinkauf, H. and Zocher, R. (1992). A highly specific D-hydroxyisovalerate dehydrogenase from the enniatin producer *Fusarium sambucinum. Journal of Biological Chemistry* 267, 11741–11744.

Levings III, C. S., Rhoads, D. M. and Siedow, J. N. (1995). Molecular interactions of *Bipolaris maydis* T-toxin and maize. *Canadian Journal of Botany* 73, S483–S489.

Levings III, C. S. and Siedow, J. N. (1992). Molecular basis of disease susceptibility in the Texas cytoplasm of maize. *Plant Molecular Biology* 19, 135–147.

Lichtenthaler, H. K., Schwender, J., Disch, A. and Rohmer, M. (1997). Biosynthesis of isoprenoids in higher plant chloroplasts proceeds via a mevalonate-independent pathway. *FEBS Letters* 400, 271–274.

Lu, S., Lyngholm, L., Yang, G., Bronson, C., Yoder, O. C. and Turgeon, B. G. (1994). Tagged mutations at the *Tox1* locus of *Cochliobolus heterostrophus* by restriction enzyme-mediated integration. *Proceedings of the National Academy of Science of the USA* 91, 12649–12653.

Martín, J. F. and Gutiérrez, S. (1995). Genes for β-lactam antibiotic biosynthesis. *Antonie van Leeuwenhoek International Journal of General Microbiology* 67, 181–200.

Mathison, L., Soliday, C., Stepan, T., Aldrich, T. and Rambosek, J. (1993). Cloning, characterization, and use in strain improvement of the *Cephalosporium acremonium* gene *cefG* encoding acetyl transferase. *Current Genetics* 23, 33–41.

Mayorga, M. E. and Timberlake, W. E. (1992). The developmentally regulated *Aspergillus nidulans wA* gene encodes a polypeptide homologous to polyketide and fatty acid synthases. *Molecular and General Genetics* 235, 205–212.

McCormick, S. P., Hohn, T. M. and Desjardins, A. E. (1996). Isolation and characterization of *Tri3*, a gene encoding 15-*O*-acetyltransferase from *Fusarium sporotrichioides. Applied and Environmental Microbiology* 62, 353–359.

Meeley, R. B., Johal, G. S., Briggs, S. P. and Walton, J. D. (1992). A biochemical phenotype for a disease resistance gene of maize. *Plant Cell* 4, 71–77.

Mende, K., Homann, V. and Tudzinski, B. (1997). The geranylgeranyl diphosphate synthase gene of *Gibberella fujikuroi*: Isolation and expression. *Molecular General Genetics* 255, 96–105.

Menne, S., Walz, M. and Kück, U. (1994). Expression studies with the bidirectional *pcbAB-pcbC* promoter region from *Acremonium chrysogenum* using reporter gene fusions. *Applied Microbiology and Biotechnology* 42, 57–66.

Noble, H. M., Langley, D., Sidebottom, P. J., Lane, S. J. and Fisher, P. J. (1991). An echinocandin from an endophytic *Cryptosporiopsis* sp. and *Pezicula* sp. in *Pinus sylvestris* and *Fagus sylvatica. Mycological Research* 95, 1439–1440.

Nozoe, S., Morisaki, M., Fufushima, K. and Okuda, S. (1968). The isolation of an acyclic C25-isoprenoid alchohol, geranylnerolidol, and a new ophiobolin. *Tetrahedron Letters* 4457–4458.

Offenzeller, M., Zhuang, S., Santer, G., Moser, H., Traber, R., Memmert, K. and Schneider-Scherzer, E. (1993). Biosynthesis of the unusual amino acid (4*R*)-4-[(*E*)-2-butenyl]-4-methyl-L-threonine of cyclosporin A; identification of 3(*R*)-hydroxy-4(*R*)-methyl-6(*E*)-octenoic acid as a key intermediate by enzymatic *in vitro* synthesis and by *in vivo* labeling techniques. *Journal of Biological Chemistry* 268, 26127–26134.

Panaccione, D. G., Scott-Craig, J. S., Pocard, J.-A. and Walton, J. D. (1992). A cyclic peptide synthetase gene required for pathogenicity of the fungus *Cochliobolus carbonum* on maize. *Proceedings of the National Academy of Science of the USA* 89, 6590–6594.

Perpetua, N. S., Kubo, Y., Yasuda, N., Takano, Y. and Furusawa, I. (1996). Cloning and characterization of a melanin biosynthetic *THR1* reductase gene essential for appressorial penetration of *Colletotrichum lagenarium*. *Molecular Plant–Microbe Interactions* 9, 323–329.

Petrini, O., Sieber, T. N., Toti, L. and Viret, O. (1992). Ecology, metabolite production, and substrate utilization in endophytic fungi. *Natural Toxins* 1, 185–196.

Pflügl, G., Kallen, J., Schirmer, T., Jansonius, J. N., Zurini, M. G. M. and Walkinshaw, M. D. (1993). X-ray structure of a decameric cyclophilin-cyclosporin crystal complex. *Nature* 361, 91–94.

Pitkin, J. W., Panaccione, D. G. and Walton, J. D. (1996). A putative cyclic peptide efflux pump encoded by the *TOXA* gene of the plant-pathogenic fungus *Cochliobolus carbonum*. *Microbiology* 142, 1557–1565.

Poch, G. K. and Gloer, J. B. (1989). Obionin A: a new polyketide metabolite from the marine fungus *Leptosphaeria obiones*. *Tetrahedron Letters* 30, 3483–3486.

Pope, M. R., Ciuffetti, L. M., Knoche, H. W., McCrery, D., Daly, J. M. and Dunkle, L. D. (1983). Structure of the host-specific toxin produced by *Helminthosporium carbonum*. *Biochemistry* 22, 3502–3506.

Proctor, R. H., Hohn, T. M., McCormick, S. P. and Desjardins, A. E. (1995). *Tri6* encodes an unusual zinc finger protein involved in regulation of trichothecene biosynthesis in *Fusarium sporotrichioides*. *Applied and Environmental Microbiology* 61, 1923–1930.

Rhoads, D. M., Levings III, C. S. and Siedow, J. N. (1995). URF13, a ligand-gated, pore-forming receptor for T-toxin in the inner membrane of *cms-T* mitochondria. *Journal of Bioengineering and Biomembranes* 27, 437–445.

Samson, S. M., Belagaje, R., Blankenship, D. T., Chapman, J. L., Perry, D., Skatrud, P. L., Van Frank, R. M., Abraham, E. P., Baldwin, J. E., Queener, S. W. and Ingolia, T. D. (1985). Isolation, sequence determination and expression in *Escherichia coli* of the isopenicillin N synthetase gene from *Cephalosporium acremonium*. *Nature* 318, 191–194.

Schmidt, B., Riesner, D., Lawen, A. and Kleinkauf, H. (1992). Cyclosporin synthetase is a 1.4 MDa multienzyme polypeptide; re-evaluation of the molecular mass of various peptide synthetases. *FEBS Letters* 307, 355–360.

Schwender, J., Seemann, M., Lichtenthaler, H. K. and Rohmer, M. (1996). Biosynthesis of isoprenoids (carotenoids, sterols, prenyl side chains of chlorophylls and plastoquinone) via a novel pyruvate/glyceraldehyde-3-phosphate non-mevalonate pathway in the green alga *Scenedesmus obliquus*. *Biochemical Journal* 316, 73–80.

Scott-Craig, J. S., Panaccione, D. G., Pocard, J.-A. and Walton, J. D. (1992). The cyclic peptide synthetase catalyzing HC-toxin production in the filamentous fungus *Cochliobolus carbonum* is encoded by a 15.7-kilobase open reading frame. *Journal of Biological Chemistry* 267, 26044–26049.

Shiotani, H. and Tsuge, T. (1995). Efficient gene targeting in the filamentous fungus *Alternaria alternata*. *Molecular and General Genetics* 248, 142–150.

Shute, R. E., Dunlap, B. and Rich, D. H. (1987). Analogues of the cytostatic and antimitogenic agents chlamydocin and HC-toxin: synthesis and biological activity of chloromethyl ketone and diazomethyl ketone functionalized cyclic tetrapeptides. *Journal of Medicinal Chemistry* 30, 71–78.

Siedow, J. N., Rhoads, D. M., Ward, G. C. and Levings III, C. S. (1995). The relationship between the mitochondrial gene T-*urf*13 and fungal pathotoxin sensitivity in maize. *Biochimica et Biophysica Acta* 1271, 235–240.

Siegel, M. R., Latch, G. C. M. and Johnson, M. C. (1987). Fungal endophytes of grasses. *Annual Review of Phytopathology* 25, 293–315.

Singh, S. B., Zink, D. L., Polishook, J. D., Dombrowski, A. W., Darkin-Rattray, S. J., Schmatz, D. M. and Goetz, M. A. (1996). Apicidins: novel cyclic tetrapeptides as coccidiostats and antimalarial agents from *Fusarium pallidoroseum*. *Tetrahedron Letters,* 37, 8077–8080.

Skatrud, P. L. and Queener, S. W. (1989). An electrophoretic molecular karyotype for an industrial strain of *Cephalosporium acremonium*. *Gene* 78, 331–338.

Smith, D. J., Burnham, M. K. R., Bull, J. H., Hodgson, J. E., Ward, J. M., Browne, P., Brown, J., Barton, B., Earl, A. J. and Turner, G. (1990). β-Lactam antibiotic biosynthetic genes have been conserved in clusters in prokaryotes and eukaryotes. *EMBO Journal* 9, 741–747.

Spencer, J. B. and Jordan, P. M. (1992). Purification and properties of 6-methylsalicylic acid synthase from *Penicillium patulum*. *Biochemical Journal* 288, 839–846.

Stein, T., Vater, J., Kruft, V., Wittmann-Liebold, B., Franke, P., Panico, M., McDowell, R. and Morris, H. R. (1994). Detection of 4′-phosphopantetheine as the thioester binding site for L-valine of gramicidin S synthetase 2. *FEBS Letters* 340, 39–44.

Takano, Y., Kubo, Y., Shimizu, K., Mise, K., Okuno, T. and Furusawa, I. (1995). Structural analysis of *PKS1*, a polyketide synthase gene involved in melanin biosynthesis in *Colletotrichum lagenarium*. *Molecular and General Genetics* 249, 162–167.

Tanaka, C., Kubo, Y. and Tsuda, M. (1991). Genetic analysis and characterization of *Cochliobolus heterostrophus* color mutants. *Mycological Research* 95, 49–56.

Thériault, Y., Logan, T. M., Meadows, R., Yu, L., Olejniczak, E. T., Holzman, T. F., Simmer, R. L. and Fesik, S. W. (1993). Solution structure of the cyclosporin A/ cyclophilin complex by NMR. *Nature* 361, 88–91.

Traber, R. and Dreyfuss, M. M. (1996). Occurrence of cyclosporins and cyclosporin-like peptolides in fungi. *Journal of Industrial Microbiology and Biotechnology* 17, 397–401.

Traber, R., Hofmann, H. and Kobel, H. (1989). Cyclosporins–new analogues by precursor directed biosynthesis. *Journal of Antibiotics* 42, 591–597.

Turner, W. B. and Aldridge, D. C. (1983). *Fungal Metabolites II*. Academic Press, New York.

Vining, L. C. (1990). Functions of secondary metabolites. *Annual Review of Microbiology* 44, 395–427.

Vining, L. C. (1992). Secondary metabolism, inventive evolution and biochemical diversity–a review. *Gene* 115, 135–140.

Walton, J. D. (1987). Two enzymes involved in biosynthesis of the host-selective phytotoxin HC-toxin. *Proceedings of the National Academy of Science of the USA,* 84, 8444–8447.

Walton, J. D. (1996). Host-selective toxins: agents of compatibility. *Plant Cell* 8, 1723–1733.

Walton, J. D. and Holden, F. R. (1988). Properties of two enzymes involved in the biosynthesis of the fungal pathogenicity factor HC-toxin. *Molecular Plant–Microbe Interactions* 1, 128–134.

Weber, G., Schmörgendorfer, K., Schneider-Scherzer, E. and Leitner, E. (1994). The peptide synthetase catalyzing cyclosporine production in *Tolypocladium niveum* is encoded by a giant 45.8 kilobase open reading frame. *Current Genetics* 26, 120–125.

Weiergang, I., Dunkle, L. D., Wood, K. V. and Nicholson, R. L. (1996). Morphogenic regulation of pathotoxin synthesis in *Cochliobolus carbonum. Fungal Genetics and Biology* 20, 74–78.

Wessel, W. L., Clare, K. A. and Gibbons, W. A. (1988). Biosynthesis of L-Aeo, the toxic determinant of the phytotoxin produced by *Helminthosporium carbonum. Biochemical Society Transactions* 16, 402–403.

Wheeler, M. H. and Klich, M. A. (1995). The effects of tricyclazole, pyroquilon, phthalide, and related fungicides on the production of conidial wall pigments by *Penicillium* and *Aspergillus* species. *Pesticide Biochemistry and Physiology* 52, 125–136.

Williams, D. H., Stone, M. J., Hauck, P. R. and Rahman, S. K. (1989). Why are secondary metabolites (natural products) biosynthesized? *Journal of Natural Products* 52, 1189–1208.

Woitek, S., Unkles, S. E., Kinghorn, J. R. and Tudzynski, B. (1997). 3-Hydroxy-3-methyl-glutaryl-CoA reductase gene of *Gibberella fujikuroi*: isolation and characterization. *Currrent Genetics* 31, 38–47.

Yang, G., Rose, M. S., Turgeon, B. G. and Yoder, O. C. (1996). A polyketide synthase is required for fungal virulence and production of the polyketide T-toxin. *Plant Cell* 8, 2139–2150.

Yang, G., Turgeon, B. G. and Yoder, O. C. (1994). Toxin-deficient mutants from a toxin-sensitive transformant of *Cochliobolus heterostrophus. Genetics* 137, 751–757.

Yoder, O. C., Yang, G., Adam, G., Diaz-Minguez, J. M., Rose, M. and Turgeon, B. G. (1993). Genetics of polyketide toxin biosynthesis by plant pathogenic fungi. *In* R. H. Baltz, G. D. Hegeman and P. L. Skatryd (eds.), *Industrial Microorganisms: Basic and Applied Molecular Genetics.* American Society for Microbiology, Washington, DC, pp. 217–225.

Yu, J.-H. and Leonard, T. J. (1995). Sterigmatocystin biosynthesis in *Aspergillus nidulans* requires a novel type I polyketide synthase. *Journal of Bacteriology* 177, 4792–4800.

Zocher, R., Keller, U. and Kleinkauf, H. (1982). Enniatin synthetase, a novel type of multifunctional enzyme catalyzing depsipeptide synthesis in *Fusarium oxysporum. Biochemistry* 21, 43–48.

Zocher, R., Salnikow, J. and Kleinkauf, H. (1976). Biosynthesis of enniatin B. *FEBS Letters* 71, 13–17.

12

Metabolic Activity, Distribution, and Propagation of Grass Endophytes *In Planta*: Investigations Using the GUS Reporter Gene System

Jan Schmid and Martin J. Spiering*

Institute of Molecular BioSciences (IMBS), Massey University, Palmerston North, New Zealand

Michael J. Christensen

Grasslands Research Centre, New Zealand Pastoral Agriculture Research Institute Limited (AgResearch), Palmerston North, New Zealand

1. INTRODUCTION

In writing this chapter we had two objectives. The first was to provide a brief overview of the results regarding the distribution and metabolic state of endophyte mycelium within perennial ryegrass (*Lolium perenne*) tillers which we have obtained using the GUS reporter gene system. The second was to share our view of the cellular biology of grass endophytes *in planta* and the evolution of some aspects of this biology—a view that has emerged from the use of the GUS system, not only from the results obtained but in some cases also by problems with the method that we encountered.

The ideas regarding the cellular biology of endophytes that we have developed explain data on two associations we have studied with the GUS system. They also explain observations on many other associations. While we cannot claim that our views are the only possible explanation for these observations we feel that we have justification to present them here, if only to draw attention to and stimulate research into important aspects of endophyte–host interactions.

---

* *Current affiliation*: University of Kentucky, Lexington, Kentucky.

2. STUDYING *IN PLANTA* ENDOPHYTE GENE EXPRESSION

An analysis of the interactions between symbionts at the molecular level (i.e., identification of molecules, their function and localization) is challenging because many of these interactions will only take place during the association of the symbionts. It is difficult to separate the contributions of the two partners or to find these molecules at all if they are produced only at specific times and locations. Molecular biology techniques can overcome this problem to a degree in that they enable us (1) to search the genomes of the partners for genes that may play a role in the symbiosis and (2) to monitor the expression of these genes. Molecular biology will not give us the entire picture- formation and final location of relevant molecules may not coincide with the where and when of expression of the genes required, but we can expect insights that we could never before have hoped to obtain.

Molecular biology techniques for ryegrass endophytes have recently become available, allowing the cloning of endophyte genes, transformation with foreign DNA, and analysis of gene expression (Herd et al., 1997; Murray et al., 1992; Schardl and An, 1993; Scott and Schardl, 1993). The time seems near when these techniques can be employed to investigate endophyte pathways of key importance. *Penicillium paxilli* (Young et al., 1998) and *Claviceps purpurea* (Tsai et al., 1995) are now successfully being used as model organisms to clone genes in the paxilline and ergot alkaloid pathways, and this should allow in the near future the cloning of the corresponding *Neotyphodium* genes. Indeed recently a gene believed to contribute to the early stages of the paxilline pathway in ryegrass endophytes, the HMGCoA reductase gene of the ryegrass endophyte Lp19, has been cloned by D. B. Scott's group (Dobson, 1997).

An easy way to assess expression of such genes is use of reporter genes such as the *Escherichia coli* β-D glucuronidase gene (GUS system), whose expression can be detected and quantified by the formation of a fluorescent product (Jefferson, 1987). When linked to the promoter of an endophyte gene of interest and inserted into the genome of an endophyte, they can give some indication of spatial and temporal distribution of the *in planta* expression of the endophyte gene, an approach whose feasibility in grass endophytes was first demonstrated by Murray *et al.* (1992).

In order to use reporter genes to learn about *in planta* regulation of endophyte gene expression it is necessary to have (1) methods that allow exact quantification of GUS activity even in small plant tissue samples, allowing mapping of tissue-specific differences in expression levels, and (2) methods for detecting the distribution of endophyte mycelium throughout the tiller and its rate of protein synthesis in order to detect upregulation and downregulation of endophyte genes.

In order to relate expression data for different genes to each other and/or to relate expression data to other measurements, it is also desirable to have a model symbiosis, maintained under constant conditions. Such a symbiosis allows one to correlate data that have been acquired sequentially in order to gain a more complete understanding of gene regulation.

Much of our recent work has focused on developing methods that fulfill the above requirements. We have recently combined the highly sensitive fluorescence assay based on the conversion of 4-methylumbelliferyl glucuronide (MUG) to methylumbelliferone (MU) (Jefferson, 1987), with a quantitative extraction method, making it possible to reliably quantify endophyte-associated GUS activity in plant tissue samples of ≤2 mg dry weight (Herd et al., 1997). Given typical ryegrass tiller dry weights of 20–400 mg, this allows a very detailed mapping of GUS activity, thus fulfilling the first requirement listed. In order to relate expression of a specific gene to the overall level of fungal gene expression in the plant tissue we have proposed the use of reporter gene constructs that place the GUS reporter gene under the control of a heterologous constitutive promoter—since this promoter is unlikely to be regulated. The promoter (*Aspergillus nidulans gpd*A promoter) is believed to be constitutive, and important regulatory sequences in the promoter are not conserved between *Aspergillus* and other fungi (Devchand and Gwyne, 1991; Jungehülsing et al., 1994; Punt et al., 1992).

The GUS activity resulting from such a construct should depend on the rate of protein synthesis (Herd et al., 1997). Actually, the measurable GUS activity depends on both the rate of synthesis and the rate of degradation of the enzyme molecules. Thus, when protein synthesis stops, GUS activity does not immediately disappear. However, we have determined a half-life of GUS activity in metabolically inactive endophyte hyphae in culture of 2 days (Spiering and Schmid, 1997). Assuming that the rate of degradation within hyphae growing in the plant is similar to the rate we have observed in culture, it should be fast enough for detecting changes in the relatively slowly developing grass system. In addition, we have developed improved staining methods for endophyte hyphae that allow the mapping of endophyte biomass concentration with a resolution comparable to that of GUS activity (Tan et al., 1997). We have also shown that it is possible to maintain a model symbiosis in a growth cabinet under constant conditions, so that measurements of endophyte biovolume concentrations and GUS activity concentrations determined several months apart are directly comparable (Tan et al., 1997). Methods are therefore now available that allow the assessment of *in planta* endophyte gene regulation.

In developing these approaches we have gained some fundamental knowledge regarding the distribution of endophyte biomass in ryegrass and the metabolic state of the mycelium.

3. DISTRIBUTION AND METABOLIC STATE OF ENDOPHYTE MYCELIUM IN RYEGRASS TILLERS

As part of our studies we have investigated for two endophyte strains, Lp1 and Lp19 (Christensen et al., 1993), the distribution of endophyte metabolic activity throughout ryegrass (cultivar Nui) tillers, i.e., expression by the endophyte of the GUS reporter gene under control of a constitutive heterologous promoter (*Aspergillus nidulans gpdA* promoter). As pointed out above, the GUS activity synthesized by a mycelium should roughly correspond to the activity of the mycelium in terms of protein synthesis, which in turn should be an indicator of its overall metabolic activity. We have verified this in culture by demonstrating a 20- to 40-fold decrease in GUS activity, in comparison to cells in log phase to early stationary phase, when the mycelium enters late stationary phase or is being exposed to metabolic inhibitors like KCN (Tan et al., 1997).

The first strain we investigated was a transformant of the *Neotyphodium* endophyte Lp1, (taxonomic group LpTG-2) (Christensen et al., 1993; Murray et al., 1992) in plants maintained in a greenhouse during the summer (Herd et al., 1997). We found that approximately 70% of all Lp1 endophyte metabolic activity in a tiller was located in the sheaths, 20% in the blades, 0–20% in immature leaves and 0.5–10% in immature tissue at the base of the tiller. In the vast majority of mature leaves, basal-apical gradients of endophyte metabolic activity concentration occurred (lower half of leaf sheath > upper half of leaf sheath > lower half of the blade > upper half of the blade). Although in immature leaves the concentration of endophyte metabolic activity was lower than in mature leaves, its distribution followed similar basal-apical gradients. We also observed that the metabolic activity concentration in the sheaths of the youngest mature leaves was usually significantly higher than the endophyte metabolic activity concentration in the older sheaths.

Endophyte metabolic activity per tiller increased at the same rate as tiller dry weight up to a tiller dry weight of 50 mg. At higher dry weights, however, each further twofold increase in tiller dry weight was only matched by a further 1.3-fold increase in endophyte metabolic activity. Larger tillers therefore had a lower concentration of endophyte metabolic activity per tiller dry weight. These larger tillers had four mature leaves, compared to two mature leaves in the tillers of less than 50 mg dry weight. However, the decrease of endophyte metabolic activity per tiller dry weight was apparently not due to an age-dependent decrease in the concentration of endophyte metabolic activity in the two older leaves. Rather, a general reduction in the concentration of endophyte metabolic activity occurred, affecting both newly forming and existing leaves to a similar degree.

Observations on the same endophyte in different Nui genotypes revealed that while the associations had similar metabolic activity distribution patterns, the concentration of endophyte metabolic activity differed between them by an order of magnitude. This seemed primarily due to variation in the number of metabolically active hyphae, as visualized by fluorescence microscopy with the GUS substrate Imagene Green (Saunders, 1997; Scott et al., 1995).

The main conclusions that we drew from these data regarding the *in planta* control of endophyte metabolic activity were (1) that basal-apical gradients of the concentration of endophyte metabolic activity were established prior to leaf maturation, (2) that endophyte metabolic activity concentration increased as the leaf matured, (3) that following maturation a decline of endophyte metabolic activity concentration in the sheath occurred, but (4) that the extent of these increases (and presumably also of the decreases) was constrained so that a predetermined threshold of endophyte metabolic activity per tiller was not exceeded and that the distribution of endophyte metabolic activity between different tissue types (blade, sheath, etc.) remained fairly constant. Our new observations on Lp19 (see below) suggest that an alternative explanation can be found for lower endophyte metabolic activity concentrations in older leaves on which the third conclusion was based.

Taken together these conclusions give the impression of a rather precise control of endophyte metabolic activity concentration in each given part of the tiller, which takes into account not only the type of tissue and its age but also the size of the tiller. The observations from this study suggested to us that the plant may actively control the endophyte.

Several open questions remained at the end of this investigation. The most important were (1) whether the apparent control of endophyte metabolic activity concentration was achieved by regulating the number of endophyte hyphae, their metabolic state, or both, and (2) whether some or all aspects of the observed patterns would also occur in other endophytes.

We therefore undertook further investigations (Tan et al., 1997, 1999) using a transformant (Saunders, 1997) of a strain of *N. lolii* (Lp19) (Christensen et al., 1993). Again we determined endophyte metabolic activity. In addition, we determined in the same sections endophyte biovolume as an indicator of endophyte biomass. Endophyte biovolume in each section was determined by counting the number of hyphae in aniline blue–stained transverse cuts taken from either ends of sections to be used for GUS assays, multiplication of this number with π, the square of the radius of the hyphae and the length of the section; the length of the section is an adequate estimate of hyphal length in the nonmeristematic tissues that we are assessing, since the endophyte shows very little branching in these tissues (Tan et al., 1998). This allowed us to determine to what degree the distribution of endophyte metabolic activity reflected differences in the endophyte mycelium's metabolic state, expressed as an endophyte metabolic state (EMS) coefficient:

$$EMS = \frac{endophyte\ metabolic\ activity}{endophyte\ biovolume}$$

We conducted these investigations in a single ryegrass genotype under controlled conditions in a growth cabinet in order to establish this association as a model system for comparisons of the expression of multiple genes and other parameters (see preceding section). The plants were maintained in growth cabinets at 15°C, 12/24 h illumination at a light intensity of 296 μmol/m²/s and watered to saturation twice weekly. Temperature and illumination were adjusted to these levels to retain the vegetative growth state of the grass tillers. Due to the growth characteristics of the "Nui" genotype used, it was difficult to obtain tillers with four mature leaves that had been part of the Lp1 investigation; only tillers with one, two, and three mature leaves could be analyzed.

The Lp19 association behaved very similar to the Lp1 association in regard to the distribution of endophyte metabolic activity between different tissue types (sheaths, blades, etc.). As with the Lp1 association, we observed basal-apical gradients in the vast majority of both mature and immature leaves. These were of comparable magnitude to those in the Lp1 association (5- to 10-fold difference between the bottom and top). Likewise, the lower part of the sheaths of the youngest mature leaf of a given tiller usually contained higher concentrations of endophyte metabolic activity than lower sheaths of the older leaves. The Lp19 association therefore behaved very similarly to tillers of less than 50 mg dry weight containing Lp1. Because the endophytes belong to different taxonomic groups and were maintained under different environmental conditions, the features common to both associations might be a general feature of grass endophytes.

There was, however, a considerable difference between the two associations in the dependence of endophyte metabolic activity per tiller on tiller dry weight. In the Lp19 association the endophyte metabolic activity per tiller increased at the same rate as tiller dry weight up to a tiller dry weight of 150 mg (the dry weight of the largest tiller analyzed). In contrast, as mentioned above, in the Lp1 association endophyte metabolic activity per tiller had only increased at the same rate as tiller dry weight up to 50 mg. We have no explanation for this difference.

Endophyte biovolume in the Lp19 association followed a distribution pattern very similar to that of endophyte metabolic activity in that (1) endophyte biovolume per tiller (like endophyte metabolic activity per tiller) increased at the same rate as tiller dry weight, (2) endophyte biovolume was distributed across the different tissue types in the same fashion as endophyte metabolic activity, (3) endophyte biovolume concentration followed basal-apical gradients comparable to those of endophyte metabolic activity concentration, and (4) the sheaths of the youngest mature leaf of a given tiller usually contained more endophyte biovolume than sheaths of older leaves.

The endophyte metabolic state values differed little between tissues. We found some slight differences that were statistically significant. However, we believe that they are of little biological significance. All endophyte metabolic state values *in planta* were similar to those of actively growing liquid cultures of Lp19 but significantly higher than endophyte metabolic state values in late stationary and KCN-treated cultures respectively of Lp19. Note that *in vitro* GUS activities of actively growing cultures varied little in different growth media and temperatures. Because it is difficult to accurately determine hyphal diameter in fixed plant sections and to determine hyphal length per milliliter liquid culture, it is difficult to accurately compare *in vitro* and *in planta* endophyte metabolic state values. However, two different methods of estimation gave very similar results.

The main contribution of the second study to our understanding of the mechanisms that regulate *in planta* endophyte levels is that in the Lp19 association studied this regulation does not seem to include curtailing of the endophyte's metabolism. Apparently a gradient of metabolically active Lp19 hyphae was established in each leaf prior to its maturation, endophyte biovolume concentration increased with leaf maturation, and following leaf maturation little or no additional proliferation of the endophyte mycelium occurred. Nevertheless, the hyphae remained metabolically highly active even in the oldest leaves.

The observation that older sheaths in most tillers contained less endophyte biovolume than younger sheaths could be an indication that hyphae disappear with increasing age of the leaf. However, we found no evidence that hyphae were disappearing. Instead we obtained evidence that on average fewer hyphae were inserted into the first leaf formed by a tiller than into subsequent leaves; the outermost sheath contained fewer hyphae because it was the first sheath that had been produced. This highlights an important point: Not only the age of a leaf itself determines how much endophyte it contains but also the age of the tiller from which it is formed.

There is an important last point we would like to bring to the reader's attention. Lp19-containing tillers behaved sufficiently uniformly in regard to endophyte biovolume and endophyte metabolic activity distribution to allow averaging and statistical analysis. However, individual tillers did not always conform to this average behavior. This indicates that an element of chance is involved in determining the concentration and possibly also the distribution of hyphae in tillers and leaves.

The above conclusions regarding the regulation of endophyte levels, based on biovolume and endophyte metabolic state measurements in Lp19, might also explain those features of the Lp1 metabolic activity distribution that it shares with Lp19. However, we emphasize that they shed no light on the general reduction of endophyte metabolic activity concentration in larger tillers of the Lp1 association. Additional studies will be needed to determine if the latter phenomenon involves changes in endophyte metabolic state or even elimination of endophyte hyphae in existing leaves.

4. HOW THE *IN PLANTA* ENDOPHYTE DISTRIBUTION PATTERN MIGHT ARISE

One of the most fascinating questions to emerge from these experiments is how the observed patterns of *in planta* endophyte metabolic activity distribution and biovolume are generated. One possibility is that the basal-apical gradients reflect the availability of nutrients within the intercellular spaces. The apparent lack of starved or semistarved hyphae in the associations we have studied, as indicated by high endophyte metabolic state values of hyphae regardless of tissue type and age, strongly argues against a restriction of *in planta* endophyte expansion by curtailed nutrient supply. However, hyphae could conceivably stop their extension as soon as they approach areas that have too low a nutrient content to sustain them (using receptors located at the very tip of hyphae). This would result in growth only into areas of adequate nutrient supply. But this idea is not supported by the reported distribution of compounds that would serve as nutrients for the endophyte: *Neotyphodium* endophytes are growing entirely in the intercellular spaces (apoplast), and, as they do not develop specialized absorption structures (e.g., haustoria or arbuscules), it seems likely that they obtain nutrients directly from the apoplast (Hinton and Bacon, 1985). The concentration of photosynthates in the apoplast is assumed to be high enough to allow growth of biotrophic fungi (Hancock and Huisman, 1981; Isaac, 1992) because sugars efflux relatively freely from mesophyll protoplasts (Guy et al., 1980; Huber and Moreland, 1980), and this efflux is essential for the functioning of the plant (phloem loading) (Giaquinta, 1976). The major nutrients translocated from host to fungus are sugars (Manners and Gay, 1983; Smith et al., 1969) and the main transport carbohydrate in the phloem is sucrose (Geiger, 1975; Mohr and Schopfer, 1992). We were not able to find data on the distribution of sugars in perennial ryegrasses but in tall fescue in which very large differences in endophyte levels between mature blade and sheath have been reported (Hinton and Bacon, 1985) no marked differences in the distribution of sugars between these two plant parts have been found (Belesky, 1989; Belesky and Fedders, 1996; Richardson et al., 1992). The same holds true for nitrogen, another essential nutrient required by the endophyte (Belesky and Fedders, 1996). We are aware of one older study that described differences in carbohydrate level between blade and sheath. Smith (1973) reported twofold higher concentrations of carbohydrate in the leaf sheath than in the blade; even if this is correct, sink effects (see above) and active uptake mechanisms in the fungal hyphae (Lam et al., 1994) would probably still allow growth in the blade at a rate similar to that found in the sheath.

Another mechanism controlling the distribution of endophyte was proposed by Hinton and Bacon to explain the absence of hyphae within leaf blades of the *N. coenophialum*/tall fescue associations they examined (Hinton and Bacon, 1985). They suggested that the ligule (or perhaps more correctly the zone of tissue in the leaf associated with the ligule) could serve as a barrier preventing hyphae growing up through the sheath into the blade. In their model, hyphae are unable to grow up from the sheath into the blade due to an absence of a continuous network of intercellular spaces in the dense layer of thick-walled parenchyma cells that form above the position of the ligule toward the end of leaf development. However, this cannot apply to the plants infected with the two transformed endophyte strains we investigated. In these associations the mature blades were colonized, and the endophyte was also present in the top parts of immature leaves (Herd et al., 1997; Tan et al., 1999), i.e., the endophyte had not only reached the blade but had done so before the ligular zone had fully differentiated, forming a thick-walled layer of cells. Note that colonization of ligules by *N. coenophialum*, other *Neotyphodium* species, and *Epichloe festucae* strains has recently been observed in tall fescue and other grasses, but only when the blades are colonized (Christensen, 1997).

Our observations on the two transformed endophytes thus indicate that early colonization of immature leaves can occur and that it may be a prerequisite for blade colonization. On this basis, we can arrive at an alternative hypothesis to explain the endophyte distributions we have observed as well as those observed by Hinton and Bacon.

In order to discuss this hypothesis we must briefly address how tillers and leaves of perennial ryegrasses are formed. Tillers arise from small meristematic regions of shoot apices, the axillary buds, which expand to become new shoot apices from whose margins the leaves (and the next generation of tillers) will be formed (Soper & Mitchell, 1956). Leaves develop from the subapical meristematic region of the shoot apex. Initially the whole developing leaf is meristematic tissue. Subsequently, a division zone is established close to the base of the newly forming leaf. This zone (and only this zone) produces new cells throughout leaf development. Because of continuing division in the division zone, cells at its distal end are pushed upward to reach the so-called elongation zone above the division zone. Here the cells elongate without further division. Following elongation, the cells are pushed further upward (as a result of continuing cell division and cell elongation below them) and differentiate after they have left the elongation zone. Initially all cells produced differentiate into blade cells, but eventually the ligule is formed at the base of the leaf and then travels upward through the elongation zone. From this point on, cells below the ligule differentiate into sheath. Finally, cell division at the base of the leaf ceases. After the last cells to be produced have elongated and differentiated, the formation of the leaf is complete; this coincides with the emergence of the ligule above the sheaths of the older leaves and the differentiation of the ligular zone (Skinner and Nelson, 1995; Soper and Mitchell, 1956). Note that as a result of this sequence of events the tip of the blade is the first part of the leaf that is formed, and the bottom of the sheath is formed last.

In order to colonize the leaves, the endophyte must first have entered the developing axillary buds from which the tiller formed. Subsequently, through branching and longitudinal extension, it must establish itself in the developing shoot apex of the tiller. The hyphae must then enter primordial leaves that develop from the subapical region of the shoot apices. Taking into consideration that the cells that will form the blades are the first to emerge from the division zone, the presence of hyphae within leaf blades as observed in our associations and their absence in the associations examined by Hinton and Bacon (1985) is then explainable as a consequence of the rate of penetration by hyphae through the tissues of the shoot apex and into the developing leaf. In the two transformed endophyte associations the division zone will have been colonized at an early stage. As a result, hyphae were already present in the division zone when the cells that were to form the blade were produced and became associated with these; this would lead to colonization of the blade. By contrast, hyphae in the tall fescue tillers examined by Hinton and Bacon would have been slow to invade the developing leaf and would have not reached the division zone in time to associate with the cells which will form the blade, leading to lack of blade colonization.

This simple model appears at first glance to explain why in some associations but not in others the blades are colonized. In addition it also explains the apical-basal gradients in endophyte concentration we have observed. If we assume that the number of hyphae in the division zone continues to increase over time, the lowest concentration of endophyte hyphae should occur in the tip of the blade since it was formed first and the highest concentration in the lower part of the sheath, which was formed last.

However, the intriguing finding that the basal apical gradients of both hyphal concentration and endophyte metabolic activity in both sheaths and blades remained relatively constant after maturation (a period of several weeks) indicates that some additional factors must be involved. As it stands, the model only explains differences in the numbers of hyphae *entering* the leaf during its development. It does not explain why additional proliferation of the hyphae during and after leaf development does not eventually lead to a uniform level of endophyte in all parts of the tiller.

Proliferation of a mycelium usually involves both the longitudinal extension of existing hyphae and the generation of additional extending tips by branching. Both of these mechanisms must therefore be suppressed in sheaths and blades for the gradients to be retained. Branching does indeed seem to be suppressed in leaf sheaths. Hyphae in sheaths of the diverse range of *Neotyphodium* and *Epichloe* endophytes that we have studied are seldom branched and run as parallel strands up through the tissue. We have made considerably fewer observations in blades, but these suggest that here too branching of hyphae is infrequent (Christensen, 1997; Tan et al., 1999). We do not have direct evidence that hyphal apical growth is suppressed in leaf sheaths and blades. However, if all hyphae inserted in a leaf continue to extend after the leaf had matured, the tips of those hyphae that were inserted last would "catch up" with those inserted first. Even in the absence of branching the number of hyphae throughout the sheath and blade would eventually be uniform. The simplest explanation as to why the basal-apical gradients we observed remain stable is therefore that hyphae within leaf sheaths and blades suppress branching and also cease apical growth once the host tissues mature.

In an extended model incorporating these findings and assumptions, only a few hyphae, if any, will be present in the division zone early in the development of the leaf, and thus few hyphae will be present in the tip region. With time, increasing numbers of hyphal tips will enter the meristematic zone from the inner regions of the shoot apex. As a result, more and more hyphal tips are inserted into the developing tissue. If all of these hyphae extend at roughly the same speed as the blade, and in the absence of both further branching and hyphal extension after blade maturation, the number of hyphae present in each section of it will be the number present when this tissue section was formed in the division zone. No increase in the number of hyphae would occur following leaf maturation, resulting in the stable basal-apical gradients of endophyte biomass (biovolume) concentration in the leaves from maturation through to senescence which we observe.

Since endophytes are often slow growers *in vitro*, readers may wonder whether fungal hyphae can extend at a speed that matches that of plant tissue, as required for the model to work. Developing perennial ryegrass leaves increase in length at a rate of 1–5 cm/day (Robson, 1973), equivalent to 7–35 μm/minute. Extension rates of fungal hyphae can reach up to 50 μm/min (Gooday and Trinci, 1980), easily matching this pace. Note that these rates are for mycelia, which, unlike the endophyte mycelium in extending leaves, branch at regular intervals as they expand so that the metabolic efforts of the mycelium must be divided between an exponentially increasing number of tips. Note also that endophyte hyphae might need specific plant signals in order to trigger extension (see below). Their growth rates *in vitro*, i.e., in the absence of such a signal, might therefore give a very misleading impression of the extension rates of which they are capable *in planta*.

If this model is correct then it was not the dense structure of the ligular zone of the tall fescue plants studied that prevented hyphae from colonizing leaf blades (Hinton and Bacon, 1985). Rather, it was that hyphae in the upper portion of the sheath had ceased apical growth when that portion of the sheath matured and hence were not able to pass through the ligular zone into the blade. However, if the cessation of hyphal apical growth does not occur at the same time as the cessation of host tissue expansion, then the ligular zone may also play an (additional) role in restricting colonization of leaf blades in some associations.

The model suggests that windows of opportunity exist during which hyphae can colonize leaf blades. Similar windows of opportunity seem to exist whenever host tissues are formed. This is supported by observations regarding the production of endophyte-free tillers in some natural (Hinton and Bacon, 1985) and novel associations (Christensen, 1995; Christensen, Ball, Bennett and Schardl, 1997), and endophyte-free seed in a novel tall fescue association (Wilson and Easton, 1997). Endophyte-free tillers would result from the failure of hyphae to grow from the meristematic tissues of the subapical region of the shoot apex into axillary buds from which new tillers are produced. Likewise, endophyte-free branches of panicles and endophyte-free embryos would result if hyphae failed to colonize any of the hierarchy of meristems that are involved in the formation of panicle and seeds. It must be noted that the model linking hyphal apical growth to that of actively growing host tissue is not the only one that could explain the window of opportunity; the observed effects could be equally caused by the physical isolation of developing tissues by the production of physical barriers to hyphal extension.

The model provided by us appears to fit the observations regarding the distribution of hyphae within the associations that we and others have studied. However, it also raises several new questions regarding the *in planta* behavior of the endophyte. The first question is why hyphae of *Neotyphodium* and *Epichloe* endophytes [and also hyphae of *Balansia* endophytes when not associated with stromata (Rykard et al., 1985)] are seldom branched in leaves. Our model itself provides a fairly convincing answer to this question (indeed one could consider the rarity of branches as indirect evidence for the model), namely, that branching would be counterproductive during the synchronous expansion of hyphae and plant tissue that we propose. It is the hallmark of filamentous fungi that their longitudinal growth occurs exclusively by addition of material into an extension zone of 2–30 μm length at the hyphal apex (Bartnicki-Garcia, 1973; Gooday and Trinci, 1980; Wessels, 1991). By contrast, as previously noted, both sheath and blade extend through addition of new cells above their basally located meristems. If plant tissue extension and endophyte hypha extension are to occur simultaneously, the entire length of the hypha must continually slide through the intercellular space in which it is located (depicted in Fig. 1; observe how the initially adjacent markers move apart as both organisms extend). As far as we are aware, the direction of endophyte growth *in planta* has never been determined, but without violating the dogma of fungal apical growth it is difficult to explain the colonization timing and pattern unless one assumes that the hyphae grow as shown in Fig. 1. If branches were formed under these conditions, they might be sheared

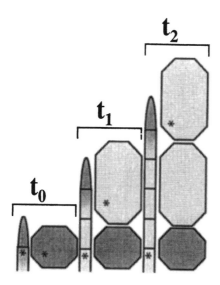

Figure 1 Apical extension of hyphae simultaneous with basal addition and elongation of new plant cells during leaf extension requires the sliding of hyphae in intercellular spaces. Asterisks mark positions in a hypha of a neighboring plant cell which are initially (at t_0) adjacent, but which slide apart over time (t_1-t_2) (see text for details).

off or could even anchor the hypha in the surrounding plant tissue, preventing the sliding motion until the hypha rips apart.

Another concern that Fig. 1 raises is whether even an unbranched hyphae would survive the stress generated by friction due to the sliding motion. This friction can be estimated in two ways. Assuming (1) that both hyphae and surrounding walls are in direct contact with each other, (2) a friction coefficient of 1.0 (the highest friction coefficient possible), (3) that the entire weight of the hypha rests on the surface of the intercellular space (which is an overestimate since the sheath is upright and the blade at an angle), a hyphal diameter of 4 μm (Tan et al., 1998), and a density of 1.1 g/cm^3 (Kubitschek, 1987), the estimated force exerted onto a hypha by friction should be less than 1.1×10^3 N/m^2 (Kuchling, 1986). Alternatively, assuming (1) that the hypha glides in a highly viscous liquid (we have chosen glycerol because it has a viscosity 1500 times that of water and thus likely to vastly exceed that of the apoplast), (2) a liquid film of 0.1 μm thickness surrounding the hypha and a speed of movement of the hypha vs. the surround plant cells of 70 μm/min (twice the maximum extension rate of the leaf), and (3) using the same dimensions of the hyphae as above, the stress generated would be 1.7×10^6 N/m^2 (Kuchling, 1986). Assuming that the tensile strength of a hypha is comparable to that of pine (Kaye and Laby, 1921), hyphae should be able to withstand 2×10^7 N/m^2, still more than 10 times higher than our worst-case assumption of the stress generated by friction.

The third question that needs to be addressed is why the endophyte apparently does not expand its territory after leaf elongation is complete. Our GUS data argue against general nutrient deprivation as a cause since we found that mycelium exhibits still appreciable metabolic activity after the extension of the leaf has stopped (see above). It has also been demonstrated that endophytes possess specific carrier-mediated transport mechanisms for sugars (Lam et al., 1994) that would allow them to accumulate nutrients from the apoplast against a concentration gradient. Furthermore, they possess proteolytic enzymes that are expressed in the mature tissue of the leaf sheath and may allow them to degrade plant proteins in this tissue (Reddy et al., 1996). One of us has also observed excessive branching and proliferation of endophyte mutants in mature plant tissue (Christensen, 1997), indicating that additional growth after maturation is possible.

We would therefore suggest that signals based on the plant cells' developmental state might be involved in determining whether hyphal extension occurs. It is well established that plant cells undergo biochemical changes over time as they become separated from their meristem or once a meristem ceases to function (Penny and Penny, 1978). If endophyte hyphae have evolved mechanisms that would allow them to sense these changes and stop apical extension once the level of these compounds has fallen below a certain threshold, we can explain why extension stops when it does. The hyphae could either sense the reduction of the concentration of meristematic compounds over time in the maturing cells with which their tips are associated or the decrease of these compounds in the meristem as it ceases to function, terminating plant extension. One could even make the additional assumption that the same molecule would regulate both apical extension and branching. High levels in the meristem allow branching and extension, lower levels in the expanding sheath and blade only apical extension, and the lowest levels may allow apical extension at a much slower rate.

The mechanism for stopping growth would also explain the discrepancy between the growth rates of endophytes *in planta* and *in vitro* as a result of a lack of extension-stimulating compounds *in vitro*.

We would add that, although we have rejected the idea that general starvation stops extension, it is possible that extension stops not because the endophyte senses a reduction in the concentration of one or a few specific substances but because these substances are required for extension. For example, Kulkarni and Nielsen (1986) have shown that while *N. coenophialum* can utilize a wide range of C and N sources, absence of a single vitamin in the growth medium, e.g., thiamine, will stop its proliferation.

The control by a chemical signal of hyphal branching and extension might also be relevant to the formation of stromata on which the sexual stage of

Epichloe endophytes is produced. The development of stromata requires a very rapid increase in the number of hyphae. Significantly, stromata are formed only above the top node of reproductive tillers, in close proximity to the developing inflorescence. The process of stromata development commences before extension of the reproductive tiller occurs, when the flowering apex is approximately 2 mm long (Kirby, 1961). Kirby postulated that stromatal development occurred only in the region of the developing inflorescence because only there was a sufficiently large volume of meristematic tissue that was active over a long enough period of time for luxuriant growth to develop. According to our model, such a mass of meristematic tissue would provide a continuous signal for hyphae within it to branch and elongate. Subsequent increases in the availability of nutrients in the region of the developing inflorescence, through changes to the host tissues associated with the developing stromata, and the invasion of vascular bundles (White et al., 1997), may allow rapid uncontrolled growth to continue.

The model we have described explains the distribution of hyphae in two associations between our transformed endophyte strains and perennial ryegrass, and the Hinton and Bacon (1985) findings with tall fescue/*N. coenophialum* associations. However, it is also indirectly supported by another set of data. According to the model, the observed differences between associations in leaf blade colonization, ranging from base to tip colonization through to blades with apparently no hyphae, would (at least partially) be a function of the balance between the expansion of the host tissue and that of the endophyte mycelium in the respective host grass. It is therefore a prediction of the model that the distribution of an endophyte in a given association should depend on both the fungal genotype and that of the host plant. Indeed, several observations made by us and others support this idea.

One example is a study in which seedlings of tall fescue and meadow fescue (*F. pratensis*) were inoculated with a wide range of *Neotyphodium* and *E. festucae* strains. The range of leaf blade colonization observed, ranging from extensive ligule to tip colonization to an apparent absence of hyphae, was influenced by both host and endophyte genotype (Christensen, 1997).

Additional evidence suggesting that endophyte distribution depends in part on the plant genotype comes from observations of dramatic differences in the concentration of hyphae when the same strain is symbiotic with different genotypes of the same cultivar (Herd et al., 1997; Saunders, 1997), with different cultivars of the same species (Christensen et al., 1997), or with different species (Christensen, Ball, Bennett and Schardl, 1997).

The influence of endophyte genotype on colonization is suggested by dramatic differences in colonization patterns when different strains are symbiotic with the same host genotype (Hiatt and Hill, 1997). Endophyte-based mechanisms involved in control are also suggested by observations of the distribution of hyphae in annual ryegrasses infected with *Neotyphodium* endophytes. In natural infections of these grasses the hyphae are largely confined to the tissues of the shoot apex, with few hyphae being found more than just 1 or 2 mm up the sheath. However, when tillers of *Lolium multiflorum* are artificially infected with strains of *N. lolii*, hyphal distribution is similar to that which occurs in the perennial ryegrass hosts of these strains (Latch et al., 1988).

Our model is thus consistent with our observations and a wide range of others regarding endophyte colonization. An appealing feature of this model is its simplicity in that it relies on only one specialized control mechanism: the ability of the endophyte to sense chemical changes that are known to occur as plant cells differentiate. Nevertheless we cannot claim that the existing evidence could not be used to devise other models. However, these would have to take into consideration (1) that colonization occurs (at least in some associations) very early in plant tissue development and (2) that unless the dogma of apical extension is untrue in endophytic fungi the extending tip of the fungal hypha may remain associated largely with the same plant cells as new leaves are formed.

5. HYPHAE REMAIN METABOLICALLY ACTIVE IN MATURE TISSUE: IMPLICATIONS FOR GENE REGULATION AND THE SYNTHESIS OF FUNGAL SECONDARY METABOLITES

Many of the compounds that play a role in enhancing the survival of endophyte-infected grasses are thought to be fungal secondary metabolites. In laboratory culture the cessation of growth usually occurs upon the depletion of nutrients and secondary metabolite production is associated with starvation and a decline in the metabolic activity of the organism. However, the term "secondary metabolism" merely implies that the pathways are not used in active growth. Indeed, our results suggest that *in planta* endophyte secondary metabolite synthesis might actually occur without starvation under conditions where the endophyte can maintain a metabolic rate similar to that during active growth.

As discussed in the preceding sections, in the one endophyte–grass association that we have studied [and which produces secondary metabolites under the conditions at which we have measured the endophyte metabolic state (Spiering and Lane, 1997)]. The endophyte metabolic state coefficient of the mycelium does not decrease noticeably even long after growth has stopped (i.e., the endophyte metabolic state coefficients in immature leaves in which the endophyte must still extend are not substantially different from the endophyte metabolic state coefficients in mature leaves, even when these are several weeks old). This is explicable as far as nutrient supply is concerned, which, as discussed earlier, should be ample. However, it is surprising if one considers the immense demands that apical extension places on the metabolism of the endophyte and that this burden is removed once growth ceases. During extension at the rate of the host tissue, each hypha would need to produce tens or even hundreds of picograms of new biomass (dry matter) per minute. This can be calculated assuming the dimensions and extension rates used earlier, as well as a dry weight content of endophyte biomass of 30% (Woldringh et al., 1993) and a buoyant density of 1.1 g/cm^3 (Kubitschek, 1987). The biomass material not only has to be synthesized but also transported to the tip, requiring the translocation of tens of thousands of vesicles every minute (Gooday and Trinci, 1980).

Once hyphal growth stops, the function of the primary biosynthetic pathways is restricted to maintenance of the existing fungal structures. With the strenuous demands on its metabolism eliminated, one would expect that even an endophyte well supplied with nutrients would now reduce its metabolic rate. However, what we observe is that the metabolism of the hyphae apparently keeps turning over at the same rate as before. Since it supposedly cannot be operating primary biosynthetic pathways at the previous rate any longer, we must assume that the metabolic activity is now directed largely to other processes. It seems likely that a major switch in endophyte gene expression must therefore occur when leaf and endophyte extension are complete, to bring about the change from biomass synthesis to postextension activities.

Some of the processes occurring after extension has stopped would involve the creation of the more complex ultrastructure characteristic of hyphae in mature leaves. Hyphae of *Neotyphodium* and *Epichloe* endophytes within mature leaf sheaths contain conspicuous crystalline inclusions and aggregates of tubules (Fineran et al., 1983; Koga et al., 1993; Siegel et al., 1987). However, nonstructural changes are also occurring as shown by the accumulation of numerous large lipid droplets that are absent or at least rare in immature and recently matured leaves, developing inflorescences and embryos (Koga et al., 1993; Philipson and Christey, 1986; (Hinton and Bacon, 1985).

It is tempting to speculate that the freeing up of resources in the hypha after cessation of apical extension might also play an important role in (or even be a prerequisite for) the synthesis of key endophyte secondary metabolites *in planta*. The resources becoming available are indeed considerable. Based on the calculations for biomass increase during extension, the several hundred hyphae in a leaf sheath should be capable of producing some tens of nanograms per minute of secondary metabolites and storage products. Foremost among these are alkaloids with activity on both mammalian and invertebrate herbivores which confer selective advantages on the grass (see Chap. 14 in this volume). The amounts of these alkaloids produced *in planta*, although they vary widely, can be considerable; loline alkaloids can occur in concentrations exceeding 2000 µg/g dry weight, while the concentrations of peramine, ergovaline and lolitrem B are typically in the range of 1–50 µg/g (Siegel et al., 1990). Another component that has recently generated much interest is a fungal serine proteinase which can comprise 1–2 % of total leaf sheath protein of *Neotyphodium*-infected *Poa ampla* (Lindstrom et al., 1993; Reddy et al., 1996).

Whether the synthesis of a given compound is linked to the cessation of apical growth is hard to determine. It is for instance difficult to measure alkaloid concentrations in very small quantities of immature expanding leaf tissue to determine if the alkaloid production is indeed low in such tissue. Furthermore, the concentration of alkaloids and other fungal products in tissue reflects both their rate of production and the stability and retention of the product within the hyphae or the surrounding plant tissue. However, for some alkaloids it is known that they are very difficult to produce in vitro (see Chap. 14). A simultaneous lack of extension in combination with high metabolic rates as a prerequisite for endophyte secondary metabolite production would provide an explanation for this; laboratory cultures either have high metabolic rates when the hyphae extend or low rates when extension stops.

We may in the future be able to determine when the synthesis of some of these components is initiated *in planta*. Once we have available endophyte genes involved in the synthesis of alkaloids, we will be able to determine whether the initiation of their expression is part of a major switch in endophyte gene expression upon leaf maturation.

6. A POSSIBLE MECHANISM FOR THE EVOLUTION OF A MECHANISM GUIDING THE DISTRIBUTION OF ENDOPHYTES

In the preceding sections we have suggested a model explaining the *in planta* distribution of endophytes and their metabolic state, which implies that today's endophytes possess the ability to alter their growth behavior in response to the type and state of the plant tissues with which they are in contact. The question that needs to be asked is whether and how the necessary biological mechanisms could have evolved. Too little is known about the evolution of the Clavicipitaceae to propose a definite sequence of events. However, using a hypothetical ancestor without such mechanisms, we can at least describe a speculative scenario which demonstrates *why* such mechanisms would have been selected for.

The phylogeny of the Clavicipitaceae, described elsewhere in this book (Chapter 8) positions the genera *Epichloe and Neotyphodium* in one branch of the Clavicipitaceae, separate from the genus *Claviceps* . Thus a common ancestor of the two branches could have been a fungus with traits common to *Epichloe* and *Claviceps* species. Furthermore, it is the prevailing opinion that parasitic fungi have evolved from saprophytes (Heath, 1986, 1997). A conceivable early ancestor in the evolution of the Clavicipitaceae could thus have been a prebio-trophic fungus that had acquired the ability to gain access from the stigmata to adjacent interior structures of the plant (style) without succumbing to host defenses. Being derived from saprophytes, this fungus might originally have simply spread from the stigmata throughout the reproductive organs of the tiller in a more or less radial fashion.

Within the reproductive structures of the plant, the ovule and the areas surrounding it are likely to be most nutrient-rich because the ovule itself behaves "parasitically" within the plant, absorbing nutrients via a well-developed vascular supply (Johri and Ambegaokar, 1984). Furthermore, biochemical and ion $[Ca^{2+}]$ gradients exist between stigma and ovule, which guide pollen tubes toward the ovule (van Went & Willemse, 1984). A parasitic fungus being able to direct its growth from the stigmata along such gradients (or using other plant tissue characteristics) would therefore quickly reach a region where it could very efficiently tap the plant (as is the case with *Claviceps* (Shaw and Mantle, 1980)). It could as a result produce more biomass and reproductive structures than competing parasites, which merely spread radially from the stigmata. Thus mutations in our hypothetical ancestor species should have been selected for that would establish biochemical mechanisms favoring (1) directed extension toward the ovule in response to characteristics of the surrounding plant tissue while (2) reducing branching and lateral growth until the ovular region has been reached. These mechanisms would form the basis of the mechanism that modern endophytes might possess.

Another selective advantage could be gained if mutations occurred that would keep the mycelium far removed from the growing tips metabolically active. This would enhance the ability of nongrowing regions of mycelium in the nutrient-rich area of the base of the ovule to supply the spore-forming reproductive structures, increasing the rate of spore production and enhancing the spread of any strain that developed such an ability. This could be the origin of high endophyte metabolic state values in today's endophytes. However, as pointed out earlier, it is difficult to investigate in the laboratory the behavior of nongrowing hyphae in the presence of adequate nutrients. It is quite conceivable that all filamentous fungi do maintain high metabolic rates in such a situation and that this mechanism is not a special adaptation.

We would emphasize again that we are not suggesting that this sequence of events describes accurately early events in the evolution of the *Clavicipitaceae*, but merely serves as a vehicle to demonstrate *why* mechanisms for sensing the plant environment and restricting radial growth might have been selected for. However, it also shows that the mechanisms that we propose exist for the guidance of today's endophytes might actually be simpler than those of their parasitic ancestors. Spreading from the core of the stem meristem, endophytes may merely need to keep pace with host cell extension and switch branching on and off according to their environment. In contrast, their parasitic ancestors had to find a particular target.

7. NATURAL SELECTION IN ENDOPHYTES: OBSERVATIONS ON THE PROPAGATION OF GENETIC MATERIAL IN ASEXUALLY REPRODUCING ENDOPHYTES

The acquisition of traits that enhance the fitness of an endophyte and the maintenance of traits that assure its survival require selection. Endophytes without a sexual cycle are believed to face a particular problem in this respect which they share with other clonally reproducing organisms. In the absence of segregation and recombination individuals cannot produce progeny with a reduced number of deleterious mutations, which leads to a progressive decline in the fitness of asexually reproducing organisms, a mechanism commonly referred to as Muller's ratchet (Lynch, 1996; Muller, 1964; Schardl, 1996).

This problem is compounded by the fact that mutations are acquired by single nuclei but that the fitness of the endophyte mycelium (and that of the tiller colonized by it) depends on the performance of the sum of all its nuclei. Thus a deleterious or a beneficial mutation in a fungal gene required for successful colonization of the host will, when first acquired by a single nucleus, not affect the mycelium's ability to penetrate the hierarchy of meristems of the tiller. It will be complemented by the surrounding nuclei (which indeed may likewise have other deleterious mutations complemented by other nuclei). Likewise a mutation in a fungal gene that confers benefits determining the survival of the plant host will initially not affect the fitness of the association. Thus in the absence of a segregation process that allows selection for or against a mutation in a single nucleus, a rapid decline in endophyte fitness should occur because in its vegetative propagation the endophyte accumulates increasingly deleterious mutations.

Schardl (1996) has postulated that the problem of accumulating mutations in asexually reproducing endophytes is overcome in two ways. One is the infusion of new genetic material into a clonally reproducing endophyte by hybridization with another endophyte species. The second 'solution' is that asexual endophytes do indeed frequently succumb to an accumulation of deleterious mutations and are then replaced by other endophytes with a sexual cycle (which will eventually likewise lose their ability for sexual recombination and share the fate of their predecessors). We believe that we may have uncovered through the use of the GUS gene a third mechanism that assists asexual endophytes in maintaining and improving fitness. It appears that asexually reproducing endophytes can use clonal segregation to select directly for or against mutations in single nuclei.

We have arrived at this conclusion in our attempts to overcome methodological problems in the use of the strain Lp19 after its genetic modification with the GUS reporter gene. Use of mycelia containing a mixture of GUS-positive and GUS-negative nuclei in the *in planta* studies could lead to misleading results, suggesting lack of expression in parts of tillers because the respective hyphae do not contain the reporter gene. It is therefore necessary that endophyte hyphae in all parts of the tiller are genetically identical as far as the reporter gene is concerned. There are two ways in which mycelia containing a mixture of GUS-positive and GUS-negative nuclei can arise. In the first, protoplasts that were transformed contained several nuclei and not all of these nuclei incorporated the GUS gene into their DNA. In the second, the reporter gene is lost in some nuclei during vegetative growth of the mycelium. Usually transformed fungi are repeatedly subcultured from spores that contain only one or a few nuclei to obtain a genetically homogeneous culture. However, Lp19 does not produce conidia. In some fungi this will make it impossible to generate genetically pure descendants because their extension involves simultaneous division of apical groups of nuclei (Trinci, 1979). However there is a second mode of extension in which the apical compartment of a hypha contains only a single nucleus whose division produces the new apical nucleus (Trinci, 1979). In the latter case every new branch of a mycelium is derived from a single nucleus. Even though mycelia of such mono-karyotic fungi can still be heterokaryons, each type of nucleus will give rise to a homokaryotic sector within the mycelium (Fig. 2a). By simultaneous staining of Lp19 with calcofluor, which reveals septae, and the nuclear stain H33258, we determined that Lp19 only had a single nucleus per apical compartment (Fig. 2b). As a result, we were able to obtain GUS-positive homokaryons from LP19 transformants by subculturing from the edges of colonies (Fig. 2c) (Tan et al., 1999). Actually, the chance of obtaining a genetically heterologous mycelium in transformation appears to be slim, the vast majority ($\geq90\%$) of protoplasts being mononucleate (Saunders, 1997; Spiering and Schmid, 1997). However, after 2 years of vegetative tiller propagation we discovered tillers containing GUS-negative endophyte—presumably due to loss of the GUS gene [the transformant used by us carried only a single copy of the GUS gene (Saunders, 1997), whose loss would immediately confer a GUS-negative phenotype. We assume that the GUS gene was lost because GUS-negative endophytes isolated from plant material showed the same morphology as the GUS-positive transformant and did not gain GUS expression in culture (Spiering et al., 1997)]. At some stage some 10–20% of our tillers contained GUS-negative endophyte. However, we never found tillers that contained a mixture of GUS-positive and GUS-negative hyphae; in every case the tiller contained either exclusively GUS-positive or GUS-negative endophyte. Based on this observation we could overcome the problem by propagating grass from single GUS-positive tillers.

What the rarity of tillers containing a mixture of GUS-positive and GUS-negative hyphae suggests is that the Lp19 endophyte was genetically "purified" when new vegetative tillers were being formed. Two similar observations are relevant to this mechanism. Both horizontal (contagious) and vertical (seed) trans-

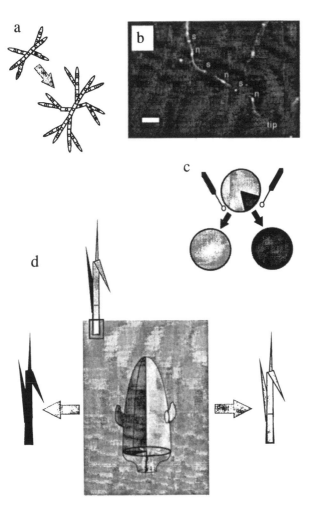

Figure 2 Clonal segregation in endophytes. (a) Schematic view of the formation of homokaryotic sectors in a heterokaryotic mycelium of a monokaryotic fungus. The two different types of nuclei are represented by white and black circles. (b) Lp19 mycelium simultaneously stained with calcofluor and H33258 to visualize septae (s) and nuclei (n) by fluorescence microscopy (bar: 20 μm); (c) Diagram showing how homokaryotic colonies can be obtained by subculturing from the edges of heterokaryotic colonies of transformed Lp19. Sectors in the colonies containing two types of genetically different nuclei are marked by light and dark shading. Note that it is not essential for obtaining homokaryotic colonies that sectors be visually distinguishable. If 1/20 of the circumference of the colony is restreaked for purification, the likelihood of obtaining another sectored colony is ≤0.1 because no more than 2 out of 20 restreaks will contain a sector border. If the sector is very small, the probability of restreaking material containing a sector border is

mission of a pleiotropic *Epichloe* endophyte occurs in *Brachypodium sylvaticum*. The dual modes of transmission have resulted in single plants being infected with more than one strain, as confirmed by allozyme analysis. However, the different allozyme genotypes were isolated from different tillers of the same *B. sylvaticum* grass clump (Leuchtmann, 1997) and dual-infected tillers were not detected. Likewise, dual-infected plants were obtained when *N. coenophialum*-infected tall fescue seedlings were inoculated with *N. lolii* strain Lp19. Although dual-infected tillers were confirmed in three of four plants, these dual-infected tillers were rare, with most tillers containing just one of the two endophytes (Christensen et al., 1999).

How could such segregation come about? We believe that it might be due to a process similar to that which we employed to obtain GUS-positive homokaryons *in vitro*: If from a few hyphal tips in the axillary bud forming a new tiller a mycelium emerges that radiates through the forming apical meristem of the new tiller, the result would resemble a "three-dimensional colony" (Fig. 2d). As a result, most axillary buds formed from this stem meristem would contain endophyte hyphae derived from a single nucleus. Only if the borderline between two sectors came to lie within an axillary bud would it contain a heterogeneous mycelium. Even then sector formation in the tiller formed from this bud would lead to third-generation tillers containing endophyte hyphae derived from a single nucleus of the mycelium present in the first-generation tiller. It might be that, as Fig. 2d suggests, the tip of the dome from which the reproductive structures will be formed would be colonized by mycelium from all founder hyphae. But even if all seeds contained a mixture of the different endophyte genotypes, the vegetative propagation of the plants derived from these seed would accomplish segregation.

We note that while the segregation mechanism suggested by our findings would offer a mechanism for selection against deleterious mutations and in favor of beneficial mutations, further research would be needed to decide whether it would suffice to overcome Muller's ratchet, especially since the small number of founder hyphae would reduce the effective population size of the endophyte.

---

reduced to 1/20. A second round of purification will lead to a homokaryotic colony with a probability of ≥99%. (d) Model of colonization of a stem meristem and clonal segregation in tillers originating from two axillary buds of a tiller colonized by a heterokaryon; in the diagram the mycelium in the stem meristem of the original tiller originates from four founder hyphae, one of which (labeled in black) has acquired a mutation and forms a sector (labeled in dark gray) from which one of the axillary buds is colonized, resulting in a tiller (labeled in dark gray) that contains mycelium homokaryotic for this mutation. A tiller formed from the second axillary bud on the right (labeled light gray) does not contain any hyphae containing the mutation.

That it might be able to do so, however, is suggested by the survival of endophytes in the Southern Hemisphere. Here asexual endophytes cannot be saved from the ratchet or be replaced as suggested by Schardl (1996) because sexually reproducing infective *Epichloe* endophytes are absent in this geographic region (White, 1997). The segregation mechanism we propose would be their only known defense.

8. CONCLUSION

The molecular tools now available open up exciting new avenues to a better understanding of the fascinating and important endophyte–ryegrass symbioses. This may make possible improved agricultural management of endophytes by understanding and influencing the regulation of their metabolic pathways. To be successful, such research must be underpinned by a solid framework of knowledge of the physiological state, distribution, and propagation of ryegrass endophytes *in planta*. This review can serve as a contribution to such a framework, or it may at least convince some of our readers that establishing it is a worthwhile and necessary enterprise.

ACKNOWLEDGMENTS

We thank Neil Pinder and Henning Klank for their assistance in the calculation of friction and Chris Schardl, Michael Lynch, Geoff Lane, Syd Easton, and Barry Scott for helpful comments on the manuscript. M.J.S. is supported by a Massey University Ph.D. scholarship. This work was supported by the New Zealand Foundation for Research Science and Technology.

REFERENCES

Bartnicki-Garcia, S. (1973). Fundamental aspects of hyphal morphogenesis. Paper presented at the 23rd Symposium of the Society for General Microbiology, Imperial College, London, 1973.

Belesky, D. P., Stringer, W. C. and Hill, N. S. (1989). Influence of Endophyte and Water Regime Upon Tall Fescue Accessions. I. Growth Characteristics. *Annals of Botany* 63, 495–503.

Belesky, D. P. and Fedders, J. M. (1996). Does Endophyte Influence Regrowth of Tall Fescue? *Annals of Botany* 78, 499–505.

Christensen, M. J. (1995). Variation in the ability of *Acremonium* endophytes of *Lolium perenne, Festuca arundinacea* and *F. pratensis* to form compatible associations in the three grasses. *Mycological Research* 99, 466–470.

Christensen, M. J. (1997). Unpublished observations.

Christensen, M. J., Ball, O. J.-P., Bennett, R. J. and Schardl, C. L. (1997). Fungal and host genotype effects on compatibility and vascular colonization by *Epichloe festucae*. *Mycological Research* 101, 493–501.

Christensen, M. J., Easton, H. S., Simpson, W. R. and Tapper, B. A. (1998). Growth of fungal endophytes in leaf blades of tall fescue; occurrence and implications for stock health. *New Zealand Journal of Agricultural Research* 41, 595–602.

Christensen, M. J., Leuchtmann, A., Rowan, D. D. and Tapper, B. A. (1993). Taxonomy of *Acremonium* endophytes of tall fescue (*Festuca arundinacea*), meadow fescue (*F. pratensis*), and perennial rye-grass (*Lolium perenne*). *Mycological Research* 97, 1083–1092.

Christensen, M. J., Simpson, W. R. and Al Samarrai, T. (1999). Infection of tall fescue and perennial ryegrass plants by combinations of different *Neotyphodium* endophytes. *Mycological Research.* In press.

Devchand, M. and Gwynne, D. I. (1991). Expression of heterologous proteins in *Aspergillus. Journal of Biotechnology* 17, 3–10.

Dobson, J. M. (1997). MSC thesis, Massey University.

Fineran, B. A., Harvey, I. C. and Ingerfeld, M. (1983). Unusual crystalloid and aggregates of tubules in the *Lolium* endophyte of ryegrass leaf sheaths. *Protoplasma* 117, 17–23.

Geiger, D. R. (1975). Phloem loading. *In* M. H. Zimmermann and J. A. Milburn (ed.), *Encyclopedia of Plant Physiology*, New Series, Vol. 1. Transport in Plants. I. Phloem Transport, Springer-Verlag, Berlin, pp. 395–343.

Giaquinta, R. T. (1976). Evidence for phloem loading from the apoplast. *Plant Physiology*, 57, 872–875.

Gooday, G. W. and Trinci, A. P. J. (1980). Wall structure and biosynthesis in fungi. Symposia of the Society for General Microbiology 30, 207–251.

Guy, M., Reinhold, L. and Rahat, M. (1980). Energisation of the sugar transport mechanism in the plasmalemma of isolated mesophyll protoplasts. *Plant Physiology* 65, 550–553.

Hancock, J. G. and Huisman, O. C. (1981). Nutrient movement in host-pathogen systems. *Annual Review of Phytopathology* 19, 309–331.

Heath, M. C. (1986). Evolution of parasitism in the fungi. Paper presented at the Evolutionary Biology of the Fungi Symposium of the British Mycological Society, University of Bristol.

Heath, M. C. (1997). Evolution of Plant Resistance and Susceptibility to Fungal Parasites. *In* K. Esser and P.A. Lemke (ed.), *The Mycota–A Comprehensive Treatise on Fungi as Experimental Systems for Basic and Applied Research*, Vol. 5. Springer-Verlag, Berlin, pp. 257–276.

Herd, S., Christensen, M. J., Saunders, K., Scott, D. B. and Schmid, J. (1997). Quantitative assessment of *in planta* distribution of metabolic activity and gene expression of an endophytic fungus. *Microbiology* 143, 267–275.

Hiatt, E. E. I. and Hill, N. S. (1997). *Neotyphodium coenophialum* mycelial protein and herbage mass effects on ergot alkaloid concentration in tall fescue. *Journal of Chemical Ecology* 23, 2721–2736.

Hinton, D. M. and Bacon, C. W. (1985). The distribution and ultrastructure of the endophyte of toxic tall fescue. *Canadian Journal of Botany* 63, 36–42.

Huber, S. C. and Moreland, D. E. (1980). Efflux of sugars across the plasmalemma of mesophyll protoplasts. *Plant Physiology* 67, 560–562.

Isaac, S. (1992). *Fungal-Plant Interactions.* Chapman and Hall, London.

Jefferson, R. A. (1987). Assaying chimeric genes in plants: the GUS gene fusion system. *Plant Molecular Biology Reporter* 5, 387–405.

Johri, B. M. and Ambegaokar, K. B. (1984). Embryology: then and now. *In* B. M. Johri (ed.), *Embryology of Angiosperms* Springer-Verlag, Berlin, pp. 1–52.

Jungehülsing, U., Arntz, C., Smit, R. and Tudzynski, P. (1994). The *Claviceps purpurea* glyceraldehyde-3-phosphate dehydrogenase gene: cloning, characterization, and use from the improvement of a dominant selection system. *Current Genetics* 25, 101–106.

Kaye, G. W. C. and Laby, T. H. (1921). *Physical and Chemical Constants and Some Mathematical Functions*, 4th ed. London: Longmans, Green.

Kirby, E. J. M. (1961). Host-parasite relations in the choke disease of grasses. *Transactions of the British Mycological Society* 44, 493–503.

Koga, H., Christensen, M. J., and Bennett, R. J. (1993). Incompatibility of some grass-Acremonium endophyte associations. *Mycological Research* 97, 1237–1244.

Kubitschek, H. E. (1987). Buoyant density variation during the cell cycle in microorganisms. *Critical Reviews in Microbiology* 14, 73–97.

Kuchling, H. (1986). Taschenbuch der Physik. Frankfurt/Main: Verlag Harri Deutsch.

Kulkarni, R. K. and Nielsen, B. D. (1986). Nutritional requirements for growth of a fungus endophyte of tall fescue grass. *Mycologia* 78(5), 781–786.

Lam, C. K., Belanger, F. C., White, J. F. and Daie, J. (1994). Mechanism and rate of sugar uptake by *Acremonium typhinum*, an endophytic fungus infecting *Festuca rubra*: evidence for presence of a cell wall invertase in endophytic fungi. *Mycologia* 86, 408–415.

Latch, G. C. M., Christensen, M. J. and Hickson, R. E. (1988). Endophytes of annual and hybrid ryegrasses. *New Zealand Journal of Agricultural Research* 31, 57–63.

Leuchtmann, A. (1997). Ecological diversity in *Neotyphodium*-infected grasses as influenced by host and fungus characteristics. Paper presented at the Third International Symposium on Neotyphodium/Grass Interactions, Athens, Georgia.

Lindstrom, J. T., Sun, S. and Belanger, F. C. (1993). A novel fungal protease expressed in endophytic infection of *Poa* species. *Plant Physiology* 102, 645–650.

Lynch, M. (1996). Mutation accumulation in transfer RNAs—Molecular Evidence For Muller's ratchet in mitochondrial genomes. *Molecular Biology and Evolution* 13, 209–220.

Manners, J. M. and Gay, J. L. (1983). The Host-Parasite Interface and Nutrient Transfer in Biotrophic Parasitism. *In* J. A. Callow (ed.), *Biochemical Plant Pathology.* John Wiley and Sons Ltd., Chichester, pp. 163–195.

Mohr, H. and Schopfer, P. (1992). *Pflanzenphysiologie*, 4th Ed. Springer-Verlag, Berlin.

Muller, H. J. (1964). The relation of recombination to mutational advance. *Mutation Research* 4, 1–9.

Murray, F. R., Latch, G. C. M. and Scott, D. B. (1992). Surrogate transformation of perennial ryegrass, *Lolium perenne*, using genetically modified *Acremonium* endophyte. *Molecular and General Genetics* 233, 1–9.

Penny, P. and Penny, D. (1978). Rapid response to phytohormones. *In* D. S. Letham, P. B. Goodwin and T. J. V. Higgins (eds.), *Phytohormones and Related Compounds: A Comprehensive Treatise*, Vol. 2. Elsevier/North-Holland, Amsterdam, pp. 537–597.

Philipson, M. N. and Christey, M. C. (1986). The relationship of host and endophyte during flowering, seed formation, and germination of *Lolium perenne*. *New Zealand Journal of Botany* 24, 125–134.

Punt, P. J., Kramer, C., Kuyvenhoven, A., Pouwels, P. H. and van den Hondel, C. A. M. J. J. (1992). An upstream activating sequence from the *Aspergillus nidulans gpdA* gene. *Gene* 120, 67–73.

Reddy, P. V., Lam, C. K. and Belanger, F. C. (1996). Mutualistic fungal endophytes express a proteinase that is homologous to proteases suspected to be important in fungal pathogenicity. *Plant Physiology* 111, 1209–1218.

Richardson, M. D., Chapman, G. W., Hoveland, C. S. and Bacon, C. W. (1992). Sugar alcohols in endophyte infected tall fescue under drought. *Crop Science* 32, 1060–1061.

Robson, M. J. (1973). The growth and development of stimulated swards of perennial ryegrass. I. Leaf growth and dry weight change as related to the ceiling yield of a seedling sward. *Annals of Botany* 37, 487–500.

Rykard, D. M., Bacon, C. W. and Luttrell, E. S. (1985). Host Relations of *Myriogenospora atramentosa* and *Balansia epichloe* (Clavicipitaceae). *Phytopathology* 75, 950–956.

Saunders, K. (1997). M.Sc. thesis, Massey University, New Zealand.

Schardl, C. L. (1996). *Epichloe* species: fungal symbionts of grasses. *Annual Reviews in Phytopathology* 34, 109–130.

Schardl, C. L. and An, Z. (1993). Molecular biology and genetics of protective fungal endophytes of grasses. *In* J. K. Setlow (ed.), *Genetic Engineering*, Vol. 15. Plenum Press, New York, pp. 191–212.

Scott, B. and Schardl, C. (1993). Fungal symbionts of grasses: evolutionary insights and agricultural potential. *Trends in Microbiology* 1, 196–200.

Scott, D. B., Saunders, K., Schmid, J. and Herd, S. (1995). Unpublished observations.

Shaw, B. I. and Mantle, P. G. (1980). Host infection of *Claviceps purpurea*. *Transactions of the British Mycological Society* 75, 77–90.

Siegel, M. R., Jarlfors, U., Latch, G. C. M. and Johnson, M. C. (1987). Ultrastructure of *Acremonium coenophialum, Acremonium lolii* and *Epichloe typhina* endophyte in host and nonhost *Festuca* and *Lolium* species of grasses. *Canadian Journal of Botany* 65, 2357–2367.

Siegel, M. R., Latch, G. C. M., Bush, L. P., Fannin, F. F., Rowan, D. D., Tapper, B. A., Bacon, C. W. and Johnson, M. C. (1990). Fungal endophyte-infected grasses: alkaloid accumulation and aphid response. *Journal of Chemical Ecology* 16, 3301–3315.

Skinner, R. H. and Nelson, C. J. (1995). Elongation of the grass leaf and its relationship to the phyllochron. *Crop Science* 35, 4–10.

Smith, D. (1973). Distribution of dry matter and chemical constituents among the plant parts of six temperate-origin forage grasses of early anthesis (Rep. R2552): Univ. Wis. Coll. Agric. Life Sci. Res. Div. Res.

Smith, D., Muscatine, L. and Lewis, D. (1969). Carbohydrate movement from autotrophs to heterotrophs in parasitic and mutualistic symbiosis. *Biological Reviews* 44, 17–90.

Soper, K. and Mitchell, K. J. (1956). The developmental anatomy of perennial ryegrass (*Lolium perenne* L.). *New Zealand Journal of Science and Technology* 37, 484–504.

Spiering, M. J., Christensen, M. J., Tan, Y. Y. and Schmid, J. (1997). Unpublished observations.

Spiering, M. J. and Lane, G. A. (1997). Unpublished observations.

Spiering, M. J. and Schmid, J. (1997). Unpublished observations.

Tan, Y. Y., Spiering, M., Christensen, M. J., Saunders, K. and Schmid, J. (1997). *In planta* metabolic state of *Neotyphodium* endophyte mycelium assessed through use of the GUS reporter gene in combination with hyphal enumeration. *In* C. W. Bacon and N. S. Hill (eds.), *Neotyphodium/Grass Interactions*. Plenum Press, New York, pp. 85–87.

Tan, Y. Y., Spiering, M., Christensen, M. J. and Schmid, J. (1999). Metabolic state of *Neotyphodium* endophyte mycelium in planta assessed using the GUS reporter gene system and hyphal enumeration, in prep).

Trinci, A. P. J. (1979). The duplication cycle and branching in fungi. *In* J. H. Burnett and A. P. J. Trinci (eds.), *Fungal Walls and Hyphal Growth*. Cambridge University Press, New York, pp. 319–358.

Tsai, H.-F., Wang, H., Gebler, J. C., Poulter, C. D. and Schardl, C. L. (1995). The *Claviceps purpurea* gene encoding dimethylallyltryptophan synthase, the committed step for ergot alkaloid biosynthesis. *Biochemical and Biophysical Research Communications* 216, 119–125.

van Went, J. L. and Willemse, T. M. (1984). Fertilization. *In* B. M. Johri (ed.), *Embryology of Angiosperms*. Springer-Verlag, Berlin, pp. 273–317.

Wessels, J. G. H. (1991). Fungal growth and development: a molecular perspective. *In* D. L. Hawksworth (ed.), Frontiers in Mycology, C.A.B. International, Regensburg, pp. 27–47.

White, J. F., Jr. (1997). Systematics of the graminocolous Clavicipitaceae. Paper presented at the Neotyphodium/Grass Interactions, Athens, Georgia.

White, J. F. Jr., Bacon, C. W. and Hinton, D. M. (1997). Modifications of host cells and tissues by the biotrophic endophyte *Epichloe amarillans* (Clavicipitaceae; Ascomycotina). *Canadian Journal of Botany 75*, 1061–1069.

Wilson, S. M. and Easton, H. S. (1997). Seed transmission of an exotic endophyte in tall fescue. Paper presented at the Third international Symposium on Neotyphodium/Grass Interactions, Athens, Georgia.

Woldringh, C. L., Huls, P. G. and Vischer, N. O. (1993). Volume growth of daughter and parent cells during the cell cycle of *Saccharomyces cerevisiae* a/alpha as determined by image cytometry. *Journal of Bacteriology* 175, 3174–81.

Young, C., Itoh, Y., Johnson, R., Garthwaite, I., Miles, C. O., Munday-Finch, S. C. and Scott, B. (1998). Paxilline negative mutants of *Penicillium paxilli* generated by heterologous and homologous plasmid integration. *Current Genetics* 33, 368–377.

13

Alkaloids of Endophyte-Infected Grasses: Defense Chemicals or Biological Anomalies?

Michael D. Richardson
University of Arkansas, Fayetteville, Arkansas

1. INTRODUCTION

Symbiotic relationships have no doubt existed for eons. In fact, it would seem almost impossible to identify a biological system in its native habitat that does not participate in some form of symbiosis. This inclusiveness is only possible because the concept of symbiosis, as originally defined by de Bary (Smith and Douglas, 1987), is very broad in scope, encompassing many different forms of interaction between two organisms. Symbiotic associations include those that are pathogenic, parasitic, mutualistic, commensalistic, and numerous overlaps between these delimiters. Although often debated, de Bary's definition has stood the test of time and is now accepted by most students in this field. Based on this broadly accepted definition, it could then be argued that the description of an in situ biological system that in no way interacts with another would seem quite illogical. Therefore, the discussion and research on symbiotic systems transcend all biological disciplines.

Although it is tempting to engage in a discussion that includes the many forms of symbiotic associations, the task at hand requires focus upon a group of symbiotic organisms that are generally characterized as mutualistic. These relationships involve a group of Ascomycete fungi (Family Clavicipitaceae, Tribe Balansieae) that form an endophytic infection in plant species from the Gramineae (Poaceae) family (Clay, 1988; Bacon and De Battista, 1990; White, 1988). Although the main focus of this chapter is on a specific group whose members are rarely considered pathogens, other species within this family possess pathogenic characteristics, especially the *Claviceps* (Luttrell, 1980), *Balansia*, and *Epichloe spp*. In fact, many of the Clavicipitaceae continue to exploit either partial or complete epiphytism during their life cycle and are fully capable of functioning as a pathogen (White et al., 1991, 1992).

The ecumenical study of grass–fungal symbiota has led to the popularization and, in some cases, misuse of the term endophyte, which in the strictest sense means "within the plant." The loose adaptation of this term to all stages and examples of these associations has principally evolved because of the complete endophytic behavior of these fungi in the most well-described and economically important endophyte-infected grass species (i.e. *Festuca arundinacea* L. *Schreb* and *Lolium perenne*) (Hinton and Bacon, 1985). However, similar fungal species, such as those that infect the fine-leaf *Festuca* sp. and grasses from the Agrostidae (White, Jr., 1993), are not entirely endophytic and the identification of those symbiotic associations as endophytes must be used with caution. Nonetheless, in order to simplify the current text, the term endophyte will again be broadly used to describe associations that express partial or complete endophytism.

At one end of the family Clavicipitaceae is *Claviceps purpurea*, a widely adapted organism that parasitizes the developing ovary of a grass floret and causes a complete loss of reproduction at the infected site (Luttrell, 1980). *C. purpurea* has been reported to infect most C3 cereal grains, as well as cool-season grasses that are used for forage, amenity or conservation purposes. At the other end of this family are the true endophytes such as *Neotyphodium coenophialum*, *N. lolli*, and *N. typhinum*. These species not only are considered nonpathogenic, but they apparently require very little energy from the host plant (Richardson, Hoveland and Bacon, 1993b), even though they are biotrophic. In addition, ecological studies continue to demonstrate that these organisms improve the fitness of the infected grasses (Hill et al., 1991; Potter et al., 1992; Prestidge et al., 1982; Funk et al., 1983). These symbiotic associations are considered mutualistic.

Clavicipitaceous fungi are commonly associated with grasses (Clay, 1988), with estimates of over 200 grass species serving as hosts for these organisms (Bacon and De Battista, 1990). The evolutionary relationship between these symbiotic forms has been established using molecular techniques (Siegel and Schardl, 1990) and appears to progress from the completely pathogenic *Claviceps sp.* to partial pathogenicity as observed in some *Balansia sp.* (Rykard et al., 1985) to complete mutualism as described for the *Neotyphodium sp.* (Glenn et al., 1996).

When considering grass–endophyte associations that exploit either a partial or complete endophytic habit, these fungi have been further classified based on their reproductive behavior, which is a reflection of their propensity to form a pathogenic or antagonistic relationship (White, 1988). *Type I* associations represent a pathogenic form of this symbiosis in which the host plant is parasitized by the mycobiont during the reproductive phase of development and seed production is either eliminated or greatly reduced. During the initiation and early development of the grass culm, the fungus differentiates into the developing floral meristem and produces an epiphytic stromata around the emerging inflorescence. This epiphytic stage is commonly referred to as "choke" because the inflorescence is aborted and reproduction of the infected tiller is eliminated. *Type I* associations have been routinely described in grasses or sedges infected by *Acremonium typhinum*. *Type II* associations are also characterized by the formation of a stromata on the host inflorescence, but stromata formation is inconsistent over years and is influenced by environmental conditions (Diehl, 1950). The primary mode of reproduction for the fungus in *type II* associations is vertical transmission of the fungus through the seed (White et al., 1992), although horizontal transmission is still possible. The fine-leafed grasses of the *Festuca spp.*, including *Festuca rubra* L. subsp. *Commutata* Gaud., *Festuca rubra* subsp. *rubra*, *Festuca longilofia*, and *Festuca ovina*, commonly contain a type II endophyte classified as *Epichloe festucae* Leuchtmann Schardl Siegel (Leuchtmann et al., 1994). The *type III* associations are the only true endophytes in that the organism completes its entire life cycle within the tissues of the grass and genetic information is vertically transferred through the embryo of the host. The type III associations, which include *Neotyphodium coenophialum* and *N. lolli,* have had the greatest economic impact on grass users and have been studied most extensively to date (Bacon and De Battista, 1990).

Two major research thrusts concerning the mutualistic group of fungi have occurred over the past 20 years. Initial interest in these symbiotic relationships was fueled by the discovery that endophyte-infected forage grasses had devastating effects on grazing animals (Fletcher and Harvey, 1981; Bacon et al., 1977). Animal toxicosis syndromes were widely reported in the scientific literature and descriptions of toxic grasses can also be found in historical records of peoples around the world (J. F. White, Jr., personal communication). Past and ongoing research in this field has shown that livestock performance is severely reduced when animals consume endophyte-infected plant materials (Stuedemann and Hoveland, 1988). These toxic properties are the result of highly bioactive compounds produced by the fungi in association with the host grass (Porter, 1994; Garner, 1984; Lyons et al., 1986; Bush et al., 1993). Because of the widespread animal poisonings that were linked to endophytes, much of the early work in this field was aimed at replacing pastures that had a history of toxicological problems with grasses free of the endophyte.

Research in the field of grass–endophyte symbiota has also been directed at exploiting these organisms for beneficial purposes. As endophytes were removed from grasses to eliminate toxicological problems, it became evident that many of the positive attributes of these grasses, especially those related to persistence and stress tolerance, were a reflection of their endophytic symbiont

(Prestidge et al., 1982; Funk et al., 1983; Arechaveleta et al., 1989). Therefore, forage and especially turfgrass researchers became keenly interested in the ecological value of endophytes (Bacon et al., 1997). Studies with forage and amenity grasses demonstrated that grasses infected by these fungi were more tolerant of many common insect pests (Potter et al., 1992; Johnson-Cicalese and White, 1990), disease (B.B. Clarke, unpublished data), and abiotic stresses (Arechaveleta et al., 1989). Although the mechanisms of drought tolerance and disease resistance in endophyte-infected grasses are far from resolved, increased insect resistance has been associated with many of the same mycotoxins that are detrimental to large herbivores (Clay and Cheplick, 1989; Rowan and Gaynor, 1986). Because of the similarities between insect resistance and animal toxicosis, much of the impetus to study grass endophytes continues to reside in the secondary metabolites produced by these fungi and the effects of these toxins on host ecology.

2. ALKALOIDS OF ENDOPHYTIC FUNGI

Over the relatively short research history of endophyte-infected grasses, more research has focused on the identification and behavior of toxins responsible for animal poisonings than any other facet of these associations. The major toxins associated with endophyte-infected grasses have been alkaloids, including ergot alkaloids, indole-diterpene tremorgens, pyrrolizidine alkaloids, and pyrrolopyrazine alkaloids (Porter, 1994). To date, the ergot alkaloids and a class of indolediterpene tremorgens appear to be most active against mammalian systems (Miles et al., 1992; Strickland et al., 1992) and have been implicated as the causal factors in the most thoroughly described toxicosis syndromes.

Although many subtle structural distinctions can be observed within the ergot alkaloids, a common characteristic of this class is the tetracyclic ergoline ring system. The most important examples of these compounds in endophyte-infected grasses are ergopeptine alkaloids (Lyons et al., 1986; Yates and Powell, 1988), in which the ergoline ring is attached to a specific tripeptide group. In addition to their toxic properties, ergot alkaloids have also been used extensively for therapeutic purposes, ranging from treatment of migraine headaches and parkinsonism, to stimulation of uterine contractions, and various types of cancer therapy (Rehacek and Sajdl, 1990). Other derivatives of ergot alkaloids, especially lysergic acid diethylamide (LSD), have been widely abused as recreational hallucinogens, while the clavine alkaloids (Porter et al., 1979) are important intermediates in the biosynthesis of lysergic acid. Although the clavines have recently received some attention because of their reported antiprolactin effects, these compounds have been considered of minor importance in the toxicosis of endophyte-infected grasses.

Ergot alkaloids appear to be the most common alkaloids associated with the

Clavicipitaceae. Not only are these compounds a primary secondary metabolite of the *Claviceps* sp., they are widely distributed in grasses infected by *Balansia*, *Epichloe*, and *Neotyphodium* spp. endophytes (Siegel et al., 1990; Porter, 1994; Lyons et al., 1986; Porter et al., 1981). Ergot alkaloids have been identified in at least 15 different grass genera found throughout the world and in almost every case where livestock poisoning has been associated with endophyte-infected grasses, ergot-type alkaloids are found.

Although the ergot alkaloids are considered economically important due to livestock toxicities (Garner, 1984), the importance of these compounds to ecological fitness of the host grass has not been well defined. The few studies that have investigated their effects on invertebrates have shown these compounds to possess lower levels of toxicity or deterrence against insects compared to other alkaloids in the symbiotum (Clay and Cheplick, 1989; Siegel et al., 1990). However, there is evidence that synergistic effects between the different classes of alkaloids may occur (Yates et al., 1989) and, therefore, the ergopeptine alkaloids may have limited insecticidal or fungicidal properties in conjunction with other compounds. This possibility needs further study.

Another group of alkaloids that have been widely associated with livestock poisonings are a class of indole diterpenes referred to as lolitrems (Miles et al., 1994). Unlike the vasoconstrictive ergot alkaloids, the lolitrems have tremorgenic activity against mammals and are generally associated with a disorder in perennial ryegrass called "staggers" (Fletcher and Harvey, 1981). The most biologically active of these compounds appears to be lolitrem B, although paxilline, a biosynthetic precursor, can also produce tremorgenic responses (Miles et al., 1992). Although these compounds were originally thought to be primarily involved in host resistance to the Argentine stem weevil, the presence of other deterrents such as peramine (see below) in this symbiosis has led to the conclusion that the lolitrems may also express limited insecticidal properties.

Pyrrolizidine (loline) alkaloids have also been routinely described in endophyte-infected grasses and are found in higher concentrations than any other alkaloid in the symbiosis (Bush et al., 1993). Nonetheless, their biological significance is probably the least understood of all the alkaloids. Over the first 20 years of research into these symbiotic grasses, it was clear that the lolines were a product of the symbiosis between endophytes and grasses. However, recent evidence has finally ascribed a fungal origin to these compounds, as they were produced by *N. coenophialum* in axenic culture (C. S. Schardl, personal communication). The loline alkaloids were initially implicated as a major factor in fescue toxicosis syndromes, but accumulating evidence would suggest that these compounds may play a very minor role in toxicosis. Endophyte-infected *Festuca pratensis*, which contains only the loline-type alkaloids, has been shown to possess few or no toxic effects (Porter, 1994). While apparently nontoxic to mammals, the loline alkaloids are believed to play a role in insect-deterring properties of these grasses.

In vitro studies using plant extracts have associated a high level of feeding deterrence to extracts containing these alkaloids (Johnson et al., 1985; Yates et al., 1989). In addition, deterrence of aphids was commonly associated with grasses containing high levels of these compounds (Siegel et al., 1990). Unfortunately, the lolines have proven to be difficult compounds to synthesize (Petroski et al., 1989) and large quantities of this material for controlled feeding trials have been limited.

The final class of alkaloid that has been widely studied in many endophyte-infected grasses is a guanidinium-containing alkaloid referred to as peramine. Peramine has been identified both in laboratory cultures of endophytes (Rowan, 1993) and in a wide range of endophyte-infected grasses (Siegel et al., 1990). Using a bioassay-directed fractionation procedure with the Argentine stem weevil as the target organism, peramine was first isolated from endophyte-infected perennial ryegrass and identified as an insect feeding deterrent (Rowan and Gaynor, 1986). Although the activity of this alkaloid against other insects and animals is largely unknown, peramine is believed to be primarily associated with insect deterrence of endophyte-infected grasses.

3. A DEFENSIVE MUTUALISM

The most common hypothesis regarding the function of endophyte-produced alkaloids is that these compounds are defense chemicals that provide an adaptive advantage to the endophyte-infected host grass (Clay, 1988; Siegel and Schardl, 1990). It has been argued that endophyte-produced mycotoxins deter mammalian (Hoveland et al., 1983), avian (Daub and Briggs, 1983), and insect herbivores (Clay and Cheplick, 1989; Prestidge et al., 1982), thus enhancing the fitness of the infected host or promoting shifts in natural grass populations toward higher endophyte levels (Clay, 1988). The advocacy of a ''defense theory'' is not surprising, as a predominant idea about secondary metabolites in many organisms is that they are defensive in nature. However, the absolute function of most secondary metabolites is still very much open for discussion.

Early research into the general field of secondary metabolites was focused on the many potent antibiotics produced by bacteria and fungi. It was logically assumed that the antibiotic properties of these compounds were ecologically relevant to securing a habitat for the producing organism. However, the production of microbial antibiotics was expressed primarily under laboratory conditions and the toxins were not found at effective levels when the organism was returned to its environment (Ciegler, 1983). Furthermore, many of the most potent, naturally occurring mycotoxins have few or no antibiotic effects against organisms considered to be natural competitors of the producing organism. For example, aflatoxins, which are common second-degree metabolites of the *Aspergillus* sp., produce devastating effects on macro-organisms, yet have little or no antibiotic activity when tested against microbiologicals (Ciegler, 1983). Since the primary competitors of a mycotoxin-producing *Aspergillus* sp. would be other microbes, it would seem logical that these compounds would function to deter competing microbes and enhance the ability of the fungus to colonize substrates. The limited antibiotic activity of aflatoxin against other microbes would suggest that the metabolites must play other important biological roles in the producing organism and their toxigenic effects are only emphasized due to the inconveniences they cause to humans. Although a large percentage of potent mycotoxins fall into this category of "minimal activity," it is noteworthy that some important mycotoxins, such as ochratoxin, are produced at lethal levels in nature and have broad-spectrum antibiotic properties (Ciegler, 1983).

In drawing an analogy to the *Aspergillus* system, I would like to address a specific question for the remainder of this chapter: Are the alkaloids of endo-phyte-infected grasses truly defense chemicals or do they also have other important biological roles in these systems? Is there overwhelming evidence that these compounds are ecologically important for the sustenance of the symbiosis? Do they play a role in host–fungal interaction, i.e., pathogenesis? Are their toxicological properties mere happenstance? Unfortunately, data to support or refute these questions are sorely lacking and much of the ensuing discussion is based on speculation derived from evidence in the field.

4. ALKALOIDS, ECOLOGY, AND EVOLUTION

The Clavicipitaceae are a highly diverse group of fungi (White, 1994). However, the evolutionary span of this group from the pathogenic *Claviceps* sp. to the mutualistic *Neotyphodium sp.* presents a relatively smooth continuum of symbiotic forms, from pathogenic to antagonistic to mutualistic. This transition would suggest a microevolutionary model for these organisms. From a biochemical standpoint, a microevolutionary model may also be synthesized, as alkaloids such as the ergot class are found across a range of symbiotic grasses (TePaske and Powell, 1993). Other bioactive chemicals, such as peramine and the loline alkaloids, are also common across many species (TePaske and Powell, 1993; Siegel et al., 1990). This conservation of specific secondary metabolites in endophyte-infected grasses clearly suggests an important role for the alkaloids, whether in the biology of the fungus or the ecology of endophytic associations.

As mentioned, the most advocated function of secondary metabolites of endophytic fungi is that they produce an ecological advantage for the host. This theory is supported by a large body of experimental evidence that demonstrates the insect-deterring properties of the endophyte-produced alkaloids, peramine and loline (Rowan and Gaynor, 1986; Siegel et al., 1990). However, the ergot alkaloids, which are probably more highly conserved across the entire family than any other compound, do not necessarily fit an ecological model in which these compounds were conserved for their benefits to the host. If these compounds have truly made an impact on the long-term selection pressures on the host, several basic criteria should be met. First, the compounds should be present in sufficient quantities within the biological system to elicit the desired effects (i.e., resistance against herbivores). Second, the presence of these compounds should be high during those times when the organism is most susceptible to selection pressure. Finally, the biological range of activity should be diverse, as a range of predators are capable of eliminating the host. Present data would suggest that the ergot alkaloids fail to meet these criteria.

The content of ergot alkaloids in endophyte-infected grass tissues generally ranges from 1–2 µg/g in leaf tissues up to 10 µg/g in seeds (Rottinghaus et al., 1991). The few studies that have tested insecticidal properties of ergot alkaloids have clearly shown that typical alkaloid levels in grass leaves would be ineffective against the most potentially damaging invertebrates. For example, fall armyworm (*Spodopter frugiperda*), a nonselective forager that is capable of decimating grass stands, was unaffected by ergopeptine and clavine alkaloids until concentrations were two to three orders of magnitude higher than plant levels (Clay and Cheplick, 1989). Similar data were reported for ergocryptine and ergonovine using the large milkweed bug (*Oncopeltus faciatus*) as the test organism (Yates et al., 1989). The combined results of these studies would suggest that ergot alkaloids are not broad enough in their activity to be effective as an insect deterrent or toxin. In addition, seasonal fluctuations in ergot alkaloids (Agee and Hill, 1994) would create scenarios where alkaloids were present in very low concentrations, further reducing their effectiveness.

Large grazers such as cattle experience reduced forage intake when exposed to endophyte-infected grasses (Schmidt et al., 1982; Aldrich et al., 1993), suggesting a deterrence mechanism against mammals. This reduction in intake has also been substantiated using laboratory animals (Neal and Schmidt, 1985). Although these observations would suggest that ergot alkaloids may satisfactorily deter macroherbivores, for the alkaloid to be considered an evolutionary asset to the host grass one must assume that those herbivores that are deterred have placed a significant selection pressure on the grasses in their natural habitat. Prior to the domestication of livestock some 9000 years ago, it is unclear as to whether macroherbivores actually placed severe selection pressure on grasses in regions of the world where many of the endophyte-infected grasses evolved. In general, morphological features of the grass plant, especially the rhizomatous growth habit and protected growing point, are considered most important to its persistence under grazing, drought, and fire. Therefore, endophyte infection may have played a very minor role in the long-term survival of a grass in its natural environment, if its primary purpose was to deter large herbivores.

The survival and evolutionary shifts of grasses in an undisturbed ecosystem are probably more affected by microherbivores such as insects and nematodes and abiotic stresses such as drought than to large grazers. Under this type of selection pressure, the presence of antibiotic secondary metabolites would certainly play an additive role in the long-term evolutionary shift toward plants that contain the symbiosis. However, many of the primary alkaloids identified in endophyte-infected grasses have shown very low levels of antibiotic activity when tested against target organisms and none of the alkaloids have proven to have a wide enough range of activity to completely protect the plant against all predators.

The direct and indirect effects of endophyte infection on drought survival of the host have been well documented (West et al., 1993) and the potential role of alkaloids in these phenomenon are certainly worth noting. Under conditions where micro- or macroherbivores might selectively avoid endophyte-infected plants, those plants sustain a greater leaf area, produce higher levels of photosynthates, and subsequently build a stronger root system. These features could have a significant effect on selection pressures in areas where limited rainfall and high temperatures are the norm during much of the growing season. Under this scenario, an indirect effect of mycotoxin production could shift populations of grasses toward those that contain the endophyte. Reduced nematode feeding would also impact the overall ability of the host to withstand environmental extremes (West et al., 1988), but the activity of alkaloids on specific nematodes is unknown. The direct effects of endophyte on such features as osmotic adjustment and stomatal regulation also indicate that metabolic activity of the host is under direct influence from the endophyte (Elmi et al., 1992). Whether these changes in host metabolism are related to alkaloids or other fungal metabolites is open for debate.

Although the ecological shifts in plant populations due to endophytic infection are well documented, it is also reasonable that the occurrence of endophyte-infected grasses in natural grasslands plays a significant role in the overall scheme of animal ecology, whether through alterations in feeding behavior, productivity, or reproductive efficiency. One of the more intriguing areas would be the effects of endophyte-infected grasses on reproduction, since many facets of animal reproduction are impacted by endophyte-produced compounds (Porter and Thompson, 1992). In ewes, beef heifers, and mares, endophytes significantly reduced conception (Bond et al., 1988), altered gestation times (Putnam et al., 1991), and increased spontaneous abortions (Garrett et al., 1980). Similar responses have been demonstrated in mice, rats, and deer (Zavos et al., 1988; Milne et al., 1990). Male animals are also sensitive to these toxins (Porter and Thompson, 1992), as sperm production, testicular growth, testosterone levels, and overall sex drive are significantly reduced in male animals that consume endophyte-infected tall fescue.

This overall reduction in reproductive efficiency that is associated with endophyte-infected grasses suggests a possible natural mechanism for balancing population size of mammalian herbivores with the rangeland food supply. Selective avoidance of endophyte-infected grasses has been demonstrated (Madej and Clay, 1991). With this in mind, it is reasonable that other herbivores might selectively consume endophyte-free forages over endophyte-infected plants, as long as forage supplies were adequate. However, if endophyte-free plants become limited due to overconsumption, animals would then be forced to consume more endophyte-infected plants. It is plausible that as animals consume higher proportions of endophyte-infected grasses, reproductive capacity would be reduced and herbivore population size would not outpace the resources of the ecosystem. The overall concept that compounds found in the natural vegetation might alter reproductive habits of wild mammals is not without basis. Reproduction in prairie voles (*Microtus ochragaster ochragaster*) of the midwestern United States is triggered by secondary metabolites found in the native grasses (Nelson and Blom, 1993; Berger et al., 1981). Because this metabolite is a component of their major food source, reproduction is initiated only when food supplies are adequate rather than by environmental factors such as photoperiod. However, we are not aware of an instance where reproduction of native herbivores is inhibited by secondary metabolites of the forage.

5. ALKALOIDS AND FUNGAL NUTRITION

An area of endophyte research that remains one of the great mysteries is how the plant and fungus interact to exchange nutrition. The study of the host–fungal interface in this symbiosis has been especially problematic, primarily because endophytes are tightly restricted to the apoplastic spaces of the grass leaf. While the transfer of nutrients between organisms may involve a range of compounds, including amino acids, ions, and secondary metabolites, it is generally assumed that carbohydrate movement is most important in maintaining a symbiosis (Smith et al., 1969).

The present concept of nutrient procurement by endophytic fungi is that it is passively accomplished in that mycelium in the intercellular spaces of plants nonaggressively survive on the few nutrients that may leak from surrounding host cells. Another hypothesis that has been put forward is that the endophyte facilitates carbohydrate movement by causing plant cells to leak (White et al., 1991). Unfortunately, very few experimental data exist to support either theory. In several pathogenic associations between plants and fungi, secondary metabolites are produced that have their primary effects on host plant physiology. For example, species of *Cercospora* are known to produce a metabolite cercosporin that kills host cells in order to make nutrients available to the fungus (Daub and Briggs, 1983). *Alternaria alternata* produces metabolites that interfere with cell membrane function of the host (Namiki et al., 1986), again making nutrients available to the parasite. It is plausible that secondary metabolites produced in a mutualistic symbiosis such as a grass–endophyte association may also play a comparable role in the infection process.

The concept that alkaloids are associated with alterations of host plant physiology is not without basis. One recent hypothesis suggests that ergot alkaloids may function to protect membranes during rapid nutrient flow (Rehacek and Sajdl, 1990). Porter et al. (1985) emphasized the similarity of the indole-containing ergot alkaloids to natural plant auxins, an observation that is supported by the fact that indole auxins are also produced by many species of the Clavicipitaceae (Porter et al., 1985; De Battista et al., 1990; Yue and Richardson, unpublished data). Ergot alkaloids or other fungal metabolites may interact with sugar absorption channels (plasmalemma transport proteins) to cause those channels to leak sugars. Recent studies of apolastic infiltrates from endophyte-infected and endophyte-free *Poa ampla* demonstrated that the apoplastic spaces of an endophyte-infected host were more nutrient-rich than the same tissues of noninfected plants (Table 1). These data would support a membrane leakage theory, but further work is needed to fully understand this phenomenon. It would be especially interesting to look at apoplastic carbohydrate levels under conditions that facilitate increased alkaloid production such as high nitrogen levels and drought.

Table 1 Apoplastic Carbohydrates ($\mu g/g$) of Endophyte-Free and Endophyte-Infected Big Bluegrass (*Poa ampla* Merrill)[a]

Carbohydrate	Endophyte-infected	Endophyte-free
Glucose	94.5*	51.5
Fructose	205.5*	132.2
Sucrose	n.d.	n.d.
Mannitol	trace	n.d.
Arabitol	n.d.	n.d.
Total	300.1*	183.8

[a] Carbohydrates were extracted using an apoplastic infiltration method and analyzed by GC analysis of the trimethylsilyl derivatives.
* Significant difference between infected and free at the 0.05 level of probability.
n.d., not detected.

6. ALKALOIDS AND ENERGY BALANCE

Another popular theory regarding secondary metabolism is that secondary pathways provide a means of dissipating intermediary compounds that accumulate under conditions of limited growth (Bu'Lock, 1980). In endophyte-infected grasses, it is well known that the presence of the fungus can have a significant effect on both nitrogen accumulation and metabolism (Lyons et al., 1990). In addition, alkaloids have been shown to accumulate under environmental conditions that limit plant growth such as drought stress and flooding (Arechaveleta et al., 1992). However, conditions that limit plant growth may not necessarily limit fungal growth, if only for the fact that energy compounds such as carbohydrates would be more available to the fungus during periods when the plant is not completely utilizing photosynthates.

Because the endophyte requires a supply of specific amino acids such as tryptophan to generate alkaloids (Bacon, 1988), fungal alkaloid synthesis may have an indirect effect on host nitrogen metabolism (Bacon and De Battista, 1990). Studies with tall fescue have suggested that the efficiency of nitrogen utilization is altered, and possibly improved, in symbiotic grasses, while other experiments have shown that nitrogen form can significantly influence the host response to nitrogen (Arechaveleta et al., 1992; Lyons et al., 1990). A recent study of *Festuca commutata* revealed that the endophyte can play a significant role in host nitrogen nutrition and have a positive effect on nitrogen use efficiency (Richardson et al., 1999). In that study, endophyte-infected grasses accumulated approximately 10% more biomass per unit of nitrogen than grasses without the infection, suggesting that absorbed nitrogenous compounds were effectively cycled within the plant. Whether this was directly related to alkaloid synthesis or some other nitrogen-fixing pathway has not been determined. Lyons (1990) reported elevated levels of glutamine synthetase activity in endophyte-infected plants and suggested that this increased activity would serve to reassimilate NH_4 compounds in the leaf tissue and effectively reduce the energy required for primary nitrogen assimilation. Lyon's study also demonstrated that although the synthesis of amino acids was enhanced in endophyte-infected plants, there was no evidence that this was a feedback response due to sink strength of the endophyte. This may suggest that alkaloid synthesis is not in itself driving enhanced NH_4 assimilation.

Historically, it has been argued that secondary metabolites may also function as storage compounds (Bu'Lock, 1980). While a storage hypothesis has limited merit, there have been very few studies of any biological system that actually demonstrated the use of secondary metabolites as storage compounds. In fact, many important secondary metabolites, especially those that are plant-derived, are present during periods of the organisms' life cycle, such as seedling development, which would in no way be associated with energy storage (Niemeyer,

1988). Although ergot alkaloids are known to accumulate under conditions that limit plant growth, such as drought and high temperatures (Arechaveleta et al., 1992; Agee and Hill, 1994), there is no evidence under these conditions that fungal growth is equally reduced. In fact, the endophyte of tall fescue has been shown to prefer a slightly osmophilic habit for optimum growth (Richardson, Bacon, Hill and Hinton, 1993). In addition, it would appear that alkaloids would play a very minor role in the overall nitrogen budget of the fungus in comparison to other nitrogen-rich compounds such as protein and amino acids.

7. SUMMARY

It has been proposed that endophytic fungi fill a unique ecological niche within grasses by providing an array of defense chemicals that are conspicuously absent in this large plant family (Clay, 1988). Unfortunately, this view does very little to explain the unmatched success of temperate and tropical grasses that possess neither a symbiotic endophyte nor an arsenal of secondary metabolites. Even within closely related members of the graminae such as the *Poa sp.*, endophyte-free grasses are as abundant as endophyte-infected grasses (White, 1987), suggesting that these important secondary metabolites have done very little to dictate the succession of this large plant family. Some of the most compelling data that supports a true ecological role for secondary metabolites were presented by Levin (1978) in a study of secondary metabolite production in higher plants. Levin found that as one moved from the temperate zones toward the equator, the number of plants containing alkaloids increased, along with an increase in alkaloid content and toxicity. This succession was looked on as an evolutionary need to defend against a more formidable population of predators as plants adapted closer to the equator. If this were the case in grasses, one would predict that endophyte-infected grasses would be much more prevalent in the tropics than in temperate climates. In reality, the presence of endophytic fungi appears to go down as one moves from temperate to tropical climates, with no evidence of increased production of other defensive systems.

Although the mycotoxic effects of the ergot alkaloids are well established, their role in the evolution of Clavicepitacious fungi may be less ecological and more biological than is now considered. It is perhaps an oversimplification to propose that a fungal endosymbiont such as *N. coenophialum* has specifically conserved this group of secondary metabolites merely to deter herbivores from its own food source. These alkaloids would more appropriately fit this role if the compounds could be shown to directly affect the competitive ability of the fungus, a feature that is not apparent based on the current body of literature. The ecological significance of endophytes to host grasses has been well defined and these fungi continue to play an important role in modern agricultural systems.

Whether these endophytes have truly blazed an evolutionary path for their host is still largely up for debate.

REFERENCES

Agee, C. S. and Hill, N. S. (1994). Ergovaline variability in *Acremonium*-infected tall fescue due to environment and plant genotype. *Crop Science* 34, 221–226.

Aldrich, C. G., Rhodes, M. T., Miner, J. L., Kerley, M. S. and Paterson, J. A. (1993). The effects of endophyte-infected tall fescue consumption and use of a dopamine antagonist on intake, digestibility, body temperature, and blood constituents in sheep. *Journal of Animal Science* 71, 158–163.

Arechaveleta, M., Bacon, C. W., Hoveland, C. S. and Radcliffe, D. E. (1989). Effect of the tall fescue endophyte on plant response to environmental stress. *Agronomy Journal* 81, 83–90.

Arechaveleta, M., Bacon, C. W., Plattner, R. D., Hoveland, C. S. and Radcliffe, D. E. (1992). Accumulation of ergopeptide alkaloids in symbiotic tall fescue grown under deficits of soil water and nitrogen fertilizer. *Applied Environmental Microbiology* 58, 857–861.

Bacon, C. W. (1988). Procedure for isolating the endophyte from tall fescue and screening isolates for ergot alkaloids. *Applied Environmental Microbiology* 54, 2615–2618.

Bacon, C. W. and De Battista, J. P. (1990) Endophytic fungi of grasses. *In* D. K. Arora (ed.), *Handbook of Applied Mycology, Vol 1, Soils and Plants.* Marcel Dekker, New York, pp. 231–244.

Bacon, C. W., Porter, J. K., Robbins, J. D. and Luttrell, E. S. (1977). *Epichloe typhina* from toxic tall fescue grasses. *Applied Environmental Microbiology* 34, 576–581.

Bacon, C. W., Richardson, M. D. and White, J. F., Jr. (1997). Modification and uses of endophyte-enhanced turfgrasses: a role for molecular biology. *Crop Science* 37, 1415–1425.

Berger, P. J., Negus, N. C., Sanders, E. H. and Gardner, P. D. (1981). Chemical triggering of reproduction in *Microtus montanus. Science* 214, 69–70.

Bond, J., Lynch, G. P., Bolt, D. J., Hawk, H. W., Jackson, C. and Wall, R. J. (1988). Reproductive performance and lamb weight gains for ewes grazing fungus-infected tall fescue. *Nutrition Report International* 37, 1099–1105.

Bu'Lock, J. D. (1980) Mycotoxins as secondary metabolites. *In* P. S. Steyn (ed.), *The Biosynthesis of Mycotoxins.* Academic Press, New York, pp. 1–17.

Bush, L. P., Fannin, F. F., Siegel, M. R., Dahlman, D. L. and Burton, H. R. (1993). Chemistry, occurrence and biological effects of saturated pyrrolizidine alkaloids associated with endophyte-grass interactions. *Agriculture Ecosystem Environment* 44, 81–102.

Ciegler, A. (1983) Evolution, ecology, and mycotoxins: Some musings. *In* J. W. Bennett and A. Ciegler (eds.), *Secondary Metabolism and Differentiation in Fungi.* Marcel Dekker, New York, pp. 364–375.

Clay, K. (1988). Fungal endophytes of grasses: a defensive mutualism between plants and fungi. *Ecology* 69, 10–16.

Clay, K. and Cheplick, G. P. (1989). Effect of ergot alkaloids from fungal endophyte-infected grasses on fall armyworm (*Spodoptera frugiperda*). *Journal of Chemical Ecology* 15, 169–182.

Daub, M. E. and Briggs, S. P. (1983). Changes in tobacco cell membrane composition and structure caused by cercosporin, toxic compound from *Cercospora nicotianae*, with *Nicotiana tabacum* as the host. *Plant Physiology* 71, 763–766.

De Battista, J. P., Bacon, C. W., Severson, R., Plattner, R. D. and Bouton, J. H. (1990). Indole acetic acid production by the fungal endophyte of tall fescue. *Agronomy Journal* 82, 878–880.

Diehl, W. W. (1950). *Balansia and Balansiae in America*. U.S. Government Printing Office, Washington, DC.

Elmi, A. A., West, C. P., Turner, K. E. and Oosterhuis, D. M. (1992) *Acremonium coenophialum* effects on tall fescue water relations. *In* S. S. Quisenberry and R. E. Joost (eds.), *Proceedings of International Symposium on Acremonium/Grass Interactions. New Orleans, LA,* Elsevier, Amsterdam, pp. 137–140.

Fletcher, L. R. and Harvey, I. C. (1981). An association of *Lolium* endophyte with ryegrass staggers. *New Zealand Veterinary Journal* 29, 185–186.

Funk, C. R., Halisky, P. M., Johnson, M. C., et al. (1983). An endophytic fungus and resistance to sod webworms. *BioTechnology* 1, 189–191.

Garner, G. (1984) Fescue foot: the search for the cause continues. *In*: *Missouri Cattle Backgrounding and Feeding Seminar*. University of Missouri Press, Columbia, p. 62.

Garrett, L. W., Heimann, E. D., Pfander, W. H. and Wilson, L. L. (1980). Reproductive problems of pregnant mares grazing fescue pastures. *Journal of Animal Science* 51, 237.

Glenn, A. E., Bacon, C. W., Price, R. and Hanlin, R. T. (1996). Molecular phylogeny of *Acremonium* and its taxonomic implications. *Mycologia* 88, 369–383.

Hill, N. S., Parrott, W. A. and Pope, D. D. (1991). Ergopeptine alkaloid production by endophyte in a common tall fescue genotype. *Crop Science* 31, 1545–1547.

Hinton, D. M. and Bacon, C. W. (1985). The distribution and ultrastructure of the endophyte of toxic tall fescue. *Canadian Journal of Botany* 63, 36–42.

Hoveland, C. S., Schmidt, S. P., King, C. C., Jr., et al. (1983). Steer performance and association of *Acremonium coenophialum* fungal endophyte of tall fescue pasture. *Agronomy Journal* 75, 821–824.

Johnson-Cicalese, J. M. and White, R. H. (1990). Effect of *Acremonium* endophytes on four species of billbugs found on New Jersey turfgrasses. *Journal of the American Society of Horticultural Science* 115, 602–604.

Johnson, M. C., Dahlman, D. L., Siegel, M. R., et al. (1985). Insect feeding deterrents in endophyte-infected tall fescue. *Applied Environmental Microbiology* 49(3), 568–571.

Leuchtmann, A., Schardl, C. L. and Siegel, M. R. (1994). Sexual compatibility and taxonomy of a new species of *Epichloe* symbiotic with fine fescue grasses. *Mycologia* 86, 802–812.

Levin, D. A. and York, B. M. Jr. (1978). The toxicity of plant alkaloids: an ecogeographic perspective. *Biochemistry Systematic Ecology* 6, 61–76.

Luttrell, E. S. (1980). Host-parasite relationship and development of the ergot sclerotium in *Claviceps purpurea*. *Canadian Journal of Botany* 58, 942–958.

Lyons, P. C., Evans, J. J. and Bacon, C. W. (1990). Effects of the fungal endophyte *Acremonium coenophialum* on nitrogen accumulation and metabolism in tall fescue. *Plant Physiology* 92, 726–732.

Lyons, P. C., Plattner, R. D. and Bacon, C. W. (1986). Occurrence of peptide and clavine ergot alkaloids in tall fescue grass. *Science* 232, 487–489.

Madej, C. W. and Clay, K. (1991). Avian seed preference and weight loss experiments: the role of fungal endophyte-infected tall fescue seeds. *Oecologia* 88, 296–302.

Miles, C. O., Munday-Finch, S. C., Wilkins, A. L., Ede, R. M. and Towers, N. (1994). Large-scale isolation of lolitrem B and structure elucidation of lolitrem E. *Journal of Agriculture and Food Chemistry* 42, 1488–1492.

Miles, C. O., Wilkins, A. L., Gallagher, R. T., Hawkes, A. D., Munday, S. C. and Towers, N. (1992). Synthesis and tremorgenecity of paxitrols and lolitrol: possible biosynthetic precursors of Lolitrem B. *Journal of Agriculture and Food Chemistry* 40, 234–238.

Milne, J. A., Loudon, A. S. I., Sibbald, A. M., Curlewis, J. D. and McNeilly, A. S. (1990). Effects of melatonin and dopamine agonist and antagonist on seasonal changes in voluntary intake, reproductive activity and plasma concentrations of prolactin and tri-iodothyronine in red deer hinds. *Journal of Endocrinology* 125, 241–251.

Namiki, F., Okamoto, H., Katou, K., et al. (1986). Studies on host-specific AF-toxins produced by *Alternaria alternata* strawberry pathotype causing *Alternaria* black spot of strawberry. 5. Effect of toxins on membrane potential of susceptible plants as assessed by electrophysiological methods. *Annals of the Phytopathology Society of Japan* 52, 610–619.

Neal, W. D. and Schmidt, S. P. (1985). Effects of feeding Kentucky-31 tall fescue seed infected with *Acremonium coenophialum* to laboratory rats. *Journal of Animal Science* 61, 603–611.

Nelson, R. J. and Blom, J. M. C. (1993). 6-Methoxy-2-benzoxazolinone and photoperiod: Prenatal and postnatal influences on reproductive development in prairie voles (*Microtus ochrogaster ochrogaster*). *Canadian Journal of Zoology* 71, 776–789.

Niemeyer, H. M. (1988). Hydroxamic acids (4-hydro-1,4-benzoxazin-3-ones), defence chemicals in the Gramineae. *Phytochemistry* 27 , 3349–3358.

Petroski, R. J., Yates, S. G., Weisleder, D. and Powell, R. G. (1989). Isolation, semisynthesis, and NMR spectral studies of loline alkaloids. *Journal of Natural Products* 52, 810–817.

Porter, J. K. (1994). Chemical constituents of grass endophytes. *In* C. W. Bacon and J. F. White, Jr. (eds.), *Biotechnology of Endophytic Fungi*. CRC Press, Boca Raton, FL, pp. 103–123.

Porter, J. K., Bacon, C. W. and Robbins, J. D. (1979). Ergosine, ergosinine, and chanoclavine I from *Epichloe typhina*. *Journal of Agriculture and Food Chemistry* 27, 595–598.

Porter, J. K., Bacon, C. W., Robbins, J. D. and Betowski, D. (1981). Ergot alkaloid identification in *Clavicipitaceae* systemic fungi of pasture grasses. *Journal of Agriculture and Food Chemistry* 29, 653–657.

Porter, J. K., Cutler, H. G., Bacon, C. W., Arrendale, R. F. and Robbins, J. D. (1985). *In vitro* auxin production by *Balansia epichloe*. *Phytochemistry* 24, 1429–1431.

Porter, J. K. and Thompson, F. N., Jr. (1992). Effects of fescue toxicosis on reproduction in livestock. *Journal of Animal Science* 70, 1594–1603.

Potter, D. A., Patterson, C. G. and Redmond, C. T. (1992). Influence of turfgrass species and tall fescue endophyte on feeding ecology of Japanese Beetle and southern masked chafer grubs (Coleopter:Scarabaeidae). *Journal of Economic Entomology* 85, 900–909.

Prestidge, R. A., Pottinger, R. P. and Barker, G. M. (1982). An association of *Lolium* endophyte with ryegrass resistance to Argentine stem weevil. *Weed Pest Control Conference*, 35, 119–122.

Putnam, M. R., Bransby, D. I., Schumacher, J., et al. (1991). Effects of the fungal endophyte *Acremonium coenophialum* in fescue on pregnant mares and foal viability. *American Journal of Veterinary Research* 52, 2071–2081.

Rehacek, Z. and Sajdl, P. (1990). *Ergot Alkaloids*. Academia, Praha.

Richardson, M. D., Bacon, C. W., Hill, N. S. and Hinton, D. M. (1993). Growth and water relations of *Acremonium coenophialum*. *In* G.C. M. Latch (eds.), *Proceedings of 2nd International Symposium on Acremonium/Grass Interactions*. Grasslands Research Centre, Palmerston North, New Zealand.

Richardson, M. D., Hoveland, C. S. and Bacon, C. W. (1993). Photosynthesis and stomatal conductance of symbiotic and nonsymbiotic tall fescue. *Crop Science* 33, 145–149.

Rottinghaus, G. E., Garner, G. B., Cornell, C. N. and Ellis, J. L. (1991). HPLC method for quantitating ergovaline in endophyte-infested tall fescue: seasonal variation of ergovaline levels in stems with leaf sheaths, leaf blades, and seed heads. *Journal of Agriculture and Food Chemistry* 39 , 112–115.

Rowan, D. D. (1993). Lolitrems, paxilline, and peramine: mycotoxins of the ryegrass/ endophyte interaction. *Agriculture Ecosystem Environment* 44, 103–122.

Rowan, D. D. and Gaynor, D. L. (1986). Isolation of feeding deterrents against Argentine stem weevil from ryegrass infected with the endophyte *Acremonium loliae*. *Journal of Chemistry Ecology* 12, 647–658.

Rykard, D. M., Bacon, C. W. and Luttrell, E. S. (1985). Host relations of *Myriogenospora atramentosa* and *Balansia epichloe* (Clavicipitaceae). *Phytopathology* 75, 950–956.

Schmidt, S. P., Hoveland, C. S., Clark, E. M., et al. (1982). Association of an endophytic fungus with fescue toxicity in steers fed Kentucky-31 tall fescue seed or hay. *Journal of Animal Science* 55, 1259–1263.

Siegel, M. R., Latch, G. C. M., Bush, L. P., et al. (1990). Fungal endophyte-infected grasses: alkaloid accumulation and aphid response. *Journal of Chemical Ecology* 16, 3301–3315.

Siegel, M. R. and Schardl, C. L. (1990). Fungal endophytes of grasses: detrimental and beneficial associations. *In* J. H. Andrews and S. S. Hirano (eds.), Microbial Ecology of Leaves, New York, Springer-Verlag, New York, pp. 198–221.

Smith, D. C. and Douglas, A. E. (1987). Introduction: the concept of symbiosis. *In* D. C. Smith and A. E. Douglas (eds.), *The Biology of Symbiosis*. Edward Arnold, London, pp. 1–10.

Smith, D. E., Muscatine, L. and Lewis, D. H. (1969). Carbohydrate movement from autotrophs to heterotrophs in parasitic and mutualistic symbiosis. *Biological Review* 44, 17–90.

Strickland, J. D., Cross, D. L., Jenkins, T. C., Petroski, R. J. and Powell, R. G. (1992).

The effect of alkaloid and seed extracts of endophyte-infected tall fescue on prolactin and secretion in an *in vitro* rat pituitary perfusion system. *Journal of Animal Science* 70, 2779–2786.

Stuedemann, J. A. and Hoveland, C. S. (1988). Fescue endophyte: history and impact on animal agriculture. *Journal Production Agriculture* 1, 39–44.

TePaske, M. R. and Powell, R. G. (1993). Analysis of selected endophyte-infected grasses for the presence of loline-type and ergot-type alkaloids. *Journal of Agriculture and Food Chemistry* 41, 2299–2303.

West, C. P., Izekor, E., Oosterhuis, D. M. and Robbins, R. T. (1988). The effect of *Acremonium coenophialum* on the growth and nematode infestation of tall fescue. *Plant and Soil* 112, 3–6.

West, C. P., Izekor, E., Turner, K. E. and Elmi, A. A. (1993). Endophyte effects on growth and persistence of tall fescue along a water-supply gradient. *Agronomy Journal* 85, 264–270.

White, J. F. (1988). Endophyte-host associations in forage grasses. XI. A proposal concerning origin and evolution. *Mycologia* 80, 442–446.

White, J. F., Breen, J. P. and Morgan-Jones, G. (1991). Substrate utilization in selected *Acremonium*, *Atkinsonella*, and *Balansia* species. *Mycologia* 83, 601–610.

White, J. F., Morgan-Jones, G. and Morrow, A. C. (1992) Taxonomy, life cycle, reproduction, and detection of *Acremonium* endophytes. *In* S. S. Quisenberry and R. E. Joost (eds.), *International Symposium on Acremonium/Grass Interactions, New Orleans, LA.* Elsevier, Amsterdam.

White, J. F., Jr. (1987). Widespread distribution of endophytes in the Poaceae. *Plant Disease* 71, 340–342.

White, J. F., Jr. (1993). Endophyte-host associations in grasses. XIX. A systematic study of some sympatric species of *Epichloe* in England. *Mycologia* 85, 444–455.

White, J. F., Jr. (1994) Taxonomic relationships among the members of the Balansieae (Clavicipitales). *In: Biotechnology of endophytic fungi of grasses*, eds. C. W. Bacon and J. F. White, Jr., Boca Raton, Florida, CRC Press, pp. 3–20.

Yates, S. G., Fenster, J. C. and Bartelt, R. J. (1989). Assay of tall fescue seed extracts, fractions, and alkaloids using the large milkweed bug. *Journal Agriculture and Food Chemistry* 37, 354–357.

Yates, S. G. and Powell, R. G. (1988). Analysis of ergopeptine alkaloids in endophyte-infected tall fescue. *Journal Agriculture and Food Chemistry* 36, 337–340.

Zavos, P. M., Varney, D. R., Jackson, J. A., Hemken, R. W., Siegel, M. R. and Bush, L. P. (1988). Lactation in mice fed endophyte-infected tall fescue seed. *Teriogen* 30, 865–875.

14

Coevolution of Fungal Endophytes with Grasses: The Significance of Secondary Metabolites

Geoffrey A. Lane and Michael J. Christensen
*Grasslands Research Centre, New Zealand Pastoral Agriculture
Research Institute Limited (AgResearch), Palmerston North,
New Zealand*

Christopher O. Miles
*Ruakura Research Centre, New Zealand Pastoral Agriculture
Research Institute Limited (AgResearch), Hamilton, New Zealand*

1. INTRODUCTION

A diverse array of biologically active secondary metabolites is found in grasses infected with *Neotyphodium* fungal endophytes. The significance of these fungal metabolites to pastoral agriculture is due as much to the prevalence of endophyte infection in grass populations and the distribution of the metabolites through plant tissues as to the yield and potency of these compounds. Alkaloids synthesized by *Neotyphodium* species in pasture grasses have been implicated in fescue toxicosis in the southern USA (Lyons et al., 1986; Yates, et al., 1985), ryegrass staggers in New Zealand (Gallagher et al., 1981), other staggers syndromes in Australia, South Africa, and Argentina (Miles et al., 1998; Towers, 1997), and "sleepygrass" syndromes of livestock in the southwestern USA (Petroski et al., 1992) and China (Miles et al., 1996). In each case, the combination of the prevalence of fungal infection, and the distribution and potency of the metabolites appears to be responsible for the impact. The related species of *Claviceps*, although well known for their production of very high concentrations of toxic alkaloids in fungal sclerotia (Flieger et al., 1997; Porter et al., 1987), do not cause problems on the same scale for pastoral agriculture, primarily because the metabolites are not spread so pervasively (in both space and time) through the pasture (Clay, 1988).

It has become clear that the prevalence of *Neotyphodium* endophyte infection in grass populations is at least in part a consequence of the contribution of the secondary metabolites of these fungi to the mutualistic relationship between fungus and host (Clay, 1988). The secondary metabolites, through their potency and distribution in the plant, help defend the host grass against a range of grass herbivores (Clay, 1988; Clement et al., 1994; Dahlman et al., 1991; Popay and Rowan, 1994), and the plant accommodates a highly efficient process of asexual transmission of the fungi, through infection of separating vegetative tillers and of seed. This life cycle is only viable for fungi that provide some advantage to the fitness of their host, as they would otherwise be eliminated by compounding transmission losses between generations (Miles et al., 1998).

While this analysis suggests that secondary metabolism must have been a very significant factor in the evolution of these symbioses, very few studies of endophyte secondary metabolism have focused on the question of how this ecologically effective combination of potency, distribution, and prevalence evolved. What evidence is there from the chemistry about the evolutionary origin of these fungi? What evidence is there that the plant and the fungus have evolved together in the production of secondary metabolites? This chapter attempts to address the present limited state of knowledge on these and related questions, and to identify key issues for future research.

The emphasis is on the secondary metabolism of the ''asymptomatic'' symbioses of the asexual seed-transmitted *Neotyphodium* endophytes within grasses, reflecting the balance of the research effort and their agricultural significance. However, we have attempted to place this in phylogenetic context, presenting the available information on the closely related *Epichloe* ''choke'' fungi, and other fungi in the tribe Balansieae (family Clavicipitaceae) that infect grasses systemically.

2. BIOLOGICALLY ACTIVE ENDOPHYTE SECONDARY METABOLITES

Endophytic *Neotyphodium* fungi synthesize a diverse range of biologically active secondary metabolites, notably alkaloids (Fig. 1). Their chemistry and activity has been extensively reviewed (Porter, 1994; Powell and Petroski, 1992b; Siegel and Bush, 1996, 1997) and is outlined only briefly here.

2.1. Pyrrolizidine Alkaloids

Pyrrolizidine alkaloids of the loline group are common in *Festuca* and annual *Lolium* spp. infected with *Neotyphodium* endophytes (Powell and Petroski,

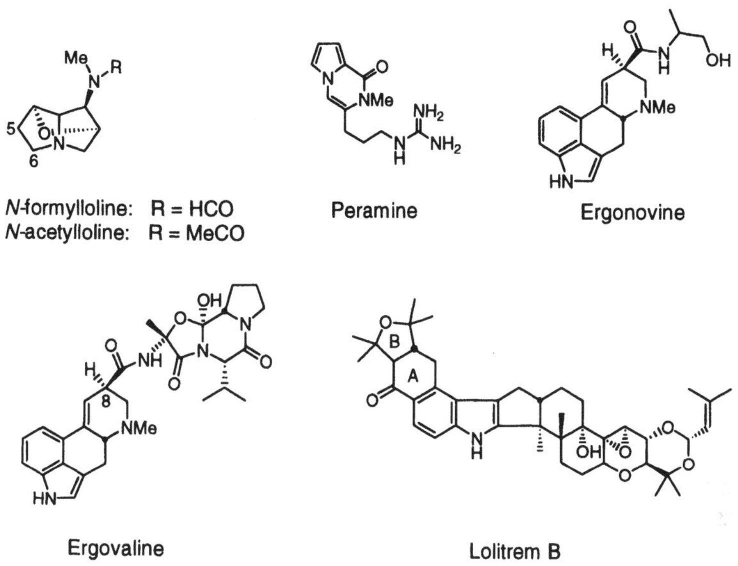

N-formylloline: R = HCO
N-acetylloline: R = MeCO

Peramine

Ergonovine

Ergovaline

Lolitrem B

Figure 1   Major bioactive alkaloids of *Epichloe/Neotyphodium* endophyte–grass associations.

1992a; TePaske et al., 1993). Unlike the common plant pyrrolizidine alkaloids, lolines contain an amino nitrogen substitution on a usually saturated pyrrolizidine ring system with a stable cyclic ether structure. In endophyte-infected grasses they occur with or without methylation and acylation of the amino nitrogen, with *N*-formyl loline and *N*-acetyl loline (Fig. 1) commonly the major components (Bush et al., 1982; Robbins et al., 1972; TePaske et al., 1993). Lolines are the most abundant of the alkaloids of endophyte-infected grasses, being present at concentrations that sometimes exceed 10,000 μg/g (Bush et al., 1993). Recently a novel ring-unsaturated variant, tentatively identified by gas chromatography–mass spectrometry (GC-MS) as 5,6-dehydro-*N*-acetyl loline, has been found in endophyte-infected *F. argentina* (Casabuono and Pomolio, 1997).

Although the significance of reported effects of lolines on grazing ruminants is unclear (Bush et al., 1993; Oliver, 1997), there are extensive data on their activity against insects (Dahlman et al., 1991; Popay and Rowan, 1994; Siegel and Bush, 1997). They may also be a factor in the allelopathic properties of endophyte-infected tall fescue (Bush *et al.*, 1993; Springer, 1996).

2.2. Pyrrolopyrazines

The unusual pyrrolopyrazine peramine (Fig. 1, in protonated form) is widespread in *Epichloe*- and *Neotyphodium*-infected grasses (Siegel et al., 1990). It is highly active as a feeding deterrent to several insect pests (Popay and Rowan, 1994; Prestidge and Ball, 1993; Rowan, 1993; Rowan and Gaynor, 1986) but has not shown toxicity to mammals or plants (Pownall *et al.*, 1995; Rowan, 1993). Although peramine remains the sole compound of this class chemically identified in such symbioses, enzyme-linked immunosorbent assay (ELISA) and chromatographic studies have indicated that a lipophilic analog of peramine is present in endophyte-infected *Achnatherum inebrians* (Stevenson, 1996).

2.3. Ergot Alkaloids

The ergopeptide ergovaline (Fig. 1), together with its 8-epimer ergovalinine— ergot alkaloids based on lysergic acid amide occur as epimeric pairs, and the presence of epimers is assumed subsequently in this discussion—is the major ergot alkaloid present in tall fescue (*Festuca arundinacea*), perennial ryegrass (*Lolium perenne*), and *Hordeum* spp. infected with *Neotyphodium* endophytes (Lyons et al., 1986; Rowan and Shaw, 1987; Siegel et al., 1990; TePaske et al., 1993; Yates et al., 1985). Other ergopeptides (Lyons et al., 1986; Shelby et al., 1997; TePaske et al., 1993), clavine alkaloids (Lyons et al., 1986; Petroski et al., 1992; Porter et al., 1979, 1981), and lysergic acid amide derivatives (Miles et al., 1996; Petroski et al., 1992; Shelby et al., 1997), have also been identified from *Neotyphodium*-infected grasses. Ergovaline is typically present at 0.5–3.0 µg/g concentrations in endophyte-infected tall fescue and ryegrass, but concentrations above 50 µg/g have been observed (Lane et al., 1997b). Lysergic acid amide derivatives predominate in *Neotyphodium*-infected sleepygrass (*Stipa robusta*) (Petroski et al., 1992) and drunken horse grass (*A. inebrians*) (Miles et al., 1996). In the latter, ergonovine (Fig. 1) concentrations exceeding 2000 µg/g were observed.

Endophyte ergopeptide alkaloids appear to be the major factor in fescue toxicosis. While direct experimental evidence of the activity of ergovaline has been limited by scarcity of the compound, the pattern of activity toward mammals appears to be similar to that of other ergot alkaloids (Berde and Schild, 1978). The characteristic symptoms of fescue toxicosis, including elevated body temperatures under external temperature stress, vasoconstriction, and reduced prolactin levels, are all consistent with ergopeptide toxicity (Cunningham, 1949; Maag and Tobiska, 1956; Oliver, 1997; Strickland and Cross, 1993), although recent oral dosing experiments with ergovaline at in planta concentrations did not entirely reproduce the effects of feeding endophyte-infected plant material (Gadberry et al., 1997; Piper et al., 1997). Ergot alkaloids are also toxic to, or inhibit the feeding of, many insects (Ball et al., 1997c; Clay and Cheplick, 1989; Popay and Rowan, 1994; Prestidge and Ball, 1993; Yates et al., 1989).

2.4. Indole–Diterpenoids

The tremorgenic indole–diterpenoid mycotoxin lolitrem B (Fig. 1) was identified initially in *N. lolii*–perennial ryegrass associations (Gallagher et al., 1981, 1982, 1984) and was subsequently found in endophyte-infected *Festuca* spp. (Christensen et al., 1993; Siegel et al., 1990). It is structurally related to the penitrems, janthitrems, paxilline, paspaline, and related mycotoxins (Gallagher et al., 1984; Steyn and Vleggaar, 1985). Lolitrem B appears to be the major factor causing ryegrass staggers in livestock grazing endophyte-infected ryegrass, although direct evidence from oral dosing experiments is currently lacking. Lolitrem B is highly tremorgenic when administered by the intravenous and intraperitoneal routes, inducing tremors lasting up to several days (Munday-Finch et al., 1996b; Towers, 1997). More recently, an extensive range of minor lolitrems and related indole–diterpenoids have been isolated from *N. lolii*-infected perennial ryegrass seed (Gatenby, 1997; Miles et al., 1994; Munday-Finch et al., 1995, 1996b, 1997, 1998). In addition to their tremorgenic effects, lolitrem B and related compounds exhibit other toxicities to mammals (Miles et al., 1992), are toxic or feeding-deterrent to insects (Prestidge and Ball, 1993), modulate calcium-activated potassium channel activity (Knaus et al., 1994), and inhibit acyl-CoA:cholesterol acyltransferase (Huang et al., 1995). Unidentified indole–diterpenoids have been detected by ELISA in endophyte-infected *A. inebrians* (Miles et al., 1996), and in the staggers-causing endophyte-infected grasses *Poa huecu*, *Echinopogon ovatus*, and *Melica decumbens* (Miles et al., 1998; Towers, 1997), as well as in *L. perenne* (Miles et al., 1993).

2.5. Antifungal Compounds

A range of fungitoxic metabolites have been isolated by Koshino and co-workers from stromata of *E. typhina* on timothy *(Phleum pratense)* during investigations of the acquired resistance of these plants to secondary infection by the leaf spot fungus *Cladosporium phlei*. These include a series of sesquiterpene alcohols (chokols A–G) (Koshino et al., 1989a; Yoshihara et al., 1985), oxygenated fatty acid derivatives (Koshino et al., 1987, 1989b), furanones (Koshino et al., 1987, 1992b), dihydropyrones and a benzopyranone (Koshino et al., 1992b). There is evidence of antifungal activity of *Neotyphodium* spp. (Christensen, 1996; Clay, 1997; Siegel and Latch, 1991), but the chemical factors involved have not been elucidated.

2.6. Other Metabolites

Several sterols have been identified in *N. coenophialum*-infected tall fescue and in culture, including ergosterol, ergosterol peroxide, and ergostatetraenone (Davis et al., 1986). Ergosterol, ergosta-5,7,22,24(28)-tetraen-3β-ol, and ergosta-7,22-dien-3β-ol were identified in cultures of *N. lolii* (Miles et al., 1992). Ergosterol has been proposed as an indicator of endophyte biomass in seeds (Richardson and Logendra, 1997). In the chemical examination of stromata of *E. typhina* on timothy, phenolic glycerides (Koshino et al., 1988), an anthrasteroid (Koshino et al., 1989c), and two sphingoid derivatives (Koshino et al., 1992a) were identified. The biological significance of these compounds is not known. In addition, plant growth regulators are produced by some endophytes (De Battista et al., 1990), and their involvement has been invoked to explain some responses of *F. arundinacea* to infection (Joost, 1995).

3. ENDOPHYTE PHYLOGENY AND EVOLUTION

Molecular genetic and biochemical studies have established that asymptomatic asexual endophytic *Neotyphodium* fungi are closely related to sexual *Epichloe* fungi ("chokes") which form stromata on the developing inflorescences of their hosts (Glenn et al., 1996; Schardl, 1996; Schardl and Clay, 1997). *Epichloe* (and *Neotyphodium*) endophytes are included within family Clavicipitaceae (White, 1997) and are closely related to *Claviceps* spp. (Glenn and Bacon, 1997), well known for the production of ergot alkaloids (Flieger et al., 1997) and indole–diterpenoids (Steyn and Vleggaar, 1985).

Fungi within Clavicipitaceae occur in obligate (usually parasitic) biotrophic associations with plants, invertebrates, or fungi (White, 1997). Those in biotrophic associations with grasses (Poaceae) and sedges (Cyperaceae) are in the subfamily Clavicipitoideae, which comprises three tribes. Tribe Clavicipiteae, with the single genus *Claviceps*, comprises fungi which infect florets of grasses and replace the host ovules with mycelium that develops into sclerotia (Tudzynski et al., 1995). Tribe Balansieae includes fungi that form stromata (structures in or on which spores are formed) around the undeveloped inflorescences or on the surface of culms (reproductive tillers) or leaves of grasses and sedges, and from which the sexual stage of the fungus is produced. Five teleomorphic genera are currently recognized in the Balansieae: *Epichloe* (and the derived anamorph *Neotyphodium*), *Balansia*, *Atkinsonella*, *Echinodothis*, and *Myriogenospora*. Tribe Ustilaginoideae comprises biotrophs of bamboos, and form hardened sclerotic stromata on stems, in the single anamorphic genus *Ustilaginoidea*. These phylogenetic relationships, and the close relationships between the asexual *Neotyphodium* fungi and the sexual *Epichloe* species, form the context in which the role of secondary metabolites in the coevolution of the endophyte symbioses with grasses must be considered.

In the sexual *Epichloe* species, reproduction takes place through the production of contagious sexual spores on a stroma formed on the flag leaf associated with the preemergent inflorescence. The formation of the stroma prevents emergence of the inflorescence, sterilizing ("choking") the reproductive tiller. Sexual recombination occurs following the transfer of spermatia from an opposite mating type. Ascospores then disseminate, infecting a new host grass via stigmata (Chung and Schardl, 1997) or through cut leaves (Western and Cavett, 1959). This route of transmission is described as horizontal. The contrasting transmission route of the asexual *Neotyphodium* endophytes is described as vertical. These latter fungi are very efficiently transferred to each new generation of plant genotypes by clonal propagation in the seed (along with the host maternal germ line), without undergoing sexual recombination.

Species with both patterns of transmission also occur, and White (1988) suggested that *Neotyphodium* and *Epichloe* species can be considered as a single complex, with three types distinguished by their mode of transmission. Although host effects can confound this classification (Leuchtmann, 1997), it provides a useful framework for discussion. For type I *Epichloe* endophytes, many of which are included within *E. typhina*, the only route of transmission is horizontal. With these fungi, even if some inflorescences emerge and develop normally to seed, any seedlings established are endophyte-free. With type II *Epichloe* endophytes, dissemination can be either through seed or through infection via ascospores. This has been described as a "balanced" association, in which substantial sexual reproductive capability is retained by both partners (Wilkinson and Schardl, 1997). Dissemination of the type III (*Neotyphodium*) endophytes is entirely via the seed, with the sexual stage absent.

Type I associations are found in a large number of grass species within at least five tribes and occur in five of the nine recognized *Epichloe* mating populations (substantially equivalent to biological species) (Schardl et al., 1997). They are very rare with *Lolium* and *Festuca* spp. Type II associations occur with grass species of at least five tribes and involve *Epichloe* endophytes in five of the mating populations. This type of association has not been reported with *Lolium* spp., and in the genus *Festuca* it appears to be most common for fine fescues (subgenus *Festuca*) infected with *E. festucae*. Type III associations (*Neotyphodium* endophytes) are found with grasses of at least six tribes (Aveneae, Bromeae, Meliceae, Poeae, Stipeae, Triticeae) and are the only type of *Epichloe/Neotyphodium* association reported in native grasses of Australasia (*Echinopogon* spp.). They are widespread in tall fescue, meadow fescue (*Festuca pratensis*) and *Lolium* spp. (Christensen et al., 1993; Latch, et al., 1987), which are important in pastoral production systems in temperate climates.

Molecular genetics and biochemical studies have shown fungi in the *Epichloe/Neotyphodium* complex to be closely related and have provided evidence that the asexual type III *Neotyphodium* endophytes evolved from type I and II sexual *Epichloe* spp. (Schardl and Clay, 1997). Gene sequence analysis and allozyme studies (Schardl et al., 1994; Tsai et al., 1994) indicate that many endophytes are of hybrid origin. Three *Neotyphodium* taxa from tall fescue, including *N. coenophialum*, appear to be interspecific hybrids (Tsai et al., 1994), as does one from perennial ryegrass (Schardl et al., 1994). Hybridization is likely to render the fungus infertile, and unable to propagate sexually. There is no evidence of hybridization for the most common endophyte species in perennial ryegrass, *N. lolii*. A sexual stage has never been observed for *N. lolii*, but gene sequence analyses indicate it to be close to *E. festucae* and it may have originated in infection of perennial ryegrass by this species (Wilkinson and Schardl, 1997).

The close relationship between the asexual *Neotyphodium* endophytes and the *Epichloe* pathogens revealed by these studies suggests a useful perspective for examining the role of secondary metabolism in the evolution of the symbiosis. While the symbiosis is arguably mutualistic, Schardl has suggested an asexual endophyte may be viewed as "a trapped pathogen whose genome and associated functions have been expropriated for the benefit of the host plant" (Schardl and Clay, 1997; Schardl et al., 1991). This perspective usefully directs attention to the genetic apparatus within the fungus coding for the synthesis of a range of biologically active secondary metabolites not otherwise available to the host plant and to how this apparatus may function for the defense of the plant.

It also raises questions about the evolution of secondary metabolism in the symbiosis. The selection pressures on both the captured, vertically transmitted, type III endophyte and its plant host should have been very different from those acting on its horizontally transmitted pathogenic relatives and their hosts, and the secondary metabolism in these associations might be expected to reflect these different pressures.

With these contrasting transmission modes in mind, we review the selection pressures acting on endophyte-infected grasses before addressing the more specific issues of secondary metabolism and coevolution in *Epichloe* and *Neotyphodium* spp. and their relatives.

4. SELECTION PRESSURES IN THE EVOLUTION OF SECONDARY METABOLISM IN ENDOPHYTE–GRASS SYMBIOSES

To understand how secondary metabolism in these symbioses may have been shaped during evolution, it is worth considering in detail the selection pressures on the association and at which stages in the life cycle of the endophyte in the plant they have greatest effect.

A prerequisite for the symbiosis is compatibility (or at least host nonresistance) between the plant and the fungus. Within this framework, the endophyte undergoes selection on two levels: within the plant, in transmission to the next plant generation, and within grass populations, in competition with endophyte-free hosts.

4.1. Endophyte–Host Compatibility

Neotyphodium and *Epichloe* endophytes in natural associations with grasses grow without eliciting any obvious defensive response from the host grass, suggesting prior selection of the endophyte and the plant for compatibility. Asexual endophytes are confined in nature by their maternal transmission. However, natural host-range boundaries can be breached through artificial transfers of endophytes into other host genotypes and related species by seedling inoculation (Christensen, 1995; Latch and Christensen, 1985; Leuchtmann and Clay, 1993), usually without symptoms of fungal pathogenicity or plant resistance.

Seedling inoculation studies have shown that the range of transfer is not unlimited and that requirements for endophyte–host compatibility are stringent. In some transfers of endophytes into nonhost species, adverse interactions have been observed, including cellular incompatibility reactions resulting in the death of either fungal hyphae (Koga et al., 1993) or host tissues (Christensen, 1995; Chung and Schardl, 1997), and invasion of vascular bundles and stunting of seedling growth (Christensen et al., 1997a). The absence of such adverse reactions provides suggestive evidence of coevolution, if only of an ancestral pathogen with its plant host.

In addition to these qualitative boundaries, endophyte–host compatibility may vary in degree. In studies of *N. coenophialum*-infected tall fescue plants, Hiatt and Hill (1997) found that compatibility, reflected in the concentration in the plant of endophyte mycelium (estimated by ELISA) and in plant herbage mass, is affected by both endophyte and plant genotype. While ELISA estimates of fungal mycelium are indirect, and plant–fungal interactions may cause antigen production to vary, it appears to be a reasonable approximation to more direct measures, e.g., hyphal counts. The data suggests that asexual transmission of the endophyte, through the sexual reproduction of the plant, continually places each endophyte genotype within a new plant genotype, with new requirements for compatibility.

This requirement for an endophyte genotype to thrive in a genetically varying sequence of hosts may be the basis for the facility of artificial cross-inoculations and the driving force favoring asexual reproduction of the endophyte. Law (1985) has argued that selection for mutual advantage of a host and symbiont favors asexual reproduction of the symbiont. Selection pressures favoring compatibility will disfavor sexual reproduction as sexual recombination of both host and symbiont will reduce the chance of favorable interactions being maintained.

4.2. Selection of the Fungus During Transmission Through the Host

The existence of the symbiosis is evidence that the barriers of the plant to fungal infection have been breached. There are further barriers to successful transmission of the fungus to a new generation of plants, and these differ with endophyte transmission type.

The *Epichloe/Neotyphodium* complex provides an unusual gradation from an antagonistic to a mutualistic relationship of the fungus to host sexual reproduction, with a corresponding gradation in selection pressures (Clay, 1988). For the type I sexual *Epichloe* species, transmission to a new host genotype entails fungal reproduction and host sterilization. For type II *Epichloe* species, the relationship is more ambiguous, as there are two routes of fungal transmission. The type III asexual *Neotyphodium* endophytes are entirely dependent on the survival and successful reproduction of their host plant for their transmission, whether by infection of host seed or of separating tillers in a vegetatively reproducing perennial host.

To ensure transmission to the next generation of plants, the type III endophyte has to continuously colonize newly formed meristematic tissue in the region of the shoot apex, so that it maintains growth within each tiller of the expanding vegetative plant. It must also successfully invade the hierarchy of meristems in each primordial inflorescence, and finally the ovule and the embryo of each seed. It must then maintain viability in the seed and repeat the colonization process as the seedling emerges and develops.

Throughout the life cycle there is effectively a competition between the growth of the plant and that of the fungus, and at each critical stage there is a possibility of endophyte loss. For stable natural associations of type III endophytes with their hosts, these losses are low, and the overall transmission process is highly efficient. Few endophyte-free tillers are found on naturally endophyte-infected plants (di Menna and Waller, 1986; Hinton and Bacon, 1985). Infection rates in seed are commonly high (Welty and Azevedo, 1993; G. C. M. Latch, personal communication), although loss of viability of endophyte infection may occur in seed prior to seed germination (Hannig, 1907, cited in Sampson 1935; Neill, 1940; Siegel et al., 1985). In some artificial associations with a higher frequency of endophyte loss, some of the critical steps in endophyte transmission are evident. For example, Wilson and Easton (1997) found that in reproductive tall fescue plants artificially infected with a *Neotyphodium* endophyte, whole panicles, or branches from infected panicles, were sometimes endophyte-free, and a proportion of the seed within infected branches was also endophyte-free.

However, the long-term persistence of asexual endophytes in host populations cannot be explained by the high efficiency of the transmission processes alone. Loss of a type III endophyte is irreversible, and 100% transmission will not occur through every cycle. A plant that loses endophyte has a chance of surviving, but the type III endophyte cannot reinfect an uninfected host plant. If selection pressures did not favor endophyte-infected plants, transmission losses, however minor, would compound. If there were no selective advantage to endophyte infection, with a 95% efficiency of transmission, the infection rate in a host population would decline from 100% to below 50% in 14 generations. Without a competitive advantage to endophyte-infected plants, the asexual endophyte would inevitably, in time, be lost from the population. Thus, the key factor in the long-term persistence of type III endophytes in host grass populations is not the high efficiency of their transmission but the ability of endophyte-infected plants to outcompete their endophyte-free counterparts.

The type I fungus faces somewhat different selection pressures. In clonally (vegetatively) propagating perennial plants, it is under the same pressures as the type III fungus to maintain infection. In the reproductive tiller, however, it must embark on rapid hyphal growth in the region of the primordial inflorescence associated with the uppermost node of the reproductive stem to produce a stroma. At this point, the relationship with the host plant is clearly antagonistic, with stroma development rendering the host infertile and the future propagation of the fungus depending on its ability to infect another host. However, in some cases contagious spread of these fungi appears to be limited, and these may persist primarily as clonally (vegetatively) propagating individual infected plants (Bradshaw, 1959; Leuchtmann and Clay, 1997; Watkins, 1987). The hostile relationship is limited to the sexual reproductive stages of plant and fungus, and the relationship of the type I endophyte and its host during vegetative growth may be mutualistic, with the type I endophyte subject to selection pressures to advantage the symbiosis similar to those for type III endophytes.

As noted above, the relationship of the type II *Epichloe* fungi with their hosts is intermediate between those of types I and III. Few individual plants show stromata, with seed transmission presumably being more efficient. The fungus can benefit by conferring selective advantage to its host, but it is not trapped by the plant. Although the type II fungus is perhaps under less pressure to meet the needs of the plant than is a type III endophyte (Leuchtmann, 1997), there is also more scope for the fungus and plant to coevolve (Schardl et al., 1997).

4.3. Selection of Endophyte-Infected Plants

The widespread occurrence of grasses infected with asexual *Neotyphodium* endophytes is evidence that natural selection in favor of endophyte-infected grasses has occurred. Indeed, there is much evidence to show that endophyte-infected plants outcompete uninfected plants in many environments, by a margin far exceeding the minimum required to offset the rate of endophyte loss. However, human selection and dissemination of endophyte-infected grasses, and modification of the natural environment, has also played a large part in the prevalence of the symbiosis today (Clay, 1997).

Toxicoses of livestock first drew scientific attention to the presence of endophyte in introduced pasture species in the USA (Bacon et al., 1977) and New Zealand (Fletcher and Harvey, 1981; Neill, 1941), where cultivars of nonnative species with high levels of endophyte infection (tall fescue with *N. coenophialum* in the USA, perennial ryegrass with *N. lolii* in New Zealand and Australia) have been propagated commercially for agricultural use. While this human involvement has confounded the picture, it has also provided experimental evidence for the ecological efficacy of endophytes. Infected grasses are highly favored in these nonnative agricultural environments. In New Zealand, *N. lolii* infection levels in old perennial ryegrass pastures are generally high (70–97%) (Hume, 1993; Wedderburn et al., 1989; Widdup and Ryan, 1992), particularly in the geographic range of the major insect pest of ryegrass, the Argentine stem weevil (*Listronotus bonariensis*). In adventive tall fescue, growing in laxly grazed situations such as road verges or stream margins, *N. coenophialum* infection levels approach 100% (M. J. Christensen, unpublished data). When New Zealand pastures are planted with low endophyte perennial ryegrass cultivars, a steady increase of the infection rate is typically observed over time (Hume and Brock, 1997; Prestidge et al., 1984, 1985). A similar pattern is seen in tall fescue pastures in the USA (Read and Walker, 1990; Shelby and Dalrymple, 1993; Thompson et al., 1989). In rather better controlled experiments, Clay (1997) has shown that endophyte-infected tall fescue is a significant ecological force when introduced into experimental ecosystems, increasing productivity but reducing diversity.

In natural environments and long-established Old World pastures, infection frequencies are also often high (Clay and Leuchtmann, 1989; Clement et al., 1997; De Battista et al., 1997; Latch et al., 1987; Leuchtmann, 1992; Lewis, 1997; Li, et al., 1997; Miles et al., 1998; Pfanmöller et al., 1997; Zabalgogeazcoa et al., 1997). The mechanisms providing this competitive advantage to endophyte-infected plants in a population have been examined in some detail, and the topic has been extensively reviewed (Clay, 1997; Leuchtmann, 1997 and references cited). A general pattern of advantage to endophyte-infected plants has been demonstrated in a range of experiments, particularly in the face of biotic and abiotic stresses. In addition to their activity against mammalian herbivores, endophytes confer advantages to the host plant in resistance to insect herbivores and nematodes, and to water stress (at least for *N. coenophialum* in tall fescue), and there is evidence of benefits in competition with other plants and some microorganisms (Clay, 1997). As indicated above, endophyte secondary metabolites are significant factors in many of these beneficial effects, particularly resistance to insect herbivores (Dahlman et al., 1991; Popay and Rowan, 1994; Prestidge and Ball, 1993; Siegel and Bush, 1997).

In view of these selection processes it is not surprising that many type III asexual endophytes are both biologically potent and much more prevalent in their host grasses than their horizontally transmitted type I relatives (Leuchtmann and Clay, 1997). Even for type II endophytes, as noted above, the vertical (seedborne) transmission route seems to be preferred to horizontal transmission, and these fungi also achieve high rates of infection. The relationship between mode of transmission, prevalence in the population, and defensive benefits to the host is worth further examination.

This outline of competition between host and endophyte within the plant, and between infected and noninfected hosts in a population, provides a framework for considering the place of secondary metabolism in the coevolution of fungus and plant. If the fungus cannot grow vigorously within the key tissues of its hosts, it will not survive to the next generation; but for the seedborne asexual transmission of *Neotyphodium* fungi to be maintained, selective advantages to the host must also accrue. Without these selective advantages, the compounding effect of transmission losses (plant escape) means that the incidence of infection will decline in the host population.

5. COEVOLUTION OF FUNGAL METABOLITE PATHWAYS

For asexual *Neotyphodium* endophytes, provision of a selective advantage to the host is vital to their survival. If endophyte secondary metabolites play an important role in plant defenses, the diversity of this metabolism should parallel phylogenetic relationships between endophytes and reflect the evolutionary processes to which they have been subjected. We examine the relationship of endophyte secondary metabolite biosynthesis: first, to that of their nonendophytic fungal relatives, and; second, to the genetic diversity of endophytic fungi, and their hosts. We then briefly consider the selection of endophytes in agricultural environments. As elsewhere, the information base is limited, and dominated by data on the agriculturally important symbioses. Care must be taken not to infer the range of distribution of particular metabolites from the absence of data. Except for the ergot alkaloids, experimental evidence of biosynthetic pathways is limited or nonexistent.

5.1. Evolution from Secondary Metabolite Pathways in Related Fungi

Despite their differences in life cycle, and the demands this may place on fungal secondary metabolism, in secondary metabolite chemistry, as in molecular genetics, *Neotyphodium* and *Epichloe* are best considered as a single complex. For some of the secondary metabolites there are obvious links to their close fungal relatives, particularly to *Claviceps* for the ergot alkaloids and indole–diterpenoids, but for the lolines and peramine there are no obvious phylogenetic connections to biosynthetic capabilities of other fungi.

5.2. Pyrrolizidines

While there is a considerable history to the understanding that the accumulation of loline alkaloids by grasses depends on fungal infection (Powell and Petroski, 1992a), it has only very recently been shown that these compounds can be synthesized independently by the fungus in culture (C. Schardl, personal communication). There have been no reports of experimental studies into the biosynthesis of these compounds, but a pathway from the amino acids ornithine and methionine, via the polyamine spermidine, has been proposed (Bush et al., 1993; Siegel and Bush, 1997).

The ancestral origin of this biosynthetic capability remains obscure. Lolines have not been reported in any other fungi. While pyrrolizidine alkaloids are not uncommon in the plant kingdom, loline alkaloids form an unusual structural class with a very restricted distribution as plant metabolites, being reported only in *Adenocarpus* and *Laburnum* spp. (family *Fabeaceae*) (Greinwald et al., 1992, and references cited). Without any prompting from phylogeny as to the occurrence of these compounds in such divergent species, either parallel evolution or horizontal gene transfer must be invoked.

5.3. Pyrrolopyrazine

The pyrrolopyrazine peramine is widely distributed throughout the *Neotyphodium/Epichloe* complex, in associations with a diverse range of hosts (Christensen et al., 1993; Siegel et al., 1990). On current evidence, it is a unique chemical marker for these fungi. It has not been reported outside these associations, and no homologous structures have been isolated, although there is ELISA evidence of an analog in endophyte-infected *A. inebrians* (Stevenson, 1996). There is no experimental evidence of the biosynthetic origin of peramine, although a pathway from proline and arginine through a diketopiperazine appears likely (Rowan, 1993). Diketopiperazine-derived alkaloids are produced by numerous fungal genera (Cordell, 1981), but analogs of peramine have not been reported in other members of the Clavicipitaceae. However, the lysergyl tripeptide lactams, thought to be intermediates in ergopeptide biosynthesis (Fig. 2) (Gröger and Floss, 1998), incorporate a proline-based diketopiperazine moiety, although free diketopiperazines do not appear to be intermediates in that pathway.

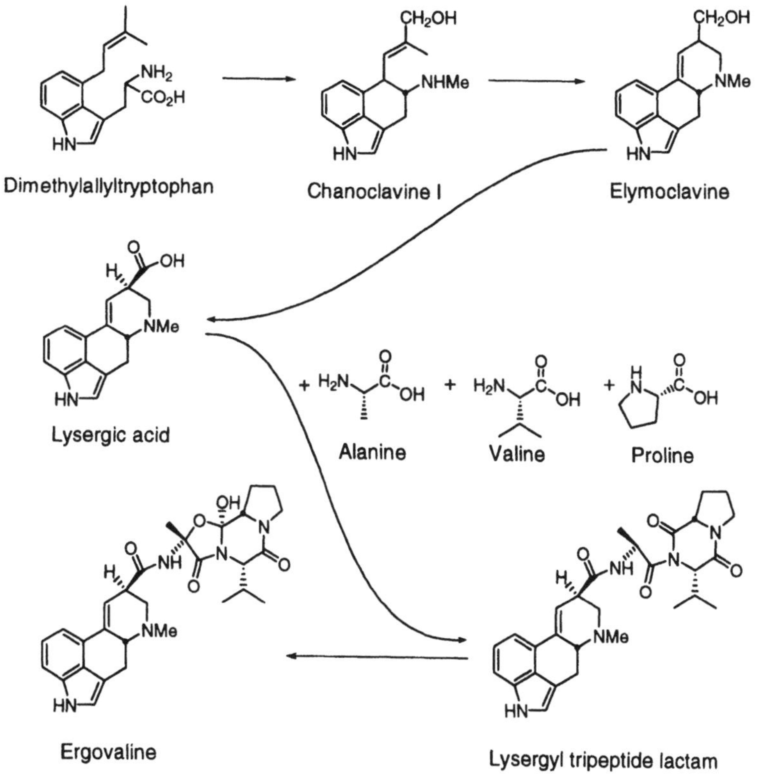

Figure 2 Proposed biosynthetic pathway to ergovaline (Gröger and Floss, 1998).

5.4. Ergot Alkaloids

The ergot alkaloids found in grasses infected with a range of *Neotyphodium* and *Epichloe* spp. (Miles et al., 1996; Siegel et al., 1990; TePaske et al., 1993) are, in almost all cases, also known as secondary products of *Claviceps* spp. (Flieger et al., 1997; Gröger and Floss, 1998). Ergovaline, the dominant ergopeptide in *Epichloe/Neotyphodium* symbioses with *Festuca*, *Lolium* and *Hordeum* spp., is a fermentation product of *C. purpurea* (Brunner et al., 1979), as are the other ergopeptides, ergosine, ergonine, ergoptine, ergocornine, and ergocryptine (Flieger et al., 1997), identified in *Neotyphodium*-infected grasses (Lyons et al., 1986; Shelby et al., 1997; TePaske et al., 1993). Shelby et al. (1997) suggest that some of the other ergot alkaloids previously identified in tall fescue seed as endophyte products may have been contaminants from *Claviceps purpurea* infections. The clavine alkaloids identified in *Neotyphodium*-infected *S. robusta* (Petroski et al., 1992) and *N. coenophialum*–infected tall fescue and in culture (Lyons et al., 1986; Porter et al., 1979, 1981), and the lysergic acid amide (ergine) (Fig. 3a) derivatives identified in *Neotyphodium*-infected tall fescue (Shelby et al., 1997), *S. robusta* (Petroski et al., 1992), and *A. inebrians* (Miles et al., 1996), are all also products of *Claviceps* species (Flieger et al., 1997).

The only ergot alkaloids to be uniquely found as natural products of a *Neotyphodium*–grass symbiosis are aci-ergovaline (Fig. 3a), the C-2′ epimer of ergovaline, and the tentatively identified didehydroergovaline (Fig. 3a, or isomer) recently identified by Shelby et al. (1997) in an LC-MS examination of *N. coenophialum*-infected tall fescue seed. There may be further examples of unique *Neotyphodium* ergot alkaloids. Shelby et al. (1997) found evidence for other unidentified ergot alkaloids in *N. coenophialum*-infected tall fescue seed, and high levels of unidentified presumptive ergot alkaloids have been de-

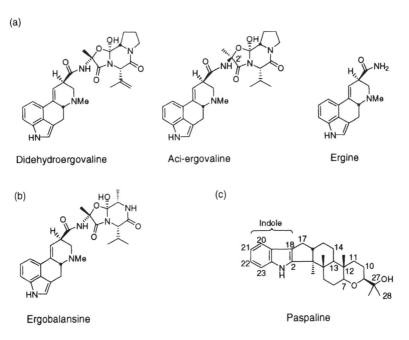

Figure 3   (a) Possible products of *in planta* ergovaline biotransformation identified in *N. coenophialum*-infected tall fescue seed (Shelby et al., 1997). (b) The nonproline ergopeptide ergobalansine, from *Balansia obtecta* and *B. cyperi* (Powell et al., 1990). (c) Sites of biosynthetic modification of paspaline.

tected by ELISA in endophyte-infected *F. argentina* (Miles et al., 1995; Towers, 1997).

The biosynthesis of ergot alkaloids in *Claviceps* species (Fig. 2) has been shown to proceed via isoprenylation of tryptophan with dimethylallylpyrophosphate to form dimethylallyltryptophan, which is then converted to lysergic acid via the clavine alkaloids, chanoclavine and elymoclavine (Gröger and Floss, 1998 and references cited). A multi-enzyme peptide synthetase complex that assembles lysergic acid, proline, and other amino acids to give D-lysergyltripeptide lactams has been isolated from *C. purpurea* (Riederer et al., 1996). Ergopeptide formation is proposed to take place by oxidation and cyclisation of the appropriate lactam (Gröger and Floss, 1998).

The gene for tryptophan prenylation in *C. fusiformis* has been cloned (Tsai et al., 1995), and a recent preliminary report indicates the presence of a homologous gene in *N. coenophialum* (Wilkinson and Schardl, 1997). Panaccione (1996) has found considerable homology in peptide synthetase gene sequences from *C. purpurea* and *N. coenophialum*, with one of the *N. coenophialum* clones hybridizing with DNA from *C. purpurea*. These data all suggest a common inheritance of ergot alkaloid biosynthesis in *Neotyphodium* and *Claviceps*.

Although ergot alkaloid synthesis is not confined to the family Clavicipitaceae, having been reported in distantly related fungi including *Aspergillus* and *Penicillium* species, and in several plant species of the Convolvulaceae (Flieger et al., 1997; Gröger and Floss, 1998), it is a characteristic of this family that *Epichloe/Neotyphodium* spp. share with other members of the tribe Balansieae. Ergot alkaloids have been identified in *Balansia* (Bacon et al., 1986), *Atkinsonella* (Leuchtmann and Clay, 1988), and *Echinodothis* (Glenn and Bacon, 1997) spp. The lysergic acid amide ergonovine, reported from two *Balansia* spp., and almost all the clavine alkaloids found in cultures of several *Balansia* spp. (Bacon et al., 1986), have also been found in *Neotyphodium*. However, ergot alkaloid synthesis in the genus *Balansia* is distinguished by the occurrence of the nonproline ergopeptide variant ergobalansine (Fig. 3b), isolated from *Balansia obtecta* and *B. cyperi* (Powell et al., 1990). This compound has not been identified as a product of *Claviceps* spp. or *Epichloe/Neotyphodium* spp., but was recently isolated from *Ipomoea piurensis* (Convulvulaceae) (Jenett-Siems et al., 1994). Recently, Glenn and Bacon (1997) suggested, on the basis of gene sequence data, that *Epichloe* is more closely related to *Claviceps* than to *Balansia*. Examination of a wider range of metabolite classes could clarify the degree of common inheritance of secondary metabolism of *Epichloe/Neotyphodium* and *Balansia*.

Shelby and co-workers (1997) have suggested that didehydroergovaline, aci-ergovaline, and ergine (Fig. 3a) in *N. coenophialum*-infected tall fescue seed are products of plant biotransformation of fungal metabolites. If confirmed, this will be the first established example of the involvement of both plant and fungal enzymes in the biosynthesis of secondary metabolites in the symbiosis.

5.5. Indole–Diterpenoids

The lolitrem indole–diterpenoids of the *Epichloe/Neotyphodium* complex are distinctive but show evidence of a common inheritance with other fungi of much of the biosynthetic pathway (Fig. 4). Although lolitrem B has not been identified in fungi other than *Epichloe/Neotyphodium* spp., the indole–diterpenoids paspaline (Gatenby, 1997), paxilline (Weedon and Mantle, 1987), and 13-desoxy-paxilline (Gatenby, 1997) identified in *N. lolii*-infected perennial ryegrass seed are also present in other fungal genera. Within the family Clavicipitaceae, paspaline occurs in *Claviceps paspali* Stevens and Hall (Springer and Clardy, 1980), the agent responsible for paspalum staggers of cattle grazing ergotized *Paspalum dilatatum* (Cysewski, 1973). However, as with ergot alkaloids, indole–diterpenoids are also produced by fungi from other families. Paspaline is also found in

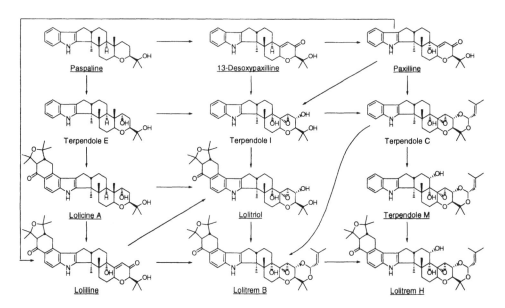

Figure 4   Possible metabolic grid for biosynthesis of lolitrems. Indole–diterpenoids with names underlined have been isolated from seed of *L. perenne* infected with *N. lolii*. Paspaline, 13-desoxypaxilline, and paxilline have been isolated from numerous other fungal genera (see text), and terpendoles E, I, and C have been isolated from *A. yamanashiensis* (Munday-Finch, 1997).

Penicillium (Munday-Finch et al., 1996a), *Emericella* (Nozawa et al., 1988a), and *Albophoma* spp. (Huang et al., 1995). Paxilline and 13-desoxypaxilline are also found in *Penicillium* (Springer et al., 1975), *Emericella* (Kawai and Nozawa, 1988; Kimura et al., 1992; Nozawa et al., 1988b), and *Eupenicillium* (Belofsky et al., 1995) spp. Related indole–diterpenoids have been identified in a wide range of fungal species (Munday-Finch, 1997; Ondeyka et al., 1997; Steyn and Vleggaar, 1985).

Most indole–diterpenoid-producing fungi appear to synthesize the toxins via paspaline (Fig. 3c), the simplest fully cyclized indole–diterpenoid. Paspaline is then modified through a number of transformations, including addition of iso-prene units to the indole unit at C-20 to C-23, often followed by cyclization; oxidative loss of the methyl at C-12; olefin formation or epoxidation at C-11 to C-12; oxidation (usually hydroxylation) at C-7, C-10, C-11, C-13, C-14, and C-17; dehydration of C-27 to C-28; addition of an isoprene unit to 27-OH to form an ether, or to 27-OH and 10-OH to form an acetal; and cyclisation of 27-OH to 7-OH with dehydration. In addition, these compounds are readily oxidized abiotically (e.g., by air) across C-2 to C-18 of the indole to give ketoamide deriva-tives (Ondeyka et al., 1997).

Many of these transformations operate in fungi in a more or less modular fashion in a metabolic grid (Mantle and Weedon, 1994), leading to a diversity of indole–diterpenoid metabolites. Approximately 70 metabolites of this class have so far been identified (Munday-Finch, 1997). Most of these transformations also appear to occur in *N. lolii* during the biosynthesis of the lolitrem neurotoxins (Miles et al., 1994; Miles et al., 1992; Munday-Finch, 1997; Munday-Finch et al., 1995, 1997, 1998), suggesting that the indole–diterpenoid biosynthetic path-way in all of these fungi has evolved from a common pathway similar to the simpler one that operates in *C. paspali*. Many of the lolitrems are analogs of simpler indole–diterpenoids found in other fungi, such as paspaline, paspaline B (Munday-Finch et al., 1998), paxilline (Munday-Finch et al., 1997), and several terpendoles (Gatenby, 1997; Munday-Finch, 1997; Munday-Finch et al., 1998), to which the endophyte has added the lolitrem A/B ring system (Fig. 4). Thus, the biosynthesis of all but the A/B rings of the lolitrems by *N. lolii* exactly paral-lels, and presumably shares a common evolutionary origin with, the indole–diter-penoid biosynthetic pathways in other fungi.

5.6. Lipids

The antifungal compounds identified in stroma of *E. typhina* stand largely in isolation. However, the antifungal hydroxylated fatty acids, particularly 12-hy-droxy-9Z,13E-octadecadienoic acid (Koshino et al., 1987), suggest a linkage with *Claviceps* spp. Ricinoleic acid (12-hydroxy-9Z-octadecenoic acid) occurs as oligotriglycerides at high concentrations in *C. purpurea* sclerotia, and is also found in submerged cultures of a range of *Claviceps* spp. (Kren, et al., 1985).

5.7. Evolutionary Diversification of Secondary Metabolite Biosynthesis in *Epichloe/Neotyphodium*

The structural diversity of secondary metabolites of *Epichloe/Neotyphodium* is not unusual in secondary metabolism, where combinations of minor variations around a structural theme are common, suggesting a modular assembly of the products. While this pattern is most evident for the indole–diterpenoids, a variety of loline and ergot alkaloid structures have also been found. Peramine variants await elucidation. Whatever the evolutionary driving force for this structural diversification, it has generated a wide range of options for the selection of endophytic fungi synthesizing diverse biologically active products.

This diversity of secondary metabolite structures is mirrored by the diversity of their accumulation in planta across the *Epichloe/Neotyphodium* complex. The range of variation in their occurrence indicates that none of the commonly examined alkaloids is essential for a successful symbiosis. There has evidently been a wide range of successful responses to the selection pressures on these fungi to provide defenses to their plant host.

Endophyte chemotypes in planta do not map cleanly on to phylogenetic categories, although there is some evidence linking alkaloid accumulation in *Epichloe/Neotyphodium*–grass associations to fungal phylogeny. The occurrence of the major alkaloids reported in natural associations of strains of *Epichloe* and *Neotyphodium* spp. with grass hosts is documented in Table 1.

For *Epichloe* spp., each of the main alkaloid classes has been reported in at least one association with a grass host (Table 1). Peramine was identified in an association with the type I endophyte *E. typhina* (Siegel et al., 1990). Ergovaline and peramine are common constituents of the associations with type II endophytes examined (Table 1), apparently irrespective of mating population [*E. amarillans* (mating population IV), *E. festucae* (MP II), and unnamed *Epichloe* spp. in *Elymus canadensis* and *Sitanion longifolium* (MP III) (Schardl et al., 1997)]. Lolines have been identified in a strain of *Epichloe festucae* in *Festuca gigantea* (Siegel et al., 1990; C. Schardl, personal communication), and lolitrem B in an *E. festucae* strain in *F. longifolia* (Siegel et al., 1990). Links between the phylogeny and chemistry of *Epichloe* spp. have yet to be established.

For *Neotyphodium* spp. associations with grasses there is wide diversity of alkaloid expression, but some links between endophyte–grass chemistry and other phylogenetic evidence can be discerned (Table 1). Lolines have been reported across the widest range of host grasses infected with *Neotyphodium* sp. endophytes, including annual *Lolium* spp., *Hordeum*, *Poa* and *Stipa* spp. (TePaske et al., 1993), *F. arundinacea*, *F. pratensis* (Siegel et al., 1990), *F. argentina*

Table 1 Secondary Metabolites of *Epichloe* and *Neotyphodium* Endophytes in Natural Host

Endophyte sp.	Host sp.	Lolines[a]	EA[b]	P[c]	LB[d]	Ref.[e]
Epichloe amarillans	*Agrostis hiemalis*	−[f]	+	+	−	(1)
E. festucae	*Festuca gigantea*	+	+	−	−	(1)
	F. glauca	−	+	+	−	(1)
	F. longifolia	−	+	+	+	(1)
	F. ovina duriuscula	+	+			(2)
	F. ovina ovina		+	Tr		(2)
	F. rubra commutata	−	+	+	−	(1)
		−	+	−	−	(1)
	F. rubra rubra	−	+	−	−	(1)
E. typhina	*Dactylis glomerata*	−	−	−	−	(3)
	Lolium perenne	−	−	+	−	(1)
Epichloe spp	*Elymus canadensis*	−	−	+	−	(1)
	Sitanion longifolium	−	+	+	−	(1)
Neotyphodium chisosum	*Stipa robusta*	+	+			(4)
N. coenophialum	*F. arundinacea*	+	+	+	−	(1, 4, 5)
		+	−	+	−	(5)
N. huerfanum	*F. arizonica*	−	−	−	−	(1)
N. lolii	*L. perenne*	−	+	+	+	(1, 5)
		−	−	+	+	(5)
		−	+	+	−	(5)
		−	−	−	−	(5)
N. starrii	*Bromus anomalous*	−	+	+	−	(1)
	F. obtusa	−	−	−	−	(1)
N. uncinatum	*F. pratensis*	+	−	−	−	(5)
Neotyphodium. sp. FaTG-2	*F. arundinacea*	−	+	+	+	(5)
		−	+	+	−	(5)
		−	+	−	−	(5)
Neotyphodium sp. FaTG-3	*F. arundinacea*	+	−	+	−	(5)
Neotyphodium sp. LpTG-2	*L. perenne*	−	+	+	−	(5)
Neotyphodium sp.	*Achnatherum inebrians*	−	+	U[g]	Tr, U[h]	(6)
	Echinopogon ovatus	+	−	−	U[h]	(7)
		−	−	−	−	(7)
	F. argentina	+	U[i]		U[h]	(8, 9, 10)
	F. paradoxa	−	−	+	−	(1)
	F. versuta	−	−	−	+	(1)
	L. multiflorum	+	−			(4)
	L. persicum	+	−			(4)
	L. rigidum	+	−			(4)
	L. temulentum	+	−			(4)
	Hordeum bogdanii	Tr	Tr			(4)
		−	−			(4)
	H. brevisubulatum	Tr	+			(4)
		−	−			(4)
	Melica decumbens				U[h]	(7, 10)
	Poa alsodes	+	+			(4)
		−	+			(4)
	P. ampla	−	−	+	−	(1)
	P. autumnalis	+	−	+	−	(1)
	P. huecu				U[h]	(7, 10)

[a] Loline alkaloids; [b] ergot alkaloids; [c] peramine; [d] lolitrem B; [e] Data compiled from (1) Siegel et al (1990), (2) Yue et al (1997), (3) Siegel and Bush (1997), (4) TePaske et al (1993), (5) Christensen et al (1993), (6) Miles et al (1996), (7) Miles et al (1998), (8) Miles et al (1995), (9) Casabuono et al (1997), (10) Towers (1997); [f] + = detected, − = not detected, Tr = possible trace, blank = not examined; [g] unidentified pyrrolopyrazine; [h] unidentified indole–diterpenoid; [i] unidentified ergot alkaloid.

(Casabuono and Pomilio, 1997), and *Echinonpogon ovatus* (Miles et al., 1998), but were not observed in *F. paradoxa*, *F. obtusa*, *F. versuta* (Siegel et al., 1990), or *Achnatherum inebrians* (Miles et al., 1996). They have also been reported from *N. lolii*-infected perennial ryegrass but only under conditions of temperature stress (Huizing et al., 1991). Ergot alkaloid synthesis is broadly distributed, with ergovaline reported in *Neotyphodium* sp.–infected *Bromus anomalus*, *F. arundinacea*, *L. perenne* (Siegel et al., 1990), and *Hordeum* spp. (TePaske et al., 1993), the ergopeptides ergosine and ergocryptine (but not ergovaline) in *P. alsodes* (TePaske et al., 1993), and lysergyl derivatives in *Stipa robusta* (Petroski et al., 1992) and *A. inebrians* (Miles et al., 1996). Antibody binding studies indicate that unidentified ergot alkaloids are present in endophyte-infected *F. argentina* (Miles et al., 1995). The occurrence of peramine and lolitrem B has been examined less extensively. Lolitrem B has been identified in *Neotyphodium* sp.–infected *F. versuta* (Siegel et al., 1990), *F. arundinacea* (Christensen et al., 1993) and *A. inebrians* (Miles et al., 1996), in addition to *L. perenne*. Evidence of indole–diterpenoids other than lolitrem B in *Neotyphodium* endophyte associations with *E. ovatus*, *Melica decumbens*, and *P. huecu* has been provided by ELISA (Miles et al., 1998; Towers, 1997). Peramine has been found in *Neotyphodium* sp.–infected *B. anomalus*, *F. arundinacea*, *F. paradoxa*, *L. perenne*, and *Poa* spp. (Siegel et al., 1990), and ELISA studies indicate an analogue in *A. inebrians* (Stevenson, 1996). Associations have also been reported with *F. arizonica* (Siegel et al., 1990), *L. perenne* (Christensen et al., 1993), and *E. ovatus* (Miles et al., 1998) in which none of these alkaloids has been detected.

The description of secondary metabolite diversity in *Epichloe/Neotyphodium* is clearly incomplete, as indicated by the uncharacterized compounds noted in Table 1. The absence of expression of one or all of the major known defensive metabolites in planta may well be the product of selection of endophyte–plant associations for expression of other unrecognized characteristics, perhaps other secondary metabolites.

If the coevolution of fungal species with their grass hosts has an ancient history [as gene sequence evidence suggests, at least for type II *Epichloe* spp. (Schardl et al., 1997)], the patterns of endophyte chemistry might be expected to reflect the host phylogeny. In practice, links between host phylogeny and chemotype are few (Table 1). Lysergyl compounds rather than ergovaline appear to be characteristic of *Neotyphodium* endophytes of the Stipeaea (*Stipa*, *Achnatherum*). *E. festucae* produces lolines in *F. gigantea*, which is the host species of *E. festucae* most closely related to tall fescue and meadow fescue, in which *Neotyphodium* sp. infections commonly produce lolines. However, the loline alkaloids occur not only with *Neotyphodium* endophytes infecting *Festuca*, *Lolium*, and *Poa* within the Poeae, but also with those infecting *Hordeum* in the Triticeae, and *Echinopogon* within the Aveneae.

Evidence of the complex phylogenetic relationships that may underlie these patterns was provided by a study of *Neotyphodium* endophyte strains in perennial ryegrass, tall fescue, and meadow fescue by isozyme phenotyping and measurement of in planta expression of alkaloids (Christensen et al., 1993). Considerable diversity was found, but with some distinctive patterns of alkaloid expression at the taxon level. Under normal growth conditions, loline alkaloids and lolitrem B did not co-occur (but note Huizing et al., 1991). Loline alkaloids were found in two of three taxa from tall fescue [*N. coenophialum* and FaTG-3 (*F. arundinacea*-taxonomic group 3)], and from the single taxon from meadow fescue (*N. uncinatum*), but not from two taxa in perennial ryegrass. Lolitrem B was found in three of five isozyme phenotypes of the third taxon found in tall fescue (FaTG-2) and in four of six isozyme phenotypes of *N. lolii*. The other taxon from perennial ryegrass (LpTG-2) produced neither lolitrem B nor lolines. Ergovaline was not found in FaTG-3 or *N. uncinatum*. Peramine production was common, but not universal, and associations lacking any of these alkaloids were also found. While the study showed associations between phylogeny and chemistry, it also clearly demonstrated that phylogenetically distinct *Neotyphodium* spp. can generate the same apparent chemotype in planta.

The genetic basis for this diversity of alkaloid synthesis in *Neotyphodium* endophytes is largely unexplored. The loss of indole–diterpenoid expression in culture by UV-induced mutation of *N. lolii* has been reported (Gurney et al., 1994). Further analysis in the asexual endophytes will require the techniques of molecular genetics. In *Epichloe* spp., gene segregation can be investigated by crossing strains of opposite mating type and differing alkaloid expression. A preliminary report indicates that loline expression in *E. festucae*, and peramine expression in *E. typhina* are in each case regulated by a single controlling locus (Bush et al. 1997).

The regulation of alkaloid biosynthesis in *Claviceps* spp. in culture can be modified by gene fusion in hybrid strains (Didek-Brumec et al., 1996), and hybridization events may have altered metabolite expression in endophytes (Bush et al., 1997). However, there is little evidence suggesting *de novo* evolution of biosynthetic capabilities in *Neotyphodium* spp. that are not already present in *Epichloe* spp. (Table 1). As noted above, gene typing shows *E. festucae* and *N. lolii* to be very similar (Wilkinson and Schardl, 1997), and the characteristic major alkaloids of *N. lolii* are all expressed in one *E. festucae* strain. The lysergyl compounds found in *Stipa* (Petroski et al., 1992) and *Achnatherum* (Miles et al., 1996) have not yet been reported in *Epichloe* but, as they occur in *Claviceps* cultures (Flieger et al., 1997), *de novo* evolution of this biosynthetic pathway is unlikely.

The simplest view is that the *Neotyphodium* fungi have inherited broad biosynthetic capabilities and that the expression of these capabilities in particular symbioses is a response to selection pressures. As Käss and Wink (1995) have pointed out *àpropos* the utility of alkaloids in legume systematics, secondary metabolite diversity may reflect the diversity of natural selection during evolution more than the phylogeny of the organisms producing them. Considering the activity of many of these compounds toward insect herbivores, this diversity of in planta accumulation may well be a product of coevolution of these symbioses with insects, generating a diversity of chemically differentiated ecological niches (see below).

5.8. Selection of Chemotypes in Agricultural Environments

In contrast to the diversity of *Neotyphodium* endophyte genotypes and chemotypes found in perennial ryegrass and tall fescue collections in Europe (Christensen et al., 1993; Schardl et al., 1994; Tsai et al., 1994), the naturalized populations of these grasses in the United States and New Zealand are remarkably uniform. In New Zealand, both of these grasses have been introduced over a period of some 150 years, presumably from Europe. However, the *Neotyphodium* endophytes identified in collections of these two grasses in New Zealand are exclusively *N. lolii* in perennial ryegrass and *N. coenophialum* in tall fescue (Latch et al., 1984), and all strains examined contain a broad range of alkaloids. Exceptions to the pattern of occurrence of lolitrem B, ergovaline, and peramine in *N. lolii*-infected perennial ryegrass, and of peramine, ergovaline, and lolines in *N. coenophialum*-infected tall fescue, have yet to be found in plants from long-established swards in New Zealand (B. A. Tapper, G. A. Lane, M. J. Christensen and G. C. M. Latch, unpublished data). It is not certain as to whether this pattern was established by importation of a limited endophyte gene pool or by selection pressures in the new environment.

In New Zealand pastures, the interaction between the *N. lolii–L. perenne* symbiosis introduced from Europe, and an insect herbivore, the Argentine stem weevil, introduced from South America, where endophyte-infected grasses are common (e.g., *P. huecu, F. argentina, F. hieronymi*) (De Battista et al., 1997), may be particularly significant. This displaced interaction may create much greater selection pressures on, and greater selective advantage for, symbiotic associations producing the full range of secondary metabolites than would be found in an undisturbed ecosystem. In a natural ecosystem, where the herbivores are likely to be better adapted to the prevailing defensive chemistry, the selective advantage provided by an endophyte may be less obvious (Faeth et al., 1997). While endophyte-infected grasses in natural ecosystems must outcompete endophyte-free grasses for infection to persist, the margin may be less extreme than that observed in New Zealand pastures, and a less comprehensive defensive repertoire may suffice. In addition, the maintenance in a natural grass population of a diversity of endophyte chemotypes, including those lacking some alkaloid classes (and indeed a proportion of endophyte-free plants), may provide an advantage in terms of long-term "resistance management," maintaining a diversity of ecological niches and a gene pool of herbivores nonadapted to each metabolite class (Bacon and Hill, 1996).

6. COADAPTATIONS OF PLANT AND FUNGUS FOR EFFECTIVE DEFENSIVE CHEMISTRY

While the potency of the secondary metabolites of *Neotyphodium* endophytes appears to derive from their ancestry rather than from coevolution with their grass hosts, there is evidence that the in planta yield and distribution of these metabolites (in time and space) is the product of mutual adaptation of plant and fungus. There is little known about the synthesis, translocation, turnover, or storage of these compounds in the fungus or plant. However, for a limited range of endophyte–grass associations, there is some useful information available about the net outcome of these processes, i.e., the accumulation of metabolites in the symbiosis.

6.1. Expression in the Symbiosis and in Culture

The biosynthesis of the distinctive secondary metabolites of the symbiosis appears to be primarily the work of the fungus. Peramine (Rowan, 1993) and lolitrem B (Penn et al., 1993) have been identified in cultures of *N. lolii*, and ergopeptides and clavines in cultures of *N. coenophialum* (Lyons et al., 1986; Porter et al., 1979, 1981). These compounds are clearly products of fungal biosynthesis, although the ergopeptide variants reported by Shelby et al. (1997) may represent plant modification of fungal metabolites. Justus et al. (1997) have reported that lolines were detected at trace levels in uninfected *F. pratensis* plants, and this calls for closer examination. However, the recent discovery of loline production by *N. uncinatum* under appropriate cultural conditions (C. Schardl, personal communication) appears to indicate a primary role for fungal enzymes in loline biosynthesis, although key substrates may be provided by the plant. Thus, all four known major groups of defensive alkaloids produced in endophyte–grass associations are primarily products of fungal metabolism.

If the biosynthetic pathways are the domain of the fungus, the plant plays a major role in governing metabolite expression, either indirectly, through the environment it provides the fungus, or by direct (and as yet unknown) regulatory mechanisms. Expression of indole–diterpenoid biosynthesis in vitro is considerably different from that in planta. Formation of paxilline analogs was found to be universal in cultures of a wide range of strains of *Neotyphodium* spp. from *L. perenne*, *F. arundinacea*, *F. pratensis*, and a *Poa* sp. and *E. festucae* strains from several *Festuca* spp. (Penn et al., 1993). The more complex indole–diterpenoids lolitrem B and lolitriol were detected at much lower concentrations in many of the cultures. Notably, lolitrem B was detected in cultures of several strains that do not accumulate it in planta, including strains of *N. uncinatum*, LpTG-2, and FaTG-3 (cf. Table 1).

A more consistent qualitative pattern of metabolite accumulation is observed when endophytic fungi are transferred between different plant species and genera, suggesting that alternative plant hosts provide similar environments for the fungus and that the diversity of metabolite expression is thus primarily a function of fungal genes. For peramine, ergovaline, and lolitrem B, the metabolite profile in the original host plant is a good predictor of the qualitative profile in an artificial association, although the relative yields may vary (Christensen et al., 1993; Siegel et al., 1990; B. A. Tapper, G. A. Lane, M. J. Christensen and G. C. M. Latch, unpublished data).

The host may play a larger role in loline biosynthesis. Although an *N. coenophialum* strain produced lolines in perennial ryegrass (as it does in tall fescue) (Siegel et al., 1990), an FaTG-3 strain and an *N. uncinatum* strain gave inconsistent loline production in *L. perenne* (D. D. Rowan, personal communication). The formation of lolines in endophyte-infected *L. perenne* (presumably with *N. lolii*) under temperature stress (Huizing et al., 1991) is further evidence that plant factors are particularly important for the accumulation of these metabolites. The role of the host genome in the formation of modified ergot alkaloids in *N. coenophialum*-infected tall fescue (Shelby et al., 1997) remains to be clarified.

The above findings suggest that the variation of in planta metabolite expression in *Epichloe/Neotyphodium* is primarily due to variation of fungal regulatory genes, rather than to the presence or absence of fungal or host genes coding for biosynthetic enzymes. With recent progress in molecular genetics, this conjecture may soon be able to be tested. If endophyte strains not expressing the common components in planta produce alternative biologically active secondary metabolites, it is likely the biosynthetic genes for these components will also be generally distributed, if less evident in other strains.

6.2. Distribution of Alkaloid Accumulation in the Plant

The distribution of toxic *Neotyphodium* metabolites throughout plant tissue is a major factor in the agricultural toxicoses (Clay, 1988; Miles et al., 1996), and in their efficacy as defenses against invertebrates, and is perhaps a key factor in the transition of endophytes from parasitism to mutualism (Clay, 1988). From the perspective of the fungus, the benefits of this metabolite distribution are indirect as, in contrast to the production of ergot alkaloids in *Claviceps* sclerotia, for example, it is not only the key tissues for fungal reproduction that are defended.

Neotyphodium endophytes maintain growth within the basal meristem tissues of the grass plant. In perennial plants, hyphae grow in the intercellular spaces of the leaf sheath and blade, typically unbranched and parallel to the leaf axis. ELISA measurements show that concentration of mycelium is usually highest in the base, declining upward in the sheath and above the ligular zone in the leaf blade (Musgrave, 1984). In the reproductive tiller, the tissues above the top node are heavily infected with endophyte, which eventually colonizes the seed embryo (Philipson and Christey, 1986). Recent studies with an endophyte strain containing a marker gene-construct indicate that the endophyte is metabolically active throughout the plant (Herd et al., 1997). The key plant tissues for propagation of the endophyte are the meristems of the shoot apices, the inflorescence, and seed, but, at least in associations with perennial grasses, neither the fungal mycelium nor the defensive metabolites are confined to these tissues.

In nonreproductive perennial ryegrass plants infected with *N. lolii*, ergovaline concentrations are highest in the crown, with a basal–apical decline up the sheath and the leaf blade above the ligular zone, paralleling the concentration of mycelium (Davies et al., 1993; Lane et al., 1997a). Lolitrem B also shows a similar basal–apical gradient, but concentrations are highest in the sheath of outer leaves and in dead leaf material, where the concentration of fungal mycelium is declining (Ball et al., 1997b; Gallagher et al., 1987; Keogh et al., 1996). The pattern of distribution of peramine is very different, with concentrations in the blade similar to, or higher than, those in the sheath, but declining in the outer leaves (Ball et al., 1997a; Keogh et al., 1996). The distribution patterns in reproductive plants are similar, except that concentrations of all three alkaloids are high in the seed, and concentrations of lolitrem B and ergovaline are also high in the upper reproductive stem (Ball et al., 1997a, b; Lane et al., 1997a; G. A. Lane, unpublished data). In seedlings, there is evidence of translocation of alkaloids from the seed (Ball et al., 1993).

In tall fescue infected with *N. coenophialum*, the distribution of ergovaline is similar to that in perennial ryegrass with the highest concentrations in the crown, seed heads and sheath material (Azevedo et al., 1993; Rottinghaus et al., 1991; Roylance et al., 1994), and concentrations are higher in mature leaves than in newly emerging or aging leaves (Belesky and Hill, 1997). Low concentrations are found in roots (Azevedo et al., 1993). Peramine is fairly evenly distributed, although at lower concentrations than in perennial ryegrass, with higher concentrations in reproductive tissues (Roylance et al., 1994). There is a declining basal–apical concentration gradient of lolines in vegetative tissues, and they are present at high concentrations in the immature inflorescence and mature seed (Bush et al., 1993; Yates et al., 1990), and at low concentrations in roots (Bush et al., 1993). A similar pattern is seen for lolines in *N. uncinatum*, with the highest concentrations in vegetative pseudostem and mature seeds (Justus et al., 1997). These authors also found evidence of translocation of alkaloids from the seed to seedlings.

There are very limited data on alkaloid distribution in other associations.

In *A. inebrians* infected with a *Neotyphodium* sp. there was no evident gradient of fungal mycelium, and the concentration of lysergyl compounds was two- to threefold fold higher in the leaf blade than in the sheath, with only relatively low concentrations in the crown (Miles et al., 1996). Concentrations in the inflorescence were also high (G. A. Lane and C. O. Miles, unpublished data). The only quantitative data to hand for a type II species is for *E. festucae*-infected *F. rubra* var. *commutata*, in which the ergovaline concentration was higher in seed than in leaf sheath, where it was in turn higher than in stroma (Yue et al., 1997). This very interesting result suggests that even in the type II association, fungal secondary metabolism has been recruited primarily to protect seedborne propagation rather than sexual reproduction of the fungus.

A conspicuous feature of the alkaloid distribution patterns (and, indeed, of the distribution of mycelium) in these perennial plants is how they appear to protect the plant and not just the fungus. Certainly the key tissues for transmission are protected (basal meristem, seed), but the overall distribution meets the requirements for survival of the symbiosis rather than that of the progeny of any individual somatic fungal cell. The distribution of peramine through the plant in perennial ryegrass and tall fescue suggests that this water-soluble alkaloid is translocated in the plant and benefits the fungus only by protecting its host. The distribution of other metabolites—especially the lipophilic (and hence, less readily translocated) indole–diterpenoids and ergopeptides—is more closely linked to that of mycelium. The growth habit of the fungus in perennial grasses appears to be determined by its defensive role in the symbiosis. In particular, the growth of mycelium into leaf sheaths and blades of perennial plants distributes defensive metabolites into plant tissues, with only indirect benefits for transmission of the fungus. In annual grasses, where maintenance of fungal infection of vegetative tissues is less important, the endophyte is found only in basal and reproductive tissues (Latch et al., 1988), and the distribution of alkaloids may well be more parsimonious.

6.3. Seasonal Variation in Concentrations

Endophyte secondary metabolites also benefit the symbiosis, rather than just the fungus, by their distribution through time as well as through space. Alkaloid synthesis in endophytes is uncoupled from the sexual life cycle of the fungus (Popay and Rowan, 1994); in *Claviceps*, alkaloid synthesis is tightly linked to fungal reproduction (Didek-Brumec et al., 1996), whereas the seasonal variation of endophyte alkaloids follows the developmental processes of the plant. The available evidence suggests that the variation of endophyte alkaloid concentrations is largely the product of the seasonal variation of endophyte mycelium in the plant (Ball et al., 1995b; di Menna and Waller, 1986), with concentrations low in winter–early spring, rising with inflorescence development in late spring, and peaking again in late summer.

Perhaps the most detailed account available of the seasonal variation of endophyte alkaloids is that of Justus and co-workers (1997). These researchers found that in meadow fescue infected with *N. uncinatum*, loline concentrations in vegetative tissues rose sharply in early spring with a subsequent decline, followed by a rise in the emerging inflorescences. High levels of lolines accumulated in mature seed heads in midsummer, and there was a renewed rise in vegetative regrowth tissues in fall.

In tall fescue infected with *N. coenophialum* in the southeastern USA, lolines often peak in early spring, then decline slightly before reaching a maximum in late summer (Bush et al., 1993). The spring rise in ergovaline concentrations occurs somewhat later, with maxima in late spring (Adcock et al., 1997a; Agee and Hill, 1994) and late fall–early winter (Belesky et al., 1988). Lower concentrations in mid-summer may be associated with temperature extremes (Adcock et al., 1997a).

A somewhat similar pattern is seen for lolitrem B and ergovaline in *N. lolii*-infected perennial ryegrass. There is a rise from low winter levels, often to a peak in late spring, and a maximum is reached in late summer–fall (Ball et al., 1995a; di Menna et al., 1992; Woodburn et al., 1993; L. R. Fletcher and G. A. Lane, unpublished data).

The seasonal variation of peramine is less marked, perhaps because it is not confined to endophyte mycelium. In perennial ryegrass a broad seasonal trend has been observed, with highest concentrations in summer and a decline in late winter (Ball et al., 1995b; L. R. Fletcher and G. A. Lane, unpublished data). Some variability, but no seasonal trend, was observed in tall fescue (Roylance et al., 1994).

It is interesting to note that concentrations of lolines in endophyte-infected tall fescue, and lolitrem B in endophyte-infected perennial ryegrass, increase after the concentration of mycelium has begun to decline at the end of summer (Ball et al., 1995b; Bush et al., 1993; di Menna et al., 1992). The seasonal patterns of alkaloid accumulation appear to serve the defensive requirements of the symbiosis. Concentrations of defensive metabolites are highest when the plant is in the reproductive phase (in late spring), and when the survival of the perennial host is most at risk—the stressful period of late summer–fall.

6.4. Effect of Growth Conditions on Concentrations

The effects of changes in plant growth conditions on endophyte metabolite concentrations are complex. However, the defensive requirements of the symbiosis appear to be a factor, as concentrations are often higher when the plant is under stress.

In *N. coenophialum*-infected tall fescue, concentrations of both lolines and ergovaline increase under conditions of water stress (Arechavaleta et al., 1992; Belesky et al., 1989; Kennedy and Bush, 1983). Ergovaline concentrations increase with applied nitrogen (Azevedo et al., 1993; Lyons et al., 1986; Rottinghaus et al., 1991), but effects on lolines are less clear-cut (Bush et al., 1993).

In field-grown *N. lolii*-infected perennial ryegrass, ergovaline concentrations increase under water stress and with applied nitrogen (Lane et al., 1997c). No consistent trends have been seen for lolitrem B and peramine concentrations. The most dramatic report of stress effects on endophyte secondary metabolism is that of Huizing and co-workers (1991), who found that *N. lolii*-infected *L. perenne* plants grown at 30°C, but not at 23°C, produced *N*-acetylloline and *N*-formylloline. This contrasts with the reported lowered concentrations of ergovaline in *N. coenophialum*-infected tall fescue at high ambient temperatures in midsummer in the southeastern USA (Adcock et al., 1997a).

Whatever the cause(s) of changes in endophyte secondary metabolite concentrations in response to changes in plant growth conditions, elevated levels of lolines and ergovaline in a water-stressed plant would presumably increase the effectiveness of its defenses at a time when compensatory growth mechanisms for coping with herbivory were not available. The reported temperature-dependent synthesis of lolines in *N. lolii–L. perenne* (Huizing et al., 1991) might be viewed in the same light. The benefits of higher alkaloid concentrations under high nitrogen growth conditions are less evident. For example, Bultmann and Conard (1997) found that endophyte infection had less effect on insect performance in fertilized plants.

6.5. Variation in Concentration with Plant and Fungal Genotype

There is a growing body of evidence indicating that if the qualitative pattern of metabolite expression in the plant is controlled by fungal genes, quantitative expression embodies a considerable plant genetic component.

Hill and co-workers (Adcock et al., 1997b; Hiatt and Hill, 1997; Roylance et al., 1994) have established that both plant and endophyte genes are important in determining the concentration of ergovaline in *N. coenophialum*-infected tall fescue plants, and that the compatibility of plant and fungus is not the only factor responsible for the variation of alkaloid concentrations in the plant. Selection of high- and low-alkaloid endophyte-infected plant lines has been demonstrated and, in a diallel crossing experiment, a significant paternal (i.e., plant only) inheritance was observed, in addition to the major maternal (endophyte + plant) effect (Adcock et al., 1997b). Different endophyte isolates in tissue culture regenerants of the same plant genotype resulted in variable ergovaline concentrations that did not correlate with mycelial mass (Hiatt and Hill, 1997). For a single endophyte genotype maternally inherited in half-sib plant genotypes, variation in endophyte mycelial mass (i.e., compatibility) affected ergovaline concentration, as did variation in plant morphology (leaf blade/sheath ratios), but other plant factors also appear to be involved (Hiatt and Hill, 1997). Peramine concentrations appear to be regulated independently from those of ergovaline (Roylance et al., 1994).

In natural *N. lolii–L. perenne* associations, a correlation was found between in planta endophyte concentrations and the concentrations of lolitrem B and peramine (Ball et al., 1995b); there was, however, a weaker correlation between the concentrations of endophyte and ergovaline (Ball et al., 1995a). A preliminary report (Fletcher and Easton, 1997) of diallel crossings of perennial ryegrass infected with a novel endophyte indicates similar findings to those of Adcock et al. (1997b).

Further evidence that the plant genome is important in regulating in planta ergovaline concentrations, independent of the amount of mycelium, has been provided by a study of several tall fescue cultivars and a meadow fescue cultivar infected with strains of *N. coenophialum* and *Neotyphodium* sp. FaTG-2 (Christensen et al., 1997b). The relationship between the ergovaline concentration and the amount of fungus in the leaf blade was found to differ with fungal strain, and with plant cultivar and species.

Thus, plant and fungal genes are involved in regulating the growth of endophyte mycelium in the plant and the accumulation of endophyte alkaloids (or at least, of ergovaline). There appears to be a wide range of variation within both plant and fungal gene pools for both characteristics (Hiatt and Hill, 1997). The apparently static gene pool of the asexual endophyte population, when superimposed on the genetic diversity generated by its sexually recombining host, creates a wide range of host–endophyte phenotypes on which selection pressures may act (Adcock et al., 1997b). Each seedling faces a new test of the defensive efficacy of the endophyte genotype in a new host genotype. The toxic alkaloid concentrations found in endophyte-infected grass populations in agricultural ecosystems may well be the product of the recent selection pressures in such ecosystems, rather than those of the natural environments in which these symbioses evolved (Adcock et al., 1997b).

Asexual *Neotyphodium* endophytes appear to be an evolutionary dead end, and coinfection by *Epichloe* sp. and the formation of new hybrid species has been suggested (Schardl, 1996; Wilkinson and Schardl, 1997) as a mechanism by which asexual endophytes can avoid Muller's ratchet—the progressive loss of fitness by the accumulation of marginally deleterious mutations in clonal lineages (Muller, 1964). However, the combination of the selection pressures imposed by the endophyte life cycle in the plant and the diverse genetic background of an outcrossing host may ensure that this is not a problem for asexual endophytes. A mutation in the endophyte genome which, in the context of a particular plant genotype, alters endophyte fitness (e.g., by altering alkaloid expression) simply adds to the range of variation available for selection. If the endophyte fails the hurdles of transmission or fails to provide selective advantage to its hosts over a period of generations (e.g., by not providing adequate defensive chemistry to the plant), it will be eliminated from the population.

The variety of potent secondary metabolites expressed in these symbioses appears to be essentially a selection from the *Epichloe/Neotyphodium* alkaloid smorgasbord. These studies of the distribution and quantitative variation of secondary metabolites in grass symbioses with type III *Neotyphodium* endophytes show evidence of coevolutionary adjustment of the yield and distribution of fungal secondary metabolites within the plant, in time and space, with mutual benefits for both partners. Further studies are required to establish whether these adjustments are peculiar to the mutualistic symbioses or common throughout the systemically infective Balansieae.

7. PROSPECTS FOR COEVOLUTIONARY STUDIES OF SECONDARY METABOLISM

Grass–endophyte symbioses provide a remarkable system for studying chemical ecology, with the ready availability of the endophyte-free control, and the ability to introduce endophytes of defined chemotype into the host plant. With the techniques of molecular biology coming to hand for fingerprinting genotypes and monitoring gene distribution and expression, there are new opportunities to investigate important biological questions. However, to establish conclusively the significance of endophyte secondary metabolites in the evolution of grass–endophyte symbioses we will need to be able to demonstrate that, for both partners, genes involved in secondary metabolite expression in planta have been shaped by the relationship. This remains a major challenge. Further progress will require a broader view of endophyte–grass associations, with greater emphasis on those in natural ecosystems.

Before tackling these issues, some basic questions about the transmission and natural selection of endophytes require answers. The role of vegetative (clonal) propagation of perennial grasses in endophyte population dynamics requires more attention. While the literature emphasizes transmission through seed, clonal propagation of the host grass, by the separation of clumps and individual tillers, will be a major route of endophyte transmission in perennial grasses, as annual reseeding usually provides only a very small proportion of the plants in a sward (Bullock et al., 1994; Hume and Barker, 1991). Endophyte–host populations may be shaped considerably by selection and clonal propagation of those combinations of endophyte and host genotype best adapted to an ecosystem. Experimental studies of variation in the efficiency of transmission of endophyte to progeny would be of value. For example, it is not known to what degree the variation found between endophyte genotypes in mycelial growth in plants (Hiatt and Hill, 1997) correlates with transmission efficiency during plant propagation (either clonally or through seed).

Quantitative studies of the distribution and variation of the secondary metabolites in grass associations with *Epichloe* endophytes could clarify some coevolutionary arguments. Defensive chemistry appears to be related to transmission mode for asexual seedborne (vertically transmitted) *Neotyphodium* endophytes, reflecting selection pressures favoring the survival and reproduction of the host. For the type I *Epichloe* endophytes, which undergo sexual (horizontal) transmission through stroma formation, and type II *Epichloe* endophytes (which utilize both horizontal and vertical modes of transmission), the secondary metabolite chemistry might be expected to benefit the fungus more directly. The findings of Yue et al. (1997) suggest that the seedborne route attracts a greater share of the defensive resource in type II endophytes. Further studies to elaborate this, and the role of secondary metabolites in the life cycle of the type I endophytes (including their defensive role in vegetatively propagating plants), could clarify the degree to which secondary metabolism has evolved in the transition from parasitism to mutualism between these fungi and their plant hosts.

Investigations of natural endophyte–grass populations could provide evidence for the efficacy of selection pressures in altering the balance of endophyte compatibility and alkaloid accumulation in plant populations. Measurements of the concentrations of endophyte mycelium and alkaloids in individual infected plants, in natural ecosystems with different levels of endophyte infection, could clarify the significance of alkaloid accumulation to the selection of endophyte-infected plants under different selection pressures. Similarly, evidence suggesting that the diversity of chemotypes is much wider in natural than in agricultural ecosystems needs to be put on a more quantitative footing.

The understanding of the factors regulating the quantitative expression of alkaloid synthesis in plants can be expected to make rapid progress with the development of marker gene technology (Herd et al., 1997). With this methodology and the likely availability in the near future of probes for expression of key genes in secondary metabolite biosynthesis, there are good prospects for clarifying whether the regulation of metabolite production in the plant is modulated by the supply to the endophyte of nutrients [e.g., sugars (Arechavaleta et al., 1992; Richardson et al., 1992; White et al., 1991), tryptophan and calcium (Hiatt and Hill, 1997)], by phytohormones (De Battista et al., 1990) or other regulatory factors, or by a combination of these. Developments in this area may shed considerable light on the degree to which fungus and plant have coevolved.

The techniques of molecular genetics can also be expected to further expedite progress towards understanding the biosynthetic pathways to endophyte alkaloids (c.f. Panaccione, 1996; Wilkinson and Schardl, 1997; Young et al., 1998).

There are also good prospects for clarifying whether plant enzymes are directly implicated in the biosynthesis of lolines (Siegel and Bush, 1997) and of modified ergopeptides (Shelby et al., 1997), issues of considerable significance in terms of coevolution and secondary metabolism. The difficulty in obtaining adequate levels of biosynthesis in culture of some toxin classes, particularly indole–diterpenoids and lolines, remains a significant obstacle in elucidating the biosynthesis of these compounds by endophytes.

There is also scope for progress toward a more complete knowledge of the suite of chemical defenses in *Neotyphodium/Epichloe*–grass associations. The unidentified compounds detected by ELISA in *A. inebrians* (Stevenson, 1996), *F. argentina* (Miles et al., 1995), *P. huecu*, *E. ovatus*, *M. decumbens* (Miles et al., 1998; Towers, 1997), and *L. perenne* (Miles et al., 1993) await identification, as do the unidentified antifungal agents of *Neotyphodium* spp. (Christensen, 1996). The chemical basis of allelopathic properties of endophyte-infected grasses (Springer, 1996; Sutherland and Hoglund, 1989) requires closer examination, and the wider distribution of the *E. typhina* metabolites identified by Koshino and co-workers (Koshino et al., 1992b and references cited) also remains to be explored. It is unlikely that all classes of secondary metabolites produced in grass symbioses with *Neotyphodium/Epichloe* spp. have yet been discovered, and a better understanding of the diversity of metabolites and their ecological roles will provide more insight into their evolutionary significance.

Genetic methods will also be of increasing value in understanding the relationships between the diversity of endophyte chemistry and fungal and host phylogeny. As probes for biosynthetic genes become available, it will be possible for the distribution of synthesis genes among *Neotyphodium/Epichloe* spp. to be directly investigated and the genetic basis of their expression in planta clarified. Gene probes will also be useful in defining the relationship between genetic and chemotypic diversity in natural populations.

The *Echinopogon* spp. of Australia and New Zealand are of particular interest in this regard (Miles et al., 1998), as their geographic isolation would suggest that *Neotyphodium* spp. infecting these grasses have had considerable opportunity to diverge from their Northern Hemisphere relatives. As infective *Epichloe* species have yet to be reported from native grasses in the Southern Hemisphere, there is no obvious candidate sexual progenitor of these asexual endophytes, raising the interesting possibility that these *Neotyphodium* endophytes may have persisted over millennia by asexual transmission. This system also provides an opportunity to investigate natural endophyte–grass associations that do not produce any of the known *Neotyphodium/Epichloe* defensive metabolites. Unless the production of defensive metabolites is less essential to the persistence of asexual endophytes in grasses than we have supposed, the absence of known metabolites is an indication to chemists of the likely presence of yet-to-be-discovered defensive components.

REFERENCES

Adcock, R. A., Hill, N. S., Boerma, H. R. and Ware, G. O. (1997a). Sample variation and resource allocation for ergot alkaloid characterization in endophyte-infected tall fescue. *Crop Science* 37, 31–35.

Adcock, R. A., Hill, N. S., Bouton, J. H., Boerma, H. R. and Ware, G. O. (1997b). Symbiont regulation and reducing ergot alkaloid concentration by breeding endophyte-infected tall fescue. *Journal of Chemical Ecology* 23, 691–704.

Agee, C. S. and Hill, N. S. (1994). Ergovaline variability in *Acremonium*-infected tall fescue due to environment and plant genotype. *Crop Science* 34, 221–226.

Arechavaleta, M., Bacon, C. W., Plattner, R. D., Hoveland, C. S. and Radcliffe, D. E. (1992). Accumulation of ergopeptide alkaloids in symbiotic tall fescue grown under deficits of soil water and nitrogen fertilizer. *Applied and Environmental Microbiology* 58, 857–861.

Azevedo, M. D., Welty, R. E., Craig, A. M. and Bartlett, J. (1993). Ergovaline distribution, total nitrogen and phosphorus content of two endophyte-infected tall fescue clones. *In* D. E. Hume, G. C. M. Latch and H. S. Easton (eds.), *Proceedings of the Second International Symposium on* Acremonium/*Grass Interactions*. AgResearch, Palmerston North, New Zealand, pp. 59–62.

Bacon, C. W. and Hill, N. S. (1996). Symptomless grass endophytes: products of coevolutionary symbioses and their role in the ecological adaptations of grasses. *In* S. C. Redlin and L. M. Carris (eds.), *Endophytic Fungi in Grasses and Woody Plants: Systematics, Ecology, and Evolution*. American Phytopathological Society (APS Press), St. Paul, MN.

Bacon, C. W., Lyons, P. C., Porter, J. K. and Robbins, J. D. (1986). Ergot toxicity from endophyte-infected grasses: a review. *Agronomy Journal* 78, 106–116.

Bacon, C. W., Porter, J. K., Robbins, J. D. and Luttrell, E. S. (1977). *Epichloe typhina* from toxic tall fescue grasses. *Applied and Environmental Microbiology* 34, 576–581.

Ball, O. J.-P., Lane, G. A. and Prestidge, R. A. (1995a). *Acremonium lolii*, ergovaline and peramine production in endophyte-infected perennial ryegrass. *Proceedings of the Forty Eighth New Zealand Plant Protection Conference*, pp. 224–228.

Ball, O. J.-P., Prestidge, R. A. and Sprosen, J. M. (1995b). Interrelationships between *Acremonium lolii*, peramine, and lolitrem B in perennial ryegrass. *Applied and Environmental Microbiology* 61, 1527–1533.

Ball, O. J.-P., Barker, G. M., Prestidge, R. A. and Lauren, D. R. (1997a). Distribution and accumulation of the alkaloid peramine in *Neotyphodium lolii*-infected perennial ryegrass. *Journal of Chemical Ecology* 23, 1419–1434.

Ball, O. J.-P., Barker, G. M., Prestidge, R. A. and Sprosen, J. M. (1997b). Distribution and accumulation of the mycotoxin lolitrem B in *Neotyphodium lolii*-infected perennial ryegrass. *Journal of Chemical Ecology* 23, 1435–1449.

Ball, O. J.-P., Miles, C. O. and Prestidge, R. A. (1997c). Ergopeptine alkaloids and *Neotyphodium lolii*-mediated resistance in perennial ryegrass against adult *Heteronychus arator* (Coleoptera: Scarabaeidae). *Journal of Economic Entomology* 90, 1382–1391.

Ball, O. J.-P., Prestidge, R. A. and Sprosen, J. M. (1993). Effect of plant age and endophyte viability on peramine and lolitrem B concentration in perennial ryegrass seedlings. *In* D. E. Hume, G. C. M. Latch and H. S. Easton (eds.), *Proceedings of the Second International Symposium on* Acremonium/*Grass Interactions*. AgResearch, Palmerston North, New Zealand, pp. 63–65.

Belesky, D. P. and Hill, N. S. (1997). Defoliation and leaf age influence on ergot alkaloids in tall fescue. *Annals of Botany* 79, 259–264.

Belesky, D. P., Stringer, W. C. and Plattner, R. D. (1989). Influence of endophyte and water regime upon tall fescue accessions. II. Pyrrolizidine and ergopeptine alkaloids. *Annals of Botany* 64, 343–349.

Belesky, D. P., Stuedemann, J. A., Plattner, R. D. and Wilkinson, S. R. (1988). Ergopeptine alkaloids in grazed tall fescue. *Agronomy Journal* 80, 209–212.

Belofsky, G. N., Gloer, J. B., Wicklow, D. T. and Dowd, P. F. (1995). Antiinsectan alkaloids shearinines A–C and a new paxilline derivative from the ascostromata of *Eupenicillium shearii*. *Tetrahedron* 51, 3959–3968.

Berde, B. and Schild, H. O. (1978). Ergot alkaloids and related compounds. *In Handbook of experimental pharmacology*, Vol. 49. Springer-Verlag. Berlin.

Bradshaw, A. D. (1959). Population differentiation in *Agrostis stolonifera* Sibth. II. The incidence and significance of infection by *Epichloe typhina*. *New Phytologist* 58, 310–315.

Brunner, R., Stutz, P. L., Tscherter, H. and Stadler, P. A. (1979). Isolation of ergovaline, ergoptine and ergonine, new alkaloids of the peptide type, from ergot sclerotia (*Claviceps purpurea*). *Canadian Journal of Chemistry* 57, 1638–1641.

Bullock, J. M., Hill, B. C. and Silvertown, J. (1994). Tiller dynamics of two grasses — responses to grazing, density and weather. *Journal of Ecology (Oxford)* 82, 331–340.

Bultmann, T. L. and Conard, N. J. (1997). Effects of endophytic fungus, nitrogen, and plant damage on performance of fall armyworm. *In* C. W. Bacon and N. S. Hill (eds.), Neotyphodium/*Grass Interactions*. Plenum Press, New York, pp. 145–148.

Bush, L. P., Cornelius, P. L., Buckner, R. C., Varney, S. R., Chapman, R. A., Burrus II, P. B., Kennedy, C. W., Jones, T. A. and Saunders, M. A. (1982). Association of *N*-acetyl loline and *N*-formyl loline with *Epichloe typhina* in tall fescue. *Crop Science* 22, 941–943.

Bush, L. P., Fannin, F. F., Siegel, M. R., Dahlman, D. L. and Burton, H. R. (1993). Chemistry, occurrence and biological effects of saturated pyrrolizidine alkaloids associated with endophyte–grass interactions. *Agriculture Ecosystems Environment* 44, 81–102.

Bush, L. P., Wilkinson, H. H. and Schardl, C. L. (1997). Bioprotective alkaloids of grass–fungal endophyte symbioses. *Plant Physiology* 114, 1–7.

Casabuono, A. C. and Pomilio, A. B. (1997). Alkaloids from endophyte-infected *Festuca argentina*. *Journal of Ethnopharmacology* 57, 1–9.

Christensen, M. J. (1995). Variation in the ability of *Acremonium* endophytes of perennial ryegrass (*Lolium perenne*), tall fescue (*Festuca arundinacea*) and meadow fescue (*F. pratensis*) to form compatible associations in the three grasses. *Mycological Research* 99, 466–470.

Christensen, M. J. (1996). Antifungal activity in grasses infected with *Acremonium* and *Epichloe* endophytes. *Australasian Plant Pathology* 25, 186–191.

Christensen, M. J., Ball, O. J.-P., Bennett, R. J. and Schardl, C. L. (1997a). Fungal and host genotype effects on compatibility and vascular colonization by *Epichloe festucae*. *Mycological Research* 101, 493–501.

Christensen, M. J., Lane, G. A., Simpson, W. R. and Tapper, B. A. (1997b). Leaf blade colonization by two *Neotyphodium* endophytes, and ergovaline distribution within leaves of tall fescue and meadow fescue. In: C. W. Bacon and N. S. Hill (eds.), Neotyphodium/*Grass Interactions*. Plenum Press, New York, pp. 149–151.

Christensen, M. J., Leuchtmann, A., Rowan, D. D. and Tapper, B. A. (1993). Taxonomy of *Acremonium* endophytes of tall fescue (*Festuca arundinacea*), meadow fescue (*F. pratensis*) and perennial ryegrass (*Lolium perenne*). *Mycological Research* 97, 1083–1092.

Chung, K. R. and Schardl, C. L. (1997). Sexual cycle and horizontal transmission of the grass symbiont, *Epichloe typhina*. *Mycological Research* 101, 295–301.

Clay, K. (1988). Clavicipitaceous fungal endophytes of grasses: coevolution and the change from parasitism to mutualism. *In* K. A. Pirozynski and D. L. Hawksworth (eds.), *Coevolution of Fungi with Plants and Animals*. Academic Press, London, pp. 79–105.

Clay, K. (1997). Consequences of endophyte-infected grasses on plant biodiversity. *In* C. W. Beacon and N. S. Hill (eds.), Neotyphodium/*Grass Interactions*. Plenum Press, New York, pp. 109–124.

Clay, K. and Cheplick, G. P. (1989). Effect of ergot alkaloids from fungal endophyte-infected grasses on fall armyworm (*Spodoptera frugiperda*). *Journal of Chemical Ecology* 15, 169–182.

Clay, K. and Leuchtmann, A. (1989). Infection of woodland grasses by fungal endophytes. *Mycologia* 81, 805–811.

Clement, S. L., Graves, W., Cunningham, P., Nebling, V., Bounejmate, W., Saidi, S., Baya, B., Chakroun, M., Mezni, A. and Porqueddu, C. (1997). *Acremonium* endophytes in Mediterranean tall fescue. *In* C. W. Bacon and N. S. Hill (eds.), Neotyphodium/*Grass Interactions*. Plenum Press, New York, pp. 49–51.

Clement, S. L., Kaiser, W. J. and Eichenseer, H. (1994). *Acremonium* endophytes in germplasms of major grasses and their utilization for insect resistance. *In* C. W. Bacon and J. F. White, Jr. (eds.), *Biotechnology of Endophytic Fungi of Grasses*. CRC Press, Boca Raton, FL, pp. 185–200.

Cordell, G. A. (1981). *Introduction to Alkaloids: A Biogenetic Approach*. John Wiley and Sons, New York.

Cunningham, I. J. (1949). A note on the cause of tall fescue lameness in cattle. *Australian Veterinary Journal* 25, 27–28.

Cysewski, S. J. (1973). Paspalum staggers and tremorgen intoxication in animals. *Journal of the American Veterinary Medical Association* 163, 1291–1292.

Dahlman, D. L., Eichenseer, H. and Siegel, M. R. (1991). Chemical perspectives on endophyte–grass interactions and their implications for insect herbivory. *In* P. Barbosa, V. A. Krischik and C. G. Jones (eds.), *Microbial Mediation of Plant–Herbivore Interactions*. John Wiley and Sons, New York, pp. 227–252.

Davies, E., Lane, G. A., Latch, G. C. M., Tapper, B. A., Garthwaite, I., Towers, N. R., Fletcher, L. R. and Pownall, D. B. (1993). Alkaloid concentrations in field-grown synthetic ryegrass endophyte associations. *In* D. E. Hume, G. C. M. Latch, and H. Easton (eds.), *Proceedings of the Second International Symposium on* Acremonium/*Grass Interactions*. AgResearch, Palmerston North, New Zealand, pp. 72–76.

Davis, N. D., Cole, R. J., Dorner, J. W., Weete, J. D., Backman, P. A., Clark, E. M., King, C. C., Schmidt, S. P. and Diener, U. L. (1986). Steroid metabolites of *Acremonium coenophialum*, an endophyte of tall fescue. *Journal of Agricultural and Food Chemistry* 34, 105–108.

De Battista, J., Altier, N., Galdames, D. R. and Dall'Agnol, M. (1997). Significance of endophyte toxicosis and current practices in dealing with the problem in South America. *In* C. W. Bacon and N. S. Hill (eds.), Neotyphodium/*Grass Interactions*. Plenum Press, New York, pp. 383–388.

De Battista, J. P., Bacon, C. W., Severson, R. F., Plattner, R. D. and Bouton, J. H. (1990). Indole acetic acid production by fungal endophyte of tall fescue. *Agronomy Journal* 82, 878–880.

di Menna, M. E., Mortimer, P. H., Prestidge, R. A., Hawkes, A. D. and Sprosen, J. M. (1992). Lolitrem B concentrations, counts of *Acremonium lolii* hyphae, and the incidence of ryegrass staggers in lambs on plots of *A. lolii*-infected perennial ryegrass. *New Zealand Journal of Agricultural Research* 35, 211–217.

di Menna, M. E. and Waller, J. E. (1986). Visual assessment of seasonal changes in amount of mycelium of *Acremonium loliae* in leaf sheaths of perennial ryegrass. *New Zealand Journal of Agricultural Research* 29, 111–116.

Didek-Brumec, M., Gaberc-Porekar, V. and Alacevic, M. (1996). Relationship between the *Claviceps* life cycle and productivity of ergot alkaloids. *Critical Reviews in Biotechnology* 16, 257–299.

Faeth, S. H., Wilson, D., Helander, M., Saikkonen, K., Schulthess, F. and Sullivan, T. J. (1997). *Neotyphodium* in native populations of Arizona fescue. *In* C. W. Bacon and N. S. Hill (eds.), Neotyphodium/*Grass Interactions*. Plenum Press, New York, pp. 165–166.

Fletcher, L. R. and Easton, H. S. (1997). The evaluation and use of endophytes for pasture improvement. *In* C. W. Bacon and N. S. Hill (eds.), Neotyphodium/*Grass Interactions*. Plenum Press, New York, pp. 209–227.

Fletcher, L. R. and Harvey, I. C. (1981). An association of a *Lolium* endophyte with ryegrass staggers. *New Zealand Veterinary Journal* 29, 185–186.

Flieger, M., Wurst, M. and Shelby, R. (1997). Ergot alkaloids—sources, structures and analytical methods. *Folia Microbiologica* 42, 3–30.

Gadberry, M. S., Denard, T. M., Spiers, D. E. and Piper, E. L. (1997). *Ovies aries*: a model for studying the effects of fescue toxins on animal performance in a heat-stress environment. *In* C. W. Bacon and N. S. Hill (eds.), Neotyphodium/*Grass Interactions*. Plenum Press, New York, pp. 429–431.

Gallagher, R. T., Campbell, A. G., Hawkes, A. D., Holland, P. T., McGaveston, D. A., Pansier, E. A. and Harvey, I. C. (1982). Ryegrass staggers: the presence of lolitrem neurotoxins in perennial ryegrass. *New Zealand Veterinary Journal* 30, 183–184.

Gallagher, R. T., Hawkes, A. D., Steyn, P. S. and Vleggaar, R. (1984). Tremorgenic neurotoxins from perennial ryegrass causing ryegrass staggers disorder of livestock: structure elucidation of lolitrem B. *Chemical Communications*, 614–616.

Gallagher, R. T., Smith, G. S. and Sprosen, J. M. (1987). The distribution and accumulation of lolitrem B neurotoxin in developing perennial ryegrass plants. *In: 4th Animal Science Congress of the Asian–Australasian Association of Animal Production Societies*. Hamilton, New Zealand, p. 404.

Gallagher, R. T., White, E. P. and Mortimer, P. H. (1981). Ryegrass staggers: isolation of potent neurotoxins lolitrem A and lolitrem B from staggers-producing pastures. *New Zealand Veterinary Journal* 29, 189–190.

Gatenby, W. A. (1997). An investigation of some mycotoxins involved in ryegrass staggers. MSc thesis, University of Waikato, Hamilton, New Zealand.

Glenn, A. E. and Bacon, C. W. (1997). Distribution of ergot alkaloids within the family Clavicipitaceae. *In* C. W. Bacon and N. S. Hill (eds.), Neotyphodium/*Grass Interactions*. Plenum Press, New York, pp. 53–56.

Glenn, A. E., Bacon, C. W., Price, R. and Hanlin, R. T. (1996). Molecular phylogeny of *Acremonium* and its taxonomic implications. *Mycologia* 88, 369–383.

Greinwald, R., Bachmann, P., Witte, L., Acebes-Grinoves, J. R. and Czygan. F.-C. (1992). Taxonomic significance of alkaloids in the genus *Adenocarpus* (Fabaceae-Genisteae). *Biochemical Systematics and Ecology* 20, 69–73.

Gröger, D. and Floss, H. G. (1998). Biochemistry of ergot alkaloids—achievements and challenges. *In* G. A. Cordell (eds.), *The Alkaloids. Chemistry and Biology*, Vol. 50. *The Alkaloids: Chemistry and Pharmacology*, Academic Press, San Diego, pp. 172–218.

Gurney, K. A., Mantle, P. G., Penn, J., Garthwaite, I. and Towers, N. R. (1994). Loss of toxic metabolites from *Acremonium lolii*, the endophyte of ryegrass, following mutagenesis. *Naturwissenschaften* 81, 362–365.

Herd, S., Christensen, M. J., Saunders, K., Scott, D. B. and Schmid, J. (1997). Quantitative assessment of *in planta* distribution of metabolic activity and gene expression of an endophytic fungus. *Microbiology* 143, 267–275.

Hiatt, E. E., III and Hill, N. S. (1997). *Neotyphodium coenophialum* mycelial protein and herbage mass effects on ergot alkaloid concentration in tall fescue. *Journal of Chemical Ecology* 23, 2721–2736.

Hinton, D. M. and Bacon, C. W. (1985). The distribution and ultrastructure of the endophyte of toxic tall fescue. *Canadian Journal of Botany* 63, 35–42.

Huang, X.-H., Tomoda, H., Nishida, H., Masuma, R. and Ōmura, S. (1995). Terpendoles, novel ACAT inhibitors produced by *Albophoma yamanashiensis* I. Production, isolation and biological properties. *The Journal of Antibiotics* 43, 1–4.

Huizing, H. J., van der Molen, W., Kloek, W. and den Nijs, A. P. M. (1991). Detection of lolines in endophyte-containing meadow fescue in The Netherlands and the effect of elevated temperature on induction of lolines in endophyte-infected perennial ryegrass. *Grass and Forage Science* 46, 441–445.

Hume, D. E. (1993). Agronomic performance of New Zealand pastures: implications of *Acremonium* presence. *In* D. E. Hume, G. C. M. Latch and H. S. Easton (eds.), *Proceedings of the Second International Symposium on* Acremonium/*Grass Interactions, Plenary Papers*. AgResearch, Palmerston North, New Zealand, pp. 31–38.

380

Lane et al.

Hume, D. E. and Barker, D. J. (1991). Natural reseeding of five grass species in summer dry hill country. *Proceedings of the New Zealand Grassland Association* 53, 97–104.

Hume, D. E. and Brock, J. L. (1997). Increases in endophyte incidence in perennial ryegrass at Palmerston North, Manawatu, New Zealand. *In* C. W. Bacon and N. S. Hill (eds.), Neotyphodium/*Grass Interactions*. Plenum Press, New York, pp. 61–63.

Jenett-Siems, K., Kaloga, M. and Eich, E. (1994). Ergobalansine/ergobalansinine, a proline-free peptide-type alkaloid of the fungal genus *Balansia*, is a constituent of *Ipomoea piurensis*. *Journal of Natural Products* 57, 1304–1306.

Joost, R. E. (1995). *Acremonium* in fescue and ryegrass: boon or bane? A review. *Journal of Animal Science* 73, 881–888.

Justus, M., Witte, L. and Hartmann, T. (1997). Levels and tissue distribution of loline alkaloids in endophyte-infected *Festuca pratensis*. *Phytochemistry* 44, 51–57.

Käss, E. and Wink, M. (1995). Molecular phylogeny of the Papilionoideae (family Leguminosae): rbcL gene sequences versus chemical taxonomy. *Botanica Acta* 108, 149–162.

Kawai, K.-I. and Nozawa, K. (1988). Novel biologically active compounds from *Emericella* species. *In Seventh International IUPAC Symposium on Mycotoxins and Phycotoxins*, Tokyo, pp. 205–212.

Kennedy, C. W. and Bush, L. P. (1983). Effect of environmental and management factors on the accumulation of *N*-acetyl and *N*-formyl loline alkaloids in tall fescue. *Crop Science* 23, 547–552.

Keogh, R. G., Tapper, B. A. and Fletcher, R. H. (1996). Distributions of the fungal endophyte *Acremonium lolii*, and of the alkaloids lolitrem B and peramine, within perennial ryegrass. *New Zealand Journal of Agricultural Research* 39, 121–127.

Kimura, Y., Nishibe, M., Nakajima, H., Hamasaki, T., Shigemitsu, N., Sugawara, F., Stout, T. J. and Clardy, J. (1992). Emeniveol; a new pollen growth inhibitor from the fungus, *Emericella nivea*. *Tetrahedron Letters* 33, 6987–6990.

Knaus, H.-G., McManus, O. B., Lee, S. H., Schmalhofer, W. A., Garcia-Calvo, M., Helms, L. M. H., Sanchez, M., Giangiacomo, K., Smith III, A. B., Kaczorowski, G. J. and Garcia, M. L. (1994). Tremorgenic indole alkaloids potently inhibit smooth muscle high-conductance calcium-activated potassium channels. *Biochemistry* 33, 5819–5828.

Koga, H., Christensen, M. J. and Bennett, R. J. (1993). Incompatibility of some grass–*Acremonium* endophyte associations. *Mycological Research* 97, 1237–1244.

Koshino, H., Terada, S.-I., Yoshihara, T., Sakamura, S., Shimanuki, T., Sato, T. and Tajimi, A. (1988). Three phenolic acid derivatives from stromata of *Epichloe typhina* on *Phleum pratense*. *Phytochemistry* 27, 1333–1338.

Koshino, H., Togiya, S., Terada, S. I., Yoshihara, T., Sakamura, S., Shimanuki, T., Sato, T. and Tajimi, A. (1989a). New fungitoxic sesquiterpenoids, chokols A–G, from stromata of *Epichloe typhina* and the absolute configuration of chokol E. *Agricultural and Biological Chemistry* 53, 789–796.

Koshino, H., Togiya, S., Yoshihara, T., Sakamura, S., Shimanuki, T., Sato, T. and Tajimi, A. (1987). Four fungitoxic C-18 hydroxy unsaturated fatty acids from stromata of *Epichloe typhina*. *Tetrahedron Letters* 28, 73–76.

Koshino, H., Yoshihara, T., Ichihara, A., Tajimi, A. and Shimanuki, T. (1992a). Two sphingoid derivatives from stromata of *Epichloe typhina* on *Phleum pratense. Phytochemistry* 31, 3757–3759.

Koshino, H., Yoshihara, T., Okuno, M., Sakamura, S., Tajimi, A. and Shimanuki, T. (1992b). Gamahonolides A, B, and gamahorin, novel antifungal compounds from stromata of *Epichloe typhina* on *Phleum pratense. Bioscience Biotechnology and Biochemistry* 56, 1096–1099.

Koshino, H., Yoshihara, T., Sakamura, S., Shimanuki, T., Sato, T. and Tajimi, A. (1989b). Novel C-11 epoxy fatty acid from stromata of *Epichloe typhina* on *Phleum pratense. Agricultural and Biological Chemistry* 53, 2527–2528.

Koshino, H., Yoshihara, T., Sakamura, S., Shimanuki, T., Sato, T. and Tajimi, A. (1989c). A ring B aromatic sterol from stromata of *Epichloe typhina. Phytochemistry* 28, 771–772.

Kren, V., Rezanka, T. and Rehacek, Z. (1985). Occurrence of ricinoleic acid in submerged cultures of various *Claviceps* sp. *Experientia* 41, 1476–1477.

Lane, G. A., Ball, O. J.-P., Davies, E. and Davidson, C. (1997a). Ergovaline distribution in perennial ryegrass naturally infected with endophyte. *In* C. W. Bacon and N. S. Hill (eds.), Neotyphodium/*Grass Interactions*. Plenum Press, New York, pp. 65–67.

Lane, G. A., Tapper, B. A., Davies, E., Christensen, M. J. and Latch, G. C. M. (1997b). Occurrence of extreme alkaloid levels in endophyte-infected perennial ryegrass, tall fescue and meadow fescue. *In* C. W. Bacon and N. S. Hill (eds.), Neotyphodium/*Grass Interactions*. Plenum Press, New York, pp. 433–436.

Lane, G. A., Tapper, B. A., Davies, E., Hume, D. E., Latch, G. C. M., Barker, D. J., Easton, H. S. and Rolston, M. P. (1997c). Effect of growth conditions on alkaloid concentrations in perennial ryegrass naturally infected with endophyte. *In* C. W. Bacon and N. S. Hill (eds.), Neotyphodium/*Grass Interactions*. Plenum Press, New York, pp. 179–182.

Latch, G. C. M. and Christensen, M. J. (1985). Artificial infection of grasses with endophytes. *Annals of Applied Biology* 107, 17–24.

Latch, G. C. M., Christensen, M. J. and Hickson, R. E. (1988). Endophytes of annual and hybrid ryegrasses. *New Zealand Journal of Agricultural Research* 31, 57–63.

Latch, G. C. M., Christensen, M. J. and Samuels, G. J. (1984). Five endophytes of *Lolium* and *Festuca* in New Zealand. *Mycotaxon* 20, 535-550.

Latch, G. C. M., Potter, L. R. and Tyler, B. F. (1987). Incidence of endophytes in seeds from collections of *Lolium* and *Festuca* species. *Annals of Applied Biology* 111, 59–64.

Law, R. (1985). Evolution in a mutualistic environment. *In* D. H. Boucher (ed.), *The Biology of Mutualism: Ecology and Evolution*, Croom Helm, London, pp. 145–170.

Leuchtmann, A. (1992). Systematics, distribution, and host specificity of grass endophytes. *Natural Toxins* 1, 150–162.

Leuchtmann, A. (1997). Ecological diversity in *Neotyphodium*-infected grasses as influenced by host and fungus characteristics. *In* C. W. Bacon and N. S. Hill (eds.), Neotyphodium/*Grass Interactions* Plenum Press, New York, pp. 93–108.

Leuchtmann, A. and Clay, K. (1988). *Atkinsonella hypoxylon* and *Balansia cyperi*, epiphytic members of the Balansiae. *Mycologia* 80, 192–199.

Leuchtmann, A. and Clay, K. (1993). Nonreciprocal compatibility between *Epichloe typhina* and four host grasses. *Mycologia* 85, 157–163.

Leuchtmann, A. and Clay, K. (1997). The population biology of grass endophytes. *In* G. C. Carroll and P. Tudzynski (eds.), *The Mycota,* Vol. 5, *Plant Relationships*, Part B. Springer-Verlag, Berlin, pp. 185–202.

Lewis, G. C. (1997). Significance of endophyte toxicosis and current practices in dealing with the problem in Europe. *In* C. W. Bacon and N. S. Hill (eds.), Neotyphodium/ *Grass Interactions*. Plenum Press, New York, pp. 377–382.

Li, B., Zheng, X. and Sun, S. (1997). A survey of endophytic fungi in some native forage grasses of Northwestern China. *In* C. W. Bacon and N. S. Hill (eds.), Neotyphodium/*Grass Interactions*. Plenum Press, New York, pp. 69–71.

Lyons, P. C., Plattner, R. D. and Bacon, C. W. (1986). Occurrence of peptide and clavine ergot alkaloids in tall fescue grass. *Science* 232, 487–489.

Maag, D. D. and Tobiska, J. W. (1956). Fescue lameness in cattle. II. Ergot alkaloids in tall fescue grass. *American Journal of Veterinary Research*, 202–204.

Mantle, P. G. and Weedon, C. M. (1994). Biosynthesis and transformation of tremorgenic indole–diterpenoids by *Penicillium paxilli* and *Acremonium lolii*. *Phytochemistry* 36, 1209–1217.

Miles, C. O., di Menna, M. E., Jacobs, S. W. L., Garthwaite, I., Lane, G. A., Prestidge, R. A., Marshall, S. L., Wilkinson, H. H., Schardl, C. L., Ball, O. J.-P. and Latch, G. C. M. (1998). Endophytic fungi in indigenous Australasian grasses associated with toxicity to livestock. *Applied and Environmental Microbiology* 64, 601–606.

Miles, C. O., Lane, G. A., di Menna, M. E., Garthwaite, I., Piper, E. L., Ball, O. J.-P., Latch, G. C. M., Allen, J. M., Hunt, M. B., Bush, L. P., Min, F. K., Fletcher, I. and Harris, P. S. (1996). High levels of ergonovine and lysergic acid amide in toxic *Achnatherum inebrians* accompany infection by an *Acremonium*-like endophytic fungus. *Journal of Agricultural and Food Chemistry* 44, 1285–1290.

Miles, C. O., Munday, S. C., Wilkins, A. L., Ede, R. M., Meagher, L. P. and Garthwaite, I. (1993). Chemical aspects of ryegrass staggers. *New Zealand Veterinary Journal* 41, 216–217.

Miles, C. O., Munday, S. C., Wilkins, A. L., Ede, R. M. and Towers, N. R. (1994). Large-scale isolation of lolitrem B and structure determination of lolitrem E. *Journal of Agricultural and Food Chemistry* 42, 1488–1492.

Miles, C. O., Wilkins, A. L., Gallagher, R. T., Hawkes, A. D., Munday, S. C. and Towers, N. R. (1992). Synthesis and tremorgenicity of paxitriols and lolitriol: possible biosynthetic precursors of lolitrem B. *Journal of Agricultural and Food Chemistry* 40, 234–238.

Miles, C. O., Wilkins, A. L., Garthwaite, I., Ede, R. M. and Munday-Finch, S. C. (1995). Immunochemical techniques in natural products chemistry: isolation and structure determination of a novel indole–diterpenoid aided by TLC–ELISAgram. *Journal of Organic Chemistry* 60, 6067–6069.

Muller, H. J. (1964). The relation of recombination to mutational advance. *Mutation Research* 1, 2–9.

Munday-Finch, S. C. (1997). Aspects of the chemistry and toxicology of indole–diterpenoid mycotoxins involved in Tremorgenic Disorders of Livestock. DPhil thesis, University of Waikato, Hamilton, New Zealand.

Munday-Finch, S. C., Miles, C. O., Wilkins, A. L. and Hawkes, A. D. (1995). Isolation and structure elucidation of lolitrem A, a tremorgenic mycotoxin from perennial ryegrass infected with *Acremonium lolii*. *Journal of Agricultural and Food Chemistry* 43, 1283–1288.

Munday-Finch, S. C., Wilkins, A. L. and Miles, C. O. (1996a). Isolation of paspaline B, an indole–diterpenoid from *Penicillium paxilli*. *Phytochemistry* 41, 327–332.

Munday-Finch, S. C., Wilkins, A. L. and Miles, C. O. (1998). Isolation of lolicine A, lolicine B, lolitriol, and lolitrem N from *Lolium perenne* infected with *Neotyphodium lolii* and evidence for the natural occurrence of 31-epilolitrem N and 31-epilolitrem F. *Journal of Agricultural and Food Chemistry* 46, 590–598.

Munday-Finch, S. C., Wilkins, A. L., Miles, C. O., Ede, R. M. and Thomson, R. A. (1996b). Structure elucidation of lolitrem F, a naturally occurring stereoisomer of the tremorgenic mycotoxin lolitrem B, isolated from *Lolium perenne* infected with *Acremonium lolii*. *Journal of Agricultural and Food Chemistry* 44, 2782–2788.

Munday-Finch, S. C., Wilkins, A. L., Miles, C. O., Tomoda, H. and Ōmura, S. (1997). Isolation and structure elucidation of lolilline, a possible biosynthetic precursor of the lolitrem family of tremorgenic mycotoxins. *Journal of Agricultural and Food Chemistry* 45, 199–204.

Musgrave, D. R. (1984). Detection of an endophytic fungus of *Lolium perenne* using enzyme-linked immunosorbent assay (ELISA). *New Zealand Journal of Agricultural Research* 27, 283–288.

Neill, J. C. (1940). The endophyte of ryegrass (*Lolium perenne*). *New Zealand Journal of Science and Technology* A 21, 280A–291A.

Neill, J. C. (1941). The endophytes of *Lolium* and *Festuca*. *New Zealand Journal of Science and Technology* A 23, 185A–193A.

Nozawa, K., Nakajima, S., Kawai, K.-I. and Udagawa, S.-I. (1988a). Isolation and structures of indoloditerpenes, possible biosynthetic intermediates to the tremorgenic mycotoxin, paxilline, from *Emericella striata*. *Journal of the Chemical Society, Perkin Transactions 1*, 2607–2610.

Nozawa, K., Yuyama, M., Nakajima, S. and Kawai, K.-I. (1988b). Studies on fungal products. Part 19. Isolation and structure of a novel indoloditerpene, emindole SA, from *Emericella striata*. *Journal of the Chemical Society, Perkin Transactions 1* 2155–2160.

Oliver, J. W. (1997). Physiological manifestations of endophyte toxicosis in ruminant and laboratory species. *In* C. W. Bacon and N. S. Hill (eds.), Neotyphodium/*Grass Interactions*. Plenum Press, New York, pp. 311–346.

Ondeyka, J. G., Helms, G. L., Hensens, O. D., Goetz, M. A., Zink, D. L., Tsipouras, A., Shoop, W. L., Slayton, L., Dombrowski, A. W., Polishook, J. D., Ostlind, D. A., Tsou, N. N., Ball, R. G. and Singh, S. B. (1997). Nodulisporic acid A, a novel and potent insecticide from a *Nodulisporium* sp. Isolation, structure determination, and chemical transformations. *Journal of the American Chemical Society* 119, 8809–8816.

Panaccione, D. C. (1996). Multiple familes of peptide synthetase genes from ergopeptine-producing fungi. *Mycological Research* 100, 429–436.

Penn, J., Garthwaite, I., Christensen, M. J., Johnson, C. M. and Towers, N. R. (1993). The importance of paxilline in screening for potentially tremorgenic *Acremonium* isolates. *In* D. E. Hume, G. C. M. Latch and H. S. Easton (eds.), *Proceedings of the Second International Symposium on* Acremonium/*Grass Interactions*. AgResearch, Palmerston North, New Zealand, pp. 88–92.

Petroski, R. J., Powell, R. G. and Clay, K. (1992). Alkaloids of *Stipa robusta* (sleepygrass) infected with an *Acremonium* endophyte. *Natural Toxins* 1, 84–88.

Pfanmöller, M., Eggestein, S. and Schoberlein, W. (1997). Occurrence of endophytes in European cultivars, seed lots and ecotypes of *Festuca* species. *In* C. W. Bacon and N. S. Hill (eds.), Neotyphodium/*Grass Interactions*. Plenum Press, New York, pp. 77–80.

Philipson, M. N. and Christey, M. C. (1986). The relationship of host and endophyte during flowering, seed formation, and germination of *Lolium perenne*. *New Zealand Journal of Botany* 24, 125–134.

Piper, E. L., Gadberry, M. S., Denard, T. M., Johnson, Z. and Flieger, M. (1997). Effect of feeding ergovaline and ergine on growing rats. *In* C. W. Bacon and N. S. Hill (eds.), Neotyphodium/*Grass Interactions*. Plenum Press, New York, pp. 437–439.

Popay, A. J. and Rowan, D. D. (1994). Endophytic fungi as mediators of plant–insect interactions. *In* E. A. Bernays (ed.), *Insect–Plant Interactions*, Vol. 5. CRC Press, Boca Raton, FL, pp. 83–103.

Porter, J. K. (1994). Chemical constituents of grass endophytes. *In* C. W. Bacon and J. F. White, Jr (eds.), *Biotechnology of Endophytic Fungi of Grasses*. CRC Press, Boca Raton, pp. 103–123.

Porter, J. K., Bacon, C. W., Plattner, R. D. and Arrendale, R. F. (1987). Ergot peptide alkaloid spectra of *Claviceps*-infected tall fescue, wheat and barley. *Journal of Agricultural and Food Chemistry* 35, 359–361.

Porter, J. K., Bacon, C. W. and Robbins, J. D. (1979). Ergosine, ergosinine, and chanoclavine I from *Epichloe typhina*. *Journal of Agricultural and Food Chemistry* 27, 595–598.

Porter, J. K., Bacon, C. W., Robbins, J. D. and Betowski, D. (1981). Ergot alkaloid identification in Clavicipitaceae systemic fungi of pasture grasses. *Journal of Agricultural and Food Chemistry* 29, 653–657.

Powell, R. G. and Petroski, R. G. (1992a). The loline group of pyrrolizidine alkaloids. *In* S. W. Pelletier (eds.), *The Alkaloids: Chemical and Biological Perspectives*, Vol. 8. John Wiley and Sons, New York, pp. 321–327.

Powell, R. G. and Petroski, R. J. (1992b). Alkaloid toxins in endophyte-infected grasses. *Natural Toxins* 1, 163–170.

Powell, R. G., Plattner, R. D., Yates, S. G., Clay, K. and Leuchtmann, A. (1990). Ergobalansine, a new ergot-type peptide alkaloid isolated from *Cenchrus echinatus* (sandbur grass) infected with *Balansia obtecta*, and produced in liquid cultures of *B. obtecta* and *Balansia cyperi*. *Journal of Natural Products* 53, 1272–1279.

Pownall, D. B., Familton, A. S., Field, R. J., Fletcher, L. R. and Lane, G. A. (1995). The effect of peramine ingestion in pen-fed lambs. *The Proceedings of the New Zealand Society of Animal Production* 55, 186.

Prestidge, R. A. and Ball, O. J.-P. (1993). The role of endophytes in alleviating plant biotic stress in New Zealand. *In* D. E. Hume, G. C. M. Latch, and H. S. Easton (eds.), *Proceedings of the Second International Symposium on* Acremonium/*Grass Interactions, Plenary Papers.* AgResearch, Palmerston North, New Zealand, pp. 141–151.

Prestidge, R. A., di Menna, M. E., van der Zijpp, S. and Badan, D. (1985). Ryegrass content, *Acremonium* endophyte and Argentine stem weevil in pastures in the Volcanic Plateau. *In:* M. J. Hartley, A. J. Popay and A. I. Popay (eds.), *Proceedings of the Thirty-eighth New Zealand Weed and Pest Control Conference,* pp. 41–44.

Prestidge, R. A., van der Zijpp, S. and Badan, D. (1984). Effects of Argentine stem weevil on pastures in the Central Volcanic Plateau. *New Zealand Journal of Experimental Agriculture* 12, 323–331.

Read, J. C. and Walker, D. W. (1990). The effect of the fungal endophyte *Acremonium coenophialum* on dry matter production and summer survival of tall fescue. *In* S. S. Quisenberry and R. Joost (eds.), *International Symposium on* Acremonium/*Grass Interactions.* New Orleans: Louisiana Agricultural Experiment Station, Baton Rouge, pp. 181–184.

Richardson, M. D., Chapman, G. W., Jr., Hoveland, C. S. and Bacon, C. W. (1992). Sugar alcohols in endophyte-infected tall fescue under drought. *Crop Science* 32, 1060–1061.

Richardson, M. D. and Logendra, S. (1997). Ergosterol as an indicator of endophyte biomass in grass seeds. *Journal of Agricultural and Food Chemistry* 45, 3903–3907.

Riederer, B., Han, M. and Keller, U. (1996). D-Lysergyl peptide synthetase from the ergot fungus *Claviceps purpurea. Journal of Biological Chemistry* 271, 27524–27530.

Robbins, J. D., Sweeny, J. G., Wilkinson, S. R. and Burdick, D. (1972). Volatile alkaloids of Kentucky 31 tall fescue seed (*Festuca arundinacea* Schreb.). *Journal of Agricultural and Food Chemistry* 20, 1040–1043.

Rottinghaus, G. E., Garner, G. B., Cornell, C. N. and Ellis, J. L. (1991). HPLC method for quantitating ergovaline in endophyte-infected tall fescue: seasonal variation of ergovaline levels in stems with leaf sheaths, leaf blades, and seed heads. *Journal of Agricultural and Food Chemistry* 39, 112–115.

Rowan, D. D. (1993). Lolitrems, peramine and paxilline: mycotoxins of the ryegrass/ endophyte interaction. *Agriculture Ecosystems and Environment* 44, 103–122.

Rowan, D. D. and Gaynor, D. L. (1986). Isolation of feeding deterrents against Argentine stem weevil from ryegrass infected with the endophyte *Acremonium loliae. Journal of Chemical Ecology* 12, 647–658.

Rowan, D. D. and Shaw, G. J. (1987). Detection of ergopeptine alkaloids in endophyte-infected perennial ryegrass by tandem mass spectrometry. *New Zealand Veterinary Journal* 35, 197–198.

Roylance, J. T., Hill, N. S. and Agee, C. S. (1994). Ergovaline and peramine production in endophyte-infected tall fescue: independent regulation and effects of plant and endophyte genotype. *Journal of Chemical Ecology* 20, 2171–2183.

Sampson, K. (1935). The presence and absence of an endophytic fungus in *Lolium temulentum* and *L. perenne. Transactions of the British Mycological Society* 19, 337–343.

Schardl, C. L. (1996). *Epichloe* species: fungal symbionts of grasses. *Annual Review of Phytopathology* 34, 109–130.

Schardl, C. L. and Clay, K. (1997). Evolution of mutualistic endophytes from plant pathogens. *In* G. C. Carroll and P. Tudzynski (eds.), *The Mycota*, Vol. 5, *Plant Relationships*, Part B. Springer-Verlag, Berlin, pp. 221–238.

Schardl, C. L., Leuchtmann, A., Chung, K.-R., Penny, D. and Siegel, M. R. (1997). Coevolution by common descent of fungal symbionts (*Epichloe* spp.) and grass hosts.
Molecular Biology and Evolution 14, 133–143.

Schardl, C. L., Leuchtmann, A., Tsai, H.-F., Collett, M., Watt, D. M. and Scott, D. B.
(1994). Origin of a fungal symbiont of perennial ryegrass by interspecific hybridization of a mutualist with the ryegrass choke pathogen, *Epichloe typhina*. *Genetics*
136, 1307–1317.

Schardl, C. L., Liu, J.-S., White, J. F., Jr., Finkel, R. A., An, Z. and Siegel, M. R. (1991).
Molecular phylogenetic relationships of nonpathogenic grass mycosymbionts and
clavicipitaceous plant pathogens. *Plant Systematics and Evolution* 178, 27–41.

Shelby, R. A. and Dalrymple, L. W. (1993). Long-term changes of endophyte infection
in tall fescue stands. *Grass and Forage Science* 48, 356–361.

Shelby, R. A., Olsovska, J., Havlicek, V. and Flieger, M. (1997). Analysis of ergot alkaloids in endophyte-infected tall fescue by liquid chromatography/electrospray ioni-
sation mass spectrometry. *Journal of Agricultural and Food Chemistry* 45, 4674–
4679.

Siegel, M. R. and Bush, L. P. (1996). Defensive chemicals in grass–fungal endophyte
associations. *In* J. T. Romeo, J. A. Saunders, and P. Barbosa (eds.), *Phytochemical
Diversity and Redundancy in Ecological Interactions*, Vol. 30. *Recent Advances in
Phytochemistry*. Plenum Press, New York, pp. 81–119.

Siegel, M. R. and Bush, L. P. (1997). Toxin production in grass/endophyte associations.
In G. C. Carroll and P. Tudzynski, eds.), *The Mycota*, Vol. 5, *Plant Relationships*,
Part A. Springer-Verlag, Berlin, pp. 185–207.

Siegel, M. R. and Latch, G. C. M. (1991). Expression of antifungal activity in agar culture
by isolates of grass endophytes. *Mycologia* 83, 529–537.

Siegel, M. R., Latch, G. C. M., Bush, L. P., Fannin, F. F., Rowan, D. D., Tapper, B. A.,
Bacon, C. W. and Johnson, M. C. (1990). Fungal endophyte-infected grasses: alkaloid accumulation and aphid response. *Journal of Chemical Ecology* 16, 3301–
3315.

Siegel, M. R., Latch, G. C. M. and Johnson, M. C. (1985). *Acremonium* fungal endophytes
of tall fescue and perennial ryegrass: significance and control. *Plant Disease* 69,
179–183.

Springer, J. P. and Clardy, J. (1980). Paspaline and paspalicine, two indole–mevalonate
metabolites from *Claviceps paspali*. *Tetrahedron Letters* 21, 231–234.

Springer, J. P., Clardy, J., Wells, J. M., Cole, R. J. and Kirksey, J. W. (1975). The structure
of paxilline, a tremorgenic metabolite of *Penicillium paxilli* Bainier. *Tetrahedron
Letters* 30, 2531–2534.

Springer, T. L. (1996). Allelopathic effects on germination and seedling growth of clovers
by endophyte-free and -infected tall fescue. *Crop Science* 36, 1639–1642.

Stevenson, G. J. (1996). Isolation of endophyte-produced metabolites using immunological detection methods. MSc thesis, University of Waikato, Hamilton, New Zealand.

Steyn, P. S. and Vleggaar, R. (1985). Tremorgenic mycotoxins. *Progress in the Chemistry
of Organic Natural Products* 48, 1–80.

Strickland, J. R., Oliver, J. W. and Cross, D. L. (1993). Fescue toxicosis and its impact on animal agriculture. *Veterinary and Human Toxicology* 35, 454–464.

Sutherland, B. L. and Hoglund, J. H. (1989). Effect of ryegrass containing the endophyte (*Acremonium lolii*), on the performance of associated white clover and subsequent crops. *Proceedings of the New Zealand Grassland Association* 50, 265–269.

TePaske, M. R., Powell, R. G. and Clement, S. L. (1993). Analyses of selected endophyte-infected grasses for the presence of loline-type and ergot-type alkaloids. *Journal of Agricultural and Food Chemistry* 41, 2299–2303.

Thompson, R. W., Fribourg, H. A. and Reddick., B. B. (1989). Sample intensity and timing for detecting *Acremonium coenophialum* incidence in tall fescue pastures. *Agronomy Journal* 81, 966–971.

Towers, N. R. (1997). Endophyte toxin research at the Ruakura Agricultural Research Centre. *In* E. Märtlbauer and E. Usleber (eds.), *Proceedings 19. Mykotoxin-Workshop.* Society for Mycotoxin Research, Munich, pp. 167–171.

Tsai, H.-F., Liu, J.-S., Staben, C., Christensen, M. J., Latch, G. C. M., Siegel, M. R. and Schardl, C. L. (1994). Evolutionary diversification of fungal endophytes of tall fescue grass by hybridization with *Epichloe* species. *Proceedings of the National Academy of Science* 91, 2542–2546.

Tsai, H.-F., Wang, H., Gebler, J. C., Poulter, C. D. and Schardl, C. L. (1995). The *Claviceps purpurea* gene encoding dimethylallyltryptophan synthase, the committed step for ergot alkaloid biosynthesis. *Biochemical and Biophysical Research Communications* 216, 119–125.

Tudzynski, P., Tenberge, K. B. and Oeser, B. (1995). *Claviceps purpurea. In* K. Kohmoto, U. S. Singh, and R. P. Singh (eds.), *Eukaryotes*, Vol. 2. *Pathogenesis and Host Specificity in Plant Disease: Histopathological, Biochemical, Genetic and Molecular Bases.* Elsevier, New York, pp. 161–187.

Watkins, A. J. (1987). *Epichloe typhina*: friend or foe? *Bulletin of the British Ecological Society* 18, 90–92.

Wedderburn, M. E., Pengelly, W. J., Tucker, M. A. and di Menna, M. E. (1989). Description of ryegrass removed from New Zealand North Island hill country. *New Zealand Journal of Agricultural Research* 32, 521–529.

Weedon, C. M. and Mantle, P. G. (1987). Paxilline biosynthesis by *Acremonium loliae*; a step towards defining the origin of lolitrem neurotoxins. *Phytochemistry* 26, 969–971.

Welty, R. E. and Azevedo, M. D. (1993). Endophyte content of seed harvested from endophyte-infected and endophyte-free tall fescue. *Journal of Applied Seed Production* 11, 6–12.

Western, J. H. and Cavett, J. J. (1959). The choke disease of cocksfoot (*Dactylis glomerata*) caused by *Epichloe typhina*. *Transactions of the British Mycological Society* 42, 298–307.

White, J. F., Jr. (1988). Endophyte–host associations in forage grasses. XI. A proposal concerning origin and evolution. *Mycologia* 80, 442–446.

White, J. F., Jr. (1997). Systematics of the graminicolous Clavicipitaceae. Applications of morphological and molecular approaches. In: C. W. Bacon and N. S. Hill, (eds.), *Neotyphodium/Grass Interactions*, pp. 27–39, New York: Plenum Press, pp. 27–39.

White, J. F., Jr., Breen, J. P. and Morgan-Jones, G. (1991). Substrate utilization in selected *Acremonium*, *Atkinsonella* and *Balansia* species. *Mycologia* 83, 601–610.

Widdup, K. H. and Ryan, D. L. (1992). Forage potential of wild populations of perennial ryegrass collected from southern New Zealand farms. *Proceedings of the New Zealand Grasslands Association* 54, 161–165.

Wilkinson, H. H. and Schardl, C. L. (1997). The evolution of mutualism in grass–endophyte associations. *In* C. W. Bacon and N. S. Hill (eds.), Neotyphodium/*Grass Interactions*. Plenum Press, New York, pp. 13–25.

Wilson, S. M. and Easton, H. S. (1997). Seed transmission of an exotic endophyte in tall fescue. *In* C. W. Bacon and N. S. Hill (eds.), Neotyphodium/*Grass Interactions*. Plenum Press, New York, pp. 281–283.

Woodburn, O. J., Walsh, J. R., Foot, J. Z. and Heazlewood, P. G. (1993). Seasonal ergovaline concentrations in perennial ryegrass cultivars of differing endophyte status. *In* D. E. Hume, G. C. M. Latch, and H. S. Easton (eds.), *Proceedings of the Second International Symposium on* Acremonium/*Grass Interactions*. AgResearch, Palmerston North, New Zealand, pp. 100–102.

Yates, S. G., Fenster, J. C. and Bartelt, R. J. (1989). Assay of tall fescue seed extracts, fractions, and alkaloids using the large milkweed bug. *Journal of Agricultural and Food Chemistry* 37, 354–357.

Yates, S. G., Petroski, R. J. and Powell, R. G. (1990). Analysis of loline alkaloids in endophyte-infected tall fescue by capillary gas chromatography. *Journal of Agricultural and Food Chemistry* 38, 182–185.

Yates, S. G., Plattner, R. D. and Garner, G. B. (1985). Detection of ergopeptine alkaloids in endophyte infected, toxic Ky-31 tall fescue by mass spectrometry/mass spectrometry. *Journal of Agricultural and Food Chemistry* 33, 719–722.

Yoshihara, T., Satoshi, T., Koshino, H., Sakamura, S., Shimanuki, T., Sato, T. and Tajimi, A. (1985). Three fungitoxic sesquiterpenes from stromata of *Epichloe typhina*. *Tetrahedron Letters* 26, 5551–5554.

Young, C., Itoh, Y., Johnson, R., Garthwaite, I., Miles, C. O., Munday-Finch, S. C. and Scott, D. B. (1998). Paxilline negative mutants of *Penicillium paxilli* generated by heterologous and homologous plasmid integration. *Current Genetics* 33, 368–377.

Yue, Q., Logendra, S., Freehoff, A. and Richardson, M. D. (1997). Alkaloids of turf-type fine fescue (*Festuca* spp.). *In* C. W. Bacon and N. S. Hill (eds.), Neotyphodium/*Grass Interactions*. Plenum Press, New York, pp. 285–287.

Zabalgogeazcoa, I., García-Cuidad, A. and García-Criado, B. (1997). Endophyte fungi in grasses from semiarid grasslands in Spain. *In* C. W. Bacon and N. S. Hill (eds.), Neotyphodium/*Grass Interactions*, Plenum Press, New York, pp. 89–91.

15
Ecology of Woody Plant Endophytes

Dennis Wilson
Arizona State University, Tempe, Arizona

1. INTRODUCTION

Relative to the enormity of the number of individuals, organismal biomass, species diversity, and the role they play in virtually any ecosystem, the attention given to microorganisms in most ecological textbooks and by macroecologists examining interactions among macroorganisms is minuscule at best. That three of the five kingdoms of organisms (bacteria, protists, and fungi) are compressed into one homogeneous group commonly and collectively referred to as microorganisms further attests to the "out of sight, out of mind" perspective that has dominated ecological studies and belied the significance of microbes.

However, microorganisms are literally everywhere and have a dynamic and important role as parasites and pathogens, symbionts, saprobes, and in nutrient recycling. Particularly intriguing are the mutualistic associations microbes form with plants and animals. The presence and nature of the microbial mutualist can affect the physiology, behavior, and outcome of the ecological interaction between the host and other organisms. In this chapter I will discuss the ecology of one group of proposed plant mutualists, the endophytic fungi that inhabit the leaves and stems of trees, although I will focus mostly on foliar endophytes. In particular, I will discuss the infection strategies, infection cycles, and temporal and spatial variation in infection levels of the host trees. In addition, I will examine the ecological roles of these ubiquitous but only relatively recently studied fungi and present some hypotheses on less studied potential roles that endophytes might have.

2. WHAT IS A TREE ENDOPHYTE?

Many authors qualify their use of the term *endophyte* and present a definition to explain their usage of the term. Discussions of the definition of the term can be found in Carroll (1991b), Petrini (1991), Stone and White (1997), Wennström (1994), Wilson (1995a) and Stone et al. (see Chap. 1). For the purposes of this chapter, I will use the definition of Wilson (1995a) as this reference is cited in the most recent Ainsworth and Bisby's dictionary of the fungi (Hawksworth et al., 1995) under the term endophyte. In brief, endophytes are fungi or bacteria (Chanway, 1996) that form symptomless infections, for part of all of their life cycle, within the healthy leaves and stems of plants. For the purposes of this chapter I will refer only to woody fungal endophytes and direct the reader interested in bacterial endophytes to the work of Chanway (1996). This definition would exclude root-inhabiting fungi, although several authors who isolated fungi from surface-sterilized root material have referred to these as endophytes (Fisher et al., 1995; Newsham, 1994). This definition would also include the symptomless or latent phase of fungi that are pathogenic for part of their life cycle. Thus, endophyte should not be used to label particular fungal species (unless they are symptomless for their entire life cycle) but can be used to describe the nature of the interaction between a fungus and its host plant. Strictly circumscribed definitions aside, for practical purposes, a fungus is typically considered an endophyte if it is cultured from surface-sterilized plant material. For a summary of different methods used to isolate endophytes from woody and similar tree materials, refer to Carroll and Carroll (1978), Stone and White (1997), and Wilson and Carroll (1994).

3. ENDOPHYTE INFECTION STRATEGIES IN TREES

In contrast to *Epichloe* and the associated anamorphic endophytes of grasses, fungal endophytes of trees are horizontally transmitted via spores and are not known to grow into the seeds and systemically infect the plant following seed germination. Although nongrass endophytes have been found in their host plant seeds (Bose, 1946; Boursnell, 1950; Petrini et al., 1992; Wilson and Carroll, 1994), where endophytes are found in tree seeds, reproductive propagules most likely infect the seeds through the seed coat rather than vegetatively grow into the seed when the seed is still attached to the parent tree. The endophyte *Discula quercina* was found in variable amounts in the cotyledons of Oregon white oak (*Quercus garryana*) in Oregon (Wilson and Carroll, 1994). While few acorns were endophyte-free, most had localized lesions that typically occupied less than 50% of one of the cotyledons and the endophyte did not infect trees grown from infected acorns under axenic conditions. However, endophytic fungal tissue in the seeds may produce spores that are capable of infecting a plant that grows from the seed. This may then ensure that the endophyte can disperse with its host if seeds are carried away from an inoculum source from adult plants.

Spores are most likely the primary reproductive propagule. Numerous studies have demonstrated contagious spread of tree endophytes and obtained infected leaves or stems by applying spore suspensions to endophyte-free tissue. Stone (1987) elegantly demonstrated conidial attachment, germination, penetration, and colonization of Douglas fir needles by the endophyte *Rhabdocline parkeri*. Toti et al. (1992b) also demonstrated the attachment and subsequent colonization of endophyte-free beech leaves by conidia of the endophyte *Discula umbrinella*. For some endophytes, presumably those that show high host fidelity, there are very specific chemical recognition cues that identify the host as compatible and initiate spore attachment and eclosion (Chapela and Boddy, 1988; Chapela et al., 1991; Toti et al., 1992a, b). For endophyte species that are more cosmopolitan in the plant hosts they infect, the recognition cues used may be much less specific. No studies have specifically examined whether hyphal fragments may initiate infection of uninfected host leaf tissue. Therefore, I will refer to spores as the infective propagules hereafter.

The origin of the spores that horizontally infect tree parts is largely unexplored. It has been shown in some studies and presumed in others that tree endophytes sporulate on abscised or senescent plant material such as leaves or twigs, since for many endophytes they are isolated from a host on which they were not previously known to sporulate (Fisher et al., 1986; Stone, 1987). Abscised plant material may become lodged in the canopy or fall to the ground. When on the ground, rainfall, wind, or insects (Malloch and Blackwell, 1992) could disperse spores onto endophyte-free plant tissue such as newly emerging leaves and twigs. When lodged in the tree canopy, rain falling through the canopy can disperse spores and carry them to susceptible tissue. Further, rain splash can be lifted up into the canopy by wind, or wind and insects alone may carry small fragments of plant material or endophyte spores up into even high tree canopies and deposit the endophyte inoculum. For deciduous trees, the endophytes would have to overwinter in leaves that would require them to be competitive saprophytes [which they might not be (Petrini et al., 1992)], or find an alternative substrate to colonize until new leaves emerge. However, endophytes that occupy stems and evergreen foliage which are present year round would not need to exist saprophytically or overwinter until new leaves are present.

Only several studies have actually examined where endophyte spores are produced, where they may overwinter, and the mode of transmission. Carroll (1995) found that the *Meria* state *R. parkeri* sporulates prolifically on *Contarinia* midge galls on Douglas fir needles and measured 1200 spores/mL in water dripping from a heavily galled branchlet. *R. parkeri* and its anamorph also sporulate in the fall on abscised needles. The spore masses of this endophyte are produced in mucilage, which is indicative of water transmission. Further, newly flushed needles in the spring do not become infected until they are rained on in the fall (Sherwood-Pike et al., 1985; Stone, 1987). The infection cycle for *Discula quercina* that infects Oregon white oak has been fully described (Wilson and Carroll, 1994). This endophyte sporulates on abscised leaves and is also found in the bark of young twigs, where presumably it also sporulates, as does a synonym of this fungus on 1-year-old *Platanus occidentalis* twigs (Neely and Himelick, 1965). These spores are also produced in mucilage and readily disperse into aqueous suspension. Rainfall collected under Oregon white oak trees in the spring just as new, endophyte-free leaves burst bud contains *D. quercina* spores (approximately 50/mL of rainwater). Both mature and newly flushed leaves on endophyte-free trees grown in the greenhouse could be infected with this rainwater, but not rainwater passed through a 1 μm pore size filter that removed fungal spores. In addition, aqueous spore suspensions obtained from *D. quercina* cultures were used to successfully infect endophyte-free leaves.

Rainfall appears to be a major vehicle of dispersal for many endophytes, so it is perhaps not a coincidence that infection (both presence and absence and relative amount) is frequently closely related to liquid precipitation (Wilson et al., 1997; Wilson and Carroll, 1994). Wilson (1996) was able to reduce *D. quercina* endophyte infection in Oregon white oak, virtually eliminate endophytic infections of *Ophiognomonia cryptica* (Diaporthales: Ascomycete) and *Plectophomella* sp. (Coelomycete), and substantially reduce infection levels of *Asteromella* sp. (Coelomycete), in the leaves of Emory oak (*Quercus emoryi*) in Arizona by shielding leaves from rainfall with bags placed over branches. The bags were constructed with a plastic tops and netting bottom allowing air flow.

Rainfall is certainly not the only mode of transport for endophyte spores. In the Emory oak study, endophytes such as *Alternaria*, *Cladosporium*, and *Aureobasidium* and other fungi that were not identified were encountered relatively infrequently individually, and so were all lumped into one group, i.e., "other." Interestingly, infection levels of these "other" endophytes (some of which do not produce spores in mucilage) were not reduced as much by the bags as three of the more common endophytes, *Asteromella* sp., *Plectophomella* sp., and *O. cryptica*, which all produce spores in mucilage. Some of these "other" endophytes, such as *Alternaria* and *Cladosporium*, are typically wind- or animal-transmitted so were likely transported onto leaves within bags through the netting bottoms (Wilson, 1996). Endophytes that produce hydrophobic spores are more likely to be wind- or animal-vectored (Malloch and Blackwell, 1992).

If a viable inoculum reaches a compatible substrate, environmental conditions may determine whether the endophyte will infect the host and some endophyte species may be more sensitive to conditions than others. *D. quercina* and *D. umbrinella* readily infected their respective hosts, Oregon white oak and *Fagus sylvatica*, under laboratory and greenhouse conditions (Viret and Petrini, 1994;

Wilson and Carroll, 1994). If spore suspensions of the former were sprayed onto leaves within protective bags under field conditions during the cool, wet, and humid months of spring in Oregon, infection levels were successfully elevated but not during the dry summer months. Infection levels of *Asteromella* sp. and *Plectophomella* sp. in Emory oak leaves were successfully elevated by spraying leaves within protective bags and on branched or unbagged trees with spore suspensions in the field, but attempts at infecting leaves in the greenhouse of 2-year-old greenhouse-grown trees were not successful (Wilson and Faeth, unpublished data).

Unsuccessful experimental infection of already infected plant tissue, such as attempted elevation of natural infection levels in the field, may be caused by plant related factors rather than unsuitable conditions. Plants may have saturation levels of infection that are not exceeded. When plants are exposed to sprayed spore suspensions at concentrations of 10^6/mL, many orders of magnitude greater than natural concentrations in rainwater, infection levels not too dissimilar to natural levels have been obtained. This indicates that the plant may have some control or have a role in regulating infection levels since environmental conditions were conducive to infection by at least some spores.

By definition endophytes inhabit healthy plant tissue, yet they may infect the plant via wounds such as insect feeding sites or from mechanical damage (Faeth and Hammon, 1997a; Faeth and Wilson, 1997; Wilson et al., 1998). Work in Arizona has shown that leaf damage to Emory oak leaves caused by a leafmining moth, or artificially with a hole punch, facilitates endophyte infection at the wound site. However, not all endophyte species responded in the same way, as some were unaffected by wounding or apparently repelled. Interestingly, the more cosmopolitan endophytes species lumped into the "other" group increased at wound sites and the newly described more specialized endophyte, *Ophiognomonia cryptica*, only known from this host, was generally less abundant at wound sites (Wilson et al., 1998). As well as providing an infection site, wounding can also trigger endophyte growth from a quiescent or dormant state into an actively growing phase where some endophytes might make the transition from latent colonizer to pathogen (Fail and Langenheim, 1990).

Generally, spores of tree endophytes are the infective propagules. They are passively horizontally transmitted via rainfall, wind, or animal vector and by chance reach a suitable host surface. Following infection, endophytes may remain dormant until triggered by natural leaf senescence, abscission, or damage to grow and perhaps sporulate. If the host tissue is not immediately available for infection, such as leaves on deciduous trees in the fall and winter, the endophyte must overwinter in abscised leaves, or may colonize the tree bark and reinfect leaves the following spring. Where leaves and stems are available year-round, they may become repeatedly colonized and the endophyte may not have to survive outside of its host for prolonged periods.

4. SPATIAL AND TEMPORAL PATTERNS OF ENDOPHYTE INFECTION

4.1. Infection Patterns Over Time

There are surprisingly few studies that have examined infection patterns within and among years, especially for deciduous trees, as many floristic studies are based on samples taken at one time in one year. However, working with evergreen trees provides a unique opportunity to examine the effects of time, since at any particular moment leaves from previous years are found on the trees; thus, one-time sampling may yield considerable among-year information assuming endophytic infection remain viable from year to year.

Leaves still within the buds are typically endophyte-free (Wilson and Carroll, 1994), although Rodrigues (1994) found that unopened palm leaves in the tropics were colonized by endophytes, albeit at much lower levels relative to opened leaves. That very young leaves still within buds are endophyte-free is not surprising as they would not have been exposed to endophyte inoculum, thus leaves are generally uninfected as leaf primordia. Infection levels then generally increase over time as the plant tissue ages: as the length of time the substrate is exposed increases so the greater the chance it has of receiving an infective propagule. Faeth and Hammon (1997b), Gaylord et al. (1996), Helander et al. (1993), Wilson et al. (1997), and Wilson and Faeth (1998) all record increases in infection levels over time within a season following bud burst. In studies where sampling was repeated over more than one season, the seasonal increase was qualitatively similar in each season.

Most, if not all, studies show important deviations from a general increase in infection levels within a season. Helander et al. (1994) found that while infection levels in young needles (1 year old) of Scotts pine increased over the season, infection levels in old needles (5 years old) did not, but needles were only sampled twice during the season. Faeth and Hammon (1997b) showed initial increases for some endophytic species in Emory oak leaves, which then reached a plateau and did not increase further, whereas some endophytes showed continued increases until leaf abscission. Similarly, Wilson and Carroll (1994) showed initial increases in infections of *D. quercina* in Oregon white oak over the first half of the leaf life, which then reached a plateau and slightly decreased until leaf abscission. Rodrigues (1994) showed variable effects of the time of season on the number of isolates recovered from a Brazilian palm with no clear pattern over time. Wilson and Faeth (1998) carried out a detailed census of endophyte infection density within leaves of Emory oak within and between seasons over a 3-year period for a randomly selected group of 20 trees. Following bud burst in April–May, infection densities for *Asteromella* sp. and endophytes grouped into the "other" category increased whereas *Plectophomella* sp. levels remained at a very low constant level, and *O. cryptica* was not detected. Then approximately 3 months after bud burst in the summer, infection levels of all endophytes present dropped significantly for a short period (less than a month), then increased slowly over the remaining summer and fall months. During late November and December, infection densities of all species increased rapidly. *O. cryptica* was first detected at this time; infection densities increased rapidly and it was typically the most abundant species at the time of leaf abscission in April. These within-season patterns were consistent between two seasons where leaves were regularly sampled during the entire season (June–April) and for corresponding times where only part of the season was sampled. The study by Gaylord et al. (1996) on Gambles oak documents a continued increase in infection levels over the season which resembles an exponential pattern of increase until leaf abscission.

Seasonal patterns are clearly variable with variations around a general trend to increase with time following bud burst. Within-season patterns vary among study systems and among endophyte species in a particular study system. However, what may appear to be highly variable results can often be explained in terms of rainfall patterns and some very basic ecological principles. Population sizes of endophytes within leaves are entirely caused by immigration and extinction. Populations will not be affected by (1) reproduction since endophytes are not known to reproduce within their live host (there may be some vegetative spread away from the infection site) or (2) emigration as they are not known to leave the live host and reinfect another site. The rate of immigration will be affected by the supply of inoculum and conditions that favor infection. I consider failure to infect the leaf that results from unfavorable conditions as an immigration phenomenon, not as an extinction event as before the fungus infects the leaf it is not actually part of the endophytic population. For rain- or water-dispersed endophytes, the seasonal occurrence of rainfall would influence seasonal patterns of endophyte infection, perhaps more accurately, the seasonal cumulative pattern of rainfall, since endophyte populations will likely consist of accumulated fungal infections within the leaf. Although the conditions that favor infection are largely unknown, for fungi in general cool humid conditions with low direct UV radiation favor fungal growth and infection. Cumulative precipitation was very closely correlated with endophyte infection levels in two studies. Wilson and Carroll (1994) showed that cumulative precipitation over the season almost perfectly mirrored endophyte infection patterns in Oregon and that endophyte infection levels stopped increasing within days of the onset of the annual summer drought. Similarly, Wilson et al. (1997) and Wilson and Faeth (1998) showed a very close match of seasonal cumulative precipitation with infection patterns of Emory oak in Arizona, and that the late season occurrence of *O. cryptica* coincides with the onset of lower winter temperatures and reduced solar radiation. Bernstein and Carroll (1977) found that endophytic infections in first-year Douglas fir needles coincided with the onset of fall rains in Oregon.

The effects of the dynamics of immigration and extinction for endophytes has never been examined [but see Andrews et al. (1987) and Kinkel et al. (1989) for epiphytes] and we have no understanding of how this might affect seasonal infection patterns. However, certain lines of indirect evidence suggest that endophyte populations may be long-lived but also subject to extinction. Wilson and Faeth (1998) found a decrease in infection levels in the summer in Arizona when daily temperature maximums can exceed 38°C and solar radiation is intense. They attributed this fall to extinction rates exceeding immigration rates, and suggested that separate endophyte infections were dying from the harsh environmental conditions and were not being replaced by new infection in the absence of rainfall during this time.

4.2. Spatial Patterns of Infection

In contrast to microbial pathogens, mycorrhizal fungi, and nitrogen-fixing bacteria, very little is known about the spatial aspects of colonization of plants by fungal endophytes. Yet considerably more studies have examined the spatial aspects of endophyte infection relative to the temporal distribution.

4.2.1. Variation Between Locations

A single host species is frequently infected with a qualitatively similar assemblage of endophytes over its natural host range; however, the relative abundance of particular species may differ depending on location (Fisher et al., 1994; Petrini et al., 1992). Rollinger and Langenheim (1993) showed that both endophyte species presence and frequency of each species in Redwood needles was remarkably similar at six sites distributed over an 850-km north–south transect along the California coast in the trees' native habitat. However, despite this similarity, using cluster and principle component analysis, the endophyte communities from each of the six sample sites formed discreet groups and so were dissimilar enough to be distinguished from one another. Similarly, Fisher et al. (1994) found that the endophytic community within the leaves of *Q. ilex* was similar for trees growing in two distantly geographically separated (the island of Majorca and Switzerland) but native host habitats, but differed to the endophyte community in a third nonnative and geographically separated area in England. However, even the similar endophyte communities from Majorca and Switzerland could be separated from each other using simple ordination analysis and were both distinguishable from the British communities. In the same way, endophytic assemblages of *Eucalyptus nitens* from their native habitat in Australia were different to the same trees growing in a nonnative habitat in England (Fisher and Petrini, 1990). Furthermore, nonnative Eucalyptus trees in England had a depauporate mycoflora and hosted a more cosmopolitan endophyte community relative to native Australian trees. In a different study of *E. nitens* endophytes in South Africa, Smith et al. (1996) found that the endophyte assemblage was similar to the same species of trees found in Australia. The endophyte assemblages of *E. grandis* trees from two geographically distinct locations in South Africa were similar at both sites.

Even in native habitats, differences in endophyte frequencies exist and can be attributed to the management practices of the area. For example, Sieber-Canavesi and Sieber (1987) showed that endophyte frequencies from European white fir needles were higher in sites where trees were naturally established following clear cutting relative to planted sites, although it is difficult to separate genotypic effects which can also alter colonization (Todd, 1988), as presumably planted trees are less genetically diverse relative to naturally established trees. Wilson and Faeth (1998) examined leaf endophytes of Emory oak trees growing in a shallow ravine alongside a wash and trees growing in an adjacent rocky hillside area and found similar species in each area, but higher infection densities of some species in trees by the wash as well as a greater number of rare species that were not identified.

Clearly, site-to-site differences in which endophyte species are present, and the abundance of those species exist. However, the causes of these differences are unclear. Several of these studies have methodological problem as they used only two trees in each site; thus site-to-site differences might be an artefact and instead be caused by tree-to-tree differences within sites, which can be great (Petrini et al., 1992). I do not wish to criticize those studies that examine isolated or few trees as the work involved to sample even a single tree can be formidable, but this should be considered in the interpretation of the results.

Site-to-site differences may be caused by a difference in native vs. non-native habitats, although with low tree number per sample site, caution is again needed when interpreting these data. However, it can be intuited that if endophyte species show high fidelity to their hosts, when the host is moved to a site where there is no inoculum of its specialized, high-fidelity endophytes it would become colonized by more generalist cosmopolitan fungi such as *Alternaria* and *Cladosporium* that are prevalent in that environment. The lower incidence of infection by the more cosmopolitan endophytes in native sites may be a result of exclusion by the more specialized endophytes as more cosmopolitan species are usually also present in the native sites. This is essentially exclusion competition and has been documented for many macroscopic organisms (Begon et al., 1990). In support of this, Wilson (1996) found that when specialist, species-specific endophytes were excluded from Emory oak leaves, they were more heavily infected with less species-specific cosmopolitan endophytes. Whether such exclusion is mediated by the host plant, fungus, or an interaction is unknown, but this would provide an interesting avenue of study.

4.2.2. Variation Among Trees in the Same Location

Variation in endophyte infection can also occur among trees in the same location; however, the variation is typically in the relative infection level of endophytes rather than the presence of particular species, at least for the more common species. Helander et al. (1994) found that endophyte colonization rates of Scotts pine needles in Finland varied widely among individual trees but that the more abundant species present were common to all trees. They explained this variation by the density of the tree stand in which the sampled tree was growing; as stand density increased so did colonization rates. Wilson and Faeth (1998) also found considerable variation in infection density but not species present among individual Emory oak trees growing at one site. Some of this variation was explained by leaf size differences among trees as small leaves had higher infection densities relative to larger leaves; thus small-leaves trees had higher endophyte infection densities. Todd (1988) showed that the genetic background of the individual tree accounted for significant among-tree variation in endophyte infection.

Many of the studies cited in this chapter that examined more than one tree show among-tree variation in endophyte infection frequency but less variation in the species present. This is not surprising as these trends are echoed by almost any parasitic group (e.g., insect, platyhelminth, etc.) of organism on its respective host. As for other macroorganisms, environmental conditions such as distance to an inoculum source (as predicted by the theory of island biogeography), weather conditions or tree microclimate, as well as humidity, temperature, and chance events are all likely to alter the tree as a host and therefore affect subsequent colonization. Genetic and presumably genotype $\times$ environment interactions can also affect the tree as a host to endophytes in an analogous way that the host genotype affects the susceptibility of a plant to a pathogen or an animal host to a parasite.

4.2.3. Within-Tree Variation

Within-tree variation in endophyte infection has been examined at different heights in the canopy, compass direction, and distance from the trunk. Where differences were found, generally infection frequencies decreased with increased height in the canopy (Johnson and Whitney, 1989), compass direction had no effect (Fisher and Petrini, 1990; Johnson and Whitney, 1989; Legault et al., 1989), and infection densities were greater in the outer canopy farther from the tree bole relative to interior leaves closer to the bole late in the season with the reverse trend in the early season (Wilson and Faeth, 1998). Although it would be premature to draw conclusions on within-tree variation based on so few studies, the most likely causal mechanisms for these results should indicate how prevalent they may be.

The higher in the canopy the further are the leaves from the inoculum source, so that lower colonization frequencies may be expected. Thus distance from the pool of colonizing propagules emerges as a general theme and is consistent with island biogeography. In addition, conditions conducive to spore deposition and infection may be less favorable higher in the canopy as the unbroken fall of rain may dislodge spores from leaves and prevent attachment, UV irradiation may be higher and humidity lower.

No effect of compass direction is somewhat surprising. The colonization of tree branches by mosses and lichens can differ depending on compass direction, so presumably similar mechanisms would impact endophyte colonization. However, in one study only one tree (balsam fir) was examined (Johnson and Whitney, 1989). In another study, 8 trees from each of two species of pine (Legault et al., 1989) were examined; this should be sufficiently large a sample size to detect differences, but none were found.

Only one study has divided trees (Emory oak) into inner or shaded leaves vs. outer or sun leaves (Wilson and Faeth, 1998). Higher infection densities in the sun leaves later in the year was explained by leaf size effects as the sun leaves are smaller relative to shade leaves, and infection density and leaf area are inversely correlated for most of the season. However, in the early season, shade leaves had higher infection densities than sun leaves. During the spring and summer, sun leaves have much higher UV exposure (which is intense in Arizona at 1200 m elevation where these trees grow) relative to the inner shaded leaves which may inhibit endophyte infection. Thus after the summer months (leaves remain on the trees until the following April–May), when UV exposure is presumably lower, infection densities may increase in sun leaves relative to shade leaves.

It is impossible to make general conclusions or suggest general patterns based on such scant data. Yet within-tree differences in colonization frequency do exist and the mechanisms that cause these differences may vary among study systems.

4.2.4. Within-Leaf Patterns of Colonization

Within-leaf colonization patterns are common but are not the same among tree species. The most common distributional pattern or infection gradient of endophyte frequency is along the axis of the midrib where the petiole end of the leaf is more heavily colonized relative to the more distal end. This trend has been observed in conifer needles (Bernstein and Carroll, 1977; Sieber-Canavesi and Sieber, 1987), as well as broadleaf trees (Wilson and Carroll, 1994). Legault et al. (1989) found this trend only in 1-year-old needles but not older needles of two pines. However, the reverse trend has also been observed (Hata and Futai, 1996). Some studies have found no within-leaf patterns (Sieber and Hugentobler, 1987), or differential within-leaf distribution of individual species (Hata and Futai, 1996). Wilson and Carroll (1994) found that infection frequency of *D. quercina* in Oregon white oak leaves also followed a lateral trend as leaf discs from the lamina were less infected relative to the closest corresponding leaf discs removed from the midrib and that the petiole-distal end gradient was present along the midrib and the lamina.

These trends might make sense in light of the mechanisms that caused them. Wilson and Carroll (1994) proposed that differential spore deposition and differential leaf expansion could explain many of the within-leaf patterns. Many endophytes are rain-dispersed, thus the affinity of different leaf parts for rainwater, the parts of the leaf where rainfall may be channeled, and the phylloplane characteristics that influence spore deposition (such as position and density trichomes) may explain infection patterns and appears to explain the within-leaf pattern in Oregon white oak. Rainwater channels into the midrib and down to the petiole end of the leaf. Trichome density is greater in the midrib relative to the lamina and greater in the petiole end relative to the distal end. Since trichomes are known to act as spore traps (Allen et al., 1991), endophyte spores will be differentially deposited in these areas and hence give rise to higher infection frequencies. In addition, endophytic infections within the leaf reflect cumulative infections over time. Thus some infections may have occurred just after bud burst when the leaf was still expanding. In this case, the relative position on the leaf will change as the leaf expands (since leaf expansion is not always uniform over the whole leaf surface). Thus in parts of the leaf which expand more than others, in the case of Oregon white oak, the distal and lamina parts of the leaf, infections will become "diluted" and infection densities will be lower. Leaves that channel rainwater to the distal segments, such as leaves with drip tips, may have higher infections in the distal parts, and leaves that do not channel rainwater or have relatively uniform trichome density over the leaf surface may show no patterns. Although never specifically investigated for endophytes, mites that inhabit leaf domatia forage over the leaf surface for fungal material and may collect endophyte spores or carry the spores on their bodies back to the domatia, which are typically located at along the midrib and closer to the petiole end. Thus these mites might influence within-leaf endophyte distribution.

At a finer within-leaf scale, endophytes tend to occupy highly localized sites and do not significantly grow away from the infection site during the symptomless endophytic phase (Stone, 1986, 1987; Wilson and Carroll, 1994). Although systemic rusts and smuts that colonize their hosts asymptotically for prolonged periods prior to causing disease symptoms are an obvious exception, but whether these fungi should be considered endophytes has been debated (Wennström, 1994; Wilson, 1995a) and they are not prevalent in trees. Following penetration, the fungi may be restricted to a single epidermal cell as shown for *R. parkeri* in Douglas fir needles, or be localized in the intercellular subcuticular and epidermal areas as seen in *D. umbrinella* from beech leaves (Viret and Petrini, 1994). Microscopic studies generally show that infection sites are discrete and highly localized. Artificial infection by *D. quercina* of attached Oregon white oak leaves also showed that the endophyte could only be reisolated from the area of the leaf where spore suspensions were applied and were not recovered from adjacent areas 1 mm away. Where endophytes colonize damaged tissue, infection foci may not be as localized as when there is infection of undamaged tissue. Furthermore, endophytic infections will spread out from the localized infection foci, but when this occurs relative to leaf abscission is unknown.

5. SAMPLING PROBLEMS

Relative to most microbial populations, it is technically relatively easy to sample and census endophyte populations. Leaves are well-defined, discrete islands that become colonized by endophytes so the population of endophytes within the leaf is finite and on a scale small enough to handle with relative ease. Since most infections are not systemic, but frequently highly localized with limited invasive colonization, the concept of the individual is preserved and allows for numbers of endophytic colonizers to be counted. Yet enumeration and quantification of endophyte infection levels or frequencies typically underestimates endophyte population sizes, especially at high colonization densities typical of older plant tissues.

Carroll (1995) has presented an elegant and detailed study of the problems associated with quantification of endophyte infection levels within leaves. The size of sampling units (whole, half, or parts of leaves) is frequently many times the size of individual fungal thalli that may be restricted to the lumens of individual cells; therefore multiple infections within a single sampling unit should be expected, which results in underestimation of the population size. Thus selection of an appropriate sampling unit size that results in no more than one infection in the sample unit is required when culturing endophytes if estimating the total population size is the goal. Bissegger and Sieber (1994) and Carroll (1995) used microdissection of bark and needles, respectively, to attempt to reduce the scale of the sampling unit and reduce multiple infections per sampling unit. The actual size of the sampling unit should depend on the density of endophytic infections and is inversely related to density. Alternatively, direct microscopic examination of stained plant material may allow each infection to be counted (Cabral et al., 1993; Stone, 1987). However, using either microdissection or direct microscopic examination for absolute quantification may become painfully laborious and prohibit sampling of large numbers of leaves.

When absolute infection levels are not required, but quantification of infection levels are needed for comparative purposes among groups (e.g., trees, branches, or treatments), it may be less important to count every infection within a leaf. Larger sample units can then be used, but the size should still be dependent on the infection density within the leaf. Typically the greater the proportion of uninfected sampling units, the better the sampling resolution. For example, if a leaf is cut into eight equally sized pieces and four pieces are uninfected, the sampling resolution might be sufficient for the purposes of comparing infection levels to other leaves even though it may have undersampled total infection level assuming that infection is random over the entire leaf. However, if infection is correlated with position within the leaf, or if infections are spatially positively correlated, undersampling for comparative purposes may still present problems. Generally, where infection levels have been reported as 100% (whatever 100% may be), or where 100% of the sampling units (leaves, leaf halves, leaf segments) are infected, comparisons among groups are meaningless as the sampling strategy has little or no resolution.

Connor et al. (1997) pointed out the importance of homogeneity of sampling unit size when using frequency counts of individuals to make comparisons. For example, a leaf (whether whole or cut into parts) is often the sampling unit in endophyte studies and variation in leaf size within and among trees can be substantial. Although it may seem typical that larger leaves may have more endophytic infections than smaller leaves (as predicted by island biogeography), infection frequency (e.g., total number of endophytic infections) is often used as the measure of endophyte infection without regard for sampling unit (leaf) size. Thus larger leaves or trees with larger mean leaf size may have higher endophyte infection frequencies relative to smaller leaves, but the infection density may be no different or even smaller on larger leaves. Connor et al. (1997) pointed out the importance of using density as a more appropriate measure of infection level, or to use frequency counts with homogeneous size of sampling unit, such as leaf discs cut with a standard size cork borer.

Petrini et al. (1992) have looked at the relationship between the number of endophytic taxa isolated and the number of trees and sampling units from each tree. They constructed discovery curves and found that, in general, 40 individuals of a given species and 30–40 sampling units from each individual will yield at least 80% of endophyte taxa assumed to be present at one site. How these numbers may change in different ecosystems or at different times of year is unknown. Sampling on this scale represents a large investment of resources, and adequately identifying and sorting the resulting slew of organisms requires a mycological expertise not present at many institutions.

Future studies need to use adequate sampling resolution and adjust the sampling unit size such that infection levels are not vastly underestimated and certainly so that less than 100% of the leaves or leaf pieces are infected. The whole leaf area can be measured, the leaf cut into appropriately sized segments, and the number of segments infected or the number of endophytic colonies growing out from each segment counted with infection levels expressed as infection densities. Alternatively, leaf discs of an appropriate size for sufficient resolution can be cut from leaves and infection levels expressed as infection densities or proportion of discs infected. Again, if 100% of the leaf discs are infected they are too large for the infection density within the leaves sampled. In addition, leaf discs should be removed from the same place on all leaves and preferably be removed from several locations within leaves. Both of these techniques will most likely underestimate endophyte levels but be sufficient for comparative purposes. For sufficient resolution I suggest that 50% or more of the leaf segments or leaf discs be uninfected.

6. ECOLOGICAL ROLES OF ENDOPHYTES

Studies on the ecological role of endophytes in trees lags far behind their seed-borne counterparts in grasses, probably because of the direct and immediate applied uses of grass endophytes and economic importance of the plants and consequences of infection. Nevertheless, the ecological role of fungal endophytes of tree leaves, in particular their role as insect antagonists, has received much discussion, debate, and speculation, yet experimental studies designed to investigate the hypotheses put forward are scant by comparison. Much of the research and the dominant hypotheses surrounding the role of tree endophytes was heavily influenced by the early finding that endophyte-infected tall fescue was responsible for maladies of cattle and that infected plants were more resistant to herbivory relative to uninfected conspecifics (see Chaps. 10, 13, and 14). After this discovery, a protective mutualistic role of endophytes for their host plant was proposed as a general hypothesis to explain why plants harbored these asymptotic organisms (Carroll, 1986). The rationale behind this hypothesis for endophytes in tree leaves is sound and experimental data support some of its predictions, yet the field has progressed significantly and hypotheses on the ecological role of tree endophytes have become more diverse and the pervasiveness of the protective role of endophytes as insect antagonists has been questioned (Faeth and Hammon, 1997c; Hammon and Faeth, 1992; Saikkonen et al., 1996).

When considering the many potential ecological roles of endophytes, it is perhaps naive to expect that all of the organisms that occupy a particular niche, i.e., the interior of plants, have the same roles or functions. Just as the gut microflora of animals have many varied functions and roles such as degrading cellulose, detoxifying allelopathic compounds (Dowd, 1991), producing vitamins (Chapman, 1971), and going along for the ride to utilize the dung (Wicklow, 1992), so different species of fungal endophytes in different hosts plants may have many and varied roles. In the rest of this chapter I will review evidence and discuss what role these ubiquitous organisms may have and how the host as well as organisms that interact with the host may be affected.

6.1. Insect Antagonism

Probably the most widely hypothesized role of endophytes in trees is as plant mutualists that act as insect antagonists (Carroll, 1986, 1988, 1991a, b; Hammon and Faeth, 1992; Wilson, 1993). Since virtually all studies on endophyte–herbivore interactions in trees involve insect herbivores, I have restricted this discussion to insects only. However, some of the arguments and rationale behind the arguments may equally apply to noninsect herbivores and some may not. Lengthy discussions on the role of endophytes as insect antagonists in trees have been presented by Carroll (1986, 1988, 1991a, b). He proposed that endophytes in trees represent an inducible mutualism (opposed to a constitutive mutualism in the case of grass endophytes), where endophytes may decrease insect fitness within a population and thereby reduce overall insect population densities. He proposed that the *modus operandum* of such antagonism was via the production and action of mycotoxins. Fungi in general produce a staggering array of diverse and metabolically active compounds, and endophytes have been shown to produce numerous insecticidal toxins. Insect antagonism could occur via the production of one mycotoxin or suites of mycotoxins that may interact synergistically. Within a population of trees where there may be a high degree of relatedness among individuals, the diffuse benefits of inducible mutualisms may be explained as a form of kin selection since antagonism by fungi usually occurs in plants that have already been attacked by insects. Endophytes would reduce the insect population size and thereby decrease the number of insects that could attack other tress. Furthermore, trees may rely on endophytes as a rapidly evolving defense that can keep pace with the rapid generation times of the insect populations and provide long-lived trees with a "moving target" of defensive mycotoxins that act similarly to plant allelochemicals. Hammon and Faeth (1992) also discussed the role of endophytes in mediating plant–insect interactions and proposed that such antagonism may be very diffuse and may be very "situation"-specific and act only under certain conditions.

If endophytes are plant mutualists that protect the plant from insect herbivory, we should expect that there is a selective advantage to the endophyte to protect its host. Presumably, the endophyte relies on its host physically as a habitat and chemically for nutrition, so when the host fitness is reduced, such as by herbivory, fitness of the endophyte is also reduced. Thus, it would be selectively advantageous for the endophyte to protect its host, and it would benefit the plant to harbor the endophytes. A similar argument is used to explain why ants defend ant acacias and why the ant acacia provides both physical and chemical resources for the insects. However, implicit in this rationale are the assumptions that (1) the endophyte is directly negatively affected by herbivory or/and (2) herbivory negatively affects the host plant, which in turn negatively impacts the endophyte. If the direct or indirect effects (via the host plant) of herbivory have a neutral impact on the endophyte, then a defensive role of the endophyte would not be selected for and protective mutualism not expected, although it is possible that insect antagonism would arise not through the evolutionary processes of selection against nonprotective endophytes but simply because endophytes are predisposed to produce insecticidal toxins or otherwise interfere with insect fitness.

How herbivory directly affects endophytes has recently been addressed by Faeth and Hammon (1997a), Faeth and Wilson (1997), and Wilson and Faeth (1998) who have examined how insect wounded and artificially damaged leaf tissue affects colonization by endophytes. Craven, Wilson, and Faeth (unpublished data) have looked at the fate of the endophyte when the leaf which it has colonized is consumed by insect herbivores. Damaged Emory oak leaf tissue is more heavily infected by most endophytes at the wound site, although one dominant endophyte was present at lower densities relative to undamaged leaf parts (Wilson et al., 1998). Furthermore, damage appears to inhibit infection in the undamaged parts of the damaged leaves relative to corresponding parts of undamaged control leaves. Whether damage alone or damage and infection cause this inhibition is unknown (Faeth and Wilson, 1997). When endophyte-colonized leaf parts were consumed by a lepidopteran herbivore, no more than 17% of *Asteromella* sp. and 0% of *O. cryptica* infections survived passage through the gut and were viable in the frass. These results indicate both positive and negative affects of herbivory on endophytes as herbivore damage facilitates infection by some newly colonizing endophytic propagules, but herbivory causes already resident organisms to suffer significant mortality. In this system, *O. cryptica* is only negatively affected by herbivory, whereas most of the other endophytes are both positively and negatively affected. Thus different endophyte species might have opposing effects on insect herbivores.

By far the majority of studies that have investigated how fungal endophytes affect insect herbivory have examined how endophytes affect insect mortality and performance, or how insect and endophyte presence are correlated. In 19 references, I found a total of 30 conclusions drawn regarding the direction of the interaction between insects and endophytes. In the correlative studies, insect and endophytes were positively associated nine times, negatively five times, and not associated seven times. In two studies (Butin, 1992; Petrini et al., 1989), the authors proposed that endophytes may cause gall insect mortality, although experimental evidence was not shown. In two studies (Wilson and Carroll, 1997; Wilson and Faeth, 1998), the negative association was explained or substantiated using manipulative experiments. In the manipulative experimental studies, endophytes had a negative effects on insects by increasing insect mortality or/and decreasing performance or/and oviposition six times and there were no reports of positive effects on insects, but three reports of neutral effects.

The results of the correlative studies should be considered with caution, however. Most obviously, correlation does not indicate causation and a correlation may have many different causes, with endophytes being one of numerous possibilities. However, a significant correlation indicates that the cause may be worth investigating with appropriate manipulative experiments to determine if there is a causal relationship. The results of correlative studies are much more meaningful if the direction of the correlation can be explained by previous results. For example, Wilson and Carroll (1997) concluded that the negative within-leaf correlation of an endophyte and a gall insect was an evolved response by the ovipositing gall wasp female to oviposit in that part of the leaf where endophyte presence was predictably low and where endophyte-caused mortality was low. They proposed that insects in general may avoid endophyte-mediated antagonism by avoiding high-endophyte space.

Correlations are typically made at only one particular spatial scale, e.g., at the whole-tree scale, or among paired leaves, whereas endophyte antagonism, if present, and like many other determinants of insect distribution (Stiling et al., 1991), may operate at some spatial scales but not others (Wilson and Faeth, 1998). Thus, insects may not necessarily select which tree to feed on based on endophyte frequency, but may select which leaves within the tree, or vice versa. Spatial correlations, or co-occurrence of endophytes and insects in the same place, such as on the same tree or on the same leaf, may be meaningless if, as for many sedentary insects such as leafminers or gall formers, the ovipositing female selects the site the larvae will occupy but the endophyte frequency is measured at a different (usually later) time when presumably endophyte frequency may have changed and may not reflect the endophyte status the ovipositing female is presented with. In addition, measuring endophyte levels for correlations after oviposition or after the insect has left the site may be more meaningless if the future endophyte distribution is unpredictable or unchangeable by the insect and does not reflect the endophyte status of the host at a time that is important for the insect.

The spatial or temporal correlation among endophytes and insects is only one component of "endophyte space." The growth activity may be an equally or more important characteristic. For example, the endophyte *D. quercina* causes mortality by growing into galls from the leaf, but it only shows high-growth activity along the leaf veins. Thus gall insects that occur on the veins suffer higher endophyte-caused mortality relative to gall insects that occur on the leaf lamina where the endophyte has very low growth activity and does not frequently grow into the galls (Wilson and Carroll, 1997).

Manipulative experiments are much more meaningful when used to investigated how endophytes affect insects and whether the relationship between the fungus and host is mutualistic. Typically these experiments have involved determining insect mortality and performance on infected and uninfected plant tissue. However, manipulating tree endophytes in the field has proven difficult, although Wilson (1996, 1998) recently used a technique to overcome this logistical problem where protective bags are placed over branches prior to bud burst. The bags prevent rain-dispersed spores from reaching the leaves, which substantially reduced or eliminated endophyte infection. Similarly bagged leaves can be sprayed with endophyte spore suspensions to create infected leaves. Some manipulative experiments have sprayed spore suspensions over leaves that are already infected with natural endophyte levels. Although this may elevate infection levels, insect performance is then measured on two groups of infected leaves, i.e., those with natural and those with elevated endophyte levels. Unless the effects of endophytes are dose-dependent, this method may not be sensitive enough to detect endophyte effects that may be present at low or natural levels with no increase in the impact on insects at elevated levels. The bagging technique has the advantage of giving the experimenter a very sensitive assay of insect performance on uninfected and infected leaves to determine how natural endophyte levels affect insects relative to the absence of endophytes. Furthermore, leaves and branches can be randomly assigned to treatments eliminating the problems of correlates of infection where treatments are not randomly assigned to replicates. However, devising controls to test the effect of the bags presents its own logistical problems.

Of the studies examined for this chapter, nine reports were based on manipulative experiments to determine how endophytes affect insects; all of the other studies are correlative and demonstrate if endophytes and insects are associated. Of the nine manipulative experiments, six indicate that endophytes had a negative affect on insect performance or survival and three that endophytes had a neutral effect. Whether six of nine is a realistic representation of how many insect–plant encounters are negatively affected by endophytes is unknown. This number likely reflects the bias in science to publish (or for journals to accept) papers showing significant effects as well as the a priori selection of systems or studies because of the promise of significant results for such high-investment experiments. Nevertheless, tree endophytes can clearly negatively affect insect survival, performance, and behavior and presumably have an important role in mediating insect–plant interactions.

Faeth and Hammon (1997b,c) and Saikkonen et al. (1996) have argued on theoretical grounds that a mutualistic relationship between tree endophytes and the tree host should not be expected, or at least should not be as strong a mutualism as between grasses and their endophytic fungi. Their arguments are based on how tightly coupled or reliant the endophyte is on its host plant, in particular how dependent the fungus is on the survival of the host plant for transmission of the endophyte. These arguments are based on the theories of Ewald (1991) who proposed that where parasites (or relevant here, fungal endophytes) are reliant on the host for transmission and dissemination, a mutualistic relationship is more likely to evolve, whereas if a parasite is not as reliant on its host for transmission, a mutualistic relationship is less likely to evolve or a mutualistic relationship would be weak. The *Neotyphodium* endophytes of grasses depend on the host for transmission via the seed or systemic spread into new tillers. Thus, by Ewald's arguments, since the endophytes depends on the host for transmission they should develop a mutualistic relationship as indicated in other chapters in this book. In contrast, tree endophytes are largely horizontally transmitted where, for example, death of a plant organ such as in leaf abscission allows the fungus to sporulate and disseminate; thus the parasite is less dependent on host survival for transmission, but transmission may be favored by host death. Thus tree endophytes would not be expected to form a mutualistic relationship or at best the relationship would be weak. Although the studies to date document clear antagonism toward insect herbivores indicative of a defensive mutualism, it is difficult to distinguish between a mutualistic relationship or fortuitous coincidence (for the plant) where these fungi act as insect antagonists. However, Ewald's arguments are not sacrosanct and do not explain why or how mycorrhizal fungi should and do form mutualistic relationships with their hosts. The transmission of many mycorrhizal fungi is horizontal from one individual to another, and, as with tree endophytes, is not totally reliant on host survival; at least transmission is not as tightly coupled to the host as it is with the grass endophytes. Yet mycorrhizal fungi clearly form often very strong mutualistic relationships with the host. Just as Ewald's theories might not adequately predict mutualistic interactions between plants and mycorrhizal fungi, they might not apply to endophyte–plant interactions. Furthermore, where Ewald's theories might predict a mutualistic relationship between seed-borne endophytes in grasses, extensive recent work in Arizona on a native fescue failed to show any antiherbivore properties of the *Neotyphodium* endophyte, although the endophyte may interact mutualistically in another way such as via increased drought tolerance.

6.2. Modes of Insect Antagonism

Several different modes of insect antagonism have emerged. The first proposed was antagonism from endophyte-produced toxins, either single toxins or the synergistic interaction of more than one toxin produced by different species or by different strains of the same species (Carroll, 1991b). Toxins would form an effective defense against both sedentary and mobile insects. In support of this hypothesis, Clark et al. (1989) showed that a small number of endophyte strains isolated from balsam fir foliage produced potent insecticidal toxins in fermentation. In addition, *R. parkeri* that infects and kills midge galls on Douglas fir has also been shown to produce potent insecticidal toxins (Carroll, 1991b). In addition to toxins, Claydon et al. (1985) found elm bark beetle boring and feeding deterrents produced by *Phomopsis* sp., an inner bark endophyte and antagonist of elm bark beetles. Considering the diverse chemical arsenal produced by the fungi (Betina, 1984), chemical defense of the host by the endophytes, as shown by grass endophytes, should not be surprising. Although no examples of endophytes acting as entomopathogens exist, in at least one example, an endophyte caused gall insect mortality by growing into the gall from the leaf and colonizing the nutritive tissue on which the insect larva feeds, thereby causing insect starvation (Wilson, 1995b). Endophytes found associated with dead gall occupants by Butin (1992) and Taper et al. (1986) may also have caused insect mortality by entering and colonizing the gall, which is the insect's food source. It would be interesting to examine the origins and functions (besides a food source) of the fungi purposely brought into ambrosia galls by the insect (Bissett and Borkent, 1988). Were these once fungal endophytes that colonized the galls and the insect adapted by feeding on them? Perhaps the fungi were used to effectively defend the gall against invasion by endophytes?

Wilson and Carroll (1997) proposed that where endophytes cause insect antagonism, insects can avoid such antagonism by spatially or temporally avoiding endophytes or by avoiding high-endophyte space in an analogous way that insects avoid natural enemies by occupying enemy-free space (Jeffries and Lawton, 1984). To avoid high-endophyte space, insects would need to have the capacity to (1) detect endophytes and actively avoid them by moving or changing oviposition or feeding site, and (2) behaviorally adapt such that they live in low-endophyte space. In the latter, the low-endophyte space would have to be consistently or predictably low from the perspective of the insect, and in the absence of an endophyte detection mechanism. The system studied by Claydon et al. (1985) described above represents an example of detection and avoidance of high-endophyte space, but I am not aware of any other nongrass examples. The infection patterns of endophytes can be highly predictable at different scales and as such insects are able to occupy predictably low-endophyte space. For instance, Wilson and Carroll (1997) showed avoidance of high-endophyte space by a gall wasp at the within-leaf scale where endophyte-caused mortality was lower. Endophyte distribution was predictable at the within-leaf scale but not between leaves, and there was no association of the insects and endophytes between leaves. Wilson and Faeth (1998) showed that endophyte density was predictable and highly inversely dependent on leaf size. Since leaf size was correlated with position, within the tree insects could avoid leaves with high endophyte densities by selecting large leaves. In fact the leafminer *Cameraria* sp. has the highest density on large leaves and is more abundant in the interior of the tree where leaves are larger compared to the exterior part of the tree (Bultman and Faeth, 1986a, b; Faeth, 1990). These as well as many other leafminers (Stiling and Simberloff, 1989) suffer mortality from density-dependent premature leaf abscission that Wilson (1998) found was caused by endophytes. *Cameraria* is most abundant on large leaves in the interior of the tree where mortality from premature leaf abscission is less relative to exterior smaller leaves because smaller leaves have higher endophyte densities and an associated higher probability of premature abscission (Bultman and Faeth, 1986b; Faeth, 1990). Thus *Cameraria* occupies low-endophyte space and as such suffers less from endophyte-caused mortality from premature leaf abscission.

It is clear that endophytes and insects interact and that endophytes may provide causal mechanisms that explain phenomena such as premature abscission of mined leaves, or distributional patterns of insects that were previously unexplained by more traditional mechanisms such as phytochemistry. However, if progress is to continue, future work in this area must consider the following: (1) manipulative experiments must be performed instead of, or in tandem with, correlative studies. (2) The effects of individual endophyte species must be examined and considered separately before they are collectively lumped together into and considered a homogeneous group of endophytes. For instance, more cosmopolitan and less host-specific endophytes should be considered separately from the more host-specific and more specialized species. (3) The effect of scale must be included into experimental designs as effect may be scale dependent as has been shown for the effects of parasitoids on phytophagous insects (Stiling et al., 1991). (4) The problem of adequate sampling and sampling resolution of endophytes must be considered as addressed above.

6.3. Leaf Senescence and Abscission

Although the interaction with insect herbivores has dominated research on the ecological role of tree endophytes, how they directly affect the plant hosts physiology, phenology, and biochemistry has been less studied. The finding that endophytes cause density-dependent premature abscission of leaves with high numbers of leafminers is indicative of a broader role endophytes might have in the phenology of leaf abscission. Fisher et al. (1986) first raised the question of whether the presence of endophytes within healthy plant tissue may hasten the onset of senescence and thus influence the lifespan of a plant. Wilson (1993) elaborated on this after observing that surface sterilized Oregon white oak leaf discs that were endophyte-free remained green and intact for over 18 months when platted out on PDA and stored in a cold room, whereas endophyte-infected discs turned brown and were rapidly degraded by the endophytes. Other lines of evidence further support the role of endophytes in leaf senescence and abscission. Dickinson (1981) discusses the acceleration of senescence of wheat and bean leaves experimentally inoculated with species of *Alternaria* and *Cladosporium* (both of which frequently form symptomless infections). Bashan (1994a, b) also found that *Alternaria* species formed endophytic infection of cotton leaves and that high spore concentrations of *A. macrospora* applied to leaves would cause accelerated abscission whereas low spores concentrations did not. Furthermore, when low spore concentrations were applied abscission could be accelerated after

A. alternata spores were also applied but *A. alternata alone* could not induce abscission, indicating a synergistic affect among endophyte species. In addition, apparently healthy nondiseased crop plants such as cereals and soybean frequently show a delay in senescence following fungicide treatment (Dickinson, 1981; Sinclair and Cerkauskas, 1996).

Faeth and Hammon (1997b) followed seasonal patterns of endophyte infection on Emory oak leaves. Most leaves of this oak are shed in April when the new leaves flush (Faeth, 1990); however, some leaves do not abscise but remain on the tree for up to an additional 7 months. They found that endophyte infection frequencies of the specialized endophyte *O. cryptica* increases from December through April but that infection frequencies in the leaves that remain on the tree are lower relative to those leaves that are shed. Although correlative, this suggests that *O. cryptica* infection loads of leaves may influence the phenology of abscission as leaves with low endophyte loads did not abscise. In this instance, it is endophyte level that better correlates with the propensity to abscise rather than leaf age.

That endophytes may influence the senescence and life of plant organs may alter the way plant physiologists investigate these processes. Most physiological investigations are in lab or greenhouse situations without the associated natural spectrum of microbes present in the field and with which the plant has evolved. If endophytes act as the gatekeepers of senescence and plant organ longevity, examining these processes in endophyte-free systems might be like examining decomposition in a microbe-free system or animal digestion in the absence of gut microbes, and the implications on the ecology of plants, nutrient reabsorption, nutrient cycling, and soil biology on the forest floor may be equally important and profound.

6.4. Induced Effects of Endophyte Infection

Induced changes in plants are typically biochemical or electrical changes in response to an external stimulus that may affect plant physiology, morphology, and phenology (Baldwin, 1988; Barker et al., 1995; Karban and Myers, 1989). Fungal infection, herbivore damage, and mechanical damage are all known to induce changes in the biochemistry of plants and ecologists have been particularly interested in how these changes affect subsequent plant–microbe or plant–insect encounters (Karban and Carey, 1984; Karban and Myers, 1989). Induced changes frequently result in greater resistance of the plant to further attack, although many studies have shown no effects or even positive effects on subsequent attackers. Matta (1971) recognized the potential interactions between temporally and spatially separated fungi colonizers of plants via induced changes in the host plant. There are many examples of nonpathogenic microbes inducing chemical changes that render the host plant more resistant to infection by pathogens (see Matta,

1971). In an elegant study by Freeman and Rodriguez (1993), a single gene mutation turned a pathogenic strain of *Colletotrichum* into a symptomless endophyte.
When the curcubit host was infected with the endophytic strain it was resistant
to infection by the pathogenic strain.

How frequently endophytes induce changes in their host plant is unknown,
but if endophytes induce changes as frequently and at the same magnitude as
infection by many pathogenic fungi, then endophytic induction of plant allelochemicals such as phytoalexins, phenolics, flavenoids, and alkaloids should be
very widespread. Given the ubiquity and abundance of endophytes, this type of
induction might be the normal situation within plants and many allelochemicals
produced constitutively might be the result of continued induced changes caused
by repeated endophytic infection. Wilson (1993) pointed out that the presence
of endophytes has never been controlled for in any study of induced responses
in plants (but see Bultman, this volume), so their effect is currently unknown.

Faeth and Wilson (1997) and Wilson et al. (1998) showed that endophyte
species in Emory oak interact via induced changes and that they respond to induced changes from insect and artificial wounding. At the whole-leaf scale in
Emory oak foliage, the presence of one endophytic species is always negatively
associated with another and usually significantly so. Although correlational, this
suggests that endophyte species might interact and that one likely cause is via
induced changes in phytochemistry, e.g., condensed tannins that are known to
be induced in Emory oak foliage (Faeth, 1992) and also to be fungistatic, although
alternative mechanisms include competition for infection sites or fungistatic metabolites produced by the endophyte.

In a series of experiments, Emory oak leaves were artificially damaged and
subsequent colonization by endophytes measured (1) at the wound site, (2) at
the undamaged part of the damaged leaf, (3) in corresponding sites of an undamaged paired control leaf. The most significant result was that endophyte infection
levels increased at the damage site relative to undamaged parts on the same or
undamaged control leaves. However, the specialized endophyte *O. cryptica* responded in the opposite way and was apparently less abundant at damage sites
relative to other endophytes, in particular more generalist species. Further, endophyte infection levels (mostly of *O. cryptica*) in the undamaged part of damaged
leaves were lower relative to a corresponding areas on undamaged control leaves.
Thus, either damage or damage and associated infection by endophytes caused
the undamaged parts of damaged leaves to be less susceptible to endophyte infection or to cause mortality of already present fungal infections. Changes in leaf
phytochemistry or the production of fungistatic metabolites most likely account
for lower infection levels in the undamaged part of the damaged leaf. Thus endophytes can interact with damage alone or damage accompanied by infection via
induced changes in the host.

Fungal endophytes can cause induced changes following infection of the
host plant and may therefore alter the host and affect subsequent interactions

with other endophytes, pathogens, or macroherbivores. However, the prevalence, magnitude, and nature of these induced changes are very poorly understood— in particular, how induced changes by endophytes compare quantitatively and qualitatively to induced changes caused by pathogens and herbivores. Edwards and Wratten (1983, 1985) suggested that insect feeding on leaves caused induced halos of defenses around the feeding sites and could explain why insect grazing damage is frequently over dispersed within leaves. Such induction around feeding sites might be explained by infection by endophytes that is facilitated at insect wound sites, assuming a negative effect of endophyte infection on the insect. Furthermore, if endophyte infection is inhibited in the rest of the leaf, avoidance of infection associated with current feeding sites and movement to low-infection sites in the rest of the leaf may be an additional explanation as to why an insect should expend time and exposure to natural enemies to change feeding sites.

6.5. Other Ecological Roles of Tree Endophytes

Except for the roles outlined above, very few other ecological roles of endophytes in trees have been examined, although other ways they might affect their host plant or other organisms that utilize the host have been proposed.

Endophytes might act as aerial counterparts of mycorrhizal fungi. Although they clearly do not have the same morphological characteristics as mycorrhizal fungi, they might have a role in nutrient acquisition from substances on the plant surface or dissolved in rainwater falling on leaves or running down stems. Plants are known to absorb certain nutrients through their foliage and endophytes might augment this process. Relatively simple experiments involving tracking radiolabeled nutrient solutions applied to greenhouse-grown plants that were endophyte-free and plants infected at known sites on the leaves or stems would address this issue.

A role of endophytes in plant decomposition has been suggested for almost a decade. Since endophytes are among the first fungi present within senescent plant tissue they may begin initial decomposition, but some evidence suggests that they are rapidly succeeded by more typical fungal saprobes (Petrini et al., 1992). Simply being the first fungus to colonize an abscised leaf with the associated access to simple carbohydrates might be a primary selecting force for endophytism. Endophytes may have been early successional saprophytes that evolved to colonize the plant material before it became senescent and so were the first to capitalize on the more easily utilized carbohydrates in the leaf. They may also compete with the leaf for mobilized nutrients at the onset of leaf senescence (Stacey, 1993).

Endophytism in trees may have evolved as a mechanism for opportunistic pathogens to infect plants, so they may already be present when conditions arise that are conducive to cause disease symptoms. These pathogens would then not have to rely on more passive processes that would bring them in contact with the host and they would reach the host before nonendophytic colonizers. Certainly, many fungi with both endophytic phases and pathogenic phases under certain conditions fit this description.

7. FUTURE DIRECTIONS AND CONCLUSIONS

Future directions in fungal tree endophyte research will most likely continue to be descriptive in nature. While these types of basic study are important, especially in light of the potentially enormous numbers of new species that await discovery and relevance to biodiversity issues (Wilson et al., 1997), they could nevertheless incorporate additional aims such as, for instance, to examine spatial and temporal variation in infection among locations, among trees within a location, etc. Such studies will need to follow rigorous sampling protocols to include adequate sampling sizes as indicated by Petrini et al. (1992) for trees and Carroll (1995) for leaves, as well as the sampling issues addressed in this chapter.

Beyond descriptive, abundance, and distribution-type studies lie the challenges of manipulative experiments to address causal relationships and ecological questions. Manipulation of tree endophytes in the field many be easier than manipulating microbes from other niches such as the soil; however, I have described at least one technique to do so. Thus endophyte systems might prove robust model systems to address some basic and fundamental ecological questions as well as the intriguing new possible roles that endophytes may have. The quality of research in this area will be increased if correlative studies are augmented with tandem experimental manipulations. Furthermore, experiments addressing ecological roles of endophytes will require interdisciplinary research efforts and the cooperation of microbiologists and macrobiologists, few of whom appreciate the subtleties of the others' fields.

Endophytes clearly have the potential to affect many different ecological, physiological, and biochemical processes in plants, yet there are very few detailed studies that have investigated much more than insect antagonism. Fruitful areas of investigation may include leaf senescence and abscission; nutrient reabsorption from leaves and absorption by leaves; induced effects on pathogens, insects, and other endophytes; and decomposition and successional trajectories of saprobes. Hopefully, the future endophyte literature on studies other than insect antagonism will blossom.

REFERENCES

Allen, E. A., Hoch, H. C., Steadman, J. R. and Stavely, R. J. (1991). Influence of leaf surface features on spore deposition and epiphytic growth of phytopathogenic fungi.

In J. H. Andrews and S. S. Hirano (eds.), *Microbial Ecology of Leaves.* Springer-Verlag, New York, pp. 87–110.

Andrews, J. H., Kinkel, L. L., Berbee, F. M. and Nordheim, E. V. (1987). Fungi, Leaves, and the theory of island biogeography. *Microbial Ecology* 14, 277–290.

Baldwin, I. T. (1988). Short-term damage-induced increases in tobacco alkaloids protect plants. *Oecologia* 75, 367–370.

Barker, A. M., Wratten, S. D. and Edwards, P. J. (1995). Wound-induced changes in tomato leaves and their effects on the feeding patterns of larval lepidoptera. *Oecologia* 101, 251–257.

Bashan, Y. (1994a). Symptom expression and ethylene production in leaf blight of cotton caused by *Alternaria macrospora* and *Alternaria alternata* alone and in combination. *Canadian Journal of Botany* 72, 1574–1579.

Bashan, Y. (1994b). Symptomless infections in alternaria leaf blight of cotton. *Canadian Journal of Botany* 72, 1580–1585.

Begon, M., Harper, J. L. and Townsend, C. R. (1990). *Ecology: Individuals, Populations, and Communities.* Blackwell Scientific, Cambridge, MA.

Bernstein, M. E. and Carroll, G. C. (1977). Internal fungi in old-growth Douglas fir foliage. *Canadian Journal of Botany* 55, 644–653.

Betina, V. (1984). *Mycotoxins: Production, Isolation, Separation, Purification.* Developments in Food Science, vol. 8. Elsevier, New York, p. 528.

Bissegger, M. and Sieber, T. N. (1994). Assemblages of endophytic fungi in copice shoots of *Castanea sativa.* Mycologia 86, 648-655.

Bissett, J. and Borkent, A. (1988). Ambrosia galls: the significance of fungal nutrition in the evolution of the Cecidomyiidae (Diptera). *In* K. A. Pirozynski and D. L. Hawksworth (eds.), *Coevolution of Fungi with Plants and Animals.* Academic Press, London, pp. 203–225.

Bose, S. R. (1946). Hereditary (seed-borne) symbiosis in *Casuarina equisetifolia* Forst. *Nature* 15, 512–514.

Boursnell, J. G. (1950). The symbiotic seed-borne fungus in the Cistaceae I. Distribution and function of the fungus in the seedling and in the tissues of the mature plant. *Annals of Botany* 14, 217–243.

Bultman, T. L. and Faeth, S. H. (1986a). Leaf size selection by leaf-mining insects on *Quercus emoryi* (Fagaceae). *Oikos* 46, 311–316.

Bultman, T. L. and Faeth, S. H. (1986b). Selective oviposition by a leaf miner in response to temporal variation in abscission. Oecologia 69, 117–120.

Butin, H. (1992). Effect of endophytic fungi from oak (*Quercus robur* L.) on mortality of leaf inhabiting gall insects. *European Journal of Forest Pathology* 22, 237–246.

Cabral, D., Stone, J. K. and Carroll, G. C. (1993). The internal mycobiota of *Juncus* spp.: microscopic and cultural observations of infection patterns. *Mycological Research,* 97, 367–376.

Carroll, G. C. (1986). The biology of endophytism in plants with particular reference to woody perennials. *In* N. J. Fokkenna and J. van den Heuvel. (eds.), *Microbiology of the Phylosphere.* Cambridge University Press, Cambridge, UK, pp. 205–222.

Carroll, G. C. (1988). Fungal endophytes in stems and leaves: from latent pathogen to mutualistic symbiont. *Ecology* 69, 2–9.

Carroll, G. C. (1991a). Beyond pest deterrence-alternative strategies and hidden costs of endophytic mutualisms in vascular plants. *In* J. A. Andrews and S. S. Hirano (eds.), *Microbial Ecology of Leaves*. Springer-Verlag, New York, pp. 358–375.

Carroll, G. C. (1991b). Fungal associates of woody plants as insect antagonists in leaves and stems. *In* P. Barbosa, V. A. Krischik, and C. G. Jones (eds.), *Microbial Mediation of Plant–Herbivore Interactions*. John Wiley and Sons, New York, pp. 253–272.

Carroll, G. C. (1995). Forest endophytes: pattern and process. *Canadian Journal of Botany* 73 (Suppl. 1), S1316–S1324.

Carroll, G. C. and Carroll, F. E. (1978). Studies on the incidence of coniferous needle endophytes in the Pacific Northwest. *Canadian Journal of Botany* 56, 3032–3043.

Chanway, C. P. (1996). Endophytes: they're not just fungi! *Canadian Journal of Botany* 74, 321–322.

Chapela, I. H., and Boddy, L. (1988). Fungal colonization of attached beech branches. II. Spatial and temporal organization of communities arising from latent invaders in bark and functional sapwood under different moisture regimes. *New Phytology* 110, 47–57.

Chapela, I. H., Petrini, O. and Hagmann, L. (1991). Monolignol glucosides as specific recognition messengers in fungus-plant symbioses. *Physiological and Molecular Plant Pathology* 39, 289–298.

Chapman, R. F. (1971). The Insects: Structure and Function. Hodder and Stroughton, London, pp. 81–82.

Clark, C. L., Miller, J. D. and Whitney, N. J. (1989). Toxicity of conifer needle endophytes to spruce budworm. *Mycological Research* 93, 508–512.

Claydon, N., Grove, J. F. and Pople, M. (1985). Elm bark beetle boring and feeding deterrents from *Phomopsis oblonga*. *Phytochemistry* 24, 937–943.

Connor, E. F., Hosfield, E., Meeter, D. M. and Niu X. (1997). Tests for aggregation and size-based sample-unit selection when sample units vary in size. *Ecology* 78, 1238–1249.

Cubit, J. D. (1974). Interactions of Seasonally Changing Physical Factors and Grazing Affecting High Intertidal Communities on a Rocky Shore. Ph.D. Dissertation, University of Oregon.

Diamandis, S. (1981). *Elytroderma torres-juanii* Diamandis and Minter. A serious attack on *Pinus brutia* L. in Greece. *In* C. S. Millar (ed.), *Current Research on Conifer Needle Diseases*. Aberdeen University Press, Aberdeen, Scotland, pp. 9–12.

Dickinson, C. H. (1981). Biology of *Alternaria alternata*, *Cladosporium cladosporioides* and *C. herbarum* in respect of their activity on green plants. *In* J. P. Blakeman (ed.), *Microbial Ecology of the Phylloplane*. Academic Press, London.

Dowd, P. F. (1991). Symbiont-mediated detoxification in insect herbivores. *In* P. Barbosa, V. A. Krischik and C. G. Jones (eds.), *Microbial Mediation of Plant–Herbivore Interactions*. John Wiley and Sons, New York, pp. 411–440.

Edwards, P. J. and Wratten, S. D. (1983). Wound induced defences in plants and their consequences for patterns of insect grazing. *Oecologia* 59, 88–93.

Edwards, P. J. and Wratten, S. D. (1985). Induced plant defences against insect grazing: fact or artifact? *Oikos* 44, 70–74.

Ewald, P. W. (1991). Culture, transmission modes and the evolution of virulence with special reference to cholera, influenza, and AIDS. *Human Nature* 2, 1–30.

Faeth, S. H. (1990). Aggregation of a leafminer, Cameraria sp. nov. (Davis): consequences and causes. *Journal of Animal Ecology* 59, 569–586.

Faeth, S. H. (1992). Do defoliation and subsequent phytochemical responses reduce future herbivory on oak trees? *Journal of Chemical Ecology* 18, 915–925.

Faeth, S. H., and Hammon, K. E. (1997a). Fungal endophytes and phytochemistry of oak foliage: Determinants of oviposition preference in leafminers. Oecologia 108, 728–736.

Faeth, S. H. and Hammon, K. E. (1997b). Fungal endophytes in oak trees: I. Long-term patterns of abundance and associations with leafminers. *Ecology* 78, 810–819.

Faeth, S. H. and Hammon, K. E. (1997c). Fungal endophytes in oak trees: II. Experimental analyses of interactions with leafminers. *Ecology* 78, 820–827.

Faeth, S. H. and Wilson, D. (1997). Induced responses in trees: mediators of interactions among macro- and micro-herbivores? *In* A. C. Gange and V. K. Brown (eds.), *Multitrophic Interactions in Terrestrial Systems.* Blackwell Scientific, London, pp. 201–215.

Fail, G. L. and Langenheim, J. H. (1990). Infection process of *Pestalotia subcuticularis* on leaves of *Hymenaea courbaril. Phytopathology* 80, 1259–1265.

Fisher, P. J., Anson, A. E. and Petrini, O. (1986). Fungal endophytes in *Ulex europaeus* and *Ulex gallii.* Transactions of the British Mycological Society 86, 153–193.

Fisher, P. J., Petrini, L. E., Sutton, B. C. and Petrini, O. (1995). A study of fungal endophytes in leaves, stems, and roots of *Gynoxis oleifolia* Muchler (Compositae) from Ecuador. *Nova Hedwigia* 60, 589–594.

Fisher, P. J., and Petrini, O. (1990). A comparative study of fungal endophytes in xylem and bark of Alnus species in England and Switzerland. *Mycological Research*, 94, 313–319.

Fisher, P. J., Petrini, O., Petrini, L. E. and Sutton, B. C. (1994). Fungal endophytes from the leaves and twigs of *Quercus ilex.* L. from England, Majorca, and Switzerland. *New Phytologist* 127, 133–137.

Freeman, S. and Rodriguez, R. J. (1993). Genetic conversion of a fungal pathogen to a nonpathogenic, endophytic mutualist. *Science* 260, 75–78.

Gange, A. C. (1996). Positive effects of endophyte infections on sycamore aphids. *Oikos* 75, 500–510.

Gaylord, E. S., Preszler, R. W. and Boecklen, W. J. (1996). Interactions between host plants, endophytic fungi, and a phytophagous insect in an oak (*Quercus grisea x Q. gambelii*) hybrid zone. *Oecologia* 105, 336–342.

Hammon, K. E., Faeth, S. H. (1992). Ecology of plant-herbivore communities: a fungal component? *Natural Toxins* 1, 197–208.

Hata, K. and K. Futai. (1995). Endophytic fungi associated with healthy pine needles and needles infected with the pine needle gall midge, *Thecodiplosis japonensis. Canadian Journal of Botany* 73, 384–390.

Hata, K. and Futai, K. (1996). Variation in fungal endophyte populations in needles of the genus *Pinus. Canadian Journal of Botany* 74, 103–114.

Hawksworth, D. L., Kirk, P. M., Sutton, B. C. and Pegler, D. N. (1995). Ainsworth and Bisby's *Dictionary of the fungi.* CAB International, Wallingford, UK.

Helander, M. L., Neuvonen, S., Sieber, T. N. and Petrini, O. (1993). Simulated acid rain affects birch leaf endophyte populations. *Microbial Ecology* 26, 227–234.

Helander, M. L., Sieber, T. N., Petrini, O. and Neuvonen, S. (1994). Endophytic fungi in Scotts pine needles: spatial variation and consequences of simulated acid rain. *Canadian Journal of Botany* 72, 1108–1113.

Jeffries, M. J. and Lawton, J. H. (1984). Enemy-free space and the structure of ecological communities. *Biological Journal of the Linnean Society* 23, 269–286.

Johnson, J. A. and Whitney, N. J. (1989). An investigation of needle endophyte colonization patterns with respect to height and compass direction in a single crown of balsam fir (*Abies balsamea*). *Canadian Journal of Botany* 56, 723–725.

Karban, R. and Carey, J. R. (1984). Induced resistance of cotton seedlings to mites. *Science* 225, 53–54.

Karban, R., and Myers, J. H. (1989). Induced plant responses to herbivory. *Annual Review of Ecology and Systematics* 20, 331–348.

Kinkel, L. L., Andrews, J. H. and Nordheim, E. V. (1989). Fungal immigration dynamics and community development on apple leaves. *Microbial Ecology* 18, 45–58.

Lappalainen, J. H. and Helander, M. J. (1997). The role of foliar microfungi in mountain birch–insect herbivore relationships. *Ecography* 20, 116–122.

Legault, D., Dessureault, M. and Laflamme, G. (1989). Mycoflore des aiguilles de *Pinus banksiana* et *Pinus resinosa* I. Champignons endophytes. *Canadian Journal of Botany* 67, 2052–2060.

Malloch, D. and Blackwell, M. (1992). Dispersal of fungal diaspores. *In* G. C. Carroll and D. T. Wicklow (eds.), *The Fungal Community: Its Organization and Role in the Ecosystem*. Marcel Dekker, New York, pp. 147–171.

Matta, A. (1971). Microbial penetration and immunization of uncongenital host plants. *Annual Review of Phytopathology* 9, 387–410.

Neely, D., and Himelick, E. B. (1965). Nomenclature of the sycamore anthracnose fungus. *Mycologia* 57, 834–837.

Newsham, K. K. (1994). First record of intracellular sporulation by a coelomycete fungus. *Mycological Research* 98, 1390–1392.

Petrini, L. E., Petrini, O. and Laflamme, G. (1989). Recovery of endophytes of *Abies balsamea* needles and galls of *Paradiplosis tumifex*. *Phytoprotection* 70. 97–103.

Petrini, O. (1991). Fungal endophytes of tree leaves. *In* J. H. Andrews and S. S. Hirano (eds.), *Microbial Ecology of Leaves*. Springer-Verlag, New York, pp. 179–197.

Petrini, O., Sieber, T. N., Toti, L. and Viret, O. (1992). Ecology, Metabolite Production, and Substrate Utilization in Endophytic Fungi. *Natural Toxins* 1, 185–196.

Preszler, R. W. and Boecklen, W. J. (1996). The influence of elevation on tri-trophic interactions: Opposing gradients of top-down and bottom-up effects on a leaf-mining moth. *Ecoscience* 3, 75–80.

Rodrigues, K. F. (1994). The foliar fungal endophytes of the Amazonian palm *Euterpe oleracea*. *Mycologia* 86, 376–385.

Rollinger, J. L., and Langenheim, J. H. (1993). Geographic survey of fungal endophyte community composition in leaves of coastal redwood. *Mycologia* 85, 149–156.

Saikkonen, K., Helander, M., Ranta, H., Neuvonen, S., Virtanen, T., Suomela, L. and Vuorinen, P. (1996). Endophyte-mediated interactions between woody plants and insect herbivores? *Entomologia Experimentalis et Applicata* 80, 269–271.

Sherwood-Pike, M., Stone, J. K. and Carroll, G. C. (1985). *Rhabdocline parkeri*, a ubiquitous foliar endophyte of Douglas fir. *Canadian Journal of Botany* 64, 1849–1855.

Sieber, T. N., and Hugentobler, C. (1987). Endophytische pilze in Blättern und Aesten gesunder und geschädigter buchen (Fagus sylvatica L.). *European Journal of Forest Pathology* 17, 411–425.

Sieber-Canavesi, F. and Sieber, T. N. (1987). Endophytische Pilze in Tanne (Abies alba Mill.).- Vergleich zweier Standorte im Schweizer Mittelland (Naturwald-Aufforstung). *Sydowia* 40, 250–273.

Sinclair, J. B. and Cerkauskas, R. F. (1996). Latent infection vs. endophytic colonization by fungi. *In* S. C. Redlin and L. M. Carris, (eds.), Endophytic Fungi in Grasses and Woody Plants: Systematics, Ecology, and Evolution. APS Press, St. Paul, MN.

Smith, H., Wingfield, M. J. and Petrini, O. (1996). *Botryosphaeria dothidea* endophytic in *Eucalyptus grandis* and *Eucalyptus nitens* in South Africa. *Forest Ecology and Management* 89, 189-195.

Stacey, L. M. (1993). Leaf senescence and the importance of fungal endophytes: a case study of *Rhytisma punctatum* and *Discula* sp. on senescing *Acer macrophyllum* leaves. M.Sc., University of Oregon, Eugene.

Stiling, P. and D. Simberloff. (1989). Leaf abscission: induced defense against pests or response to damage. *Oikos* 55, 43–49.

Stiling, P., Throckmorton, A., Silvanima, J. and Strong, D. R. (1991). Does spatial scale affect the incidence of density dependence? a field test with insect parasitoids. *Ecology* 72, 2143–2154.

Stone, J. K. (1986). Foliar endophytes of Douglas fir: cytology and physiology of the host-endophyte relationship. Ph.D., University of Oregon, Eugene.

Stone, J. K. (1987). Initiation and development of latent infections by *Rhabdocline parkeri* on Douglas-fir. *Canadian Journal of Botany*, 65, 2614–2621.

Stone, J. K. and White, J. F. (1997). Biodiversity of endophytic fungi. *In* G. M. Mueller, G. F. Bills, A. Y. Rossman and H. H. Burdsall. *Measuring and Monitoring Biological Diversity: Standard Methods for Fungi*. Smithsonian Institution Press, Washington, DC.

Taper, M. L., Zimmerman, E. M. and Case, T. J. (1986). Sources of mortality for a cynipid gall-wasp (Dryocosmus dubiosus [Hymenoptera: Cynipidae]): the importance of the tannin/fungus interaction. *Oecologia* (Berlin) 68, 437–445.

Todd, D. (1988). The effects of host genotype, growth rate, and needle age on the distribution of a mutualistic, endophytic fungus in Douglas-fir plantations. *Canadian Journal of Forest Research* 18, 601–605.

Toti, L., Chapela, I. H., and Petrini, O. (1992a). Morphometric evidence for host-specific strain formation in *Discula umbrinella*. *Mycological Research* 96, 420–424.

Toti, L., Viret, O., Chapela, I. H. and Petrini, O. (1992b). Differential attachment by conidia of the endophyte, *Discula umbrinella* (Berk. & Br.) Morelet, to host and non-host surfaces. *New Phytologist* 121, 469–475.

Viret, O., and Petrini, O. (1994). Colonization of beech leaves (*Fagus sylvatica*) by the endophytes *Discula umbrinella* (Teleomorph: *Apiognomonia errabunda*). *Mycological Research* 98, 423–432.

Webber, J. (1981). A natural biological control of Dutch elm disease. *Nature* 292, 449–450.

Wennström, A. (1994). Endophyte—the misuse of an old term. *Oikos* 71, 535–536.

Whitham, T. G. (1978). Habitat selection by Pemphigus aphids in response to resource limitation and competition. *Ecology* 56, 1164–1176.

Wicklow, D. T. (1992). The coprophilous fungal community: an experimental system. *In* G. C. Carroll, and D. T. Wicklow (eds.), *The Fungal Community: Its Organization and Role in the Ecosystem*. Marcel Dekker, New York, pp. 715–728.

Wilson, D. (1993). Fungal endophytes: out of sight but should not be out of mind. *Oikos* 68, 379–384.

Wilson, D. (1995a). Endophyte—the evolution of a term, and clarification of its use and definition. *Oikos* 73, 274–276.

Wilson, D. (1995b). Fungal endophytes which invade insect galls: Insect pathogens, benign saprophytes, or fungal inquilines? *Oecologia* 103, 255–260.

Wilson, D. (1996). Manipulation of infection levels of horizontally transmitted fungal endophytes in the field. *Mycological Research* 100, 827–830.

Wilson, D. (2000). Endophyte-driven premature abscission of insect damaged leaves. *Ecology* (in press).

Wilson, D., Barr, M. E. and Faeth, S. H. (2000). Ecology and description of a new species of *Ophiognomonia* endophytic in the leaves of *Quercus emoryi*. *Mycologia* (submitted).

Wilson, D. and Carroll, G. C. (1994). Infection studies of *Discula quercina*, an endophyte of *Quercus garryana*. *Mycologia* 86, 635–647.

Wilson, D., and Carroll, G. C. (2000). Avoidance of high-endophyte space by gall forming insects. *Ecology* (in press)

Wilson, D. and Faeth, S. F. (2000). Do fungal endophytes result in selection for leafminer ovipositional preference? *Ecology* (submitted).

Wilson, D., Faeth, S. F. and Anderson, R. A. (2000). Fungal endophyte mediated induced changes in oak leaves and infection at wound sites. *Ecology* (submitted).

16
Do Fungal Endophytes Mediate Wound-Induced Resistance?

Thomas L. Bultman and John Charles Murphy
Truman State University, Kirksville, Missouri

1. INTRODUCTION

Microorganisms associated with plants are generally diverse and abundant (Dickenson and Preece, 1976; Blakeman, 1981; Wicklow and Carroll, 1981; Fokkema and van den Hueval, 1986; Andrews and Minao, 1991; Bills and Polishook, 1991). Despite this, entomologists studying ecological interactions between plants and insect herbivores have until recently overlooked the potential influence of microbes (Price et al., 1986; Letourneau, 1988; Berenbaum, 1988; Marquis and Alexander, 1992). Only recently has the role of microbes in modulating plant–insect interactions been addressed empirically. For example, fungi may indirectly affect seed mortality in shrub-steppe communities through reducing predation by granivores of fungal-infested seeds (Crist and Friese, 1993). Microorganisms likely mediate interactions between insect herbivores and plants in diverse ways (Hatcher, 1995; Barbosa et al., 1991).

Fungal symbionts are common associates of vascular plants. Many recent studies have documented the taxonomy, distribution, abundance, and ecological importance of endo- and ectomycorrhizal fungi (Ingham and Molina, 1991; Read, 1991). In addition to these below-ground fungal associates, an array of fungal endophytes also interact with above-ground portions of many plants (Petrini, 1986). One group of fungal endophytes (Ascomycota: Clavicipitaceae) forms intimate associations with forage grasses. The fungi intercellularly infect aerial shoot tissue of many grass species (White, 1987). Many endophytes that infect agronomically important species, like tall fescue and perennial ryegrass, never leave their host and can only reproduce by invading seed tissue of the host. These endophytes have diverse impacts on their host grasses, ranging from enhanced growth, vigor, and drought tolerance to enhanced resistance to herbivores (Clay, 1990a; West and Gwinn, 1993). Herbivore resistance is apparently mediated by mycotoxins produced by the fungus (Dahlman et al., 1991) and this resistance to herbivores has often been invoked as the centerpiece of hypotheses aimed at explaining the evolution of fungal endophytism in grasses (Clay, 1988, 1990a, b, 1991; Breen, 1994; Popay and Rowan, 1994).

We review the evidence for this hypothesis and conclude that the results are far from completely supportive. We propose a modification of the hypothesis—that in addition to constitutive resistance, fungal endophytes also mediate induced resistance. We suggest that induced resistance is as important, or perhaps even more important, than the constitutive resistance provided by grass endophytes and more fully explains the selective advantage of infection to grasses. Finally, we present preliminary empirical work designed to specifically test the hypothesis of endophyte-mediated induced resistance and discuss potential future directions in this area of research.

2. HISTORY OF ECOLOGICAL WORK ON ENDOPHYTE-INSECT INTERACTIONS

The presence of endophytic fungi within grasses was established several decades ago (Sampson, 1933, 1935; Neill, 1941; Diehl, 1950). Yet an ecological role of the fungi was not apparent until the association between the tall fescue endophyte [*Neotyphodium* (formerly *Acremonium*—see Glenn et al., 1996) *coenophialum*] and livestock toxicity was presented (Bacon et al., 1977) and substantiated (Hoveland et al., 1980). A similar association holds for the endophyte (*Neotyphodium lolii*) infecting perennial ryegrass and toxicoses of sheep that graze infected grass (Fletcher and Harvey, 1981). Researchers quickly discovered lolitrem toxins present in endophyte-infected ryegrass that cause ryegrass staggers in sheep (Gallagher et al., 1984). Similarly, ergot alkaloids appear to be the toxins responsible for tall fescue toxicity to cattle (Yates et al., 1985; Lyons et al., 1986). With these discoveries, efforts were made to remove endophytes from forage grasses. One way this was achieved was by prolonged seed storage (Rolston et al., 1986; Siegel et al., 1984). Pastures were replanted with endophyte-free grasses. However, it soon became apparent that endophyte-free grass was heavily damaged by insects (Prestidge et al., 1982) and that pastures quickly again became dominated by endophyte-infected individuals (Clay, 1990b). Indeed, the observation that endophyte-infected ryegrass plants predominate in seminatural, native habitats in Europe indicates that infected plants are at a selective advantage compared to their endophyte-free counterparts (Latch et al., 1987).

3. THE IMPACT OF ENDOPHYTES ON INSECT
HERBIVORES: THE PROBLEM

The prevailing hypothesis to explain the predominance of endophyte-infected grass in pastures is the defensive mutualism hypothesis (Clay, 1988). It is hypothesized that endophytes protect their hosts from insect herbivores through the production of alkaloids that confer resistance to the grass. Due to this protection, endophyte-infected plants are expected to dominate over endophyte-free plants. While some previous work supports this hypothesis, substantial variation exists in published impacts of endophyte-infected grasses on insect herbivores (Hammon and Faeth, 1992; Clay, 1996). Moreover, there is not a thorough understanding of the mechanisms by which fungal grass endophytes might defend their hosts.

In the last 12 years, several researchers have assessed the insect resistance reported for endophyte-infected grasses. For example, some insects tend to avoid endophyte-infected grasses in laboratory choice experiments. Latch et al. (1985) and Eichenseer and Dahlman (1992) found that the aphid *Rhopalosiphum padi* strongly preferred uninfected plants of tall fescue over infected plants. However, Latch et al. (1985) found no preference by the aphid for uninfected plants of perennial ryegrass. Similarly, results for the Russian wheat aphid (*Diuraphis noxia*) are mixed. Clement et al. (1990) found it did not select endophyte-free over endophyte-infected leaf sheaths and stems of perennial ryegrass. In contrast, Kindler et al. (1991) found that *D. noxia* preferred uninfected over infected tall fescue plants.

Workers have also investigated the potential effects of endophytes on chewing insects. Considering stem-feeding insects, Breen (1994) reported that endophyte-infected blue, Chewings, and hard fescues were resistant to chinch bug (*Blissus leucopterous*). Billbug (Johnson-Cicalese and White, 1990) and sod webworm (Funk et al., 1983) both preferred endophyte-free perennial ryegrass over uninfected plants in the field. Results of laboratory choice experiments with Argentine stem weevil (*Listronotus bonariensis*) showed that weevils strongly preferred endophyte-free perennial ryegrass over plants containing the endophyte (Barker, et al. 1984a). The folivore fall armyworm preferred leaves of uninfected tall fescue over infected tall fescue, but only when leaves were old; larvae showed no preference when younger leaves were used (Hardy et al., 1986). Similarly, work with root-feeding insects has also produced variable results. Both the Japanese beetle (*Popillia japonica*) and southern masked chafer (*Cyclocephala lurida*) were not affected by endophyte-infected tall fescue. Neither density nor mass of grubs differed between field plots of infected and uninfected grass (Potter et al., 1992). In sum, reactions of insects to endophytes has ranged from avoidance to neutrality.

Effects of grass endophytes on performance of insect herbivores have also received attention from investigators. For example, crickets restricted to infected perennial ryegrass died within 84 h, yet they survived when fed uninfected plants (Ahmad et al., 1985). Russian wheat aphid experienced decreased survival when feeding on infected tall fescue (Kindler et al., 1991) and perennial ryegrass (Kindler et al., 1991; Clement et al., 1992) compared to uninfected plants.

Some studies have also shown negative impacts on lepidopterous herbivores. For example, fall armyworm showed reduced larval and pupal mass and delayed development when reared on infected compared to uninfected tall fescue and perennial ryegrass (Clay et al., 1985). Yet for some experiments in that same study the authors reported no impact of fungal infection on pupal mass, survival, or developmental time to adult eclosion, and even faster development of fall armyworm reared on infected compared to uninfected tall fescue. Overall, it appears that endophyte infection has a more strongly negative impact on fall armyworm performance in perennial ryegrass than in tall fescue (Hardy et al., 1985, 1986; Clay et al., 1985). Eichenseer (1992) investigated utilization of tall fescue (both with and without endophyte) by fall armyworm and the orthopteran grasshopper *Melanoplus differentialis*. Comparing growth and consumption rates and digestive efficiencies, he found no effect of endophyte on either insect.

The impact of fungal endophytes on the true armyworm *Pseuduletia unipuncta* has also been evaluated (Eichenseer and Dahlman, 1993). Results showed no difference in development or survival rates and pupal mass between insects fed endophyte-infected or uninfected tall fescue. Similarly, Breen (1993) reported that both fall armyworm and southern armyworm (*Spodoptera eridania*) were unaffected by endophyte-infected tall fescue. Additionally, both Johnson et al. (1985) and Dubis et al. (1992) found southern armyworm to be unaffected by endophyte-infected tall fescue.

While endophyte-infected perennial ryegrass has a strong negative impact on Argentine stem weevil in New Zealand (see above), some other insects are insensitive to perennial ryegrass that contains endophyte. These include black field cricket (*Teleogryllus commodus*), porina (*Wisenana* spp.; Lepidoptera: Hepialidae), Tasmanian grass grub (*Aphodius tasmaniae*), several hemipteran species, and the dipteran *Cerodontha australis* (Argompyzidae) (Barker et al., 1984b; Prestidge and Ball, 1993). Prestidge (1989) also listed several homopteran leafhoppers in New Zealand that had distributions not related to endophyte infection. Additionally, endophyte in perennial ryegrass in field plots in the United Kingdom did not influence infestation of the frit fly (*Oscinella*), a major pest of seedling ryegrass plants (Lewis and Clements, 1986).

In striking departure from nearly all published work on insect resistance in endophyte–grass associations, Lopez et al. (1995) studied a native noncommercial grass occurring naturally in Arizona. Using a native grasshopper, *Melanoplus femurrubrum*, they found that endophyte infection did not deter herbivory and insects actually showed increased assimilation efficiencies when feeding on infected plants. Furthermore, similar results have been obtained for another native grasshopper, *Xanthipus corallipes*, feeding on Arizona fescue (S. Faeth, personal communication).

In summary, while several studies have shown a negative impact of grass endophytes on insect herbivore performance, substantial variability exists among published reports.

4. CHEMICAL BASIS FOR REPORTED INSECT HERBIVORE RESISTANCE

Alkaloids produced by fungal endophytes appear to be the source of resistance to at least some insects. Concentrations of the alkaloids, like endophyte hyphae, are generally greater in leaf sheaths and seeds, and lower in leaf blades (Gallagher et al., 1985; Siegel et al. 1987). An exception appears to be the alkaloid peramine, which is as concentrated in leaf blades as it is in leaf sheaths (Ball et al., 1995; Keogh et al., 1996). Alkaloids are heterocyclic ring compounds that contain an amino group. Not surprisingly, increased nitrogen fertilizer application can result in elevated concentrations of at least some alkaloids associated with grass endophytes (Archavaleta et al., 1992). Given the close phylogenetic relationship of *Neotyphodium* with alkaloid-producing *Claviceps*, the ergot fungus (Groger, 1972), it is not surprising that *Neotyphodium* produces similar toxins (Clay, 1986). The ergopeptine alkaloids are synthesized by the fungus (Gwinn and Savary, 1990). The most prevalent ergopeptine alkaloid, ergovaline, occurs at concentrations of 0.3–6 µg/g dry mass of tall fescue and perennial ryegrass shoot tissue (Siegel et al., 1990). Lolitrems are lipophilic complex–substituted indole compounds that occur in endophyte-infected perennial ryegrass. Lolitrem B is the major component and is produced by the fungus (Gallagher et al., 1984). Concentration of lolitrem B ranges from 5 to 10 µg/g dry mass of perennial ryegrass shoot tissue (Gallagher et al., 1985; Siegel et al., 1990). Peramine is a guanidinium alkaloid containing a unique pyrrolopyrazine structure that is produced by the fungus. It occurs at concentrations ranging from 10 to 42 µg/g dry mass of tall fescue and perennial ryegrass shoot material (Siegel et al., 1990). The loline or pyrrolizidine alkaloids have not been detected in perennial ryegrass but are common in endophyte-infected tall fescue. The two most common are *N-formyl* loline and *N-acetyl* loline, which occur at combined concentrations of 1000–3000 µg/g dry mass of shoot tissue. While the loline alkaloids were thought to be produced by the plant, not the fungal endophyte (Bacon and Siegel, 1988), recent work shows that the fungus does in fact produce the lolines in vitro and hence are a product of the fungus (H. Wilkinson and C. Schardl, personal communication). Fungal infection is necessary for loline production; uninfected tall fescue plants produce <20 µg/g dry mass of shoot tissue (Eichenseer et al., 1991).

Concentrations of alkaloids associated with grass endophytes generally follow a seasonal pattern with lowest levels in winter and highest levels in late spring and summer (Belesky et al., 1987, 1989; Rottinghaus et al., 1991; Rowan et al., 1990; Woodburn et al., 1993; Kennedy and Bush, 1983; Ball et al., 1995). Ball et al. (1995) also followed concentrations in *N. lolii* in perennial ryegrass over a year and found (using ELISA) that fungal hyphae were least concentrated in the winter.

While ergot alkaloids produced by *Neotyphodium* spp. decrease growth of fall armyworm (Clay and Cheplick, 1989) and the large milkweed bug (Yates et al. 1989), concentrations of the compounds necessary for these effects were greater than those naturally found in grass tissue. While probably not involved in insect resistance, the ergot alkaloids do have negative impacts on livestock (Lyons et al., 1986).

More likely mediators of insect resistance may be the lolines and peramine. A survey of tall fescue, perennial ryegrass, and other grasses showed that survival of the aphid *R. padi* was negatively correlated with grasses containing lolines (Siegel et al., 1990). *N*-Formyl loline, one of the two major lolines present in endophyte-infected tall fescue (Dahlman et al., 1991), added to artificial diet was quite toxic to milkweed bugs at concentrations naturally occurring in plant tissue (Yates et al., 1989). Yet lolines cannot be responsible for the negative effects on the insect of endophyte-infected perennial ryegrass since lolines are absent from this grass-endophyte association (Popay and Rowan, 1994). *N*-Formyl loline also has insecticidal activity against a number of nonherbivorous insects (Dahlman et al., 1991). The presence of peramine at naturally occurring concentrations deters feeding by Argentine stem weevils (Rowan, 1993). Additionally, the survival of the aphid *Schizaphis graminum* was negatively correlated with grasses containing peramine (Siegel et al., 1990).

In summary, while many studies at least partially support the hypothesis of defensive mutualism via alkaloids, considerable variation exists among published reports. Some insects appear unaffected by the presence of endophytes in grass, and effects of specific chemicals produced in endophyte-infected grass on insect performance are not clearly and consistently negative. While some of this variation might be due to differences in host–plant and endophyte genotype, as well as variation in environmental factors, such as soil fertility (Breen, 1994), one might expect considerable variation in these factors in natural and even agricultural settings. Yet, even in these circumstances, endophyte-infected plants dominate uninfected plants (Clay, 1990b; Shelby and Dalrymple, 1993). If the observed dominance of endophyte-infected plants is due to a defensive mutualism [as the hypothesis is currently articulated (Clay, 1988)], one would expect consistently negative impacts of endophytes on insect herbivores under a wide range of experimental conditions; this expectation is not met by the published literature.

5. INDUCED RESPONSES IN PLANTS

Rapid (within days or even shorter time frames) changes in plants following damage or induced responses appear to be widespread (Karban and Myers, 1989; Karban and Baldwin, 1997). These responses include chemical, physical, and phenological changes (Myers and Bazely, 1991; Tallamy and Raupp, 1991; Karban and Myers, 1989). Initially, many entomologists suggested that damage-induced changes were defensive in nature, playing a role in protecting the plant from further damage. Yet a defensive interpretation depends on two important criteria (Fowler and Lawton, 1985). First, does the induced response decrease herbivore preference or performance? If so, the change is referred to as induced resistance (Karban and Myers, 1989). Second, does the response result in increased plant reproductive fitness? If so, it can accurately be called an induced defense (Karban and Myers, 1989). Tests of induced responses often employ artificial damage, like clipping, and while many plants respond to artificial damage much like they do to natural damage, responses of some plants to simulated and natural damage are not identical (Baldwin, 1990; Stout et al., 1994).

Many entomologists have tested for, and demonstrated, induced resistance in agronomic plants (Karban, 1991). For example, Edwards et al. (1985) showed strong preference of *Spodoptera littoralis* larvae for undamaged tomato plants compared to damaged ones. Cotton seedlings damaged either mechanically or by mites subsequently supported fewer mites than undamaged plants (Karban and Carey, 1984). Recent work with tomato shows that plants respond differentially to different types of damage (Stout et al., 1994). Investigators followed effects of mechanical, chemical, and biotic damage on concentrations of proteinase inhibitors and oxidative enzymes, and found that different subsets of the foliar chemicals were induced by different types of damage. The expression of induced resistance can depend on other factors as well, like plant genotype, plant age, and abiotic conditions (Coleman and Jones, 1991); hence, these factors need to be controlled in experimental studies.

The mechanism of induced resistance can be active or passive (Karban and Myers, 1989). Active resistance involves de novo synthesis of energetically costly enzymes, whereas passive resistance results from passive rearrangement of resources within the plant. Obviously, augmentation of mycotoxins by endophyte-infected grass that provides resistance for the grass would be regarded as an active response.

Some microorganisms induce responses in plants that result in elevated resistance to insect herbivores. For example, Karban et al. (1987) showed infection of cotton by a pathogenic fungus induced resistance in the plant to spider mites. In contrast, cucumbers that experienced a localized infection from tobacco necrosis virus showed elevated resistance to anthracnose fungus, but were not protected against insect or mite attack (Apriyanto and Potter, 1990). Wound-induced resistance by plants that is mediated by a purported microbial mutualist of the plant, like endophytic fungi, has not been investigated until very recently (see below).

Induced resistance can be viewed within the framework of plant defense theory. Several different but overlapping ideas of plant defense have emerged within the last 20 years (Stamp, 1992). The optimal defense theory assumes that there are limited resources that plants can devote to defense and that plants will optimally allocate these toward alternative demands (i.e., growth and defense) (Rhoades, 1979). The optimal defense theory assumes that defenses are costly. This provides a potential explanation for the widespread occurrence of induced resistance. Employing defense only after initial damage may help plants optimally allocate limited resources (Karban and Myers, 1989). In a different approach, Bryant et al. (1983) and Coley et al. (1985) suggested that phenotypic responses of secondary metabolism are governed by the plant's ability to replace tissue lost to herbivores. The authors assumed that two factors should operate here: the resources available to the plant and the plant's growth rate. For example, slow-growing plants living in resource-limited soils should have high levels of defense due to the high cost of replacing lost tissue. An extension of the resource availability theory makes predictions concerning the balance of resources available to plants. For example, it is argued that moderate nutrient stress should enhance carbon-based defenses (in contrast to the predictions of the optimal defense theory which predicts reduced defenses due to their inherent cost). More recently, Herms and Mattson (1992) proposed the plant growth–differentiation balance theory which provides a framework that places plants along a growth–differentiation continuum and attempts to explain patterns in defense from the level of cellular processes within plants. When growth declines more than photosynthesis, the plant should switch to differentiation (and the accumulation of carbon-based allelochemicals). In contrast, growth-dominated plants should favor the use of induced resistance to herbivores. For the most part, predictions of the growth–differentiation balance hypothesis and carbon–nutrient hypothesis are similar to one another (Tuomi, 1992).

Grasses have long been considered to primarily exhibit morphological adaptations to herbivory, particularly that caused by vertebrates (Barnard and Frankel, 1964; Stebbins, 1981). In particular, the basal intercalary meristems of grasses allow them to tolerate some herbivory. Yet, cultivated species, primarily maize and wheat, do possess some chemical protection from insect herbivores. The aglycone versions of the cyclic hydroxamic acid derivatives offer resistance to some insects (Klun et al., 1967; Russel et al., 1975) and increase in concentration following wounding of the plant (Gutierrez et al., 1988). Seedling grasses may also possess antifeeding alkaloids (Bernays and Chapman, 1976). Nonetheless, compared to most other angiosperms, grasses appear to possess relatively few allelochemicals.

6. EXPECTATIONS OF INDUCED RESISTANCE IN GRASS-ENDOPHYTE ASSOCIATIONS

The defensive mutualism hypothesis states that grasses acquire their defense through toxins produced by fungal endophytes (Clay, 1988; Cheplick and Clay, 1988). The alkaloids present in endophyte-infected grasses are interpreted as constitutive (fixed) defenses for the otherwise poorly defended grasses. Yet support for this hypothesis (reviewed above) is somewhat mixed. Lack of consistent support for the defensive mutualism hypothesis suggests that additional mechanisms associated with endophyte infection are involved to confer the strong advantage infected grasses appear to have in natural and agricultural settings.

The real advantage of infection to plants may be that endophytes are stimulated to increase production of mycotoxins after damage to the plant has occurred. The potential adaptive significance of inducible resistance in the grass–endophyte system is readily apparent. *Neotyphodium* spp. live exclusively within their host and the fungi have no mechanism of proliferation except through reproduction of their host. Hence, any feature of the fungi that promotes reproductive success of their host should be selected. Based on both carbon–nutrient balance and growth–differentiation balance models, one might predict that fast-growing herbaceous plants, like grasses, should employ induced resistance to herbivores, particularly under nutrient-rich growing conditions. The idea of endophyte-mediated resistance represents an extension of the defensive mutualism hypothesis, i.e., plants might acquire protection through constitutive *and* induced resistance.

The expectation of induced resistance in infected plants is strengthened by reports that environmental stresses can stimulate the production of mycotoxins by endophyte-infected grasses. For example, endophyte-infected tall fescue damaged by nematodes produced greater amounts of chitinase than undamaged, uninfected plants (Roberts et al., 1992). The study is noteworthy in that it shows that endophytes may mediate chemical changes following damage to the host, even if the damage is to parts of the plant (roots) not containing the endophyte. A finding even more relevant to the question of induced resistance is that elevated temperatures (30°C compared to 23°C) stimulated the production of lolines in endophyte-infected perennial ryegrass (Huizing et al., 1991). Similarly, water deficits imposed on endophyte-infected perennial ryegrass resulted in a greater than twofold increase in levels of ergovaline, whereas no difference was seen in watered plants over the same time period (Barker et al., 1993). Even greater increases in ergovaline were found in drought-stressed tall fescue (Arechevaleta et al., 1992). Drought induced similar increases in loline alkaloids in tall fescue (Kennedy and Bush, 1983).

The expectation of induced resistance in infected plants is further strengthened by reports that loline alkaloids were substantially elevated following cutting of tall fescue (Eichenseer et al., 1991). Five-week-old plants that had been fertilized were clipped 1.5 cm above the soil and then assessed for N-formyl and N-acetyl loline in regrowth tissue 1 and 4 weeks following damage. Elevated levels of both lolines were found in previously clipped, compared to unclipped, plants. No change in loline levels (which are negligible) were found in uninfected plants. Thus, loline levels are rather labile and their production is stimulated by damage to the plant, i.e., grass endophytes may mediate an induced response to damage that enhances resistance by the plant to further insect damage. Given the tie between clipping and alkaloids, one might expect that stocking rates of livestock would influence alkaloid concentrations in endophyte-infected plants. Interestingly, recent work with endophyte-infected perennial ryegrass (L. Fletcher, personal communication) shows damage to plants can influence alkaloid concentrations. Grass from paddocks with continuous sheep grazing had lower concentrations of ergovaline and peramine than grass from paddocks with rotational grazing by sheep. Hence, it appears that grazing intensity can influence alkaloid production.

In an experiment designed to test the defensive mutualism hypothesis, Clay et al. (1993) allowed free-ranging fall armyworm larvae to forage on potted plants of tall fescue, red fescue, and perennial ryegrass in the greenhouse. Plants either infected or uninfected with fungal endophytes were grown within pots in single- and mixed-species arrangements. Their results showed that, in general, infected plants were at a competitive advantage when herbivores were present. Yet, because insect herbivores were left on plants for 11–14 weeks (with one or two reintroductions during that period), their study does not distinguish between constitutive and induced resistance. Rapid induction of resistance during the time frame of their experiment is also a viable explanation for their results. It is also unclear if endophyte-infected plants might have exhibited faster regrowth from herbivore damage and that this alone could explain their experimental results.

7. EMPIRICAL TESTS OF THE INDUCED RESISTANCE HYPOTHESIS

7.1. Tests Involving Shoot Damage

Several experiments in our laboratory have recently been conducted to specifically test for mediation of induced resistance by grass endophytes. In one experiment, perennial ryegrass, both infected and uninfected with its fungal endophyte

(*N. lolii*), was artificially damaged by clipping to simulate herbivory 4 weeks after germination (Bultman and Ganey, 1995). Plants were fertilized weekly with a water-soluble Peter's solution (20–10–20 with micronutrients, diluted to 150 ppm N). Fall armyworm larvae (*Spodoptera frugiperda*) were reared by feeding them shoot material of damaged and undamaged plants in a two-factor (infection status and damage) design. Feeding neonate larvae fresh shoot material began 2 days after the initial damage. Larvae were kept in individual Petri dishes within an environmental chamber under a 12:12 L D cycle at 26°C. Fall armyworm has frequently been used in past work on grass endophyte-insect interactions (reviewed in Breen, 1994) because it is often an important pest of many grasses (Luginbill, 1928).

Experimental results showed that the interaction between damage and infection status affected pupal mass of fall armyworm (Fig. 1). Pupae produced from larvae reared on damaged, infected plants weighed less than those produced from larvae reared on undamaged, infected plants. The reverse was true for uninfected plants. Mass of 8-day-old larvae and developmental rate qualitatively followed the results for pupal mass but lacked statistical significance. It is also apparent that the mass of pupae reared on undamaged, infected plants was not different from the mass of those reared on undamaged, uninfected plants (Fig. 1). Hence, this experiment failed to give strong support for the hypothesis that endophyte-infected perennial ryegrass confers constitutive resistance to fall armyworm (the overall main effect of infection status, summed over all damaged treatments, was,

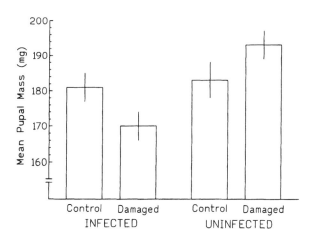

Figure 1 Mean (+/− SEM) mass of fall armyworm pupae reared from larvae fed perennial ryegrass with or without infection of the endophyte *N. lolii*. Control refers to grass lacking prior damage (Bultman and Ganey, 1995).

however, significant). These results are exciting in that they are consistent with the hypothesis that grass endophytes mediate induced resistance. Yet, a potential problem with the study is that plants were damaged by clipping off 25% of the apical shoot material. Because grass endophytes and their associated allelochemicals are generally more concentrated in the leaf sheaths than in the leaf blades (Gallagher et al., 1984; Lyons et al., 1986; Hinton and Bacon, 1985), the experimental results may been due to the removal of apical and therefore less chemically defended tissue. Yet control for this possibility was made by feeding larvae roughly equal proportions of leaf blade and leaf sheath material across all treatments.

In a second experiment, an alternative type of damage was employed that is not subject to the criticism of the damage protocol used in the first experiment (Boning and Bultman, 1996). Other than this modification, all aspects of the study were identical to the previous experiment with the exception of the grass and fungal species used. Tall fescue plants that were either infected or uninfected with the endophyte (*N. coenophialum*) were grown in greenhouse flats. Damage was accomplished by removing a tiller from each plant 4 weeks after germination. Four days after damage, fresh plant material (stems and blades) was fed to neonate fall armyworm larvae. In this way, potential problems of the first experiment were avoided because fall armyworm larvae in all treatment groups were fed intact tillers. As with the first experiment, the impacts of infection status and damage on pupal mass interacted; pupae reared from larvae fed infected, damaged plants were lighter than those reared from larvae fed infected, undamaged plants (Figure 2). Note that pupae actually weighed more when reared from infected compared to uninfected plants. If there is a protective advantage for grass of harboring the endophyte, these results indicate that it should be most (and perhaps only) apparent after plants have been damaged.

We draw five important conclusions from the two studies. First, damage induces fungal-mediated resistance to a generalist lepidopteran herbivore. Second, in both experiments, support for constitutive resistance in endophyte-infected plants was weak and variable. These studies add to the growing body of information that grass endophytes have variable effects on insect herbivores (see above). Third, induced resistance appears to occur in at least two different systems: tall fescue and perennial ryegrass. Fourth, both clipping blades and removing whole tillers appears to elicit induced resistance in infected plants. Fifth, induced resistance in tall fescue appears to be systemic, i.e., damage to one tiller can result in induced resistance in adjoining tillers of the same plant.

A third experiment repeated the second experiment using the same fungus–plant–insect system, but with three modifications (Bultman and Conard, 1998). Herbivory was simulated when plants were 4 weeks old using a fabric tracing wheel that produced small punctures in the leaf blades (Ohnmeiss and Baldwin, 1994). This mode of damage did not remove leaf tissue; therefore, it did not

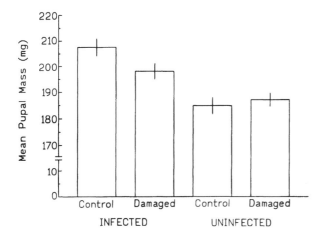

Figure 2 Mean $(+/- \text{SEM})$ mass of fall armyworm pupae reared from larvae fed tall fescue with or without infection of the endophyte *N. coenophialum*. Control refers to grass lacking prior damage (Boning and Bultman, 1996).

influence the portion of the plant subsequently fed to fall armyworm larvae—a potential complication of the two experiments described above. In addition, plants were maintained at two rates of 20–10–20 liquid fertilizer: 150 ppm nitrogen (high) and 50 ppm nitrogen (low). A third difference was that the time between damage and insect feeding was held constant throughout the 14-day period of fall armyworm larval development. Seeds were planted and foliage damaged sequentially so that all insects received plant tissue 4 days after wounding occurred with the tracing wheel. Plants were fertilized weekly. Hence, the factorial experiment had three treatment effects: infection status, fertilizer level, and wounding. A prediction from the hypothesis of induced resistance, that the main effects of infection status and wounding should interact, was not supported (pupal mass: $F_{1,280} = 1.76$, $p > 0.10$; similar results were obtained for 8-day larval mass and development time). Failure to find evidence of inducible resistance in the experiment may be due to the nature of the damage; pricking leaf blades of tall fescue may be insufficient to elicit a response that affects the performance of fall armyworm. In addition, a fungal-mediated inducible response may require a period longer than the 4 days allowed in this experiment.

As with the previous two experiments, infection status had a variable effect on insect performance. Endophyte reduced pupal mass ($F_{1,280} = 23.0$, $p < 0.05$), but also reduced development time ($F_{1,280} = 4.5$, $p < 0.05$), and increased 8-day larval mass ($F_{1,280} = 7.3$, $p < 0.01$). A reason for including a fertilizer treatment was to test if inducible resistance waned as nutrient availability to the plant dropped. While this prediction was not supported, plants receiving low fertilizer did show a stronger negative impact of the endophyte on pupal mass than insects reared from plants receiving high fertilizer (reduction in mass of pupae reared from plants receiving high vs. low fertilizer: 25.3% for insects feeding on infected plants and 4.0% for insects feeding on uninfected plants; $F_{1,280} = 13.5, p < 0.001$). In nitrogen-poor soils the fungus may become an important nitrogen sink and reduce the suitability of the grass for fall armyworm larvae. This hypothesis is supported by the report that endophyte infection tends to reduce the concentration of nitrate in tall fescue leaf blades and sheaths (Lyons et al., 1990).

We have begun to test the generality of induced resistance by conducting an investigation with a nonlepidopteran—the aphid, *R. padi*. If endophytes also mediate induced resistance to a phloem feeder, this would suggest that results obtained with fall armyworm are indicative of a more general phenomenon characteristic of grass endophytes.

Aphids were obtained from the Illinois Natural History Survey in Champaign, Urbana, IL, and were reared on barley sprouts in an environmental chamber set at 21°C and a 14/10 hr light/dark cycle. *Rhopalosiphum padi*, a vector of the costly barley yellow dwarf virus (Power, 1991; Villanueva and Strong, 1964), is a host-alternating species. Overwintering eggs are laid on bird cherry trees, and after hatching in the spring, alate (winged) individuals disperse to infest herbaceous vegetation such as grasses. While on grasses, most *R. padi* are apterous and parthenogenetic until the autumn when crowding and day length changes initiate aphid migration back to cherry trees where sexual morphs are produced and eggs are laid (Dixon, 1973). Several workers have utilized *R. padi* to investigate insect resistance in endophyte-infected grasses and results have generally shown strong negative effects on aphid reproduction (reviewed in Breen, 1994).

Tall fescue (KY 31) seeds (both uninfected and infected with *N. coenophialum*) were sown into root-master boxes filled with Pro-Mix potting soil (Premier Brands, Red Hill, PA). Microscopic inspection of seeds stained with aniline blue revealed an infection frequency of 87% in the infected group. Plants were fertilized weekly with a solution of 1.33 g/L of 20–20–20 (diluted to 260 ppm nitrogen) fertilizer. Ten root-master boxes (each with 28 plants) were used with one box assigned to each treatment combination (see below). Plants were rotated weekly within the greenhouse. At five weeks of age, plants were moved to light-tables and kept on a 14/10 h light/dark schedule. At 7 weeks of age, plants were damaged according to the following treatments: low damage, in which the second to largest tiller was removed from the plant near the base; medium damage, in which the second to largest tiller was removed, along with 50% of the remaining leaf material; and high damage, in which all but the largest tiller was removed along with 50% of each leaf blade on that tiller, except the blade on which the aphids were placed (see below). Artificial damage was performed by clipping with scissors. In each treatment, additional plants were used for estimating the percentage of tissue removed by artificial damage treatments. Oven-dry weight of plant material removed was compared to the oven-dry weight of the shoot material that remained following damage. The artificial damage procedure resulted in different ($F_{2,9} = 79.7$, $p < 0.0001$; analysis on arcsin transformed data) levels of tissue removal among treatments: low ($\bar{x} = 19\%$, $n = 28$ for both endophyte-infected and uninfected), medium ($\bar{x} = 49\%$, $n = 28$ and 25 for endophyte-infected and uninfected, respectively), and high ($\bar{x} = 70\%$, $n = 28$ and 27 for endophyte-infected and uninfected, respectively) (Fig. 3). A natural damage ($n = 27$ and 28 for endophyte-infected and uninfected, respectively) was also imposed by placing two fourth-instar fall armyworm larvae in a mesh envelope (2.6 × 10.3 cm) that enclosed the second to largest tiller of the plant. Larvae

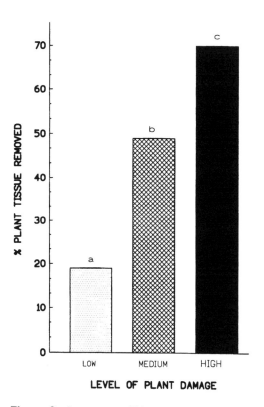

Figure 3 Percentage of biomass removed from tall fescue plants in artificial damage treatments. Error bars are too small to be visible.

were allowed to remain (2–3 days) on the plant until the tiller was removed at its base. Control plants ($n = 28$ for both E+ and E−) were not damaged.

Aphids were confined to the undamaged, second leaf from the base of the largest tiller and is referred to as the "aphid leaf." We adopted this protocol to minimize differences in leaf age which can affect insect herbivore growth and survivorship (Eichenseer et al., 1991; Hardy et al., 1986). Enclosure bags were made of fine (0.33 mm), translucent, fabric mesh. The 2.6 × 5.1 cm enclosures were sewn lengthwise closing the enclosure and forming a tube with open ends. Aphids were unable to cross the mesh barrier. Aphid enclosures were placed on leaves with the bottom of the enclosures bordering the base of the leaf blade. The ends of the enclosures were sealed with thin strips of paper shims that were compressed by large paper-clips.

Seven days after plants were damaged, apterous aphids were placed on the aphid leaves of plants. Two aphids were placed, by paintbrush, into each enclosure bag. Four days after aphids were introduced, the apterous aphids were counted under a dissecting scope at 10×. Results were analyzed with two-factor ANOVA (SAS 1985). The number of apterous aphids was not normally distributed; therefore, data were ranked prior to analysis (Conover and Iman, 1981).

After 4 days of containment on plants, higher numbers of apterous aphids were found on uninfected plants (Fig. 4; Table 1). The number of aphids did not differ among damage treatments (Table 1). However, as predicted by the induced resistance hypothesis, there was a significant interaction between infection status and damage, signifying that these two factors were not independent of one another. Aphids feeding on endophyte-infected plants experienced reduced reproduction; however, damaging uninfected plants did not have this effect on aphids.

The negative impact on aphids of damaging infected plants was only observed in the low and medium, artificial damage treatments (Fig. 4). The low-damage treatment, in which one tiller was removed from each plant, was similar to the damage level imposed in the experiment by Boning and Bultman (1996) with tall fescue and fall armyworm (see above). By contrast, the high-damage treatment was much more severe, involving removal of nearly all of the shoot tissue from plants. A plausible explanation for the lack of induced resistance in the high-damage treatment group is that the magnitude of damage was so great that the plant–fungus symbiotum was unable to channel resources into defense. Lack of correspondence between the extent of initial plant damage and plant responses to damage has been reported for several plant species (Karban, 1991) and also appear to operate in the tall fescue–endophyte system.

Others (Siegel et al., 1990; Eichenseer et al., 1991) have found that R. padi reproduces much faster on endophyte-free than on endophyte-infected tall fescue. While we did find a significant reduction in aphid numbers on infected plants,

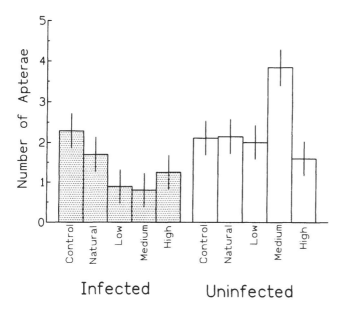

Figure 4 Mean number of apterous *R. padi* aphids contained on tall fescue plants infected with or free of the endophyte *N. coenophialum*. Damage to plants prior to introduction of aphids included low, medium, and high levels of artificial damage and natural damage by fall armyworm larvae. Control plants were not damaged. Error bars = 1 SE.

Table 1 Two-Way ANOVA Results Using Type III Sums of Squares (SAS 1985) of Apterous Aphid Numbers After Placement for 4 Days on Tall Fescue Plants[a]

Source	df	ss	MS	F	P
Infection	1	72206.7	72206.7	13.22	<0.001
Damage	4	39162.3	9790.6	1.79	n.s.
Infection/damage	4	60734.1	15183.5	2.78	<0.05
Error	265	1447955.6	5464.0		
Total	274	1617144.5			

[a] Data for aphid numbers were rank transformed prior to analysis because they were not normally distributed (Conover and Iman 1981). n.s. = not significant.

the difference in aphid numbers between the two groups was not as large as has been found by others. Several factors may explain this. First, unlike designs employed by previous workers, all aphids in our experiment were confined within enclosure bags that constrained the movements and perhaps reproduction of the aphids. However, because this effect was constant across all treatment groups, observed differences can be attributed to the manipulated factors. Restricting aphids to leaf blades may have reduced their reproduction in other ways as well. For example, if free-ranging aphids feeding on uninfected plants select grass stems and in so doing experience high reproduction, then confining aphids to blades would result in lower reproduction. Also, another difference between our study and that by previous workers that may explain the low reproduction on uninfected plants is that we left aphids on plants for only 4, rather than 7, days. Finally, there are likely genetic differences between the population of *R. padi* we used and those used by other workers; these differences may, in part, explain the differences observed in reproductive capacities.

Our results also showed that endophyte-infected tall fescue did not respond to damage by the chewing folivore, fall armyworm (Fig. 4). Even though the level of damage caused by the insect (removal of one tiller per plant) was similar to that in the artificial, low-damage treatment, there was a distinct difference in the time frame within which the damage took place. In contrast to our clipping, larvae took 2–3 days to remove tillers. Clipping more closely simulated damage from vertebrate, rather than invertebrate, herbivores. Given this, negative interactions between vertebrate and invertebrate herbivores feeding on grasses may occur that are modulated by fungal endophytes. Vertebrates grazing on endophyte-infected grass could, through inducing resistance in the grass, have negative impacts on the performance of invertebrates that subsequently feed on the grass. Similar interactions between herbivores that share a common host plant have been noted (Faeth, 1986) and may be more prevalent than commonly thought. If this interaction occurs one would predict vertebrate grazing to influence alkaloid levels in endophyte-infected grasses. While information on effects of grazing on mycotoxins in endophyte-infected grasses is scarce, ongoing work shows that stocking rates can influence alkaloid levels in perennial ryegrass (L. Fletcher, personal communication).

To summarize, preliminary experiments tend to support the hypothesis that damage to endophyte-infected grass shoots stimulates elevated levels of resistance to insects. Mediation of induced resistance by endophytes occurs in both tall fescue and perennial ryegrass, and is effective against lepidopteran and homopteran insects. These exciting results are the first to show wound-induced responses that are mediated by purported microbial mutualists of plants. Clearly, fungal endophytes modulate plant–insect interactions through multiple pathways.

7.2. A Test Involving Root Damage

While grass endophytes are primarily confined to shoot tissue (Latch et al., 1984, but see also Azevedo and Welty, 1995), it has become apparent that endophytes can influence their hosts below ground. For example, tall fescue infected with *N. coenophialum* consistently allocated more resources to roots than did tall fescue that lacked endophyte infection (Kelrick et al., 1990, 1995). In addition, endophytes appear to offer some protection against root herbivores (West et al. 1988, 1990; Pedersen, 1988). Hence, understanding below-ground processes may be necessary for a full understanding of how grass endophytes influence their hosts and insect herbivores. Given this, Nyree Conard and Ben Nomann while students at Truman State University conducted an experiment under the direction of Michael Kelrick in which they tested for the mediation of resistance to a folivore induced by below-ground herbivory.

Tall fescue seeds both infected and free of endophyte were planted in 10.2-cm-diameter pots filled with a 1:1 mixture of fine-grained sand and Fison's LG3 germinating mix (ground peat and vermiculite). Half of the pots had been cut cross-sectionally and taped back together prior to planting. When plants in these pots were 4 weeks old, the tape was removed and a metal blade used to slice through the pot and the soil it contained. This manipulation simulated root herbivory by removing on 60–80% of the total root biomass at the time of wounding (estimated from additional plants not used in the experiment).

In addition to infection status and artificial root herbivory, a treatment of nitrogen (N) fertilizer level was also imposed: plants were maintained with weekly applications of low, medium, and high levels of N (25%, 75%, and 125% of Hoaglunds's No. 2 solution, respectively). Nitrogen concentrations per 50 mL of these treatment solutions were 4 mM, 12 mM, and 20 mM, respectively. Each treatment combination contained 15 replicate plants. As a bioassay of response to treatment effects, one third-instar fall armyworm larva was placed on each plant a week after roots were damaged. All plants were fitted with a gauze canopy that prevented larvae from escaping. Two response variables were measured: weight gained by fall armyworm larvae after 4 days on plants and activity of larvae (recorded daily each of the 4 days).

Results showed no effect of root herbivory on the growth (Table 2) or activity (Table 3) of the folivore. Furthermore, infection status and simulated root herbivory did not interact (Table 2), and therefore the hypothesis of induced resistance was not supported. The behavior of the larvae was also not associated with simulated root herbivory (Table 3). Of the three main effects, only N level supplied to the plants influenced larval growth; larvae placed on plants receiving high N gained more weight than those placed on plants receiving low nitrogen (Fig. 5). Once again, infection status did not influence insect performance (Table

Table 2 Effects of Endophyte Infection Status, Fertilization Level, and Simulated Root Herbivory on Mass of Fall Armyworm Larvae

Source	df	SS	MS	F	p
Infection	1	1817.1	1817.1	1.23	n.s
Fertilizer	2	13321.1	6660.5	4.51	<0.01
Damage	1	2831.5	2831.55	1.93	n.s.
Infection × fertilizer	2	2106.6	1053.3	0.71	n.s.
Infection × damage	1	4500.7	4500.7	3.05	n.s.
Fertilizer × damage	2	6747.7	3373.9	2.29	n.s.
Infect. × fert. × damage	2	13064.3	6532.1	4.43	<0.05
Initial mass	1	57802.2	57802.2	39.2	<0.01
Error	153	225721.0	1475.3		
Total	165	382289.2			

Analysis was conducted with a three-factor ANCOVA using the type III sums of squares in PROC GLM (SAS, 1985). Initial mass of larvae when placed on plants was used as the covariate. Because the distribution of the data was not normal, values were rank-transformed prior to analysis (Conover and Iman, 1981).

2). These results suggest that in the presence of simulated root herbivory, endophyte-infected tall fescue is no more resistant to an above-ground herbivore than is tall fescue that lacks endophytic infection. So, in contrast to experiments with shoot damage, there is no evidence that root damage induces insect resistance in endophyte-infected grass. However, work in this area needs to be repeated and expanded before any general conclusions can be made.

Table 3 Behavior of Fall Armyworm on Tall Fescue Infected with *N. coenophialum* or Free of Infection[a]

	Behavior of fall armyworm larvae		
	No. feeding	No. not feeding	No. off plant
Infected	131 (2.1)	124 (0.1)	150 (2.8)
Uninfected	183 (1.9)	141 (0.2)	126 (2.6)

[a] Summed over three fertilizer application and two root herbivory levels (see text). Contingency table analysis showed that insect behavior depended upon infection status of plants ($G^2 = 9.46$, df $= 2$, $p < 0.01$). Numbers in cells represent frequencies and those with parentheses are the contributions of the cell to the overall G^2 statistic. There was no association between larval behavior and root damage ($G^2 = 0.72$, df $= 2$, $p > 0.9$).

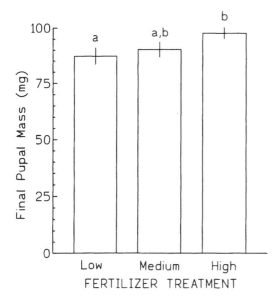

Figure 5 Weight gain by fall armyworm larvae after 4 days on tall fescue plants fertilized at low, medium, and high rates of nitrogen (see text). Data presented are summed over the treatments of endophyte infection status and simulated root herbivory. Histograms with the same letter are not significantly different as determined by Tukey's studentized range test. Error bars = 1 SE.

8. CONCLUSIONS AND FUTURE DIRECTIONS

The recent empirical tests of shoot herbivory coupled with reports of elevated alkaloids in infected plants following clipping and other environmental stresses (see above) lend support to the idea that grass endophytes mediate induced resistance. Nonetheless, we caution that acceptance of the induced resistance hypothesis is certainly premature.

The work to date is laboratory-based and may not reflect patterns that would be found in the field. More variable results would be expected from work conducted in the field where conditions often vary in time and space. Another concern is that tall fescue and perennial ryegrass originated in Europe (Gleason and Cronquist, 1963) and are now widely used commercially for forage and turf. As such, they have been subject to intense breeding programs. Whether endophytes in native, noncommercial grasses provide induced or constitutive resistance to insects remains largely unexplored. In one of the few studies of a noncommercial grass, Lopez et al. (1995) found no evidence of constitutive resistance to a native grasshopper. A similar criticism of the tests of endophyte-mediated induced resistance, and nearly all endophyte–insect herbivore work in general, is the use of polyphagous pest species that may have little evolutionary history with the grass cultivars used.

While the defensive mutualism hypothesis has received the most attention as an explanation for the predominance of endophyte-infected grasses, it remains only one of several alternative hypotheses. Other ideas that should be more closely examined include the hypotheses that endophytes promote general plant vigor and competitive ability (Clay, 1990a), and that endophytes are beneficial to their hosts due to enhanced drought tolerance of the host (West and Gwinn, 1993 and others).

Even if subsequent work establishes that endophyte-mediated induced resistance is widely present in grasses, it may have no effect on the fitness of the endophyte-infected plants. That is, effects on individual herbivores may not be translated into effects on herbivore populations and may not factor back to provide plants with defense against herbivores (Fowler and Lawton, 1985). It is also widely accepted that understanding plant-insect interactions requires knowledge of the third trophic level (Price et al., 1980). A full understanding of the role endophytes play in modulating interactions between grasses and herbivorous insects must incorporate these tritrophic level influences. The impact of constitutive and induced resistance on insect herbivores could be reduced if endophytes have a negative effect on natural enemies of the herbivores. Three recent studies have focused on this question.

Bultman et al. (1997) fed fall armyworm larvae tall fescue with or without endophyte and allowed eulophid parasitoids to parasitize the larvae. They found that the presence of endophyte-infected plants in the diet of fall armyworm had a negative impact on the pupal mass of parasitoids. Furthermore, they also fed larvae artificial diets to which the *Neotyphodium*-associated alkaloids, *N-acetyl* and *N-formyl* loline, were added and they found that both resulted in reduced survival of the parasitoids. A similar interaction was also observed in a much different plant–endophyte system under field conditions. Preszler et al. (1996) found lower parasitism of a leaf-mining moth on Gambel's oak trees with high infection levels of endophyte, suggesting parasitoids avoided hosts feeding on trees heavily infected by endophyte. These two studies suggest that the benefits to plants of constitutive and/or induced resistance that endophytes may provide could be canceled or at least reduced by negative effects endophytes have on the third trophic level. However, others have found that endophytes can have the opposite effect on natural enemies. Japanese beetles feeding on roots of endophyte-infected fescue were more susceptible to an entomopathogenic nematode than beetles feeding on endophyte-free plants (Grewal et al., 1995). The investigators attributed this effect to reduced activity and vigor of the larval beetles when feeding on endophyte-infected plants. It is clear from these three studies that the pathway of interaction between endophytes and the third trophic level can fluctuate from positive to negative effects. Further work is needed to determine if general patterns emerge and to determine the relative impacts of endophytes on herbivores and their natural enemies and how the balance of these two influences the fitness of the plant.

Given that alkaloids contain nitrogen and soil fertility is positively correlated with alkaloid concentrations in endophyte-infected plants (Archavaleta et al., 1992), further study of the interaction between nitrogen availability and inducible resistance should help illuminate our understanding of the defensive roles grass endophytes play. This line of work would also have the advantage of utilizing theoretical frameworks on plant defense (see above) for the interpretation of results.

Another feature that has often been lacking in past tests of endophyte-mediated constitutive and induced resistance is the impact on insect feeding efficiency. Various measurements, like assimilation efficiency and growth (Scriber and Slansky, 1981; Horton and Redak, 1993), could provide insights on the mechanisms responsible for reduction in development by insect herbivores feeding on endophyte-infected grasses.

The repeated observation that individuals of endophyte-infected grasses dominate their uninfected conspecifics (Clay, 1990b) suggests that endophyte infection often enhances the fitness of grasses. The primary hypothesis to explain the enhanced fitness is that endophytes form a protective mutualism with their hosts by providing constitutive resistance (Clay, 1988). Our work shows that in some situations, endophytes also provide their hosts with inducible resistance. Further work is needed to determine if inducible responses are general in grass endophytes and if they are important in the evolution of the symbiosis between clavicipitaceous fungi and many grasses.

ACKNOWLEDGMENTS

Earlier versions of the manuscript benefited from comments by K. Craven, S. Faeth, M. Kelrick, F. Schulthess, T. J. Sullivan, and D. Wilson. H. Eichenseer, L. Fletcher, H. Wilkinson, and C. Schardl all kindly offered results of unpublished work. A portion of the manuscript was written during a sabbatical leave granted by Truman State University to T.L.B. and spent at the Canterbury Agriculture and Science Center in Lincoln, New Zealand. The wonderful AgResearch staff in Lincoln provided computer and library resources and excellent hospitality. Funding for the stay came from an NSF International Travel Grant and the Science Division at Truman State University. Experimental results with *R. padi* and tall fescue are part of the work completed by J.C.M. for the degree of Master of Science at Truman State University and was funded by NSF Grant DEB-9527600.

REFERENCES

Ahmad, S., Govindarajan, J. M. Funk, C. R. and Johnson-Cicalese, J. M. (1985). Fatality of house crickets on perennial ryegrass infected with fungal endophyte. *Entomologia experimentalis et applicata* 39, 183–190.

Andrews, J. H. and Minao, S. S. (1991). *Microbial Ecology of Leaves.* Springer-Verlag, New York.

Apriyanto, D. and Potter, D. (1990). Pathogen-activated induced resistance of cucumber: response of arthropod herbivores to systemically protected leaves. *Oecologia* 85, 25–31.

Arechavaleta, M., Bacon, C. W., Plattner, R. D., Hoveland, C. S. and Radcliffe, D. E. (1992). Accumulation of ergopeptide alkaloids in symbiotic tall fescue grown under deficits of soil water and nitrogen fertilizer. *Applications in Environmental Microbiology* 58, 857–861.

Azevedo, M. D. and Welty, R. E. (1995). A study of the fungal endophyte *Acremonium coenophialum* in the roots of tall fescue seedlings. *Mycologia* 87, 289–297.

Bacon, C. W. and Siegel, M. R. (1988). Endophyte parasitism of tall fescue. *Journal of Production Agriculture* 1, 45–55.

Bacon, C. W., Porter, J. K., Robbins, J. D. and Luttrell, E. S. (1977). *Epichloe typhina* from toxic tall fescue grasses. *Applications in Environmental Microbiology* 34, 576–581.

Baldwin, I. T. (1990). Herbivory simulations in ecological research. *Trends in Ecology and Evolution* 5, 91–93.

Ball, O. J.-P., Prestidge, R. A. and Sprosen, J. M. (1995). Interrelationships between *Acremonium lolii*, peramine, and lolitrem B in perennial ryegrass. *Applications in Environmental Microbiology* 61, 1527–1533.

Barbosa, P., Krischik, V. A., and Jones, C. G. (eds.). (1991). Microbial Mediation of Plant-Herbivore Interactions. John Wiley and Sons, New York.

Barker, D. J., Davies, E., Lane, G. A., Latch, G. C. M., Nott, H. M. and Tapper, B. A. (1993). Effect of water deficit on alkaloid concentrations in perennial ryegrass endophyte associations. *In* D. E. Hume, G. C. M. Latch and H. S. Easton (eds.), *Proceedings of the 2nd International Symposium on* Acremonium/*Grass Interactions.* AgResearch, Palmerston North, New Zealand, pp. 67–71.

Barker, G. M., Pottinger, R. P., Addison, P. J. and Prestidge, R. A. (1984a). Effect of Lolium endophyte fungus infections on behaviour of adult Argentine stem weevil. *New Zealand Journal of Agricultural Research* 27, 271–277.

Barker, G. M., Pottinger, R. P., Addison, P. J. and Oliver, E. H. A. (1984b). Pest status of *Ceredontha spp.* and other shoot flies in Waikato pasture. *Proceedings of the 37th New Zealand Weed and Pest Control* 37, 96–100.

Barnard, C. and Frankel, O. H. (1964). Grass, grazing animals, and man in historic perspective. *In* C. Barnard (ed.), *Grasses and Grasslands.* Macmillan, New York, pp. 1–12.

Belesky, D. P., Robbins, J. D., Stuedemann, J. A., Wilkinson, S. R. and Devine, O. J. (1987). Fungal endophyte infection-loline derivative alkaloid concentration of grazed tall fescue. *Agronomy Journal* 79, 217–220.

Belesky, D. P., Stringer, W. C. and Plattner, R. D. (1989). Influence of endophyte and water regime upon tall fescue accessions. II. Pyrrolizidine and ergopeptine alkaloids. *Annals of Botany* 64, 343–349.

Berenbaum, M. R. (1988). Allelochemicals in insect-microbe-plant interactions: agents provocateurs in the coevolutionary arms race, *In* P. Barbosa and D. K. Letourneau (eds.). *Novel Aspects of Insect–Plant Interactions.* John Wiley and Sons, New York, pp. 97–124.

Bernays, E. A. and Chapman, R. F. (1976). Antifeedant properties of seedling grasses. *In* T. Jermy (ed.), *The Host Plant in Relation to Insect Behavior and Reproduction.* Akademiai Kiado, Budapest, pp. 41–46.

Bills, G. F. and Polishook, J. D. (1991). Microfungi from *Carpinus caroliniana. Canadian Journal of Botany* 69, 1477–1487.

Blakeman, J. P. (ed.) (1981). *Microbial Ecology of the Phylloplane.* Academic Press, London.

Boning, R. A. and Bultman, T. L. (1996). A test for constitutive and induced resistance by tall fescue (*Festuca arundinacea*) to an insect herbivore: impact of the fungal endophyte, *Acremonium coenophialum. American Midland Naturalist* 136, 328–335.

Breen, J. P. (1993). Enhanced resistance to fall armyworm (Lepidoptera: Noctuidae) in *Acremonium* endophyte-infected turfgrasses. *Journal of Economic Entomology* 86, 621–629.

Breen, J. P. (1994). *Acremonium* endophyte interactions with enhanced plant resistance to insects. *Annual Review of Entomology* 39, 401–423.

Bryant, J. P., Chapin, F. S. III and Klein, D. R. (1983). Carbon/nutrient balance of boreal plants in relation to vertebrate herbivory. *Oikos* 40, 357–368.

Bultman, T. L., Borowicz, K. L., Schneble, R. M., Caudron, T. A., Crowder, R. J. and Bush, L. P. (1997). Effect of a fungal endophyte and loline alkaloids on the growth and survival of two *Euplectrus* parasitoids. *Oikos* 78, 170–176.

Bultman, T. L. and Conrad, N. S. (1998). Effects of endophytic fungus, nutrient level, and plant damage on performance of fall armyworm (Lepidoptera: Noctuidae). *Environmental Entomology* 27, 631–635.

Bultman, T. L. and Ganey, D. T. (1995). Induced resistance to fall armyworm (Lepidoptera: Noctuidae) mediated by a fungal endophyte. *Environmental Entomology* 24, 1196–1200.

Cheplick, G. P. and Clay, K. (1988). Acquired chemical defenses in grasses: the role of endophytes. *Oikos* 52, 309–319.

Cheplick, G. P., Clay, K. and Marks, S. (1989). Interactions between infection by endophytic fungi and nutrient limitation in the grasses *Lolium perenne* and *Festuca arundinacea. New Phytologist* 111, 89–97.

Clay, K. (1986). Grass endophytes. *In* J. Fokkema and J. van den Heuvel (eds.), *Microbiology of the Phyllosphere.* Cambridge Univ. Press, London, pp. 188–204.

Clay, K. (1988). Fungal endophytes of grasses: a defensive mutualism between plants and fungi. *Ecology* 69, 10–16.

Clay, K. (1990a). Fungal endophytes of grasses. *Annual Review of Ecology & Systematics* 21, 275–297.

Clay, K. (1990b). Insects, endophytic fungi and plants. *In* J. J. Burdon and R. S. Leather (eds.), *Pests, Pathogens and Plant Communities. Blackwell Scientific, Oxford, pp. 111–130.*

Clay, K. (1991). Fungal endophytes, grasses, and herbivores. *In* P. Barbosa, V. A. Krischik, and C. G. Jones (eds.), *Microbial Mediation of Plant-Herbivore Interactions.* John Wiley and Sons, New York, pp. 194–226.

Clay, K. 1996. Fungal endophyte, herbivores, and the structure of grassland communities. *In* A. C. Gange (ed.), *Multitrophic Interactions in Terrestrial Systems.* British Ecological Society Symposium, Blackwell Scientific, pp. 151–169.

Clay, K., Hardy, T. N. and Hammond, A. M. Jr. (1985). Fungal endophytes of grasses and their effects on an insect herbivore. *Oecologia* 66, 1–6.

Clay, K. and Cheplick, G. P. (1989). Effect of ergot alkaloids from fungal endophyte-infected grasses on fall armyworm (*Spodoptera frugiperda*). *Journal of Chemical Ecology* 15, 169–182.

Clay, K., Marks, S. and Cheplick, G. P. (1993). Effects of nutrient availability and fungal endophyte infection on competitive interactions among grasses. *Ecology* 74 1767–1777.

Clement, S. L., Pike, K. S., Kaiser, W. J. and Wilson, A. D. (1990). Resistance of endophyte-infected plants of tall fescue and perennial ryegrass to the Russian wheat aphid (Homoptera: Aphididae). *Journal of the Kansas Entomological Society* 63, 646–648.

Clement, S. L., Lester, D. G., Wilson, A. D. and Pike, K. S. (1992). Behavior and performance of *Diuraphis noxia* (Homoptera: Aphididae), on fungal endophyte-infected and uninfected perennial ryegrass. *Journal of Economic Entomology* 85, 583–588.

Coleman, J. S. and Jones, C. G. (1991). A phytocentric perspective of phytochemical induction by herbivores. *In* D. W. Tallamy and M. J. Raupp (eds.), *Phytochemical Induction by Herbivores.* John Wiley and Sons, New York, pp. 3–41.

Coley, P. D., Bryant, J. P. and Chapin, F. S. III. (1985). Resource availability and plant antiherbivore defense. *Science* 230, 895–899.

Conover, W. J. and Iman, R. L. (1981). Rank transformations as a bridge between parametric and nonparametric statistics. *American Statistician* 66, 1–6.

Crist, T. O. and Friese, C. F. (1993). The impact of fungi on soil seeds: implication for plants and granivores in a semiarid shrub-steppe. *Ecology* 74, 2231–2239.

Dahlman, D. L., Eichenseer, H. and Siegel, M. R. (1991). Chemical perspectives on endophyte grass interactions and their implications to insect herbivory. *In* P. Barbosa, V. A. Krischik and C. G. Jones (eds.), *Microbial Mediation of Plant-Herbivore Interactions.* John Wiley and Sons, New York, pp. 227–252.

Dickenson, C. H. and Preece, T. F. (eds.). (1976). *Microbiology of Aerial Plant Surfaces.* Academic Press, London.

Diehl, W. W. (1950). Balansia *and the Balansiae in America.* Agric. Monogr. 4, USDA, Washington, DC.

Dixon, A. F. G. (1973). *Biology of Aphids*, 5th ed. Edward Arnold, London.

Dubis, E. N., Brattsten, L. B. and Dungan, L. B. (1992). Effects of endophyte-associated alkaloid peramine on southern armyworm cytochrome P-450. *In* C. V. Mullin and J. G. Scott (eds.), *Molecular Mechanisms of Insecticide Resistance: Diversity Among Insects.* Am. Chem. Soc. Symp. Ser. 505, Washington, DC, pp. 125–136.

Edwards, P. J., Wratten, S. D. and Cox, H. (1985). Wound-induced changes in the accept-ability of tomato to larvae of *Spodoptera littorialis*: a laboratory bioassy. *Ecological Entomology* 10, 155–158.

Eichenseer, H. (1992). Behavioral and physiological modification of selected insects medi-ated by lolines, endophyte-associated alkaloids in tall fescue. Ph.D. Dissertation, University of Kentucky, Lexington.

Eichenseer, H. and Dahlman, D. L. (1992). Antibiotic and deterrent qualities of endophyte-infected tall fescue to two aphid species (Homoptera: Aphididae). *Environmental Entomology* 21, 1046–1051.

Eichenseer, H. and Dahlman, D. L. (1993). Survival and development of the true armyworm *Pseudaletia unipuncta* (Haworth) (Lepidoptera: Noctuidae), on endophyte-infected and endophyte-free tall fescue. *Journal of Entomological Science* 28, 462–467.

Eichenseer, H., Dahlman, D. L. and Bush, L. P. (1991). Influence of endophyte infection, plant age and harvest interval on *Rhopalosiphum padi* survival and its relation to quantity of N-formyl and N-acetyl loline in tall fescue. *Entomologia experimentalis et applicata*, 60, 29–38.

Faeth, S. H. (1986). Indirect interactions between temporally separated herbivories medi-ated by the host plant. *Ecology,* 870–875.

Fletcher, L. R. and Harvey, I.C. (1981). An association of a *Lolium* endophyte with rye-grass staggers. *New Zealand Veterinary Journal* 29, 185–186.

Fokkema, N. J. and van den Hueval, J. (eds.). (1986). Microbiology of the Phylloplane. Cambridge Univ. Press, New York.

Fowler, S. V. and Lawton, J. H. (1985). Rapidly induced defenses and talking trees: the devil's advocate position. *American Naturalist* 126, 181–195.

Funk, C. R., Halisky, P. M., Johnson, M. C., Siegel, M. R., Stewart, A. V., Ahmad, S., Hurley, R. H. and Harvey, I. C. (1983). An endophytic fungus and resistance to sod webworms: association in *Lolium perenne* L. *Biotechnology* 1, 189–191.

Gallagher, R. T., Hawkes, A. D., Steyen, P. S. and Vleggar, R. (1984). Tremorgenic neuro-toxins from perennial ryegrass causing ryegrass staggers disorder in livestock: struc-ture and elucidation of lolitrem B. *Chemical Communications*, pp. 614–616.

Gallagher, R. T., Hawkes, A. D., Steyen, P. S. and Vleggar, R. (1985). Tremorgenic neuro-toxins from perennial ryegrass causing ryegrass staggers disorder of livestock: structure elucidation of lolitrem B in perennial ryegrass by high performance liquid chromatography with fluorescence detection. *Journal of Chromatography* 321, 217–226.

Gleason, H. A. and Cronquist, A. (1963). *Manual of Vascular Plants of Northeastern United States and Adjacent Canada.* D. Van Nostrand, New York.

Glenn, A. E., Bacon, C. W., Price, R. and Hanlin, R. T. (1996). Molecular phylogeny of *Acremonium* and its taxonomic implications. *Mycologia* 88, 369–383.

Grewal, S. K., Grewal, P. S. and Gaugler, R. (1995). Endophytes of fescue grass enhance susceptibility of *Papilla japonica* larvae to an entomopathogenic nematode. *Ento-mologic experiemtalis et applicator* 74, 219–224.

Groger, D. (1972). Ergot. *In* S. Kadis, A. Ciegler, and S. J. Ajl (eds.), *Microbial Toxins*, Vol. 7. Academic Press, New York, pp. 321–373.

Guppy, J. C. (1961). Life history and behaviour of the armyworm, *Pseudaletia unipuncta* (Haw.) (Lepidoptera: Noctuidae), in eastern Ontario. *Canadian Entomologist* 93, 1141–1153.

Gutierrez, C., Castanera, P. and Torres, V. (1988). Wound-induced changes in DIMBOA (2,4 dihydroxy-7-methoxy-2H-1, 4 benzoxazin-3(4H)-one) concentration in maize plants caused by *Sesamia nonagriodes* (Lepidoptera: Noctuidae). *Annals of Applied Biology* 113, 447–454.

Gwinn, K. D. and Savary, B. J. (1990). Production of ergovaline alkaloids in both liquid and solid media by *Acremonium coenophialum* isolated from tall fescue seed. *Abstract of the Annual Meeting of APS & CPS, Phytophathology* 80, 1031.

Hammon, K. E. and Faeth, S. H. (1992). Ecology of plant-herbivore communities: a fungal component? *Natural Toxins* 1, 197–208.

Hardy, T., Clay, K. and Hammond, A. M. Jr. (1985). Fall armyworm (Lepidoptera: Noctuidae): a laboratory bioassay and larval performance study for the fungal endophyte of perennial ryegrass. *Journal of Economic Entomology* 78, 571–574.

Hardy, T., Clay, K. and Hammond, A. M. Jr. (1986). Leaf age and related factors affecting endophyte-mediated resistance to fall armyworm (Lepidoptera: Noctuidae) in tall fescue. *Environmental Entomology* 15, 1083–1089.

Hatcher, P. E. (1995). Three-way interactions between plant pathogenic fungi, herbivorous insects and their host plants. *Biological Reviews* 70, 639–694.

Herms, D. A. and Mattson, W. J. (1992). The dilemma of plants: to grow or defend. *Quarterly Review of Biology* 67, 283–335.

Hinton, D. M. and Bacon, C. W. (1985). The distribution and ultrastructure of the endophyte of toxic tall fescue. *Canadian Journal of Botany* 63, 36–42.

Horton, D. R. and Redak, R. A. (1993). Further comments on analysis of covariance in insect dietary studies. *Entomologia experimentalis et applicata* 69, 263–275.

Hoveland, C. S., Schmidt, S. P., King, C. C. Jr., Odom, J. W., Clark, E. M., McGuire, J. A., Smith, L. A., Grimes, H. W. and Holliman, J. L. (1980). Association of *Epichloe typhina* fungus and steer performance on tall fescue pasture. *Agronomy Journal* 72, 1064–1065.

Huizing, H. J., Van der Molen, W., Kloek, W. and Den Nijs, A. P. M. (1991). Detection of lolines in endophyte-containing meadow fescue in the Netherlands and the effect of elevated temperature on induction of lolines in endophyte-infected perennial ryegrass. *Grass Forage Science* 46, 441–445.

Ingham, E. R. and Molina, R. (1991). Interactions among mycorrhizal fungi, rhizosphere organisms, and plants. *In* P. Barbosa, V. A. Krischik, and C. G. Jones (eds.), *Microbial Mediation of Plant-Herbivore Interactions.* John Wiley and Sons, New York, pp. 169–198.

Johnson, M. C., Dahlman, D. L., Siegel, M. R., Bush, L. P. and Latch, G. C. M. (1985). Insect feeding deterrents in endophyte-infected tall fescue. *Applied Environmental Microbiology* 49, 568–571.

Johnson-Cicalese, J. M. and White, R. H. (1990). Effect of *Acremonium* endophytes on four species of billbug (Coleoptera: Curculionidae) found on New Jersey turfgrasses. *Journal of the American Society of Horticultural Science* 115, 602–604.

Karban, R. (1991). Inducible resistance in agricultural systems. *In* D. W. Tallamy and M. J. Raupp (Eds.). Phytochemical induction by herbivores. John Wiley & Sons, New York, pp. 403–419.

Karban, R. and Carey, J. R. (1984). Induced resistance of cotton seedlings to mites. *Science* 225, 53–54.

Karban, R. and Baldwin, I. T. (1997). *Induced Responses to Herbivory.* University of Chicago Press, Chicago.

Karban, R. and Myers, J. H. (1989). Induced plant responses to herbivory. *Annual Review of Ecology and Systematics* 20, 331–348.

Karban, R., Adamchak, R. and Schnathorst, W. C. (1987). Induced resistance and interspecific competition between spider mites and a vascular wilt fungus. *Science* 235, 678–680.

Kelrick, M. I., Kasper, N. A., Bultman, T. L. and Taylor. S. (1990). Direct interactions between infected and uninfected individuals of *Festuca arundiancea*: differential allocation to shoot and root biomass. In S. S. Quisenberry and R. E. Joost (eds.), *Proceedings of the First International Symposium on* Acremonium/*Grass Interactions.* Louisiana Agricultural Experiment Station, New Orleans, pp. 21–29.

Kelrick, M. I., Maas, D. S. and Nomann, B. N. (1995). The *Acremonium* endophyte and its tall fescue host: nuances of a contingent mutualism. *Bulletin of the Ecological Society of America* 76, 140.

Kennedy, C. W. and Bush, L. P. (1983). Effect of environmental and management factors on the accumulation of N-acetyl and N-formyl loline alkaloids in tall fescue. *Crop Science* 32, 547–552.

Keogh, R. G., Tapper, B. A. and Fletcher, R. H. (1996). Distributions of the fungal endophyte *Acremonium lolii*, and of the alkaloids lolitrem B and peramine, within perennial ryegrass. *New Zealand Journal of Agricultural Research* 39, 121–127.

Kindler, S. D., Breen, J. P. and Springer, T. L. (1991). Reproduction and damage by Russian wheat aphid (Homoptera: Aphididae) as influenced by fungal endophytes and cool-season turfgrasses. *Journal of Economic Entomology* 84, 685–692.

Klun, J. A., Tipton, C. L. and Brindley, T. A. (1967). 2,4-dihydroxy-7-methoxy-1, 4-benzoxazin-3-one (DIMBOA), an active agent in the resistance of maize to the European corn borer. *Journal of Economic Entomology* 60, 1529–1533.

Latch, G. C. M., Christensen, M. J. and Samuels, G. J. (1984). Five endophytes of Lolium and Festuca in New Zealand. *Mycotaxon* 20, 535–550.

Latch, G. C. M., Christensen, M. J. and Gaynor, D. L. (1985). Aphid detection of endophytic infection in tall fescue. *New Zealand Journal of Argicultural Research* 28, 129–132.

Latch, G. C. M., Potter, L. R. and Tyler, B. F. (1987). Incidence of endophytes in seeds from collections of *Lolium* and *Festuca* species. *Annals of Applied Biology* 111, 59–64.

Letourneau, D. K. (1988). Microorganisms as mediators of intertrophic and intratrophic interactions. *In* P. Barbosa and D. K. Letourneau (eds.), *Novel Aspects of Insect-Plant Interactions.* John Wiley and Sons, New York, pp. 91–95.

Lewis, G. C. and Clements, R. O. (1986). A survey of ryegrass endophyte (*Acremonium loliae*) in the U.K. and its apparent ineffectuality on a seedling pest. *Journal of Agricultural Science* 107, 633–638.

Lopez, J. E., Faeth, S. H. and Miller, M. (1995). The effect of endophytic fungi on herbivory by redlegged grasshoppers (Orthoptera: Acrididae) on Arizona fescue. *Environmental Entomology* 24, 1576–1580.

Luginbill, P. (1928). The fall armyworm. USDA Techn. Bull. 34.

Lyons, P. C., Platner, R. D. and Bacon, C. W. (1986). Occurrence of peptide and clavine ergot alkaloids in tall fescue. *Science* 232, 487–489.

Lyons, P. C., Evans, J. J. and Bacon, C. W. (1990). Effects of the fungal endophyte *Acrem-*

onium coenophialum on nitrogen accumulation and metabolism in tall fescue. *Plant Physiology* 92, 726–732.

Marquis, R. J. and Alexander, F. M. (1992). Evolution of resistance and virulence in plant-herbivore and plant-pathogen interactions. *Trends in Ecology and Evolution* 7, 126–129.

Mattson, W. J. (1980). Herbivory in relation to plant nitrogen content. *Annual Review of Ecology and Systematics* 11, 119–161.

Myers, J. H. and Bazely, D. (1991). Thorns, spines, prickles, and hairs: are they stimulated by herbivory and do they deter herbivores? *In* D. W. Tallamy and M. J. Raupp (eds.). Phytochemical induction by herbivores. John Wiley and Sons, New York, pp. 325–344.

Neill, J. C. (1941). The endophytes of *Lolium* and *Festuca*. *New Zealand Journal of Science and Technology* 23, 185–193.

Ohnmeiss, T. E. and Baldwin, I. T. (1994). The allometry of nitrogen allocation to growth and an inducible defense under nitrogen-limited growth. *Ecology* 75, 995–1002.

Pedersen, J. F., Rodriguez-Kabana, R. and Shelby, D. B. (1988). Ryegrass cultivars and endophyte in tall fescue affect nematodes in grass and succeeding soybean. *Agronomy Journal* 80, 811–814.

Petrini, O. (1986). Taxonomy of endophytic fungi of aerial plant tissues. *In* N. J. Fokkema and J. van den Heuvel (eds.), *Microbiology of the Phyllosphere*. Cambridge Univ. Press, London, 175–187.

Popay, A. J. and Rowan, D. D. (1994). Endophytic fungi as mediators of plant–insect interactions. *In* E. A. Bernays (ed.), Insect-Plant Interactions, Vol. V. CRC Press, Boca Raton, FL, pp. 83–103.

Potter, D. A., Patterson, C. G. and Redmond, C. T. (1992). Influence of turfgrass species and tall fescue endophyte on feeding ecology of Japanese beetle and southern masked chafer grubs (Coleoptera: Scarabaeidae). *Journal of Economic Entomology* 85, 900–909.

Power, A. G. (1991). Virus spread and vector dynamics in genetically diverse plant populations. *Ecology* 72, 232–241.

Prestidge, R. A. (1989). Preliminary observations on the grassland leafhopper fauna of the central North Island volcanic plateau. *New Zealand Entomology* 12, 54–57.

Prestidge, R. A. and Ball, O. J.-P. (1993). The role of endophytes in alleviating plant biotic stress in New Zealand. *In* D. E. Hume, G. C. M. Latch, and H. S. Easton (eds.), *Proc. Second Int. Symp.* Acremonium/*Grass Interactions*. AgResearch, Palmerston North, New Zealand, pp. 141–151.

Prestidge, R. A., Pottinger, R. P. and Barker, G. M. (1982). An association of *Lolium* endophyte with ryegrass resistance to Argentine stem weevil. *Proceedings of the New Zealand Weed and Pest Control Conference* 35, 119–122.

Preszler, R. W., Gaylord, E. S. and Boecklon, W. J. (1996). Reduced parasitism of a leaf-mining moth on trees with high infection frequencies of an endophytic fungus. *Oecologia* 108, 159–166.

Price, P. W., Bouton, C. E., Gross, P., McPheron, B. A., Thompson, J. N. and Weis, A. E. (1980). Interactions among three trophic levels: influence of plants on interactions between insect herbivores and natural enemies. *Annual Review of Ecology and Systematics* 11, 41–65.

Price, P. W., Westoby, M., Rice, B., Atsatt, P. R., Fritz, R. S., Thompson, J. N. and Mobley, K. (1986). Parasite mediations in ecological interactions. *Annual Review of Ecology and Systematics* 17, 487–505.

Read, D. J. (1991). Mycorrhizas in ecosystems. *Experientia* 47, 376–391.

Rhoades, D. F. (1979). Evolution of plant chemical defense against herbivores. *In* G. A. Rosenthal and D. H. Janzen (eds.) *Herbivores: Their Interaction with Secondary Plant Metabolites.* Academic Press, Orlando, FL, pp. 3–54.

Roberts, C. A., Marek, S. M., Niblack, T. L. and Karr, A. L. (1992). Parasitic *Meloidogyne* and mutualistic *Acremonium coenophialum* increase chitinase in tall fescue. *Journal of Chemical Ecology* 18, 1107–1116.

Rolston, M. P., Hare, M. D., Moore, K. K. and Christensen, M. J. (1986). Viability of *Lolium* endophyte fungus in seed stored at different moisture contents and temperatures. *New Zealand Journal of Experimental Agriculture* 14, 297–300.

Rottinghaus, G. E., Garner, G. B., Cornell, C. N., and Ellis, J. L. (1991). An HPLC method for quantitating ergovaline in endophyte infected fall fescue: seasonal variation of ergovaline levels in stems, leaves and seed heads. *Journal of Agriculture and Food Chemistry* 39, 112–115.

Rowan, D. D. (1993). Lolitrems, peramine and paxilline: mycotoxins of the ryegrass/ endophyte interaction. *Agriculture, Ecosystems and Environment* 44, 103–122.

Rowan, D. D., Tapper, B. A., Sergejew, N. L. and Latch, G. C. M. (1990). Ergopeptine alkaloids in endophyte-infected ryegrass and fescues in New Zealand. *In* S. S. Quisenberry and R. E. Joost (eds.), *Proceedings of the 1st International Symposium* Acremonium/*Grass Interactions.* Louisiana Agricultural Experiment Station, New Orleans, pp. 97 -99.

Russell, W. A., Guthrie, W. D., Klun, J. A. and Grindeland, R. (1975). Selection for resistance in maize to first-brood European corn borer by using leaf-feeding damage of the insect and chemical analysis for DIMBOA in the plant. *Journal of Economic Entomology* 68, 31–34.

Sampson, K. (1933). The systemic infection of grasses by *Epichloe typhina* (Pers.) Tul. *Transactions of the British Mycological Society* 18, 30–47.

Sampson, K. (1935). The presence and absence of an endophytic fungus in *Lolium tementulum* and *L. perenne. Transactions of the British Mycological Society* 20, 337–343.

SAS Institute. (1985). *SAS User's Guide: Statistics, Version 5.* SAS Institute, Cary, NC.

Scriber, M. R. and Slansky, F., Jr. (1981). The nutritional ecology of immature insects. *Annual Review of Entomology* 26, 183–211.

Shelby, R. A. and Dalrymple, L. W. T. (1993). Long-term changes of endophyte infection in tall fescue stands. *Grass and Forage Science* 48, 356–361.

Siegel, M. R., Varney, D. R., Johnson, M. S., Nesmith, W. C., Buckner, R. C., Bush, L. P., Burrus, P. B. II and Hardison, J. R. (1984). A fungal endophyte of tall fescue: evaluation of control methods. *Phytopathology* 74, 937–941.

Siegel, M. R., Latch, G. C. M. and Johnson, M. C. (1987). Fungal endophytes of grasses. *Annual Review of Phytopathology* 25, 293–315.

Siegel, M. R., Latch, G. C. M., Bush, L. P., Fannin, F. F., Rowan, D. D., Tapper, B. A., Bacon, C. W. and Johnson, M. C. (1990). Fungal endophyte-infected grasses: alkaloid accumulation and aphid response. *Journal Chemical Ecology* 16, 3301–3315.

Stamp, N. E. (1992). Theory of plant-insect herbivore interactions on the inevitable brink of re-synthesis. *Bulletin of the Ecological Society of America* 73, 28–34.

Stebbins, G. L. (1981). Coevolution of grasses and herbivores. *Annals of the Missouri Botanical Gardens*, 68, 75–86.

Stout, M. J., Workman, J. and Duffey, S. S. (1994). Differential induction of tomato foliar proteins by arthropod herbivores. *Journal of Chemical Ecology* 20, 2575–2594.

Strauss, S. Y. (1991). Indirect effects in community ecology: their definition, study and importance. *Trends in Ecology and Evolution* 6, 206–210.

Tallamy, D. W. and Raupp, M. J. (eds.). (1991). *Phytochemical Induction by Herbivores.* John Wiley and Sons, New York.

Tapper, B. A., Rowan, D. D. and Latch, G. C. M. (1989). Detection and measurement of the alkaloid peramine in endophyte-infected grasses. *Journal of Chromatography* 463, 133–138.

Tuomi, J. (1992). Toward integration of plant defense theories. *Trends in Ecology and Evolution* 7, 365–367.

Villanueva, J. R. and Strong, F. E. (1964). Laboratory studies on the biology of *Rhopalosiphum padi* (Homoptera: Aphidae). *Annals of the Entomological Society of America* 57, 609–613.

West, C. P., Izekor, E., Oosterhuis, D. M., and Robbins, R. T. (1988). The effect of *Acremonium coenophialum* on the growth and nematode infestation of tall fescue. *Plant and Soil* 112, 3–6.

West, C. P., Izekor, E., Robbins, R. T., Gergerich, R. and Mahmood, T. (1990). *Acremonium coenophialum* effects on infestations of barley yellow drawf virus and soil-borne nematodes and insects in tall fescue. *In* S. S. Quisenberry and R. E. Joost (eds.), *Proceedings of the 1st International Symposium* Acremonium/*Grass Interactions.* Louisiana Agricultural Experiment Station, New Orleans, pp. 196–198.

West, C. P. and Gwinn, K. D. (1993). Role of *Acremonium* in drought, pest, and disease tolerances of grasses. *In* D. E. Hume, G. C. M. Latch, and H. S. Easton (eds.), *Proc. Second Int. Symp.* Acremonium/*Grass Interactions.* AgResearch, Palmerston North, New Zealand, pp. 131–140.

White, J. F., Jr. (1987). The widespread distribution of endophytes in the Poaceae. *Plant Disease* 71, 340–342.

Wilson, D. (1993). Fungal endophytes: out of sight but should not be out of mind. *Oikos* 68, 379–384.

Wicklow, D. T. and Carroll, G. C. (eds.) (1981). *The Fungal Community, Its Organization and Role in the Ecosystem.* Mycology Series Vol 2. Marcel Dekker, New York.

Woodburn, O. J., Walsh, J. R., Foot, J. Z., and Heazlewood, P. G. (1993). Seasonal ergovaline concentrations in perennial ryegrass cultivars of diffing endophyte status. *In* D. E. Hume, G. C. M. Latch, and H. S. Easton (eds.), *Proceedings of the 2nd International Symposium on* Acremonium/*Grass Interactions.* AgResearch, Palmerston North, New Zealand, pp. 100–102.

Yates, S. G., Plattner, R. D. and Garner, G. B. (1985). Detection of ergopeptine alkaloids in endophyte infected, toxic Ky-31 tall fescue by mass spectrometry/mass spectrometry. *Journal Agricultural and Food Chemistry* 33, 719–722.

Yates, S. G., Fenster, J. C. and Bartlet, R. J. (1989). Assay of tall fescue seed extracts, fractions and alkaloids using the large milkweed bug. *Journal of Agricultural and Food Chemistry* 37, 354–357.

IV
ECOLOGY OF ENDOPHYTES

17

Abiotic Stresses and Morphological Plasticity and Chemical Adaptations of *Neotyphodium*-Infected Tall Fescue Plants

David P. Belesky
Appalachian Farming Systems Research Center, Agricultural Research Service, U.S. Department of Agriculture, Beaver, West Virginia

Dariusz P. Malinowski
Texas A&M University, Vernon, Texas

1. INTRODUCTION

Pasture plant communities respond to biotic, abiotic, and management factors at community, species, and individual plant levels (Turkington and Mehrhoff, 1990). Each level of response, in turn, can modify the other so that a resilient and dynamic community occurs. For example, one can deduce from a large body of agronomic and applied botanical research that botanically complex swards respond to external factors and their interactions by shifts in species, lines within species, and reallocation of resources within individual plants. Our interest as agriculturalists is in plant adaptability that sustains productive and persistent swards under dynamic conditions in a given production system, often with disregard to floristic diversity. Pastures are usually relegated to marginal sites on the landscape imposing additional challenges to reliable production of quality herbage. This chapter focuses on morphological and chemical plasticity within *Festuca* species, a major component of humid-temperate region pastures, attributable to a mutualistic symbiosis between host grasses and endophyte infection. We consider genotypic responses of particular endophyte–tall fescue associations to

455

abiotic stresses common to marginal soil resources (e.g., acidity, water deficit, limited P availability), which should enable us to establish reference points from which we can develop plant resources to suit specific production system needs.

Tall fescue (*Festuca arundinacea* Schreb.) is a cool-temperate grass grown extensively in the eastern United States. Vigorous growth and productivity under a range of edaphic, environmental, and management conditions contribute to widespread agronomic, conservation, and amenity plantings. Tall fescue tolerates drought and high temperatures, which may be attributable, in part, to the presence of a fungal endophyte, *Neotyphodium coenophialum* (West, 1994). Report of a symptomless fungal endophyte associated with tall fescue (Neill, 1941) was verified by Bacon et al. (1977).

Despite agronomic versatility, endophyte-infected tall fescue is often toxic to livestock. Toxicity symptoms in livestock vary with environmental conditions, e.g., summer syndrome (Robbins, 1983) and fescue foot in winter (Garner and Cornell, 1978), and may occur as distal tissue necrosis, necrotic fat lesions, depressed weight gain, impaired reproductive efficiency, and low milk production (Stuedemann and Hoveland, 1988). A similar relationship was found between endophyte-infected perennial ryegrass (*Lolium perenne* L.) and livestock health disorders in New Zealand (Latch and Christensen, 1982). Research conducted since the late 1970s has focused on the economically important effects of endophyte–grass interactions on livestock production and grassland management.

Environment and endophyte–tall fescue associations interact to influence stand persistence as well as livestock performance. Several examples of infected plant superiority in highly stressful environments suggest that endophyte–grass associations can alter biodiversity in a pasture system (Hill, et al., 1991; Clay et al., 1993; Fribourg and MacLaren, 1987). Resultant changes in nutrient and environmental resource use in production situations dominated by endophyte-infected grasses remain to be resolved. The presence of an endophyte in a perennial temperate grass such as tall fescue enables it to thrive outside of typical adaptation zones. Endophyte-infected tall fescue grows and persists in the southeastern coastal plain of Georgia, USA whereas noninfected isogenic material succumbed to stresses associated with the environment, e.g., water deficit, insects and nematodes, high temperature, soil acidity (Bouton et al., 1993). Substantial stand loss occurred in tall fescue paddocks with low levels of endophyte infection in midsummer while infected plants became the dominant contributor to herbage production in northern Texas (Read and Camp, 1986). This suggests a competitive advantage for endophyte-infected plants growing under water deficit and associated high-temperature conditions. Livestock selectivity and preference for noninfected plants could be assumed to influence botanical dynamics as well.

Endophytes of the economically important grasses including tall fescue, perennial ryegrass, and meadow fescue (*F. pratensis* Huds.) are specific to the host grass (Christensen, 1997). These fungi belong to the Balansieae in the family Clavicipitaceae (Ascomycetes) (Diehl, 1950) and complete their life cycle within the above-ground portions of the grass host forming nonpathogenic, systemic, and usually intercellular associations (Bacon and De Battista, 1991). Currently, six genera are classified in the Balansieae: *Atkinsonella*, *Balansia*, *Echinodothis*, *Epichloe*, *Myriogenospora* (White, 1994), and *Parepichloe* (White and Reddy, 1998). Morgan-Jones and Gams (1982) erected the section *Albo-lanosa* section of the genus *Acremonium* to accommodate the anamorphic (imperfect) states of *Epichloe typhina* and the tall fescue endophyte. The tall fescue endophyte is distinguished from the *Epichloe* endophytes by existing as a symptomless infections of grasses in the subfamily Pooideae (Siegel, 1993). Recently, Glenn et al. (1996) reclassified fungi in the *Acremonium* sect. *Albo-lanosa* into a new genus, *Neotyphodium* (e-endophytes) to separate endophytic and nonendophytic *Acremonium* species. Thus, the *Albo-lanosa* section was abolished. The e-endophytes occur intercellularly in leaves, stems, and reproductive structures, but not in roots of soil-grown grasses (Hinton and Bacon, 1985).

Subsequent studies on fungal endophytes of C_3 grasses led to the discovery of at least two other groups of endophytes with symptomless colonization of host shoots. One group (p-endophytes) consists of closely related *Gliocladium*-like endophytes reported in perennial ryegrass (Latch et al., 1984; Phillipson, 1991) and *Phialophora*-like endophytes detected in meadow fescue, *F. gigantea*, and Arizona fescue *F. arizonica* (Schmidt, 1991; An et al., 1993). *Gliocladium*-like and *Phialophora*-like endophytes are classified as Eurotiales (Ascomycetes) (Siegel et al., 1995) and are not related to clavicipitaceous grass endophytes (Leuchtmann, 1992). These often occur cosymbiotically with *Neotyphodium* species endophytes in the same grass hosts (Latch et al., 1984; Phillipson, 1989; Schmidt, 1991). The p-endophytes differ from e-endophytes in patterns of mycelium growth, leaf and root colonization, ability to sporulate on the host grass (thus being infective), lack of alkaloid production (Siegel et al., 1995), and negligible effects on host in terms of tolerance to biotic and abiotic stresses (Malinowski et al., 1997a; Schmidt, 1993). The other group of grass endophytes not related to e-endophytes corresponds to *A. chilense*-like endophytes of annual ryegrass (*L. multiflorum*) and *F. paniculata* (L.) Schinz & Thell (a-endophytes) (Naffaa et al., 1998). The a-endophytes belong to the section *Simplex* of *Acremonium* genus and are represented by pathogenic species of *Acremonium* that are similar to *A. chilense*, an endophyte of orchardgrass *(Dactylis glomerata)* (Morgan-Jones et al., 1990). Knowledge is limited on the effects of p- and a-endophytes on physiology and ecology of cool-temperate origin grasses, regardless of e-endophyte status.

Circumstantial evidence suggests that intercellular endophytes, despite their symbiotic association with grasses, elicit a chemical response in the host plant. For example, resveratrol, a phenolic compound with antifungal activity, is produced by plants in response to fungal infection (Sylvia and Sinclair, 1983). Resveratrol concentration was greater in *Neotyphodium*-infected than in noninfected grasses (Powell et al., 1994), and Malinowski et al. (1998a) found that concentrations of phenolic-like compounds were greater in all parts of endophyte-infected than in endophyte-free (hence referred to as noninfected) tall fescue. Endophyte-infected tall fescue also contained chitinase, a pathogenesis-related protein (Roberts et al., 1992). The occurrence of phenolic compounds and chitinase suggests that *Neotyphodium*–grass associations create a less than benign environment for the fungal component, as well as for other fungal infections, i.e., grass pathogens. This evidence appears to support the asymptomatic association premise of Wilkinson and Schardl (1997) wherein the host benefit is obvious and costs imposed by the fungal symbiont are negligible. Our knowledge of interactions between species of *Neotyphodium* endophytes and grass hosts is developing rapidly, but many aspects remain obscure.

Our ability to explore the effects of endophyte infection on host fitness developed in the early 1980s when isogenic lines devoid of endophyte were generated. Investigations in the past 20 years focused on canopy responses to external factors including defoliation, nutrient input, and water deficit, whereas rhizosphere responses, which are more difficult to quantify, have been overlooked. We present a synopsis of abiotic stress tolerance of *Neotyphodium*–grass associations, emphasizing the influence of mineral nutrition and rhizosphere conditions on the growth and composition of endophyte-infected tall fescue. While abiotic factors often have substantial influence on endophyte–grass associations, we will present evidence that host plant responses to *N. coenophialum* infection interact with and in some instances affect the abiotic environment through localized changes in nutrient availability and acquisition, and pH. Biotic factors are considered briefly since these factors interact with abiotic features to affect partitioning of photosynthate within plants as well as within canopies as a function of selective herbivory. Our discussion will focus primarily on *N. coenophialum*–tall fescue associations with pertinent discussion of other endophyte–host associations where appropriate.

2. BIOTIC STRESS TOLERANCE OF ENDOPHYTE-INFECTED GRASSES

Biotic factors interact with abiotic factors to influence persistence of individuals and ultimately populations under field conditions (Ayres, 1991). For example, herbivory can create a dynamic canopy environment (altered light quantity and quality, nutrient cycling dynamics, and water demand) that can affect productivity and resource allocation of endophyte–grass associations. Nutrients, such as high tissue N concentrations, can also affect herbage preference to certain insect herbivores (Davidson and Potter, 1995), although this is complicated by the likelihood of increased alkaloid production in endophyte-infected grasses (Bush et al., 1993).

There is a strong selective advantage for endophyte-infected grasses in populations that include noninfected plants (Clay, 1998). Some of this advantage is attributable to an array of bioprotective alkaloids in the host grass (Porter, 1994; Bush et al., 1997) which deter livestock and insect herbivores (Ball, 1997). The alkaloids include ergopeptine alkaloids in tall fescue which cause a syndrome in livestock termed "fescue toxicosis" (Thompson and Stuedemann, 1993); lolitrem alkaloids in endophyte-infected perennial ryegrass causing "ryegrass staggers" (Prestige, 1993); loline alkaloids that are toxic to insects (Dahlman et al., 1997); and peramine, which deters feeding by insects (Rowan, 1993). Infected grasses produce specific flavonoids (nonalkaloid compounds) that are toxic to insects (Ju et al., 1998). Indirectly, endophytes can reduce insect-transmitted viral diseases in grasses (West et al., 1990b). Endophyte-infected grasses are more resistant than noninfected grasses to soilborne nematodes (West et al., 1988; Eerens et al., 1997), with resistance probably attributable to alkaloids (Malinowski, et al., 1999c; Malinowski et al. 1998b). Increased secretion of phenolic-like compounds from roots of infected tall fescue (Malinowski, et al., 1998a) may act as feeding deterrents in the rhizosphere, although this remains to be resolved. The reader is referred to Bacon, et al. (1997) for a thorough review of endophyte–grass association effects on biotic factors including bacteria, fungi, nematodes, insects, and nondomesticated herbivores.

Endophyte-infected grasses, in general, compete successfully with noninfected conspecifics as well as other plant species and in time can dominate a sward. The competitive advantage arises from phenotypic responses such as larger and more numerous tillers (Hill et al., 1991; Belesky et al., 1989a), greater leaf elongation (Belesky and Fedders, 1996), and altered root architecture (Malinowski et al., 1999d). For example, white clover (*Trifolium repens* L.) declines in pastures dominated by endophyte-infected compared to noninfected perennial ryegrass (Percival and Duder, 1983; Sutherland and Hoglund, 1989). The decline could be a function of more efficient nutrient assimilation by endophyte-infected plants, direct chemical effects on the competitor (allelopathic factors), chemically induced selection factors related to herbivory deterrence or preference, and resistance to disease in the infected plant. Petroski et al. (1990) noticed that synthetic forms of loline alkaloids (e.g., *N*-formyl loline, NL) affected germination and seedling development of annual ryegrass, and alfalfa (*Medicago sativa* L.). Bush et al. (1997) reported that NL detected in soils below endophyte-infected tall fescue plants was four times the amount required to inhibit germination of annual ryegrass. Seed extracts of endophyte-infected tall fescue inhibited germination of *Trifolium* species (Springer, 1997). Suppression of red clover (*T. pratense* L.) appeared closely related to loline alkaloid concentration in roots of endophyte-infected plants (Malinowski et al., 1999c). Strong suppression of leguminous companion species by endophyte-infected tall fescue in a mixed sward could have significant effects on productivity and N economy of a sward. This would also create a situation where a high degree of N-use efficiency would be advantageous.

Current efforts focus on the development of benign or "friendly" endophytes that, in association with grass hosts, produce little or no ergot alkaloids toxic to livestock yet retain pest and drought resistance attributes of symbiotic plants. The impact of endophyte-mediated surrogate transformation (Bacon and Hinton, 1998) on mechanisms of adaptation and host survival in a competitive environment under marginal edaphic conditions is unknown and warrants investigation.

3. ABIOTIC STRESS TOLERANCE OF ENDOPHYTE-INFECTED GRASSES

3.1. Drought Stress Tolerance

The benefit of endophyte infection in terms of drought stress tolerance in host grasses is well known. Endophyte-infected tall fescue (Buck et al., 1997; Bacon, 1993; West, 1994) and meadow fescue (Malinowski et al., 1997a, b) showed greater drought stress tolerance than noninfected conspecifics, although mechanisms and modes of action are not well understood. Host genotype, endophyte strain, and environmental conditions interact to influence water deficit effects on endophyte–grass associations (Elbersen & West, 1996; Hill et al., 1996).

Endophyte-infected compared to noninfected plants often have lower stomatal conductance and greater negative water potential in leaf blades, and exhibit osmotic adjustment in leaf bases (Elmi et al., 1990; West et al., 1990a; Elmi and West, 1995; Elbersen and West, 1996). Endophyte may induce an internal stress in the host plant that could precondition the host to drought, thereby permitting the infected plant to adapt more rapidly than noninfected conspecifics (West, 1994). Belesky et al. (1987a) suggested that endophyte might have some effect on the hormonal control of stomata in infected plants. Joost et al. (1993) showed that endophyte infection indeed affected concentration of abscisic acid (ABA) in leaves of drought-stressed tall fescue, suggesting that ABA is involved in the rapid response and recovery of endophyte-infected tall fescue in water deficit situations. Osmotically active substances, which enable the plant to maintain turgor under water deficit, accumulate in endophyte-infected grasses (Lyons et al., 1990; Richardson et al., 1992); however, none has been specifically identified or implicated in turgor maintenance in infected tall fescue.

Bacon and Hinton (1998) suggest that loline alkaloids may be involved in water deficit tolerance of endophyte-infected grasses. Loline alkaloids are water-soluble and may therefore act as osmoregulators (Bacon, 1993). Belesky et al. (1989b) showed that tall fescue accessions infected with their respective endophytes had significant increases in the actual amount of loline alkaloid production and occurrence in leaf tissues when plants were subjected to water deficit. High levels of loline alkaloids (up to 0.8% of foliar dry matter mass) occurred in endophyte-infected tall fescue and meadow fescue, and as such were likely to affect osmotic potential in cells (Kennedy and Bush, 1983; Bush et al., 1993). Loline alkaloids are the only alkaloids detected in endophyte-infected meadow fescue (Bush and Schmidt, 1994). Their presence in leaves and roots of tall fescue (Bush et al., 1993; Malinowski et al., 1999c) suggests that lolines may be translocated within plants. Interestingly, loline alkaloids are uncommon in endophyte-infected perennial ryegrass (Bush et al., 1993) where endophyte-related effects on drought tolerance are absent (Eerens et al., 1998).

Endophyte-infected grasses often have larger root systems than noninfected plants (Latch et al., 1985; Belesky et al., 1989a; De Battista et al., 1990; Malinowski et al., 1997a, b). While changes in root mass and architecture can enhance plant access to water in soils, a direct relationship between root system size and drought tolerance has not been validated for endophyte-infected grasses. Malinowski et al. (1999d) were the first to show that *N. coenophialum*–endophyte modified tall fescue root morphology. Roots of endophyte-infected compared to noninfected plants were smaller in diameter and had longer root hairs, resulting in increased root surface area. Greater root mass and increasing root density can not only lead to enhanced water acquisition but can change the soil structure and associated properties (like water holding capacity) affecting plant growth (Schulze, 1991).

Endophyte-infected tall fescue adjusts to water deficit by reducing leaf area and thus area contributing to transpirational water loss (West et al., 1990a; Belesky et al., 1989a). Laminar area reduction can occur through senescence as well as leaf rolling (Arachevaleta et al., 1989; Hill et al., 1990). Water-deficit induced senescence can lead to accumulation of polyamines which appear to be direct precursors of pyrrolizidine alkaloids (Bush et al., 1993). Pyrrolizidine alkaloids increase as a function of water deficit and may function in osmotic adjustment of infected plants. Bacon (1993) addresses the issue of osmotic adjustment in a review of endophyte-infected tall fescue response to abiotic factors.

An additional soil-water-related phenomenon involves morphological adjustment of infected tall fescue to water-logged soil conditions. Arechavaleta et al. (1992) found that infected tall fescue grew and accumulated ergopeptine alkaloids when exposed to water-logged soil conditions, whereas noninfected plants were not as productive and did not produce ergoalkaloids, even when grown under waterlogged conditions. Arechavaleta and co-workers noticed that infected plants produced rhizomes to circumvent water-logged soil conditions unlike noninfected plants. However this observation has not been substantiated in the genotypes used in our studies.

Responses of endophyte-infected grasses to water deficit are a combination of drought stress avoidance, tolerance, and recovery mechanisms (Clarke and Durley, 1981). Considering this we see that endophyte-infected tall fescue adjusts physiologically (e.g., increased osmotic adjustment) and morphologically (greater apparent root surface area) and may, as a function of modified root architecture, have some effect on soil physical properties associated with structure and aggregate stability. The sum of responses of individual endophyte–grass associations determines the ecological success of a given grass population (West, 1994), so it is likely that specific endophyte–grass associations are adjusting to a given set of conditions with a range of responses. As a result, the range of adaptation of endophyte-infected cool-season grasses can be extended into areas dominated by warm-season grasses (Bouton et al., 1993; Cunningham et al., 1994). In addition to a soil–water deficit, high temperature (above 35°C) seems to have a more deleterious effect on physiological processes (photosynthesis and carbon exchange rate) in noninfected than endophyte-infected grasses (Marks and Clay, 1996).

3.2. Light and Temperature

Very little information is available concerning the photosynthetic response of endophyte–grass associations in response to increasing light intensity (Belesky et al., 1987a), water deficit (Richardson et al., 1993), and air temperature (Marks and Clay, 1996). In each situation, results were influenced by interaction of environmental conditions with endophytegrass association. Photosynthetic rate was greater at high light intensity for endophyte-infected than noninfected plants (Marks and Clay, 1996) at leaf temperatures above 35°C. Richardson et al. (1993) found that the photosynthetic rate was greater at moderate to low water potential in infected plants. Belesky et al. (1987a) found that net photosynthesis of infected plants was depressed relative to that of noninfected plants growing at high light intensity. In both cases where photosynthetic rate increased (Marks and Clay, 1996; Richardson et al., 1993), measurements were made on individual leaves of defined age, whereas Belesky et al. (1987) determined net photosynthetic rate on the entire canopy composed of mixed leaf ages. Consider that high temperatures and high light intensities are likely to occur simultaneously, especially in the southern portions of the United States at the extremes of the tall fescue adaptation zone (Shelby and Dalrymple, 1987). Ability to tolerate water deficit and maintain photosynthetic capability at high leaf temperature and light intensity might enable cool-temperate-origin species to persist in warmer reaches of the humid-temperate climatic zone. These attributes contribute to the versatility of tall fescue and provide pasture managers with a forage resource that can extend the herbage production interval over a range of climatic and short-term weather conditions.

Competition for light is a deciding factor in the composition of plant communities and influences response characteristics of individuals in the community (Donald, 1963). Clay (1987), and Marks and Clay (1996) showed that infected clones of fescue and ryegrass were more productive than noninfected clones when grown at high light intensities. The opposite response occurred under low-light conditions. Low-light-intensity conditions in a sward could be caused by a competitive canopy environment that can influence the quality of light as well as quantity. Changes in light quality (e.g., red/far-red ratio, R/FR) affect tiller expression and elongation (Casal et al., 1987, 1990) and leaf development (Skinner and Nelson, 1994) in grasses. Endophyte effects on grass development are well documented but data on interactions between light quality and endophyte infection on development are scarce. Gautier and Gaborcik (1993) investigated end-of-day light effects on morphogenesis of tall fescue and perennial ryegrass to address the role of light in competitive situations involving endophyte-infected grasses. They found that neither tall fescue nor perennial ryegrass responded as expected to end-of-day far-red light enrichment. Tillering rate was not affected nor was leaf sheath or lamina length when comparing infected to noninfected plants.

We conducted experiments to determine the interaction of red light enrichment (far-red depletion) with endophyte on the growth and development of infected and noninfected tall fescue isolines (Belesky, unpublished data). Red light–emitting diodes were positioned at the base of plants for a factorial arrangement of tall fescue genotype (DN2, DN9, DN11, and DN15), endophyte infection status (E+, E−), and end-of-day light condition (red light–enriched; ambient). In two separate experiments, the response of tall fescue to end-of-day red light enrichment was dependent upon host genotype. For example, DN9 had a greater increase in tiller number than DN15 as a function of end-of-day red light compared to their respective noninfected isolines (Table 1). Other measures of response to light quality like pseudostem length, leaf area, and leaf thickness were affected by end-of-day light conditions in DN9 but not DN15.

The interaction of light quality and endophyte infection could influence the competitive ability of the plant host in a range of canopy environments. This could be manifested as pliant canopy architecture or through altered herbage dry matter production and interplant allocation. Herbage production or plant size is linked to competitiveness of plants (Harper, 1977). Dense swards influence light quality by decreasing the R/FR within the canopy. The altered light environment can influence tiller production and leaf elongation and ultimately phytomass or photosynthate available for growth, storage or use in secondary metabolite production.

End-of-day light quality can influence alkaloid production in certain plants (Tso et al., 1970), but light quality effects on alkaloid production of tall fescue remain largely unexplored. Bransby et al. (1988) suggested that close grazing

Table 1 Interaction of End-of-Day Light Quality and Endophyte on Some Developmental Features of Tall Fescue[a]

Genotype	Endophyte status	End-of-day light regime	Tiller number	Leaf area (cm²)	Pseudostem length (cm)
9	+	ambient	30 ± 11	366 + 67	7.1 ± 1.9
9	+	red-light	38 ± 9	424 + 133	5.5 ± 1.3
9	−	ambient	12 ± 5	237 + 88	8.0 ± 2.2
9	−	red-light	10 ± 5	265 + 78	9.0 ± 0.8
15	+	ambient	33 ± 4	286 + 135	6.0 ± 0.8
15	+	red-light	33 ± 5	390 + 83	6.0 ± 1.2
15	−	ambient	31 ± 5	430 + 54	6.0 ± 0.8
15	−	red-light	25 ± 4	392 + 83	6.3 ± 1.0

[a] Plants were clipped to a 10-cm residue every 10 days. Each value is the mean of four replicates. (From D. Belesky, unpublished data.)

of infected tall fescue minimized deleterious effects on the grazer. Subsequent experiments in field and laboratory showed that leaf area and ergoalkaloid production were highly correlated (Hill et al., 1990) and that defoliation affected ergoalkaloid production (Belesky and Hill, 1997). Preliminary data (Belesky, unpublished data) suggests that red-light-enriched conditions (representing closely grazed canopies) decrease ergoalkaloid yield (alkaloid concentration × dry matter) of infected tall fescue relative to dense canopies (lower R/FR) and that the response is genotype-dependent (Table 2).

Low R/FR (greater far-red light which may occur in dense or lightly grazed canopies) stimulates accumulation of phenolics in some plant species (Kandeler, 1958). Phenolics can act as feeding deterrents (Scehovic and Jadas-Hecart, 1989) Phenolic-like compounds also contribute to the remarkable adaptability of tall fescue to mineral stress conditions (Malinowski et al., 1998a), which we present in the following sections. Understanding the interaction of management and endophyte on production of phenolics in tall fescue would be useful when designing associations devoid of ergoalkaloids, but with persistence (insect deterrence) and enhanced mineral acquisition capabilities under low-input (marginal resource) conditions.

Light quality can be affected by changes in atmospheric conditions as well as changes associated with canopy structure. Newsham et al. (1998) showed that exposure of a clone of endophyte-infected perennial ryegrass to elevated UV-B radiation reduced fertility (seed set) compared to noninfected plants. The potentially deleterious impact of UV-B radiation on the regenerative propagation of endophyte-infected perennial ryegrass could be offset by enhanced tiller production in infected plants. The authors concluded that UV-B radiation might have

Table 2 Pyrrolyzidine (Loline) and Ergopeptine Alkaloid Yield in Endophyte Infected Tall Fescue as a Function of End-of-Day Light Quality (μg alkaloid per plant)[a]

Genotype	Light regime	Leaf		Pseudostem		Root	
		Lolines	Ergopeptines	Lolines	Ergopeptines	Lolines	Ergopeptines
9	Ambient	2524	<1	4275	9.16	389	<1
9	Red light	2109	<1	3202	7.55	69	<1
15	Ambient	2628	<1	4154	8.04	304	<1
15	Red light	2522	1.88	4124	7.98	302	<1

[a] Each value is the mean of four replicates. Pyrrolizidine alkaloid analysis conducted by L. P. Bush (University of Kentucky) and ergopeptine analysis conducted by N. S. Hill (University of Georgia.) (From D. Belesky, unpublished data.)

a negative impact on endophyte-infected grasses that rely to a great extent on seed production for propagation.

3.3. Soil Acidity

Knowledge of mechanisms of soil acidity tolerance in perennial grasses, particularly endophyte-infected grasses, is very limited. Low soil pH and high soluble Al concentrations can restrict root growth, and affect N uptake and N supply to growing grass leaves (Thornton, 1998). Soil acidity affects plant growth through a complex of chemical changes in the rhizosphere involving increased H^+, Al, and Mn; inhibition of metal cation (Ca, Mg) uptake; a decrease in P and Mo solubility; and increased leaching of nutrients and metabolites from roots (Marschner, 1991). The net effect is modified nutrient activity in the root zone and physiological damage to root tissue. Plants adapted to acidic soils possess a variety of mechanisms that enable them to cope with adverse chemical conditions.

Cheplick (1993) found that infected tall fescue was less productive than noninfected plants under conditions of simulated acid rain. In contrast, Belesky and Fedders (1995) reported that some endophyte-infected tall fescue genotypes (DN2 and DN7) produced more massive root systems than noninfected plants grown in acidic, high-aluminum-content soil, while other genotypes (DN4, DN5, and DN11) were not influenced by infection. Very high P (800 mg P kg⁻¹) was applied, which may have affected the response to acidity conditions. As shown by Malinowski et al. (1998b) for the same soil type, applying more than 50 mg P kg⁻¹ soil depressed root and shoot DM of infected compared to noninfected tall fescue plants. Malinowski and Belesky (1999a) monitored soil pH as a function of tall fescue endophyte status. In control (nonfertilized) and soil supplied with a low level (13 mg P kg⁻¹) of commercial P fertilizer, soil pH increased more rapidly and to a greater extent when associated with infected compared to noninfected plants. No endophyte-influenced differences in soil pH were observed when P was supplied as phosphate. Changes in nutrient solution pH were also observed in experiments with hydroponically grown tall fescue, especially under limited P and high Al, as a function of endophyte infection (Malinowski and Belesky, 1999a; Malinowski et al., 1999d).

3.4. Mineral Stress Tolerance

Gentry et al. (1969) reported what appears to be the first documented evidence that mineral nutrition affected production of certain alkaloids in field-grown tall fescue. Results indicated that concentrations of perloline and "unknown" alkaloids were increased by nitrogen (N) but decreased by phosphorus (P) and potassium (K) when compared to nonfertilized controls. Occurrence of endophyte in tall fescue was not known at the time. Based largely on this early work and the occurrence of toxicosis symptoms in livestock on well-managed tall fescue pastures, research over the next 20 years focused on the involvement of N in tall fescue toxicosis. We now have evidence (Malinowski et al., 1998b) that the elicitor(s) of livestock disorders, ergopeptide alkaloids, might be influenced by phosphorus nutrition as well in a production system.

3.4.1. Nitrogen

Secondary metabolites such as alkaloids produced by endophyte-infected grasses (Porter, 1994) are nitrogen-rich compounds. Consequently, most of the early work on developing a management solution for tall fescue toxicosis focused on N as a factor in and control point of toxicity. Lyons, et al. (1990) found that N form (NH_4^+ or NO_3^-) in plant tissue was affected by endophyte infection status in both leaf sheaths (localized site of endophyte infection) and blades (tissues devoid of endophyte) of infected tall fescue. Infected plants had greater ammonium and amino acid levels and less nitrate than noninfected plants. Infected plants were also more N-use-efficient than noninfected plants when N was limited in the environment, which is a competitive advantage for infected plants. The form of N in pasture situations is probably less important than the quantity. Belesky et al. (1987b) reported an increase in pyrrolizidine alkaloid concentration in response to N fertilization on a unit of endophyte-infection-rate basis from grazed tall fescue. Concentrations of ergopeptine alkaloids also increased in response to N fertilization under controlled environment (Lyons and Bacon, 1984) and field conditions (Belesky et al., 1988). Other metabolites containing N, e.g., proline (Belesky et al., 1982) and amino acids (Lyons et al., 1990; Belesky et al., 1984), increased in endophyte-infected tall fescue receiving high N levels.

Nitrogen and photosynthate partitioning are linked and might influence endophyte–host responses. In general non-structural carbohydrate accumulation decreases as N supply increases. Belesky et al. (1991) showed that, on a population level, endophyte did not affect carbohydrate accumulation in grazed tall fescue supplied with either a modest (134 kg N ha^{-1}) or high (336 kg N ha^{-1}) amount of fertilizer. Alkaloid concentrations decreased as defoliation frequency increased and did so as a function of nonstructural carbohydrate concentration (Belesky and Hill, 1997), suggesting that management and environmental conditions that allow nonstructural carbohydrate to accumulate can lead to increased ergoalkaloid production.

Belesky et al. (1989a) found that dry matter yield increased an average of 15% probably because of more numerous and massive tillers on infected compared to noninfected plants, when N (150 mg kg^{-1} soil) and water supply was adequate. Cheplick et al. (1989) reported that endophyte infection differentially affected growth of perennial ryegrass and tall fescue seedlings as well as adult plants, under varying nutrient supply. Endophyte-infected seedlings of both species were more productive than noninfected seedlings at intermediate and high nutrient levels. Depressed seedling growth of infected plants at low nutrient level suggests that endophyte might impose a metabolite cost to the developing plant as a result of competition for nutrients and energy. Mature plants of endophyte-infected tall fescue were more productive than noninfected plants at all nutrient levels, whereas no such mineral nutrition related response was observed for perennial ryegrass.

Arachevaleta et al. (1989) found that at low N (11 mg N pot^{-1}), endophyte-infected plants produced as much herbage DM as noninfected plants supplied with 220 mg N pot^{-1}, with productivity increasing as N rates increased. These results are further evidence that endophyte-infected plants used N more efficiently than noninfected plants. Meadow fescue nitrogen use efficiency was not affected by endophyte (*N. uncinatum* and *Phialophora*-like endophyte) when under moderate N fertilization (Malinowski, 1995). Endophyte was beneficial to growth of fully developed tillers of Arizona fescue genotypes at low N level, but as N availability increased, noninfected plant productivity improved (Louis and Faeth, 1997). Results obtained with Arizona fescue support the premise of increased N-use efficiency in infected plants.

3.4.2. Phosphorus

Widespread use of tall fescue in disturbed land reclamation efforts suggests that the species persists under marginal soil conditions (e.g., high exchangeable Al, low pH, shallow and eroded) to a much greater extent than many other cool-season grasses. Apparently, tall fescue has a means to avoid stresses and acquire nutrients under challenging chemical and physical conditions. Acidic soils limit phosphorus availability which can influence botanical diversity and consequently the type of agricultural activity conducted on a soil resource. Herbaceous and woody species often benefit from association with mycorrhizae in acidic or marginal nutrient supply conditions (Wilcox, 1996; also see review by Smith and Gianinazzi-Pearson, 1988). Could endophytes confined to the shoot, such as *Neo-typhodium* species, have a similar influence on nutrient acquisition by the symbiotum in cool-temperate origin grasses?

Endophyte (*N. coenophialum*) hyphae contained inorganic P structures (Azevedo and Welty, 1995) similar to those found in cytoplasm of mycorrhizal fungi (White and Brown, 1979; Lapeyrie et al., 1984). Azevedo and Welty (1995) found that endophyte hyphae were a sink for P in media-grown tall fescue plants. Phosphorus localized in granules might be a source of P in endophyte-infected grasses when soil P availability is low, although this remains to be resolved. The first obstacle to overcome would be acquisition of P from soil.

Agronomic practices and growing conditions that influence P availability become important considerations when managing infected tall fescue. Malinowski et al. (1998b) found that P uptake and herbage production were greater in infected than to noninfected plants when soil P availability was low; however, herbage mass was less in infected than noninfected plants at high soil P availability. Infected plants grown at low P availability had greater specific root length (*i.e.*, finer roots) and concentrations of P, Mg, and Ca in roots and shoots than did noninfected plants. At the same time, Al concentration was greater in roots than shoots of infected plants, suggesting a mechanism to exclude Al from shoots.

Phosphorus source affected P uptake rate by endophyte-infected plants in P-deficient soils. Phosphorus uptake rates were greater in infected plants fertilized with phosphate rock compared to commercial P fertilizer (Malinowski and Belesky, 1999a). Infected plants had more root DM (10%) and greater relative growth rate (16%) than noninfected plants when P was supplied as phosphate rock, but endophyte-mediated responses were minimal when conventional P fertilizer was applied. We attributed differential response to P source as a function of root exudates, especially from endophyte-infected plants.

Two endophyte-related mechanisms for P uptake appear to be operating in tall fescue grown in P-deficient soils: (1) altered root morphology and (2) increased activity of root exudates. Malinowski et al. (1999d) found that infected tall fescue grown in nutrient solution produced roots with smaller diameter and longer root hairs than noninfected plants regardless of P level (Fig. 1). Using a simple color-reaction technique, we (Malinowski et al., 1998a) showed that more reducing activity was generated by roots of endophyte-infected than by noninfected tall fescue in P-deficient growth media (Fig. 2). Reducing activity was associated with phenolic-like compounds occurring in greater amounts in roots (as well as shoots) of infected compared to noninfected plants when P was limited (Fig. 3). Since phenolic-like compounds were detected in noninfected plants as well, they might not be endophyte-specific but might be produced by the symbionts under stress.

Phosphorus is involved in ergoalkaloid synthesis in *Claviceps* species. (Robbers, 1984; Flieger et al., 1991), controlling dimethylallyl tryptophan synthase (DMATase) activity, the first enzyme in the ergoalkaloid biosynthetic pathway. Garner et al. (1993) proposed a similar mechanism of ergoalkaloid biosynthesis for tall fescue endophytes, linking P nutrition with ergoalkaloid production in infected grasses. Ergovaline concentration in tissues increased as P level in nutrient solution increased (Azevedo et al., 1993), clearly demonstrating that P nutrition was involved in ergoalkaloid production in infected tall fescue. Ergoalkaloid concentrations increased as P availability increased in endophyte-tall fescue associations with low ergoalkaloid production capability (e.g., DN4) (Malinowski et al., 1998b). Genotypes with greater ergoalkaloid production potential, such as DN7 and DN11, reached the greatest ergoalkaloid concentration at moderate P (50 mg P kg^{-1}) in the soil and reflected the feedback inhibition caused by relatively high P levels noticed for media grown *Claviceps* species.

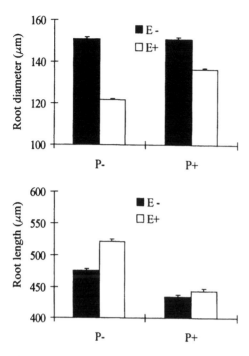

Figure 1 Average root diameter and root hair length of endophyte infected (E+) and noninfected (E−) tall fescue (pooled for two genotypes DN2 and DN4) in response to P concentrations in nutrient solution. Bars indicate standard errors. (From Malinowski et al., 1999d).

Host plant adaptation to marginal soil nutrient conditions can be mediated by endophyte, where in some instances the symbiosis enhances acquisition of nutrients like P, while in other circumstances ergoalkaloid production can be altered to affect herbivory. Despite this, the influence of fertilizer P on tall fescue toxicosis symptoms in livestock has been overlooked. Consider that much of the tall fescue grown in the southeastern United States is associated with the combination of beef and poultry operations. Manure generated from confinement-grown poultry operations is applied on adjacent pastures, which are often dominated by tall fescue. The P (as well as N) associated with poultry waste could lead to increased ergoalkaloid production in endophyte-infected tall fescue exacerbating fescue toxicosis symptoms.

3.4.3. Aluminum

Phenolic-like compounds capable of reducing Fe^{3+} can chelate Al (Marschner, 1986) and may facilitate Al transport within plants in nontoxic forms (Taylor,

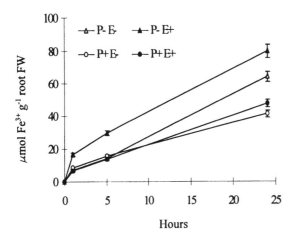

Figure 2 Fe^{3+} reducing activity of intact roots of endophyte infected (E+) and noninfected (E−) tall fescue grown in P− and P+ nutrient solutions after 1, 5, and 24 h. Data represent values averaged over four tall fescue genotypes (DN2, DN4, DN7, and DN11). Bars indicate standard errors ($n = 8$). (From Malinowski et al., 1998a.)

1988). Foy and Murray (1998) showed that tall fescue accumulated much less Al in shoots than in roots, suggesting an efficient mechanism for Al sequestration at the roots. Endophyte-infected fine fescues (*Festuca* species) were more tolerant of elevated Al levels than were noninfected plants (Liu et al., 1996) but a mechanism for the enhanced tolerance was not proposed.

Phenolic-like compounds exuded from roots of endophyte-infected tall fescue might be involved in Al tolerance. More Al (47%) and P (49%) was desorbed from root surfaces of infected than noninfected plants (Malinowski, unpublished data). Amounts of Al desorbed from root surfaces depended on root DM of plants, suggesting an additional Al sequestration mechanism located on roots of endophyte-infected plants. (Fig. 4). More Al (35%) but less P (10%) was found in root tissues as a function of endophyte infection, suggesting that Al is sequestered in roots and P is translocated to the shoot. Resveratrol, a phenolic substance produced in fungus-infected plants, can chelate Cu in vitro (Belguendouz et al., 1998). Is it possible that resveratrol is involved in Al sequestration in infected plants?

3.4.4. Other Macro- and Microelements

Very little is known about the influence of minerals other than N and P on the growth and chemical composition of endophyte-infected grasses. Likewise, knowledge of endophyte effects on uptake and allocation of microelements is limited. Along with N and P, Ca is involved in ergoalkaloid biosynthesis, being

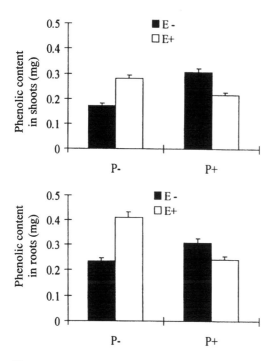

Figure 3 Total phenolic content (based on ethanol/water-soluble phenolic) in roots and shoots of endophyte infected (E+) and noninfected (E−) tall fescue (averaged over four genotypes DN2, DN4, DN7, and DN11) in response to phosphorus concentration in nutrient solution. Bars indicate standard errors. (From Malinowski et al., 1998a.)

a coenzyme for DMATase (Lee et al., 1976; Cress et al., 1981). Hill (1995) found that Ca fertilization increased ergoalkaloid concentration in infected tall fescue independent of soil pH. We have evidence (Malinowski and Belesky, unpublished data) that endophyte infection increased specific absorption rate for Ca as well as P, Mg, K, and several microelements. Endophyte-related mineral absorption rate differences were especially pronounced under P deficiency. Dennis et al. (1998) reported that Cu concentrations decreased in herbage of infected Kentucky-31 tall fescue grown under greenhouse conditions. Subsequent field experiments with tall fescue in Virginia and Mississippi showed that Cu concentration behaved in the same way. Steers consuming infected tall fescue in Virginia had lower serum Cu levels than steers fed noninfected plants (Saker et al., 1998), while no differences in serum Cu were observed in steers fed tall fescue in Mississippi (Allen et al., 1997). Apparently, Cu uptake is influenced by endophyte as well as by soil parent material. Endophyte appears to benefit the host by re-

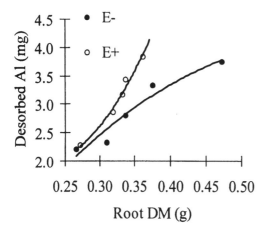

Figure 4 Relationship between root dry matter and aluminum desorbed by citric acid from root surfaces and free spaces of noninfected (E−) ($y = -15.8 \chi^2 + 19.8x -2.1$; $R^2 = 0.93$) and endophyte infected (E+) ($y = 87.6 \chi^2 - 37.7x + 6.0$; $R^22 = 0.98$) tall fescue. Data were pooled for two genotypes (DN2 and DN4) (from Malinowski and Belesky, 1999b.)

stricting potentially phytotoxic metal ion accumulation in above-ground tissues; however, restricted levels of essential micronutrients and possible bioavailability factors may present additional nutritive quality concerns in extensively managed tall fescue–based pasture systems where endophyte is present.

Increased release of exudates with Fe^{3+} reducing capability from infected tall fescue (Malinowski et al., 1998a) may present an interesting question should Fe uptake become an issue. Grasses possess a unique mechanism of Fe uptake that involves nonproteinogenic amino acid (phytosiderophores) carriers (Tagaki, 1976; Marschner, 1986). The relationship between release of reductants into the rhizosphere and Fe uptake by roots of tall fescue as a function of the endophyte's siderophore and mineral stress requires further study.

4. SUMMARY AND CONCLUSIONS

Mutualistic symbiosis of grasses with specific shoot-localized endophytes contributes to adaptability and subsequent widespread persistence of the associations in challenging edaphic and environmental conditions. Shoot-localized endophytes, specifically *Neotyphodium* species, induce changes in root morphology and root function. These changes may be a result of intricate signaling and biofeedback systems involving the symbionts. Phytohormones are produced by mycorrhizae (Dannenberg et al., 1992) and *Neotyphodium* species (Joost, 1993). Compounds produced as a function of endophyte infection, such as pyrrolozidine alkaloids in tall fescue, may have dual functions as antiherbivory (biotic) and as osmoregulator agents (abiotic). Endophyte infection–induced root exudates appear to have both a biotic and abiotic stress resistance function in tall fescue, such as, nematode resistance and apparent allelopathic effects in mixed swards (biotic), and metal ion chelation affecting sequestration and availability (abiotic). This facet of the symbiotic response requires further investigation to realize the full benefit from endophyte–tall fescue associations surrogately transformed for non ergot–alkaloid production. The nature of communication between endophyte and host grass remains to be determined.

Benefits to grass hosts arising from infection with *Neotyphodium* species endophytes appear similar to well-known benefits attributable to mutualistic symbioses involving mycorrhizae, i.e., enhanced acquisition of phosphorus (Read, 1991), and a range of other functions especially under nutrient-stress conditions (Newsham, et al., 1995). A number of cosymbiotic associations with mycorrhizae and *Neotyphodium* species occur; however, these duel associations seem to be antagonistic in cool-season grasses (Barker, 1987; Chu-Chou et al., 1992).

Defining the responses of endophyte–host associations to various extrinsic factors is fundamental to understanding pasture community structure and, ultimately, success of the population. Furthermore, investigation of surrogately transformed endophyte–host associations is imperative if we are to capitalize on the valuable attributes of such associations. The close linkage between shoot–root function and our evolving knowledge of how endophytes mediate the relationship complicates our extrapolation from an individual behavior to community level responses. Modifying endophyte–host relationships for a specific characteristic may have unknown consequences for survival.

We have shown that *N. coenophialum*-infected grasses adapt to abiotic factors through morphological, physiological and chemical modifications. Observed changes in rhizosphere chemistry, root morphology, and root function in endophyte–grass associations present exciting new opportunities for future research addressing endophyte–host communication, mineral uptake mechanisms and use efficiency, and persistence in production systems. Once the multifaceted extrinsic factors and trophic levels have been determined, the evolutionary strategies used by both organisms may be addressed.

ACKNOWLEDGMENTS

We thank our many colleagues and co-workers, both supporters and critics, who helped us carry out our research efforts. Special thanks to Dr. G. A. Alloush for assistance with rhizosphere chemistry; Dr. A. Leuchtmann for helpful suggestions with our statements regarding taxonomic classification of endophytes; Dr. C. M. Willmer for discussions on carbon exchange and stomatal function; Mr. J. M. Fedders for help with light quality and other experiments over the past 10 years; and Dr. J. D. Robbins who, though now departed, still inspires me (D.P.B.) to probe the tall fescue–endophyte mystery.

REFERENCES

Allen, V. G., Fontenot, J. P., Bagley, C. P., Ivy, R. L. and Evans, R. R. (1997). Effects of seaweed treatment of tall fescue on grazing steers. *In* M. J. Williams (ed.), *Proceedings of the 1997 American Forage and Grassland Council Conference*. Georgetown (TX): American Forage and Grassland Council, Georgetown, TX, pp. 168–172.

An, Z. Q., Siegel, M. R., Hollin, W., Tsai, H. F., Schmidt, D., Bunge, G and Schardl, C. L. (1993). Relationships among non-*Acremonium* sp. fungal endophytes in five grass species. *Applied Environmental Microbiology* 59, 1540–1548.

Arachevaleta, M., Bacon, C. W., Hoveland, C. S. and Radcliffe, D. E. (1989). Effect of the tall fescue endophyte on plant responses to environmental stress. *Agronomy Journal* 81, 83–90.

Arechavaleta, M. Bacon, C. W. Plattner, R. D. Hoveland, C. S. and Radcliffe, D. E. (1992). Accumulation of ergopeptide alkaloids in symbiotic tall fescue grown under deficits of soil water and nitrogen fertilizer. *Applied and Environmental Microbiology* 58, 857–861.

Ayres, P. G. (1991) Growth responses induced by pathogens and other stresses. *In* H. A. Mooney, W. E. Winner and E. J. Pell (eds.), *Response of Plants to Multiple Stresses*. Academic Press, San Diego, pp. 227–248.

Azevedo, M. D. and Welty R. E. (1995). A study of the fungal endophyte *Acremonium coenophialum* in the roots of tall fescue seedlings. *Mycologia* 87, 289–297.

Azevedo, M. D., Welty, R. E., Creaig, A. M. and Bartlett, J. (1993). Ergovaline distribution, total nitrogen and phosphorus content of two endophyte-infected tall fescue clones. *In* D. E. Hume, G. C. M. Latch, and H. S. Easton (eds.), *Proceedings of the Second International Symposium on Acremonium/Grass Interactions*. AgResearch, Palmerston North, New Zealand, pp. 59–62.

Bacon, C. W. (1993). Abiotic stress tolerances (moisture, nutrients) and photosynthesis in endophyte-infected tall fescue. *Agriculture, Ecosystems and Environment* 44, 123–141.

Bacon, C. W. and Hinton, D. M. (1998). Tall fescue: Symbiosis, and surrogate transformations for increased drought tolerance. *In* G. E. Brink (ed.), *Proceedings of the 54th Southern Pasture and Forage Crop Improvement Conference*. Lafayette, LA, pp. 17–33.

Bacon, C. W. and De Battista, J. (1991). Endophytic fungi of grasses. *In* D. K. Arora, B. Rai, K. G, Mukerji, and G. R. Knudsen (eds.), *Handbook of Applied Mycology, Vol. 1, Soil and Plants*. New York, Marcel Dekker, pp. 231–256.

Bacon, C. W. Richardson, M. D. and White, J. F., Jr. (1997) Modification and uses of endophyte–enhanced turfgrasses: A role for molecular technology. *Crop Science* 37, 1415–1425.

Bacon, C. W., Porter, J. K., Robbins, J. D. and Luttrell, E. S. (1977). *Epichloe typhina* from toxic tall fescue grasses. *Applied Environmental Microbiology* 35, 576–581.

Ball, D. M. (1997). Significance of endophyte toxicosis and current practices in dealing with the problem in the United States. *In* C. W. Bacon and N. S. Hill (eds.), *Neotyphodium/Grass Interactions*. Plenum Press, New York, pp. 395–410.

Barker, G. M. (1987). Mycorrhizal infection influences *Acremonium*-induced resistance to Argentine stem weevil in ryegrass. *In Proceedings of the 40th New Zealand Weed and Pest Control Conference* 40:199–203.

Belesky, D. P. and Hill, N. S. (1997). Defoliation and leaf age influence on ergot alkaloids in tall fescue. *Annals of Botany* 79, 259–264.

Belesky, D. P. and Fedders, J. M. (1996). Does endophyte influence regrowth of tall fescue? *Annals of Botany* 78, 499-505.

Belesky, D. P. and Fedders, J. M. (1995). Tall fescue development in to *Acremonium coenophialum* and soil acidity. *Crop Science* 35, 529–533.

Belesky, D. P., Wilkinson, S. R. and Stuedemann, J. A. (1991). The influence of nitrogen fertilizer and *Acremonium coenophialum* on the soluble carbohydrate content of grazed and-non-grazed tall *Festuca arundinacea*. *Grass and Forage Science*, 46, 159-66.

Belesky, D. P., Stringer, W. C. and Hill, N. S. (1989a). Influence of endophyte and water regime upon tall fescue accessions. I. Growth characteristics. *Annals of Botany* 63, 495–503.

Belesky, D. P., Stringer, W. C. and Plattner, R. D. (1989b). Influence of endophyte and water regime upon tall fescue accessions. II. Pyrrolizidine and ergopeptine alkaloids. *Annals of Botany* 64, 343–349.

Belesky, D. P., Stuedemann, J. A., Plattner, R. D. and Wilkinson, S. R. (1988). Ergopeptine alkaloids in grazed tall fescue. *Agronomy Journal* 80, 209–212.

Belesky, D. P., Devine, O. J., Pallas, J. E., Jr., and Stringer, W. C. (1987a). Photosynthetic activity of tall fescue as influenced by a fungal endophyte. *Photosynthetica* 21, 82–87.

Belesky, D. P., Robbins, J. D., Stuedemann, J. A., Wilkinson, S. R. and Devine, O. J. (1987b). Fungal endophyte infection-loline derived alkaloid concentration of grazed tall fescue. *Agronomy Journal* 79, 217–220.

Belesky, D. P., Wilkinson, S. R. and Evans, J. J. (1984). Amino acid composition of fractions of 'Kentucky-31' tall fescue as affected by N fertilization and mild water stress. *Plant and Soil* 81, 257–267.

Belesky, D. P., Wilkinson, S. R. and Pallas, J. E., Jr. (1982). Response of four tall fescue cultivars grown at two nitrogen levels to low soil water availability. *Crop Science* 22, 93–97.

Belguendouz, L., Fremont, L. and Gozzelino, M. T. (1998) Interaction of transresveratrol with plasma lipoproteins. *Biochemical Pharmacology*, 55, 811–816.

Bouton, J. H., Gates, R. N., Belesky, D. P. and Owsley M. (1993). Yield and persistence of tall fescue in the Southeastern Coastal plain after removal of its endophyte. *Agronomy Journal* 81, 220–223.

Bransby, D. I., Schmidt, S. P., Griffey, W. and Eason, J. T., (1988). Heavy grazing is best for infected fescue. *Alabama Agricultural Experiment Station. Highlights of Agricultural Research* 35, 12.

Buck, G. W., West, C. P. and Elbersen, H. W. (1997). Endophyte effect on drought tolerance in diverse *Festuca* species. *In* C. W. Bacon and N. S. Hill (eds.), *Neotyphodium/Grass Interactions*. Plenum Press, New York, pp. 141–143.

Bush, L. P. and Schmidt, D. (1994). Alkaloid content of meadow fescue and tall fescue with their natural endophytes. *International Organization for Biological and Integrated Control, West Palaearctic Regional Section (IOBC/WPRS) Bulletin* 17, 259–265.

Bush, L. P., Wilkinson, H. H. and Schardl, C. L. (1997). Bioprotective alkaloids of grass-fungal endophyte symbioses. *Plant Physiology* 114, 1–7.

Bush, L. P., Fannin, F. F., Siegel, M. R., Dahlman, D. L. and Burton, H. R. (1993). Chemistry, occurrence and biological effects of saturated pyrrolizidine alkaloids associated with endophyte-grass interactions. *Agriculture, Ecosystems and Environment* 44, 81–102.

Casal, J. J., Sanchez, R. A., and Gibson, D. (1990). The significance of changes in the red/far-red ratio, associated with either neighbour plants or twilight, for tillering in *Loium multiflorum* Lam. *New Phytologist* 116, 565–572.

Casal, J. J., Sanchez, R. A. and Deregibus, D. A. (1987) the effect of light quality on shoot extension growth in three species of grasses. *Journal of Applied Ecology* 59, 1–7.

Cheplick, G. P. (1993). Effect of simulated acid rain on the mutualism between tall fescue (*Festuca arundinacea*) and an endophytic fungus (*Acremonium coenophialum*). *International Journal of Plant Science* 154, 134–143.

Cheplick, G. P., Clay, K. and Marks S. (1989). Interactions between infection by endophytic fungi and nutrient limitation in the grasses *Lolium perenne* and *Festuca arundinacea*. *New Phytologist* 111, 89–97.

Christensen, M. J. (1997). Endophyte compatibility in perennial ryegrass, meadow fescue, and tall fescue: a short review. *In* C. W. Bacon and N. S. Hill (eds.), *Neotyphodium/Grass Interactions*. Plenum Press, New York, pp. 45–48.

Chu-Chou, M., Guo, G., An, Z.Q., Henrix, J. W., Ferris, R. S., Siegel, M. R., Dougherty, C. T. and Burrus, P. B. (1992). Suppression of mycorrhizal fungi in fescue by the *Acremonium coenophialum* endophyte. *Soil Biology and Biochemistry*, 24, 633–637.

Clarke, J. M. and Durley, R. C. (1981). The responses of plants to drought stress. *In* G. M. Simpson (ed.), *Water Stress on Plants*. Praeger Publishers, New York, pp. 89–139.

Clay, K. (1998). Fungal endophytes of grasses: a defensive mutualism between plants and fungi. *Ecology* 69, 10–16.

Clay, K. (1987). Effects of fungal endophytes on seed and seedling biology of *Lolium perenne* and *Festuca arundinacea*. *Oecologia* 73, 358–362.

Clay, K., Marks, S. and Cheplick, G. P. (1993) Effects of insect herbivory and fungal endophyte infection on competitive interactions among grasses. *Ecology* 74, 1767–1777.

Cunningham, P. J., Blumenthal, M. J., Anderson, M. W., Prakash, K. S. and Leonforte,

A. (1994). Perennial ryegrass improvement in Australia. *New Zealand Journal of Agricultural Research*, 37, 295–310.

Cress, W. A., Chayet, L. T. and Rilling, H. C. (1981). Crystallization and partial characterization of dimethylallyl pyrophosphate: L-Tryptophan dimethylallyl transferase from *Claviceps* sp. SD58. *Journal of Chemistry* 256, 10917–10923.

Dahlman, D. L., Siegel, M. R. and Bush, L. P. (1997). Insecticidal activity of *N*-formylloline. *Proceedings of the XVIII International Grasslands Congress* 8–10 June 1997, Winnipeg, Canada, pp. 13/5–13/6.

Dannenberg, G., Latus, C., Zimmer, W., Hundeshagen, B., Schneider-Poetsch H. J. and Bothe, H. (1992). Influence of vesicular-arbuscular mycorrhiza on phytohormone balances in maize (*Zea mays* L.). *Journal of Plant Physiology* 141, 33–39.

Davidson, A. W. and Potter, D. A. (1995). Response of plant-feeding, predatory, and soil-inhabiting invertebrates to Acremonium coenophialum and nitrogen fertilization in turf fescue turf. *Journal of Economic Entomology* 88, 367–379.

De Battista, J. P., Bouton, J. H., Bacon, C. W. and Siegel, M. R. (1990). Rhizome and herbage production of endophyte removed tall fescue clones and populations. *Agronomy Journal* 82, 651–654.

Dennis, S. B., Allen, V. G., Saker, K. E., Fontenot, J. P., Ayad J. Y. M., and Brown, C. P. (1998). Influence of *Neotyphodium coenophialum* on copper concentration in tall fescue. *Journal of Animal Science*, 76, 2687–2693.

Diehl, W. W. (1950). *Balansia* and the Balansiae in America. U. S. Department of Agriculture, Washington, DC.

Donald, C. M. (1963). Competition among crop and pasture plants. *Advances in Agronomy*, 15, 1–118.

Eerens, J. P. J., Lucas, R. J., Easton, S. and White, J. G. H. (1998). Influence of the endophyte *(Neotyphodium lolii)* on morphology, physiology, and alkaloid synthesis of perennial ryegrass during high temperature and water stress. *New Zealand Journal of Agricultural Research* 41, 219–226.

Eerens, J. P. J., Visker, M. H. P. W., Lucas, R. J., Easton, H. S. and White, J. G. H. (1997). Influence of the ryegrass endophyte on phyto-nematodes. *In* C. W. Bacon and N. S. Hill (eds.), *Neotyphodium/Grass Interactions*, Plenum Press, New York, pp. 153–156. New York: Plenum Press.

Elbersen, H. W. and West C. P. (1996). Growth and water relations of field-grown tall fescue as influenced by drought and endophyte. *Grass and Forage Science* 51, 333–342.

Elmi, A. A. and West, C. P. (1995). Endophyte infection effects on stomatal conductance, osmotic adjustment and drought recovery of tall fescue. *New Phytologist* 131, 61–67.

Elmi, A. A., West, C. P., Turner, K. A. and Oosterhuis, D. M. (1990). *Acremonium coenophialum* effects on tall fescue water relations. *In* S. S. Quisenberry and R. E. Joost (eds.), *Proceedings of the International Symposium on Acremonium/Grass Interactions*. Louisiana Agricultural Experimental Station, Baton Rouge, pp. 137–140.

Flieger, M., Sedmera, P., Novak, J., Cvak, L., Zapletal, J. and Stuchlik, J. (1991). Degradation products of ergot alkaloids. *Journal Natural Production*, 54, 390–395.

Foy, C. D. and Murray, J. J. (1998). Developing aluminum tolerant strains of tall fescue for acid soils. *Journal of Plant Nutrition* 21, 1301–1325.

Fribourg, H. A. and MacLaren, J. B. (1987). Interactions between presence of *Acremonium coenophialum* in tall fescue, clover stand and pasture forage consumption, pp. 42–43. *In: Progress Report on Clovers and Special Purpose Legumes Research*, Vol. 20. U.S. Dairy Forage Research Center, University of Wisconsin. Madison.

Garner, G. B. and Cornell, C. N. (1978) Fescue foot in cattle. *In* T. D. Wylie & L. G. Morehouse (eds.), *Mycotoxic Fungi, Mycotoxins, Mycotoxicoses, Vol. 1.* Marcel Dekker, New York, pp. 45–62.

Garner, G. B., Rottinghaus, G. E., Cornell, C. N. and Testereci H. (1993). Chemistry of compounds associated with endophyte/grass interaction: ergovaline- and ergopeptine-related alkaloids. *Agriculture, Ecosystems and Environment*, 44, 65–80.

Gautier, H. and Gaborcik, N. (1993) The effect of red and far-red end of day treatment on perennial ryegrass and tall fescue morphogenesis. *In* J. Neuteboom, K. Wind, E. Lantinga, and R. van Loo (eds.), *Working Group on Pasture Ecology*. Paper IV8. Wageningen, Netherlands.

Gentry, C. E., Chapman, R. A., Henson, L. and Buckner, R. C. (1969). Factors affecting the alkaloid content of tall fescue. *Agronomy Journal*, 61, 313–316.

Glenn, A. E., Bacon, C. W., Price, R. and Hanlin, R. T. (1996). Molecular phylogeny of *Acremonium* and its taxonomic implications. *Mycologia* 88, 369–383.

Harper, J. L. (1977). *Population Biology of Plants*, Academic Press, London.

Hill, N. S. (1995). Progress report—the University of Georgia. *In* F. T. Withers (ed), *Proceedings of the Tall Fescue Workshop*, Southern Extension and Research Activity Information Exchange Group 8, Mississippi Agricultural and Forestry Experiment Station, Mississippi State, Mississippi, pp. 14–16.

Hill, N. S., Pachon, J. G. and Bacon C. W. (1996) *Acremonium coenophialum*-mediated short- and long-term drought acclimation in tall fescue. *Crop Science* 36, 665–672.

Hill, N. S., Belesky, D. P. and Stringer, W. C. (1991) Competitiveness of tall fescue as influenced by *Acremonium coenophialum*. *Crop Science* 31, 185–190.

Hill, N. S., Stringer, W. C., Rottinghaus, G. E., Belesky, D. P., Parrott, W. A. and Pope, D. D. (1990). Growth, morphological and chemical component responses of tall fescue *to Acremonium coenophialum*. *Crop Science* 30, 156–161.

Hinton, D. M and Bacon, C. W. (1985). The distribution and ultrastructure of the endophyte of toxic tall fescue. *Canadian Journal of Botany* 63, 36–42.

Joost, R. E., Sharp, R.E. and Holder, T. L. (1993). Involvement of the *Acremonium*-endophyte in ABA-mediated gas exchange responses in tall fescue. Tall Fescue Toxicosis Workshop, Atlanta, GA. *SERAIEG*, 8, 41–42.

Ju, Y., Sacalis, J. N., and Still, C. C. (1998) Bioactive flavonoids from endophyte-infected blue grass (*Poa ampla*). *Journal of Agriculture and Food Chemistry* 46, 3785–3788.

Kandeler, R. (1958). Dir Wirkung von farbigem und weissem licht auf die Anthocyanbildung bei Cruciferen-Keimlingen. *Deutsche Botanische Gesellschaft* 71, 34–44.

Kennedy, C. W. and Bush, L. P. (1983). Effect of environmental and management factors on the accumulation of *N*-acetyl and *N*-formyl loline alkaloids in tall fescue. *Crop Science* 23, 547–552.

Lapeyrie, F. F., Chilvers, G. A. and Douglass, P. A. (1984). Formation of metachromatic granules following phosphate uptake by mycelial hyphae of an ectomycorrhizal fungus. *New Phytologist* 98, 345–360.

Latch, G. C. M. and Christensen, M. J. (1982). Ryegrass endophyte, incidence and control. *New Zealand Journal of Agriculture Research*, 25, 443–448.

Latch, G. C. M., Hunt, W. F. and Musgrave, D. R. (1985). Endophytic fungi affect growth of perennial ryegrass. *New Zealand Journal of Agricultural Research*, 28, 165–168.

Latch, G. C. M., Christensen, M. J. and Samuels, G. J. (1984). Five endophytes of *Lolium* and *Festuca* in New Zealand. *Mycotaxon* 20, 535–550.

Lee, S. L., Floss, H. G. and Heinstein, P. (1976). Purification and properties of dimethylallylpyrophosphate: Tryptophan dimethylallyl tryptophan transferase, the first enzyme of ergot alkaloid biosynthesis in *Claviceps* sp. SD 58. *Archives of Biochemistry and Biophysiology*, 177, 84–94.

Leuchtmann, A. (1992). Systematics, distribution, and host specificity of grass endophytes. *Natural Toxins* 1, 150–162.

Liu, H., Heckman, J. R. and Murphy, J. A. (1996). Screening fine fescues for aluminum tolerance. *Journal of Plant Nutrition*, 19, 677–688.

Louis, M. St. and Faeth, S. H. (1997). The effect of endophytic fungi on the fitness of Arizona fescue (*Festuca arizonica*) under varying nitrogen levels. *The Fourth Annual Undergraduate Research Poster Symposium*, Arizona State University.

Lyons, P. C. and Bacon, C. W. (1984). Ergot alkaloids in tall fescue infected with *Sphacelia typhina*. *Phytopathology*, 75, 501.

Lyons, P. C., Evans, J. J. and Bacon, C. W. (1990). Effect of the fungal endophyte *Acremonium coenophialum* on nitrogen accumulation and metabolism in tall fescue. *Plant Physiology*, 92, 726–732.

Malinowski, D. (1995). Rhizomatous ecotypes and symbiosis with endophytes as new possibilities of improvement in competitive ability of meadow fescue (*Festuca pratensis Huds.*). Swiss Federal Institute of Technology (ETH), Zurich. Diss. ETH No. 11397.

Malinowski, D. P. and Belesky, D. P. (1999a). *Neotyphodium coenophialum*-endophyte infection affects the ability of tall fescus to use sparingly soluble phosphorus. *Journal of Plant Nutrition* 22:835–853.

Malinowski, D. P. and Belesky, D. P. (1999b). Tall fescue aluminum tolerance is affected by *Neotyphodium coenophialum*-endophyte. *Journal of Plant Nutrition* 22:1335–1349.

Malinowski, D. P., Belesky, D. P. and Fedders, J. M. (1999c). Endophyte infection may affect the competitive ability of tall fescue grown with red clover. *Journal of Agronomy and Crop Science* 183:91–102.

Malinowski, D. P., Brauer, D. K. and Belesky, D. P. (1999d). *Neotyphodium coenophialum*-endophyte affects root morphology of tall fescue grown under phosphorus deficiency. *Journal of Agronomy and Crop Science* 183:53–60.

Malinowski, D. P., Alloush, G. A. and Belesky, D. P. (1998a). Evidence for chemical changes on the root surface of tall fescue in response to infection with the fungal endophyte *Neotyphodium coenophialum*. *Plant & Soil*, 205, 1–12.

Malinowski, D. P., Belesky, D. P., Hill, N. S., Baligar, V. C. and Fedders, J. M. (1998b). Influence of phosphorus on the growth and ergot alkaloid content of *Neotyphodium coenophialum*-infected tall fescue (*Festuca arundinacea* Schreb.). *Plant & Soil* 198, 53–61.

Malinowski, D., Leuchtmann, A., Schmidt, D. and Nösberger, J. (1997a). Growth and water status in meadow fescue is affected by *Neotyphodium* and *Phialophora* species endophytes. *Agronomy Journal*, 89, 673–678.

Malinowski, D., Leuchtmann, A., Schmidt, D. and Nösberger, J. (1997b). Symbiosis with *Neotyphodium uncinatum* endophyte may increase the competitive ability of meadow fescue. *Agronomy Journal*, 89, 833–839.

Marschner, H. (1991). Mechanisms of adaptation of plants to acid soils. *In* R. J. Wright, V. C. Baligar and R. P. Murrmann (eds.), *Plant–Soil Interactions at Low pH: Proceedings of the Second International Symposium on Plant–Soil Interactions at Low pH*. Kluwer Academic, Netherlands, pp. 683–702.

Marschner, H. (1986). *Mineral Nutrition of Higher Plants*. Academic Press, Orlando, FL.

Marks, S. and Clay, K. (1996). Physiological responses of *Festuca arundinacea* to fungal endophyte infection. *New Phytologist* 133, 727–733.

Morgan-Jones, G. and Gams, W. (1982). Notes on Hyphomycetes. XLI. An endophyte of *Festuca arundinacea* and the anamorph of *Epichloe typhina*, new taxa in one of two new sections of *Acremonium*. *Mycotaxon* 15, 311–318.

Morgan-Jones, G., White, J. F. and Piontelli, E. L. (1990). Endophyte-host associations in forage grasses. XIII. *Acremonium chilense*, an undescribed endophyte occurring in *Dactylis glomerata* in Chile. *Mycotaxon* 15, 311–318.

Naffaa, W., Ravel, C. and Guillaumin, J. J. (1998). Nutritional requirements for growth of fungal endophytes of grasses. *Canadian Journal of Microbiology*, 44, 231–237.

Neill, J. C. (1941). The endophyte of *Lolium* and *Festuca*. *New Zealand Journal of Science and Technology, Section A*, 23, 185–193.

Newsham, K. K., Lewis, G. C., Greenslade, P. D. and McLeod, A. R. (1998). *Neotyphodium lolii*, a fungal leaf endophyte, reduces fertility of *Lolium perenne* exposed to elevated UV-B radiation. *Annals of Botany*, 81, 397–403.

Newsham, K. K., Fitter, A. H. and Watkinson, A. R. (1995) Multi-functionality and biodiversity in arbuscular mycorrhizas. *Trends in Ecology and Evolution*, 10, 407–411.

Percival, N. S. and Duder, F. R. (1983). A comparison of perennial grasses under sheep grazing on the Central Plateau. *Proceedings of the New Zealand Grassland Association* 44, 81–90.

Petroski, R. J., Dornbos, D. L., Jr. and Powell, R. G. (1990). Germination and growth inhibition of annual ryegrass (*Lolium multiflorum* L.) and alfalfa (*Medicago sativa* L.) by loline alkaloids and synthetic *N*-acylloline derivates. *Journal of Agriculture and Food Chemistry* 38, 1716–1718.

Phillipson, M. N. (1989). A symptomless endophyte of ryegrass (*Lolium perenne*) that spores on its host–a light microscope study. *New Zealand Journal of Botany* 27, 513–519.

Phillipson, M. N. (1991). Ultrastructure of the *Gliocladium*-like endophyte of perennial ryegrass (*Lolium perenne* L.). *New Phytologist* 117, 271–280.

Porter, J. K. (1994). Chemical constituents of grass endophytes. *In* C. W. Bacon and J. F. White, Jr. (eds.), *Biotechnology of Endophytic Fungi of Grasses*. CRC Press, Boca Raton, FL, pp. 103–123.

Powell, R. G., Te Paske, M. R., Plattner, R. D. White, J. F. and Clement, S. L. (1994).

Isolation of resveratrol from *Festuca versuta* and evidence for the widespread occurrence of this stilbene in the Poaceae. *Phytochemistry* 35, 335–338.

Prestige, R. A. (1993). Causes and control of perennial ryegrass staggers in New Zealand. *Agriculture, Ecosystems and Environment*, 44, 283–300.

Read, D. J. (1991). Mycorrhizas in ecosystems. *Experientia*, 47, 376–391.

Read, J. C. and Camp, B. J. (1986). The effect of the fungal endophyte *Acremonium coenophialum* in tall fescue on animal performance, toxicity, and stand maintenance. *Agronomy Journal* 78, 848–850.

Richardson, M. D., Hoveland, C. S. and Bacon C. W. (1993). Photosynthesis and stomatal conductance in symbiotic and nonsymbiotic tall fescue. *Crop Science* 33, 145–149.

Richardson, M. D., Chapman, G. W., Jr., Hoveland, C. S. and Bacon, C. W. (1992). Sugar alcohol in endophyte-infected tall fescue under drought. *Crop Science* 32, 1060–1061.

Robbers, J. E. (1984). The fermentative production of ergot alkaloids. *In* A. Mizrahi and A. L. van Wezel (eds.), *Advances in Biotechnological Processes, Vol. 3.* A. R. Liss, New York, p. 197.

Robbins, J. D. (1983). The tall fescue toxicosis problem. *In Proceedings of the Tall Fescue Toxicosis Workshop*. Georgia Agricultural Extension Service, Athens, pp. 1–4.

Roberts, C. A., Marek, S. M., Niblack, T. L. and Karr A.L. (1992). Parasitic *Meloidogyne* and mutualistic *Acremonium* increase chitinase in tall fescue. *Journal of Chemistry and Ecology* 18, 1107–1116.

Rowan, D. D. (1993). Lolitrems, peramine and paxilline: mycotoxins of the ryegrass/endophyte interaction. *Agriculture, Ecosystems and Environment*, 44, 103–122.

Saker, K. E., Allen, V. G., Kalnitsky, J., Thatcher, C. D., Swecker, W. S., Jr. and Fontenot, J. P. (1998). Monocyte immune cell response and copper status in beef steers that grazed-endophyte-infected tall fescue. *Journal of Animal Science*, 76, 2694–2700.

Scehovic, J. and Jadas-Hecart, J. (1989). La qualite des hybrides festulolium comparée à celle de la fétuque elevée. *Revue Suisse d'Agriculture* 21, 345–349.

Schmidt, D. (1991). Les endophytes de la fétuque des prés. *Revue Suisse Agriculture*, 23, 369–375.

Schmidt, D. (1993). Effects of *Acremonium uncinatum* and a *Phialophora*-like endophyte on vigour, insect and disease resistance of meadow fescue. *In* D. E. Hume, G. C. M. Latch, and H. S. Easton (eds.), *Proceedings of the Second International Symposium on Acremonium/Grass Interactions*. AgResearch, Palmerston North, New Zealand, pp. 185–187.

Schulze, E.-D. (1991) Water and nutrient interactions with plant water stress. p. 89–101. *In* H. A. Mooney et al. (eds.). *Response of Plants to Multiple Stresses*. Academic Press, San Diego, pp. 89–101.

Shelby, R. A. and Dalrymple, L. W. (1987). Incidence and distribution of tall fescue endophytes in the United States. *Plant Diseases* 71, 783–786.

Siegel, M. R. (1993). *Acremonium* endophytes: our current state of knowledge and future directions for research. *Agriculture, Ecosystems and Environment* 44, 301–321.

Siegel, M. R., Schardl, C. L. and Phillips, T. D. (1995). Incidence and compatibility of nonclavicipitaceous fungal endophytes in *Festuca* and *Lolium* grass species. *Mycologia* 87, 196–202.

Skinner, R. H. and Nelson, C. J. (1994) Effect of tiller trimming on phyllochron and tillering regulation during tall fescue development. *Crop Science* 34, 1267–1273.

Smith, S. E. and Gianinazzi-Pearson, V. (1988). Physiological interactions between symbionts in vesicular-arbuscular mycorrhizal plants. *Annual Reviews of Plant Physiology and Plant Molecular Biology* 39, 221–244.

Southerland, B. L. and Hoglund, J. H. (1989). Effect of ryegrass containing the endophyte (*Acremonium lolii*) on the performance of associated white clover and subsequent crops. *In: Proceedings of the New Zealand Grassland Association* 50, 265–269.

Springer, T. L. (1997). Allelopathic effects of tall fescue. *Proceedings of the 53rd Southern Pasture and Forage Crops Improvement Conference*, ed. G. E. Brink. pp. 25–33.

Stuedemann, J. A. and Hoveland, C. S. (1988) Fescue endophyte: history and impact on animal agriculture. *Journal of Production Agriculture* 1, 39–44.

Sylvia, D. M. and Sinclair, W. A. (1983). Phenolic compounds and resistance to fungal pathogens induced in primary roots of Douglas-fir seedlings by the ectomycorrhizal fungus *Laccaria laccata*. *Phytopathology* 73, 390–397.

Tagaki, S. (1976). Naturally occurring iron-chelating compounds in oat and rice-root washings. 1. Activity, measurement and preliminary characterization. *Soil Science and Plant Nutrition*, 22, 423–433.

Taylor, G. J. (1988). The physiology of aluminum tolerance in higher plants. *Communication in Soil Science and Plant Analyses*, 19, 1179–1194

Thompson, F. N. and Stuedemann, J. A. (1993). Pathophysiology of fescue toxicosis. *Agriculture, Ecosystems and Environment* 44, 263–281.

Thornton, B. (1998). Influence of pH and aluminum on nitrogen partitioning in defoliated grasses. *Grass and Forage Science* 53, 170–178.

Tso, T. C., Kasperbauer, M. J. and Sorokin, T. P. (1970). Effect of photoperiod and end-of-day light quality on non-alkaloids and phenolic compounds of tobacco. *Plant Physiology*, 45, 330–333.

Turkington, R. and Mehrhoff, L. (1990). The role of competition in structuring pasture communities. p. 307–340. *In* J. B. Grace and D. Tilman (eds.), *Perspectives on Plant Competition*. Academic Press, San Diego, pp. 307–340.

West, C. P. (1994). Physiology and drought tolerance of endophyte-infected grasses. *In* C. W. Bacon and J. F. White (eds.), *Biotechnology of Endophytic Fungi of Grasses*. CRC Press, Boca Raton, FL, pp. 87–99.

West, C. P., Oosterhuis, D. M. and Wullschleger, S. D. (1990a). Osmotic adjustment in tissues of tall fescue in response to water deficit. *Environmental Experimental Botany*, 30, 149–156.

West, C. P., Izekor, E., Robbins, R. T., Gergerich, R. and Mahmood, T. (1990b). *Acremonium coenophialum* effects on infestations of Barley Yellow Dwarf Virus and soil-borne nematodes and insects in tall fescue. *In* S. S. Quisenberry and R. E. Joost (eds.), *Proceedings of the International Symposium on Acremonium/Grass Interactions*. Louisiana Agricultural Experimental Station, Baton Rouge, pp. 196–198.

West, C. P., Izekor, E., Oosterhuis, D. M. and Robbins, R. T. (1988). The effect of *Acremonium coenophialum* on the growth and nematode infestation of tall fescue. *Plant and Soil*, 112, 3–6.

White, J. A. and Brown, M. R. (1979). Ultrastructure and x-ray analysis of phosphorus granules in a vesicular-arbuscular mycorrhizal fungus. *Canadian Journal of Botany* 57, 2812–2818.

White, J. F., Jr. (1994). Taxonomic relationships among the members of the Balansiae (Clavicipitales). *In* C. W. Bacon and J. F. White, Jr. (eds.), *Biotechnology of Endophytic Fungi of Grasses*. CRC Press, Boca Raton, FL, pp. 3–20.

White, J. F., Jr. and Reddy, P. V. (1998). Examination of structure and molecular phylogenetic relationships of some graminicolous symbionts in genera *Epichloe* and *Parepichloe*. *Mycologia*, 90, 226–234.

Wilcox, H. E. (1996) Mycorrhizae. *In* Y. Waisel, A. Eschel, and U. Kafkafi (eds.), *Plant Roots: The Hidden Half*. 2nd Ed. Marcel Dekker, New York, pp. 689–721.

Wilkinson, H. H. and Schardl, C. L. (1997). The evolution of mutualism in grass-endophyte associations. *In* C. W. Bacon and N. J. Hill (eds.), *Neotyphodium/Grass Interactions*. Plenum Press, New York, pp. 13–25.

Index

Printed and bound by CPI Group (UK) Ltd, Croydon, CR0 4YY

23/10/2024

01778237-0018